T0211061

Lecture Notes in Computer Science　　10156

Commenced Publication in 1973
Founding and Former Series Editors:
Gerhard Goos, Juris Hartmanis, and Jan van Leeuwen

More information about this series at http://www.springer.com/series/7407

Daya Gaur · N.S. Narayanaswamy (Eds.)

Algorithms and Discrete Applied Mathematics

Third International Conference, CALDAM 2017
Sancoale, Goa, India, February 16–18, 2017
Proceedings

 Springer

Editors
Daya Gaur
University of Lethbridge
Lethbridge, AB
Canada

N.S. Narayanaswamy
Indian Institute of Technology Madras
Chennai
India

ISSN 0302-9743   ISSN 1611-3349   (electronic)
Lecture Notes in Computer Science
ISBN 978-3-319-53006-2   ISBN 978-3-319-53007-9   (eBook)
DOI 10.1007/978-3-319-53007-9

Library of Congress Control Number: 2016963600

LNCS Sublibrary: SL1 – Theoretical Computer Science and General Issues

Printed on acid-free paper

This Springer imprint is published by Springer Nature
The registered company is Springer International Publishing AG
The registered company address is: Gewerbestrasse 11, 6330 Cham, Switzerland

Preface

This volume contains the papers presented at CALDAM 2017: the Third International Conference on Algorithms and Discrete Applied Mathematics held during February 16–18, 2017 in Goa. CALDAM 2017 was organized by the Department of Mathematics, Birla Institute of Technology and Science, Pilani (BITS Pilani), K. K. Birla Goa Campus, Goa. The conference had papers in the areas of algorithms, graph theory, codes, polyhedral combinatorics, computational geometry, and discrete geometry. The 103 submissions had authors from 18 different countries. Each submission received at least one detailed review and nearly all were reviewed by three Programme Committee members. The committee decided to accept 32 papers. The program also included four invited talks by Sumit Ganguly, Martin C. Golumbic, Günter Rote, and Ola Svensson.

The first CALDAM was held in February 2015 at the Indian Institute of Technology, Kanpur, and had 26 papers selected from 58 submissions from 10 countries. The second edition was held in February 2016 at Thiruvananthapuram (Trivandrum), India, and had 30 papers selected from 91 submissions from 13 countries.

We would like to thank all the authors for contributing high-quality research papers to the conference. We express our sincere thanks to the Program Committee members and the external reviewers for reviewing the papers within a very short period of time. We thank Springer for publishing the proceedings in the *Lecture Notes in Computer Science* series. We thank the invited speakers Martin C. Golumbic, Günter Rote, Ola Svensson, and Sumit Ganguly for accepting our invitation. We thank the Organizing Committee chaired by Tarkeshwar Singh from Birla Institute of Technology and Science, Pilani (BITS Pilani), K. K. Birla Goa Campus, Goa, for the smooth functioning of the conference. We thank the chair of the Steering Committee, Subir Ghosh, for his active help, support, and guidance throughout. We thank our sponsors Google Inc., Microsoft Research India, and National Board of Higher Mathematics, Department of Atomic Energy, for their financial support. We also thank Springer for its support for the two Best Paper Presentation Awards. Last and definitely most importantly, we thank the EasyChair conference management system, which was very effective in handling the entire reviewing process.

December 2016

Daya Gaur
N.S. Narayanaswamy

Organization

Program Committee

John Augustine	Indian Institute of Technology Madras, India
Amitabha Bagchi	Indian Institute of Technology, Delhi, India
Niranjan Balachandran	Indian Institute of Technology, Bombay, India
Bostjan Bresar	University of Maribor, Slovenia
Subramanian C.R.	The Institute of Mathematical Sciences, India
L. Sunil Chandran	Indian Institute of Science, India
Manoj Changat	University of Kerala, India
Sandip Das	Indian Statistical Institute, Kolkata, India
Fabrizio Frati	Roma Tre University, Italy
Zachary Frigstaad	University of Alberta, Canada
Sumit Ganguly	Indian Institute of Technology, Kanpur, India
Daya Gaur (Co-chair)	University of Lethbridge, Canada
Partha Goswami	University of Calcutta, India
Sathish Govindarajan	Indian Institute of Science, India
R. Inkulu	Indian Institute of Technology, Guwahati, India
Erik Jan van Leeuwen	MPI Saarbrücken, Germany
Subrahmanyam Kalyanasundaram	Indian Institute of Technology, Hyderabad, India
Matya Katz	Ben-Gurion University, Israel
Sandi Klavzar	University of Ljubljana, Slovenia
Ramesh Krishnamurti	Simon Fraser University, Canada
Stefano Leonardi	University of Rome La Sapienza, Italy
Andrzej Lingas	Lund University, Sweden
Anil Maheshwari	Carleton University, Canada
Kaz Makino	Kyoto University, Japan
Bodo Manthey	University of Twente, The Netherlands
Rogers Mathew	Indian Institute of Technology, Kharagpur, India
Bojan Mohar	Simon Fraser University, Canada
Apurva Mudgal	Indian Institute of Technology Ropar, India
Narayanaswamy N.S. (Co-chair)	Indian Institute of Technology Madras, India
Reza Naserasr	IRIF Paris-Diderot, France
Gyula O.H. Katona	Renyi Institute, Budapest, Hungary
Sudebkumar Pal	Indian Institute of Technology, Kharagpur, India
Abraham Punnen	Simon Fraser University, Canada
Venkatesh Raman	The Institute of Mathematical Sciences, India
Abhiram Ranade	Indian Institute of Technology, Bombay, India
Bhawani S. Panda	Indian Institute of Technology, Delhi, India
Jiří Sgall	Computer Science Institute of Charles University, Czech Republic
Michiel Smid	Carleton University, Canada
Dorothea Wagner	Karlsruhe Institute of Technology (KIT), Germany

Organizing Committee (BITS Pilani K. K. Birla Goa Campus, Goa, India)

Souvik Bhattacharyya (Vice Chancellor (Cheif Patron))
G. Raghurama (Director (Patron))
Sasikumar Punnekkat (Former Director and Senior Professor)
Ashwin Srinivasan (Deputy Director)
Tarkeshwar Singh (Organizing Chair)
Prasannakumar N.
Bharat M. Deshpande
Veeky Baths
Ramprasad Joshi
P. Dhanumjaya
Anil Kumar
Manoj Kumar Pandey
Amit Kumar Setia
J.K. Sahoo
Mayank Goel
Alpesh M. Dhorajia
Prabal Paul
Gauranga Samanta
Himadri Mukharjee
Soumyadip Bandyopadhyay
Jessica Pereira
Bijil Prakash

Steering Committee

Subir Kumar Ghosh (Chair)	Ramakrishna Mission Vivekananda University, India
János Pach	École Polytechnique Fédérale De Lausanne (EPFL), Switzerland
Nicola Santoro	Carleton University, Canada
Swami Sarvattomananda	Ramakrishna Mission Vivekananda University, India
Peter Widmayer	ETH Zurich, Switzerland
Chee Yap	Courant New York University, USA

Additional Reviewers

Atre, Medha
Balakrishnan, Kannan
Banerjee, Niranka
Basavaraju, Manu
Basu Roy, Aniket
Bhattacharya, Pritam
Brückner, Guido
Das, Syamantak
Dhannya, S.M.
Estrada-Moreno,
    Alejandro
Francis, Mathew
Gologranc, Tanja
González Yero, Ismael
Gregor, Petr
Hamann, Michael
Jakovac, Marko
Jansson, Jesper
Johannsen, Jan
Kare, Anjeneya Swami
Kern, Walter
Kothapalli, Kishore
Kowaluk, Miroslaw
Krithika, R.

Levcopoulos, Christos
Lust, Thibaut
Majumdar, Diptapriyo
Mehta, Shashank K.
Meirun, Chen
Michael, Henning
Mishra, Tapas Kumar
Mohammad, Meesum
Molla, Anisur Rahaman
Moses Jr., William K.
Mukherjee, Joydeep
Mulder, Henry Martyn
Muthu, Rahul
N., Sharmili
Nagar, Mukesh Kumar
Nandakumar, Satyadev
Nandi, Soumen
Nasre, Meghana
Natarajan, Aravind
Niedermann, Benjamin
Nilsson, Bengt J.
Padinhatteeri, Sajith
Pandey, Arti
Pandurangan, Ragukumar

Panigrahi, Pratima
Philip, Geevarghese
Pradhan, D.
Ramamoorthi,
    Vijayaragunathan
Rao, M.V. Panduranga
Redlich, Amanda
Rezapour, Mohsen
Rollova, Edita
Roy, Sasanka
Sahoo, Uma
Saptharishi, Ramprasad
Sen, Sagnik
Severin, Daniel
Strasser, Ben
Surynek, Pavel
Tewari, Raghunath
Tratnik, Niko
Wood, David
Zemljic, Sara Sabrina
Zündorf, Tobias

Abstracts of Invited Talks

Sketching for Data Streams and Numerical Linear Algebra

Sumit Ganguly

Indian Institute of Technology, Kanpur, India
sganguly@cse.iitk.ac.in

Abstract. In the data stream model, data arrives in high volume and speed and there is not enough storage space to hold all the input. The input data is examined one record at a time upon arrival and processed. The typical processing done is randomized sketching that allows for sublinear storage, and often uses only poly-logarithmic space and time (per record). We present some interesting sketching solutions that have appeared in the literature for several statistical problems over data streams. We will also discuss some recent advances in algorithms for numerical linear algebra obtained using linear sketching. In this technique, an input matrix is compressed into a much smaller matrix by multiplying it with a random matrix chosen from certain distributions. The original problem is now solved, approximately to within factors of $1\pm\epsilon$ with high probability, by computing over the smaller matrix. We will illustrate this method using the least squares regression problem.

The Elusive Nature of Intersecting Paths on a Grid

Martin Charles Golumbic

University of Haifa, Haifa, Israel
golumbic@cs.haifa.ac.il

Abstract. In this lecture, we will survey the mathematical and algorithmic results on the *edge intersection graphs of paths in a grid* (EPG) together with several restrictions on the representations. Two important restrictions that are motivated by network and circuit design problems are (1) allowing just a single bend in any path, and (2) limiting the paths within a rectangular grid with the endpoints of each path on the boundary of the rectangle.

Golumbic, Lipshteyn and Stern introduced EPG graphs in 2005, proving that every graph is an EPG graph, and then turning their attention to the subclass of graphs that admit an EPG representation in which every path has at most a single bend, called B_1-EPG graphs. They proved that any tree is a B_1-EPG graph and gave a structural property that enables generating non B_1-EPG graphs. A characterization of the representation of cliques and chordless 4-cycles in B_1-EPG graphs was given, and also prove that single bend paths on a grid have Strong Helly number 4, and when the paths satisfy the usual Helly property, they have Strong Helly number 3. Subsequent results by our colleagues will be surveyed, as well as open problems and future work.

We then present our new work on boundary generated B_1-EPG graphs together with Gila Morgenstern and Deepak Rajendraprasad. For two boundary vertices u and v on two adjacent boundaries of a rectangular grid $\mathcal{G}$, we call the unique single-bend path connecting u and v in $\mathcal{G}$ using no other boundary vertex of $\mathcal{G}$ as the path *generated* by (u,v). A path in $\mathcal{G}$ is called *boundary-generated*, if it is generated by some pair of vertices on two adjacent boundaries of $\mathcal{G}$. In this work, we study the edge-intersection graphs of boundary-generated paths on a grid or ∂EPG graphs.

We show that ∂EPG graphs can be covered by two collections of vertex-disjoint cobipartite chain graphs. This leads us to a linear-time testable characterization of ∂EPG trees and also a tight upper bound on the equivalence covering number of general ∂EPG graphs. We also study the cases of two-sided ∂EPG and three-sided ∂EPG graphs, which are respectively, the subclasses of ∂EPG graphs obtained when all the boundary vertex pairs which generate the paths are restricted to lie on at most two or three boundaries of the grid. For the former case, we give a complete characterization.

We do not know yet whether one can efficiently recognize ∂EPG graphs. Though the problem is linear-time solvable on trees, we suspect that it might be NP-hard in general.

Keywords: Edge intersection graphs · Paths in a grid · Strong Helly number · Boundary generated EPG graphs

Minimal Dominating Sets in Trees: Counting, Enumeration, and Extremal Results

Günter Rote

November 29, 2016

Abstract. A tree with n vertices has at most $95^{n/13}$ minimal dominating sets. The corresponding growth constant $\lambda = \sqrt[13]{95} \approx 1.4194908$ is best possible.

These results are obtained more or less automatically by computer, starting from the dynamic-programming recursion for computing the number of minimal dominating sets of a given tree. This recursion defines a bilinear operation on sixtuples, and the growth constant arises as a kind of "dominant eigenvalue" of this operation.

We also derive an output-sensitive algorithm for listing all minimal dominating sets with linear set-up time and linear delay between reporting successive solutions. It is open whether the delay can be reduced to a constant delay, for an appropriate modification of the problem statement.

Strong Convex Relaxations
for Allocation Problems

Ola Svensson

EPFL Lausanne, Lausanne, Switzerland
ola.svensson@epfl.ch

Abstract. Allocation problems, such as machine scheduling and distributing indivisible goods, are classic problems in combinatorial optimization. The study of their approximability goes back to the late 60's and, since then, impressive progress has led to an (almost) complete understanding of natural linear programming relaxations referred to as Assignment-LPs.

However, for most allocation problems, these natural relaxations fail to give tight approximation guarantees. This motivates the study of stronger convex relaxations such as Configuration-LPs. Although these relaxations are believed to give much better (and even tight) guarantees for several allocation problems, our current techniques are unable to fully exploit their strength.

In this talk, we give an overview of recent progress. In particular, we show algorithmic techniques that better exploit the power of Configuration-LPs, yielding improved approximation algorithms for the restricted assignment problem, budgeted allocation, and max-min fair allocation.

Contents

Optimal Embedding of Locally Twisted Cubes into Grids

Jessie Abraham[✉] and Micheal Arockiaraj

Department of Mathematics, Loyola College, Chennai 600034, India
jessie.abrt@gmail.com, marockiaraj@gmail.com

Abstract. The hypercube has been used in numerous problems related to interconnection networks due to its simple structure and communication properties. The locally twisted cube is an important class of hypercube variants with the same number of nodes and connections per node, but has only half the diameter and better graph embedding capability as compared to its counterpart. The embedding problem plays a significant role in parallel and distributed systems. In this paper we devise an optimal embedding of the n-dimensional locally twisted cube onto a grid network.

Keywords: Locally twisted cube · Embedding · Edge congestion · Optimal set

1 Introduction

In a parallel distributed system, the execution of parallel algorithms devised for a particular network on other networks in such a way that communication overhead could be minimized when they are run concurrently, can be modeled as a graph embedding problem.

Let $G = (V(G), E(G))$ and $H = (V(H), E(H))$ be undirected connected graphs on n nodes each known as a *guest graph* and *host graph*, representing the network underlying the algorithm and the network on which it is to be embedded respectively. A *graph embedding* is an ordered pair $\prec f, p \succ$ of injective maps where f maps $V(G)$ onto $V(H)$ and p maps the edges of G into simple paths of H such that if $e = (a, b) \in E(G)$, then $p(e)$ is a simple path in H with $f(a)$ and $f(b)$ as endpoints [11].

The *edge congestion* $EC_{\prec f, p \succ}(e)$ [3,10] of an edge $e \in H$ denotes the maximum number of edges of the guest graph that are embedded on e and is denoted as

$$EC_{\prec f, p \succ}(e) = |\{(u, v) \in E(G) : e \in E(p(u, v))\}|.$$

For any set $S \subseteq V(G)$, let $I_G(S) = \{(u, v) \in E(G) : u \in S \,\&\, v \in S\}$ and for $1 \le k \le |V(G)|$, let $I_G(k) = \min_{S \subseteq V(G), |S| = k} |I_G(S)|$. The *maximum subgraph problem* is to find $S \subseteq V(G)$ with $|S| = k$ such that $I_G(k) = |I_G(S)|$. Such a set S is called an *optimal set* [8].

© Springer International Publishing AG 2017
D. Gaur and N.S. Narayanaswamy (Eds.): CALDAM 2017, LNCS 10156, pp. 1–11, 2017.
DOI: 10.1007/978-3-319-53007-9_1

Lemma 1 *(Congestion Lemma).* *[10] Let G be an r-regular graph and $\prec f, p \succ$ be an embedding of G into H. Let S be an edge cut of H such that the removal of edges of S splits H into 2 components H_1 and H_2 and let $G_1 = f^{-1}(H_1)$ and $G_2 = f^{-1}(H_2)$. The following conditions are sufficient for $EC_{\prec f, p \succ}(S)$ to be minimum where $EC_{\prec f, p \succ}(S)$ denotes the sum of edge congestion over all the edges in S.*

1. *For every edge $(a, b) \in G_i$, $i = 1, 2$, $p(a, b)$ has no edges in S.*
2. *For every edge (a, b) in G with $a \in G_1$ and $b \in G_2$, $p(a, b)$ has exactly one edge in S.*
3. *G_1 or G_2 is an optimal set.*

Let $\{S_i\}_{i=1}^m$ be a partition of $E(H)$ such that $EC_{\prec f, p \succ}(S_i)$ is minimum over all embeddings for every i. An embedding $\prec f, p \succ$ from G into H for which such a partition can be determined is known as an *optimal embedding* [10]. Finding an optimal embedding helps in solving the layout problem [12–14], which finds application in VLSI circuit design [4], graph drawing [2], crossing number problem [5] and structural engineering [9].

The rest of the paper is organized as follows. Some fundamental definitions and preliminary results for locally twisted cube and grid are given in the next section. In Sect. 3, we devise a labeling algorithm and prove that it yields the optimal embedding for locally twisted cubes into grids. In Sect. 4, we conclude the paper.

2   Basic Definitions and Terminologies

The binary hypercube is one of the most popular, versatile and efficient topological structures of interconnection networks having simple deadlock-free routing, a small diameter, bounded link traffic density and a good support for parallel algorithms.

Definition 1. *[8] For $n \geq 1$, the node set of an n-dimensional hypercube Q_n is made up of n-bits binary strings labeled in order using $\{0, 1, \ldots, 2^n - 1\}$ beginning with 0 at $\underbrace{00 \ldots 00}_{n \text{ times}}$ and ending at $\underbrace{11 \ldots 11}_{n \text{ times}}$ with $2^n - 1$. Two nodes $x, y \in V(Q_n)$ are adjacent if and only if their corresponding binary strings differ in exactly one bit. An incomplete hypercube on i nodes of Q_n is the subgraph induced by $L_i = \{0, 1, \ldots, i - 1\}$, $1 \leq i \leq 2^n$, and is denoted by $Q_n[L_i]$.*

Theorem 1. *[7] For $1 \leq i \leq 2^n$, L_i is an optimal set in Q_n.*

Definition 2. *[15] For $n \geq 2$, an n-dimensional locally twisted cube is defined recursively as follows:*

1. *LTQ_2 is a graph consisting of four nodes labeled with $00, 01, 10, 11$ respectively, connected by four edges $(00, 01), (00, 10), (01, 11)$ and $(10, 11)$.*

2. *For $n \geq 3$, LTQ_n is built from two disjoint copies of LTQ_{n-1} as follows: Let $0LTQ_{n-1}$ denote the graph obtained by prefixing the binary representation of each node of one copy of LTQ_{n-1} with 0. Let $1LTQ_{n-1}$ denote the graph obtained by prefixing the binary representation of each node of another copy of LTQ_{n-1} with 1. Connect each node $0x_2x_3 \ldots x_n$ of $0LTQ_{n-1}$ to the node $1\,(x_2 \oplus x_n)\,x_3 \ldots x_n$ of $1LTQ_{n-1}$ by an edge, where $\oplus$ denotes addition modulo 2.*

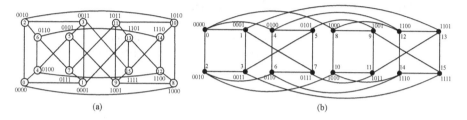

(a)   (b)

Fig. 1. Isomorphic 4-dimensional locally twisted cubes with decimal and binary labeling

Figure 1 depicts a 4-dimensional locally twisted cube. LTQ_n can be equivalently defined non-recursively as follows.

Definition 3. *[15] For $n \geq 2$, an n-dimensional locally twisted cube is a graph with node set of the form $\{0,1\}^n$. Two nodes $x = x_1x_2x_3 \ldots x_n$ and $y = y_1y_2y_3 \ldots y_n$ of LTQ_n are adjacent if and only if either of the following conditions is satisfied.*

1. *(a) There is an integer i with $1 \leq i \leq n-2$ such that $x_i = \overline{y}_i$ and $x_{i+1} = y_{i+1} \oplus x_n$.*
   (b) All remaining bits of x and y are identical.
2. *There is an integer $i \in \{n-1, n\}$ such that x and y differ only in the ith bit.*

For $1 \leq i \leq 2^{n-1}$, let $E_i = \{0, 2, \ldots, 2i-2\}$ and let $(TO)_i = \{a_t : 1 \leq t \leq i\}$ where $a_1 = 1$, $a_2 = 3$ and for $2 \leq k \leq n-1$, $1 \leq j \leq 2^{k-1}$,

$$a_{2^{k-1}+j} = \begin{cases} a_j + 2^k + 2^{k-1} : 1 \leq j \leq 2^{k-2} \\ a_j + 2^{k-1} \quad\;\; : 2^{k-2} < j \leq 2^{k-1}. \end{cases}$$

Lemma 2. *[1] For $1 \leq i \leq 2^n$,*

$$(ETO)_i = \begin{cases} E_i & : 1 \leq i \leq 2^{n-1} \\ E_{2^{n-1}} \cup (TO)_{i-2^{n-1}} : 2^{n-1} < i \leq 2^n. \end{cases}$$

is an optimal set in LTQ_n.

In what follows, $A \simeq B$ represents an isomorphism between A and B.

Lemma 3. *[1] For $1 \leq i \leq 2^{n-1}$, $LTQ_n[E_i] \simeq LTQ_n[(TO)_i] \simeq Q_n[L_i]$.*

Definition 4. *[6] An $n \times m$ grid $G[n \times m]$ is a graph with node set $V(G[n \times m]) = \{\alpha_{ij} \mid 1 \leq i \leq n, 1 \leq j \leq m\}$ and edge set $E(G[n \times m]) = \{(\alpha_{ij}, \alpha_{i(j+1)}) \mid 1 \leq i \leq n, 1 \leq j \leq m - 1\} \cup \{(\alpha_{kp}, \alpha_{(k+1)p}) \mid 1 \leq k \leq n - 1, 1 \leq p \leq m\}$. A $2^a \times 2^b$ grid is of the form $G[2^a \times 2^b]$, where $a \leq b$, $a + b = n$.*

3 Optimal Embedding Methodology

The optimal embedding proof comprises of three steps namely, node labeling, embedding proposal and proof of optimization for the proposed embedding.

3.1 Node Labeling Algorithm

In this section we label the nodes of LTQ_n and $G[2^a \times 2^b]$ in a particular pattern and prove that this labeling pattern gives the optimal embedding of LTQ_n into $G[2^a \times 2^b]$.

LTQ_n **Labeling:** Label the nodes of LTQ_n by lexicographic order [10] using $\{0, 1, \ldots, 2^n - 1\}$ starting with 0 at $\underbrace{00 \ldots \ldots 00}_{n \; times}$ and ending at $\underbrace{11 \ldots \ldots 11}_{n \; times}$ with $2^n - 1$. A clear illustration for this decimal labeling is given in Fig. 1(b).

Grid Labeling: For $1 \leq k \leq 2^a$, $1 \leq l \leq 2^b$, let $g(k, l)$ denote the node located in the kth row and l^{th} column of the grid $G[2^a \times 2^b]$. We shall split the grid labeling into two cases.

Case 1. For $1 \leq k \leq 2^a$, $1 \leq l \leq 2^{b-1}$,

$$g(k, l) = 2^b(k - 1) + 2(l - 1).$$

Case 2. For $1 \leq k \leq 2^a$, $2^{b-1} < l \leq 2^b$, labeling $g(k, l)$ is divided into three sub-cases.

Sub-case 2(a). When $l = 2^{b-1} + 1$, let

$$g(k, l) = \begin{cases} 2^b(k - 1) + 1 & : k = 1 \\ 2^b(k - 1) + 2^{b-1} + 1 : k = 2. \end{cases}$$

For $3 \leq k \leq 2^a$, let p be a positive integer such that $2^{p-1} < k \leq 2^p$. Then

$$g(k, l) = \begin{cases} 2^b(k - 1) + 1 & \& : g(k - 2^{p-1}, l) = 2^b(k - 2^{p-1} - 1) + 2^{b-1} + 1 \\ 2^b(k - 1) + 2^{b-1} + 1 & \& : g(k - 2^{p-1}, l) = 2^b(k - 2^{p-1} - 1) + 1. \end{cases}$$

Sub-case 2(b). When $l = 2^{b-1} + 2$,

$$g(k, l) = g(k, 2^{b-1} + 1) + 2.$$

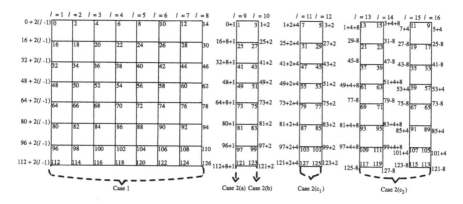

Fig. 2. The different cases of node labeling of $G[2^3 \times 2^4]$

Sub-case 2(c). For $2^{b-1} + 3 \leq l \leq 2^b$, we shall represent l in the form $l = 2^{b-1} + 2^{i-1} + j$, where $2 \leq i \leq b-1, 1 \leq j \leq 2^{i-1}$. The labeling of $g(k,l)$ for this case is subdivided into two cases.

Case 2c_1: Suppose $l = 2^{b-1} + 2^{i-1} + j, 1 \leq i \leq b-1, 1 \leq j \leq 2^{i-1}$,

$$g(k,l) = \begin{cases} g(k, 2^{b-1} + j) + 2^{i-1} + 2^i : & 1 \leq j \leq 2^{i-2} \\ g(k, 2^{b-1} + j) + 2^{i-1} & : 2^{i-2} < j \leq 2^{i-1}. \end{cases}$$

Case 2c_2: Suppose $l = 2^{b-1} + 2^{b-2} + j, 1 \leq j \leq 2^{b-2}$,

$$g(k,l) = \begin{cases} g(k, 2^{b-1} + j) + 2^{b-2} + 2^{b-1} : g(k, 2^{b-1} + 1) = 2^b(k-1) + 1 \\ \qquad\qquad\qquad\qquad\qquad\qquad\qquad\qquad\qquad\qquad\qquad \& \; j \leq 2^{b-3} \\ g(k, 2^{b-1} + j) + 2^{b-2} : g(k, 2^{b-1} + 1) = 2^b(k-1) + 1 \; \& \; j > 2^{b-3} \\ g(k, 2^{b-1} + 2^{b-2} - j + 1) - 2^{b-1} : g(k, 2^{b-1} + 1) = 2^b(k-1) \\ \qquad\qquad\qquad\qquad\qquad\qquad\qquad\qquad\qquad\qquad\qquad + 2^{b-1} + 1. \end{cases}$$

Figure 2 illustrates the node labeling pattern of a $2^3 \times 2^4$ grid.

3.2   Embedding Optimization

The following results are used to obtain the proof of optimization for the embedding.

For $1 \leq i \leq 2^a$, let $D_i = \{(i-1)2^b, (i-1)2^b + 1, \ldots, (i-1)2^b + (2^b - 1)\}$ and $R_i = \{D_1, D_2, \ldots, D_i\}$.

Lemma 4. *For $1 \leq i \leq 2^a$, R_i is an optimal set in LTQ_n.*

Proof. We prove this result by induction on i. By the recursive definition of locally twisted cubes, $LTQ_n[R_1] \simeq LTQ_b$ and hence $|E(LTQ_n[R_1])| = b.2^{b-1}$.

By Lemmas 2 and 3, R_1 is an optimal set in LTQ_n. Assuming that the result is true for $i = k - 1$, we prove that R_k is optimal in LTQ_n.

We first show that $LTQ_n[D_k] \simeq LTQ_b$ for every k, $1 \le k \le 2^a$. We have that $LTQ_n[D_1] \simeq LTQ_b$. We prove the hypothesis for every other k by showing that $LTQ_n[D_k]$ is isomorphic to $LTQ_n[D_{k-2^{p-1}}]$, where $1 \le p \le a$ and $2^{p-1} < k \le 2^p$. Define a mapping $\varphi : V(LTQ_n[D_{k-2^{p-1}}]) \to V(LTQ_n[D_k])$ such that $\varphi(x) = x + 2^{b+p-1}$.

Let the binary representation of x be $\underbrace{00\ldots00}_{(a-p)\ times} 0x_{a-p+2}\ldots x_n$. Then the binary representation of $\varphi(x)$ is $\underbrace{00\ldots00}_{(a-p)\ times} 1x_{a-p+2}\ldots x_n$.

For $j \in \{a-p+2,\ldots,n-2\}$, the bits variation of $(x,y) \in E(LTQ_n[D_{k-2^{p-1}}])$ and the corresponding $(\varphi(x),\varphi(y)) \in E(LTQ_n[D_k])$ is as follows.

$$x = \underbrace{000\ldots\ldots00}_{(a-p)\ times} 0x_{a-p+2}\ldots x_j\ldots x_{n-1}x_n \text{ and}$$

$$y = \begin{cases} \underbrace{00\ldots00}_{(a-p)\ times} 0\ldots\overline{x}_j(x_{j+1} \oplus x_n)\ldots x_{n-1}x_n : j \in \{a-p+2,\ldots,n-2\} \\ \underbrace{00\ldots00}_{(a-p)\ times} 0x_{a-p+2}\ldots\overline{x}_jx_n \qquad\qquad : j = n-1 \\ \underbrace{00\ldots00}_{(a-p)\ times} 0x_{a-p+2}\ldots x_{n-1}\overline{x}_j \qquad\qquad : j = n \end{cases}$$

then the binary representations of $\varphi(x)$ and $\varphi(y)$ are

$$\varphi(x) = \underbrace{000\ldots\ldots00}_{(a-p)\ times} 1x_{a-p+2}\ldots x_j\ldots x_{n-1}x_n \text{ and}$$

$$\varphi(y) = \begin{cases} \underbrace{00\ldots00}_{(a-p)\ times} 1\ldots\overline{x}_j(x_{j+1} \oplus x_n)\ldots x_{n-1}x_n : j \in \{a-p+2,\ldots,n-2\} \\ \underbrace{00\ldots00}_{(a-p)\ times} 1x_{a-p+2}\ldots\overline{x}_jx_n \qquad\qquad : j = n-1 \\ \underbrace{00\ldots00}_{(a-p)\ times} 1x_{a-p+2}\ldots x_{n-1}\overline{x}_j \qquad\qquad : j = n \end{cases}$$

Hence $(x,y) \in E(LTQ_n[D_{k-2^{p-1}}]) \Leftrightarrow$ for $j \in \{a-p+2,a-p+3,\ldots,n-2\}$, the binary representations of x and y differ in the jth bit and have either identical or different $(j+1)^{th}$ bit depending on x_n, the remaining bits being identical or their binary representations differ only in the $(n-1)^{th}$ or n^{th} bit $\Leftrightarrow$ the binary representations of $\varphi(x)$ and $\varphi(y)$ differ in the jth bit and have either identical or different $(j+1)^{th}$ bit depending on x_n, the remaining bits being identical or their binary representations differ only in the $(n-1)^{th}$ or n^{th} bit $\Leftrightarrow (\varphi(x),\varphi(y)) \in E(LTQ_n[D_k])$. Therefore φ is an isomorphism and hence $LTQ_n[D_k] \simeq LTQ_b$.

Let $E(LTQ_n[D_k] \wedge LTQ_n[R_{k-1}]) = \{(x,y) : x \in V(LTQ_n[D_k]) \text{ and } y \in V(LTQ_n[R_{k-1}])\}$. Let k be represented as $k = 2^{r_1} + 2^{r_2} + \ldots + 2^{r_q} + 1$ such that $r_1 > r_2 > \ldots > r_q \geq 0$. Each node in $LTQ_n[D_k]$ is adjacent to q nodes in $LTQ_n[R_{k-1}]$. For $m \in \{1, 2, \ldots, a\}$, the binary representation of x and y such that $(x, y) \in E(LTQ_n[D_k] \wedge LTQ_n[R_{k-1}])$ are of the form $x = x_1 x_2 \ldots x_{m-1} x_m \ldots x_{n-1} 0$ and $y = x_1 x_2 \ldots x_{m-1} \overline{x}_m x_{m+1} \ldots x_{n-1} 0$, when $x \in \{(k-1).2^b, (k-1).2^b+2, \ldots, (k-1).2^b+(2^b-2)\}$, $y \in \{0, 2, \ldots, (k-2).2^b+(2^b-2)\}$ and $x = x_1 \ldots x_{m-1} 1 x_{m+1} \ldots x_{n-1} 1$ and $y = x_1 \ldots x_{m-1} 0 \overline{x}_{m+1} \ldots x_{n-1} 1$, when $x \in \{(k-1).2^b+1, (k-1).2^b+3, \ldots, (k-1).2^b+(2^b-1)\}$, $y \in \{1, 3, 5, \ldots, (k-2).2^b+(2^b-1)\}$. Clearly, there is no edge (x, y) where $x \in \{(k-1).2^b, (k-1).2^b+2, \ldots, (k-1).2^b+(2^b-2)\}$ and $y \in \{1, 3, 5, \ldots, (k-2).2^b+(2^b-1)\}$ and vice versa. Therefore, $|E(LTQ_n[R_k])| = |E(LQ_n[R_{k-1}])| + |E(LTQ_n[D_k])| + |E(LTQ_n[D_k] \wedge LTQ_n[R_{k-1}])| = |E(LTQ_n[(ETO)_{k-1.2^b}])| + |E(LTQ_b)| + q(2^b) = |E(LTQ_n[(ETO)_{(k).2^b}])|$ and hence by Lemma 2, R_k is an optimal set in LTQ_n. $\qquad \square$

For $1 \leq j \leq 2^{b-1}$, let $S_j = \{o_{1j}, o_{2j}, \ldots, o_{2^a j}\}$ be the set of all nodes in the $(2^{b-1}+j)^{th}$ column of $G[2^a \times 2^b]$ taken in a distinct order, where o_{ij} is defined as follows.

1. For $i = 1, 2, 1 \leq j \leq 2^{b-1}$,

$$o_{ij} = \begin{cases} a_j & : i = 1 \\ a_{2^{b-1}+j} & : i = 2. \end{cases}$$

where a_k denotes the kth element of $(TO)_{2^b-1}$.

2. For $3 \leq i \leq 2^a$, $1 \leq j \leq 2^{b-1}$, let t be a positive integer such that $t \leq n$ and $2^{t-1} < i \leq 2^t$. Then

$$o_{ij} = \begin{cases} o_{(i-2^{t-1})j} + 2^{b+t-1} + 2^{b+t-2} : 2^{t-1} < i \leq 2^{t-1} + 2^{t-2} \\ o_{(i-2^{t-1})j} + 2^{b+t-2} \qquad : 2^{t-1} + 2^{t-2} < i \leq 2^t. \end{cases}$$

Let $CO_j = \{S_{2^{b-1}}, S_{2^{b-1}-1}, \ldots, S_{2^{b-1}-j+1}\}$.

Lemma 5. For $1 \leq j \leq 2^{b-1}$, CO_j is an optimal set in LTQ_n.

Proof. The proof consists of two parts. First we prove that for $1 \leq j \leq 2^{b-1}$, $LTQ_n[S_j]$ is isomorphic to a copy of Q_a. In the second part we prove that $|E(LTQ_n[CO_j])| = |E(Q_n[L_{j.2^a}])|$ for all j, which asserts the optimality of CO_j, according to Lemma 3.

For $1 \leq j \leq 2^{b-1}$, let $C_{2^1}^j = \{o_{1j}, o_{2j}\}$. For $2 \leq r \leq a$, $1 \leq j \leq 2^{b-1}$, let $C_{2^r}^j = C_{2^{r-1}}^j \cup X$, where $C_{2^{r-1}}^j = \{o_{ij} : 1 \leq i \leq 2^{r-1}\}$ and $X = \{o_{ij} : 2^{r-1} < i \leq 2^r\}$. We prove the first part by induction on r. By verification, $LTQ_n[C_{2^1}^j] \simeq Q_1$. Assuming that the result is true for $r = k-1$, we show that $LTQ_n[C_{2^k}^j] \simeq Q_k$. For this we first prove that $LTQ_n[X] \simeq LTQ_n[C_{2^{k-1}}^j]$.

Define a function $f : V(LTQ_n)[C_{2^{k-1}}^j] \rightarrow V(LTQ_n[X])$ such that

$$f(x) = \begin{cases} x + 2^{b+k-1} + 2^{b+k-2} : x \in C_{2^{k-2}j} \\ x + 2^{b+k-2} : x \in C_{2^{k-1}j} \setminus C_{2^{k-2}j}. \end{cases}$$

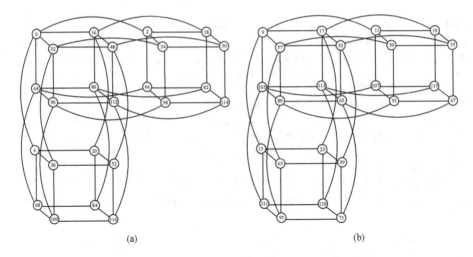

Fig. 3. (a) $LTQ_7[CE_3]$ (b) $LTQ_7[CO_3]$ both isomorphic to $Q_7[L_{24}]$

If the binary representation of x is $\underbrace{00\ldots00}_{(a-k)times} 0x_{a+k+2}\ldots x_{n-1}1$, then the binary representation of $f(x)$ is $\underbrace{00\ldots00}_{(a-k)times} 1\overline{x}_{a+k+2}x_{a+k+3}\ldots x_{n-1}1$. The bits variation of $(x,y) \in E(LTQ_n[C^j_{2^k-1}])$ and the corresponding $(f(x),f(y)) \in E(LTQ_n[X])$ for $i \in \{a-k+3, a-k+4\ldots, n-3\}$ is as follows.

The binary representations of x and y are

$$x = \underbrace{00\ldots00}_{(a-k)times} 0x_{a-k+2}\ldots x_i x_{i+1}\ldots x_{n-1}1 \text{ and}$$

$$y = \underbrace{00\ldots00}_{(a-k)times} 0x_{a-k+2}\ldots x_{i-1}\overline{x}_i\overline{x}_{i+1}x_{i+2}\ldots x_{n-1}1.$$

Then the binary representations of $f(x)$ and $f(y)$ are

$$f(x) = \underbrace{00\ldots00}_{(a-k)times} 1\overline{x}_{a-k+2}\ldots x_i x_{i+1}\ldots x_{n-1}1 \text{ and}$$

$$f(y) = \underbrace{00\ldots00}_{(a-k)times} 1\overline{x}_{a-k+2}\ldots x_{i-1}\overline{x}_i\overline{x}_{i+1}x_{i+2}\ldots x_{n-1}1.$$

Hence $(x,y) \in E(LTQ_n[C^j_{2^k-1}]) \Leftrightarrow$ there exists an $i \in \{a-k+3, a-k+4,\ldots, n-3\}$ such that the binary representations of x and y differ only in the ith and $(i+1)^{th}$ bits $\Leftrightarrow$ the binary representations of $f(x)$ and $f(y)$ differ only in the same i^{th} and $(i+1)^{th}$ bits $\Leftrightarrow (f(x),f(y)) \in E(LTQ_n[X])$. Therefore f is an isomorphism.

Next we prove that there is a perfect matching between $LTQ_n[C^j_{2^k-1}]$ and $LTQ_n[X]$. For any $o_{ij} \in V(LTQ_n[X])$ and $o_{(i-2^{k-1})j} \in V(LTQ_n[C^j_{2^k-1}])$, let the binary representation of o_{ij} be

$$o_{ij} = \alpha_1\alpha_2\ldots\alpha_{a-k}\alpha_{a-k+1}\ldots\alpha_{n-1}1$$

Then the binary representation of $o_{(i-2^p-1)j}$ is

$$o_{(i-2^{k-1})j} = \alpha_1\alpha_2\ldots\alpha_{a-k}\overline{\alpha}_{a-k+1}\overline{\alpha}_{a-k+2}\alpha_{a-k+3}\ldots\alpha_{n-1}1.$$

The binary representations of o_{ij} and $o_{(i-2^{k-1})j}$ differ only in the $(a-k+1)^{th}$ and $(a-k+2)^{th}$ bits. Hence $(o_{ij}, o_{(i-2^{k-1})j})$ is an edge in LTQ_n for $2^{k-1} < i \leq 2^k$, implying that there is a perfect matching between the two isomorphic components, similar to a hypercube connectivity. Therefore $LTQ_n[C_{2^k}^j] \simeq Q_k$ and $LTQ_n[S_j]$ is isomorphic to Q_a for all $j \in \{1, 2, \ldots, 2^{b-1}\}$.

We prove the second part by induction on j. Clearly $LTQ_n[CO_1] \simeq Q_a$. Hence $|E(LTQ_n[CO_1])| = |E(Q_n[L_{1.2^a}])|$. By Lemma 3, it is optimal. Assuming that the postulate is true for $j = q - 1$, we prove that $|E(LTQ_n[CO_q])| = |E(Q_n[L_{q.2^a}])|$.

$V(LTQ_n[CO_q])$ can be represented as $V(LTQ_n[CO_q]) = V(LTQ_n[CO_{q-1}]) \cup V(LTQ_n[S_{2^{b-1}-q+1}])$. Let $E(LTQ_n[CO_{q-1}] \wedge LTQ_n[S_{2^{b-1}-q+1}]) = \{(u, v) : u \in V(LTQ_n[CO_{q-1}]), v \in V(LTQ_n[S_{2^{b-1}-q+1}])\}$. Let q be represented as $q = 2^{s_1} + 2^{s_2} + \ldots + 2^{s_m} + 1$ such that $s_1 > s_2 > \ldots > s_m \geq 0$. For any q, there are $m.2^a$ edges in $E(LTQ_n[CO_{q-1}] \wedge LTQ_n[S_{2^{b-1}-q+1}])$.

For $t \in \{a+1, a+2, \ldots, n-1\}$, the binary representations of u and v such that (u, v) is an edge in $E(LTQ_n[CO_{q-1}] \wedge LTQ_n[S_{2^{b-1}-q+1}])$ are of the form

$$u = u_1u_2\ldots u_tu_{t+1}\ldots u_{n-1}1 \text{ and}$$

$$v = \begin{cases} u_1u_2\ldots\overline{u}_t\overline{u}_{t+1}\ldots u_{n-1}1 : t \in \{a+1, a+2\ldots, n-2\} \\ u_1u_2\ldots u_{t-1}\overline{u}_t1 \qquad : t = n-1. \end{cases}$$

Hence the number of edges in $LTQ_n[CO_q]$ is given by $|E(LTQ_n[CO_q])| = |E(LTQ_n[CO_{q-1}])| + |E(LTQ_n[S_{2^{b-1}-q+1}])| + |E(LTQ_n[CO_{q-1}] \wedge LTQ_n[S_{2^{b-1}-q+1}])| = |E(Q_n[L_{q-1.2^a}])| + |E(Q_a)| + m.2^a = |E(Q_n[L_{q.2^a}])|$. Figure 3(b) depicts $LTQ_7[CO_3]$. □

Lemma 6. *For $1 \leq j \leq 2^{b-1}$,*

$$CE_j = \begin{cases} 0, & 1 \times 2^b, & \ldots & , (2^a - 1) \times 2^b, \\ 2, & 1 \times 2^b + 2, & \ldots & , (2^a - 1) \times 2^b + 2, \\ & & \ldots & \\ & \& \ldots & & \\ & \& \ldots & & \\ 2(j-1), & 1 \times 2^b + 2(j-1), & \ldots, & (2^a - 1) \times 2^b + 2(j-1) \end{cases}$$

is an optimal set in LTQ_n.

Proof. We have to prove that $LTQ_n[CE_j]$ is isomorphic to $LTQ_n[(ETO)_{j.2^a}]$. But $j.2^a \leq 2^{n-1}$ and hence by Lemma 3, it is enough to show that $LTQn[CE_j]$ is isomorphic to $Q_n[L_{j.2^a}]$. Define a function $\pi : V(LTQ_n[CE_j]) \to V(Q_n[L_{j.2^a}])$ such that for $0 \leq g \leq 2^{a-1}, 0 \leq h \leq j-1, \pi(g \times 2^b + 2h) = h \times 2^a + g$. Let the binary representation of $g \times 2^b + 2h$ be $\alpha_1\alpha_2\ldots\alpha_{b-1}\beta_1\ldots\beta_{a+1}$. Then the binary representation of $h \times 2^a + g$ is $\beta_1\beta_2\ldots\beta_{a+1}\alpha_1\alpha_2\ldots\alpha_{b-1}$.

Two nodes $x = x_1x_2\ldots x_n$ and $y = y_1y_2\ldots y_n$ are adjacent in $LTQ_n[CE_j]$ $\Leftrightarrow$ there exists an integer $i \in \{1,2,\ldots,n-1\}$ such that $x_i = \overline{y}_i \Leftrightarrow \pi(x)$ and $\pi(y)$ differ only in the ith bit $\Leftrightarrow (\pi(x),\pi(y)) \in E(Q_n[L_{j\cdot 2^a}])$. Hence $LTQ_n[CE_j]$ is isomorphic to $Q_n[L_{j\cdot 2^a}]$. Figure 3(a) illustrates this result. $\square$

Proposed Embedding. Define an embedding f from LTQ_n into $G[2^a \times 2^b]$ such that $f(x) = x$, together with $p(u,v)$, a shortest path irrespective of choice in $G[2^a \times 2^b]$ between $f(u)$ and $f(v)$ for every $(u,v) \in E(LTQ_n)$.

Theorem 2. *The embedding $\prec f,p \succ$ from LTQ_n into $G[2^a \times 2^b]$ is optimal.*

Proof. Table 1 gives a list of certain edge cuts covering the entire edge set of $G[2^a \times 2^b]$ and the node set of the components obtained by the removal of these edge cuts, which is illustrated in Fig. 4.

Table 1. Edge cuts of $G[2^a \times 2^b]$

Edge Cuts	Type of Cuts	Components	V(Component)
X_i $i = 1,2,\ldots,2^a-1$	Horizontal cut	$A_i, \overline{A}_i$	$V(A_i) = R_i$
Y_{ej} $j = 1,2,\ldots,2^{b-1}-1$	Left to middle vertical cut	$B_{ej}, \overline{B}_{ej}$	$V(B_{ej}) = CE_j$
Y_{oj} $j = 1,2,\ldots,2^{b-1}-1$	Right to middle vertical cut	$B_{oj}, \overline{B}_{oj}$	$V(B_{oj}) = CO_j$
Z	Middle vertical cut	P_1, P_2	$V(P_1) = E_{2^n-1}$

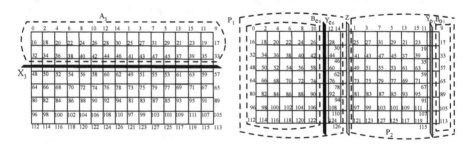

Fig. 4. Edge cuts along the rows and columns of $G[2^3 \times 2^4]$.

All the edge cuts satisfy conditions $(i) - (iii)$ of Lemma 1. In addition, $\{X_i : i = 1,2,\ldots,2^a-1\}\cup\{Y_{ej} : j = 1,2\ldots,2^{b-1}-1\}\cup\{Y_{oj} : j = 1,2,\ldots 2^{b-1}-1\}\cup Z$ is a partition of $E(G[2^a \times 2^b])$. Therefore by the definition of optimal embedding, $\prec f,p \succ$ is optimal. $\square$

4   Conclusion

In this paper we have devised a rigorous node labeling algorithm and proved that this labeling gives the optimal embedding of an n-dimensional locally twisted cube into a $2^a \times 2^b$ grid structure, where $a \leq b < n$, using Congestion Lemma and edge partitioning techniques. It would be an interesting line of research to find another elegant and simple node labeling pattern which induces an optimal embedding.

Acknowledgement. This work was supported by Project No. 5LCTOI14MAT002, Loyola College - Times of India, Chennai, India.

References

1. Arockiaraj, M., Abraham, J., Quadras, J., Shalini, A.J.: Linear layout of locally twisted cubes. Int. J. Comput. Math. (2015). doi:10.1080/00207160.2015.1088943
2. Battista, G.D., Eades, P., Tamassia, R., Tollis, I.G.: Algorithms for drawing graphs: an annotated bibliography. Comput. Geom. **4**, 235–282 (1994)
3. Bezrukov, S.L., Chavez, J.D., Harper, L.H., Röttger, M., Schroeder, U.-P.: The congestion of n-cube layout on a rectangular grid. Discret. Math. **213**(1), 13–19 (2000)
4. Bhatt, S.N., Leighton, F.T.: A framework for solving VLSI graph layout problems. J. Comp. Syst. Sci. **28**, 300–343 (1984)
5. Djidjev, H.N., Vrto, I.: Crossing numbers and cutwidths. J. Graph Algorithms Appl. **7**, 245–251 (2003)
6. Han, H., Fan, J., Zhang, S., Yang, J., Qian, P.: Embedding meshes into locally twisted cubes. Inform. Sci. **180**, 3794–3805 (2010)
7. Harper, L.H.: Optimal assignments of numbers to vertices. J. Soc. Ind. Appl. Math. **12**, 131–135 (1964)
8. Harper, L.H.: Global Methods for Combinatorial Isoperimetric Problems. Cambridge University Press, London (2004)
9. Lai, Y.L., Williams, K.: A survey of solved problems and applications on bandwidth, edgesum, and profile of graphs. J. Graph Theory **31**, 75–94 (1999)
10. Manuel, P., Rajasingh, I., Rajan, B., Mercy, H.: Exact wirelength of hypercube on a grid. Discret. Appl. Math. **157**(7), 1486–1495 (2009)
11. Opatrny, J., Sotteau, D.: Embeddings of complete binary trees into grids and extended grids with total vertex-congestion 1. Discret. Appl. Math. **98**, 237–254 (2000)
12. Rajasingh, I., Arockiaraj, M.: Linear wirelength of folded hypercubes. Math. Comput. Sci. **5**, 101–111 (2011)
13. Rajasingh, I., Arockiaraj, M., Rajan, B., Manuel, P.: Minimum wirelength of hypercubes into n-dimensional grid networks. Inform. Process. Lett. **112**, 583–586 (2012)
14. Rajasingh, I., Rajan, R.S., Parthiban, N., Rajalaxmi, T.M.: Bothway embedding of circulant network into grid. J. Discret. Algorithms. **33**, 2–9 (2015)
15. Yang, X., Evans, D.J., Megson, G.M.: The locally twisted cubes. Int. J. Comput. Math. **82**(4), 401–413 (2005)

Polynomial Time Algorithms
for Bichromatic Problems

Sayan Bandyapadhyay[1] and Aritra Banik[2(✉)]

[1] Computer Science, University of Iowa, Iowa City, USA
`sayan-bandyapadhyay@uiowa.edu`
[2] Department of Computer Science and Engineering,
Indian Institute of Technology, Jodhpur, India
`aritrabanik@gmail.com`

Abstract. In this article, we consider a collection of geometric problems involving points colored by two colors (red and blue), referred to as bichromatic problems. The motivation behind studying these problems is two fold; (i) these problems appear naturally and frequently in the fields like Machine learning, Data mining, and so on, and (ii) we are interested in extending the algorithms and techniques for single point set (monochromatic) problems to bichromatic case. For all the problems considered in this paper, we design low polynomial time exact algorithms. These algorithms are based on novel techniques which might be of independent interest.

1 Introduction

In discrete and computational geometry one of the most important classes of problems are those involving points colored by two colors (red and blue), referred to as bichromatic problems. These problems have vast applications in the fields of Machine learning, Data mining, Computer graphics, and so on. For example, one natural problem in learning and clustering is separability problem [19,21], where given a red and a blue set of points, the goal is to separate as many red (desirable) points as possible from the blue (non-desirable) points using geometric objects. Another motivation to study bichromatic problems is to extend the algorithms and techniques in discrete and computational geometry for single point set problems to bichromatic case (see the survey by Kaneko and Kano [25] and some more recent works [2–4,10,11]).

In this paper we consider two bichromatic problems that arise naturally in practice. Throughout the paper as coloring we refer to a function that maps points to the range $\{red, blue\}$. In the first problem, we are given two finite disjoint sets of points R and B in the plane, colored by red and blue, respectively. Denote the respective cardinality of R and B by n and m. Also let $T = R \cup B$. In the second problem, we are given a finite collection Q of pairs of points in the plane and we need to find a coloring of the points using red and blue such that certain optimality criterion is satisfied. For the sake of simplicity of exposition we assume that all the points are in general position. Assuming this, we formally define the two problems mentioned before.

© Springer International Publishing AG 2017
D. Gaur and N.S. Narayanaswamy (Eds.): CALDAM 2017, LNCS 10156, pp. 12–23, 2017.
DOI: 10.1007/978-3-319-53007-9_2

Maximum Red Rectangle Problem. A rectangle (of arbitrary orientation) is called *red* if it does not contain any blue points in its interior[1]. The *size* of a red rectangle is defined as the number of red points contained in it. The *maximum red rectangle problem* (MRR) is to find a red rectangle of maximum size. Such a rectangle will be referred to as a *maximum red rectangle*. We note that the version where the rectangles are constrained to be axes parallel is a special case of MRR.

Liu and Nediak [27] considered the axes parallel version of MRR and designed an algorithm that runs in $O(n^2 \log n + nm + m \log m)$ time and $O(n)$ space. Backer and Keil [8] improved this time bound to $O((n + m)\log^3(n + m))$ using a divide-and-conquer approach due to Aggarwal and Suri [1]. However, their algorithm uses $O(n \log n)$ space. For axes parallel squares, they designed an $O((n + m)\log(n + m))$ time algorithm. As far as we are concerned the general version of MRR was not considered before. However, two related problems involving rectangles of arbitrary orientation have been studied before. The first one is the largest empty rectangle problem (LER) where given a point set P, the goal is to find a rectangle of the maximum area which does not contain any point of P in its interior [1,14,15,28,29]. This problem can be solved in $O(|P|^3)$ time with $O(|P|^2)$ space [14]. The second problem is a bichromatic problem, where the goal is to find the rectangle that contains all the red points, the minimum number of blue points and has the largest area. This problem can be solved in $O(m^3 + n \log n)$ time [5].

Several variants of MRR have been studied in the past. Aronov and Har-Peled [6] considered the problem of finding a maximum red disk and gave a $(1-\epsilon)$-factor approximation algorithm with $O((n+m)\epsilon^{-2}\log^2(n+m))$ expected running time. Eckstein *et al.* [22] considered a variant of MRR for axes parallel hyperrectangles in high dimensions. They showed that, if the dimension of the space is not fixed, the problem is NP-hard. However, they presented an $O(n^{2d+1})$ time algorithm, for any fixed dimension $d \geq 3$. Later, Backer and Keil [7] have significantly improved this time bound to $O(n^d\log^{d-2}n)$. Bitner *et al.* [12] have studied the problem of computing all circles that contain the red points and as few blue points as possible in its interior. In [18] Cortes *et al.* have considered the following problem: find a largest subset of $R \cup B$ which can be enclosed by the union of two not necessarily disjoint, axes-aligned rectangles R_1 and R_2 such that R_1 (resp. R_2) contains only red (resp. blue) points.

Maximum Coloring Problem. We are given a collection $Q=\{(a_1,b_1),\ldots,(a_n,b_n)\}$ of pairs of points. Let $P = \cup_{i=1}^n\{a_i,b_i\}$. A coloring of the points in P is *valid* if for each pair of points in Q, exactly one point is colored by blue and the other point is colored by red. Now consider any valid coloring C. For any halfplane h, we denote the set of red points and blue points in h by $h_r(C)$ and $h_b(C)$, respectively. Let $\mathcal{H}(C)$ be the set of all halfplanes h such that $|h_b(C)| = 0$. Define $\eta(C) = \max_{h \in \mathcal{H}(C)} |h_r(C)|$. The *Maximum Coloring Problem* (MaxCol) is to find a valid coloring C that maximizes $\eta(C)$.

---

[1] The red rectangle may contain blue points on its boundary.

Problems related to coloring of points have been studied in literature. Even *et al.* [23] defined Conflict-free coloring where given a set X of points and a set of geometric regions, the goal is to color the points with minimum number of colors so that for any region r, at least one of the points of X that lie in r has a unique color. This problem have further been studied in offline [16,26] and online [9,17] settings.

Our Results. We give exact algorithms for both of the problems considered in this paper. In Sect. 2, we design an algorithm for MRR that runs in $O(g(n, m) \log(m + n) + n^2)$ time with $O(n^2 + m^2)$ space, where $g(n, m) \in O(m^2(n + m))$ and $g(n, m) \in \Omega(m^2 + mn)$. To solve the problem in general case, we need to solve an interesting subproblem which can be of independent interest. The nontriviality of this problem arises mainly due to the arbitrary orientations of the rectangles.

In Sect. 3, we show that MaxCol can be solved in $O(n^{\frac{4}{3}+\epsilon} \log n)$ time. In particular, we show a linear time reduction from MaxCol to a problem and design an algorithm for the latter problem with the desired time complexity.

2   Maximum Red Rectangle Problem

Before stepping into the general case let us consider the axes parallel version. We note that the ideas mentioned here for the axes parallel case have been noted before in [8].

2.1   The Axes Parallel Case

Note that the number of red rectangles can be infinite. But for the sake of computing a maximum red rectangle we can focus on the following set. Consider the set S of red rectangles such that each side of any such rectangle contains a point of B or is unbounded. We note that the candidate set for the axes parallel version of Largest Empty Rectangle problem (LER) on B is exactly S. Using this connection we will use several results from the literature of LER. Namaad *et al.* [15] proved that $|S| = O(m^2)$ and in expected case $|S| = O(m \log m)$. Orlowski [29] has designed an algorithm that computes the set S in $O(|S|)$ time. The algorithm requires two sorted orderings of B, one with respect to x-coordinates and the other with respect to y-coordinates.

To compute a maximum red rectangle we use Orlowski's algorithm. In each iteration when the algorithm computes a rectangle, we make a query for the number of red points inside the rectangle. Lastly, we return the rectangle that contains the maximum number of red points. Now, using $O(n \log n)$ preprocessing time and space, one can create a data structure that can handle orthogonal rectangular query in $O(\log n)$ time. Hence the axes parallel version of MPP can be solved in $O(|S| \log n + n \log n + m \log m)$ time with $O(n \log n + m)$ space required.

2.2 The General Case

For a point p, we denote its x and y coordinate by x_p and y_p, respectively. A rectangle is said to be *anchored* by a point set P if it contains the points of P on its boundary.

Like in the axes parallel case, in general case also the number of red rectangles can be infinite. However, we will define a finite subset of those rectangles such that the subset contains a maximum red rectangle. We note that this is a key step based on which we design our algorithm. The approach is not that subtle like the one for axes parallel case.

Consider the set of red rectangles C with the following two properties. For any rectangle $T \in C$, (i) at least one side of T contains two points p, q such that $p \in B$ and $q \in R \cup B$; and (ii) each of the other sides of T either contains a point of B or is unbounded. In the full version of this paper, we prove the following two lemmas:

Lemma 1. $|C| = O(m^2(n+m))$ *and* $|C| = \Omega(m^2 + mn)$.

Lemma 2. *The set C as defined above contains a maximum red rectangle.*

We compute all the rectangles of C and return one that contains the maximum number of red points. Given two points p, q such that $p \in B$ and $q \in R \cup B$, we design a subroutine to compute the rectangles of C such that each of them contains p and q on a single side.

2.2.1 The Subroutine

We are given two points p, q such that $p \in B$ and $q \in R \cup B$. We would like to compute the set of rectangles C_{pq} anchored by p and q. Without loss of generality, suppose the line through p and q is on the x-axis. We show how to compute the subset C' of rectangles of C_{pq} whose interior are lying above the x axis. By symmetry, using a similar approach we can compute the rectangles of C_{pq} whose interior are lying below the x axis. We consider all the blue points strictly above the x axis, and we denote this set by B_1.

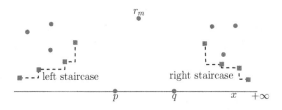

Fig. 1. Staircase points are shown by squares. (Color figure online)

Now consider the points of B_1 whose x coordinates are between x_p and x_q (if any). Let r_m be a point having minimum y coordinate among those points.

Let B' be the subset of points of B_1 having y coordinate less than y_{r_m}. Define the *left staircase* B_l to be the subset of B', such that for any point $r_2 \in B'$, r_2 belongs to B_l if $x_{r_2} < x_p$ and $\nexists r_1 \in B'$ such that $x_{r_2} < x_{r_1} < x_p$ and $y_{r_2} \geq y_{r_1}$ (see Fig. 1). Similarly, we define the *right staircase* B_r as follows. For any r_2 of B', $r_2 \in B_r$ if $x_q < x_{r_2}$ and $\nexists r_1 \in B$ such that $x_q < x_{r_1} < x_{r_2}$ and $y_{r_1} \leq y_{r_2}$ (see Fig. 1). Also let $B_a = B_l \cup B_r \cup \{r_m\}$. Now we have the following observation.

Observation 3. *For any rectangle $T \in C'$, T contains p, q in one of its sides and each of the other sides of T either contains a point of B_a or is unbounded.*

Orlowski [29] designed a linear time algorithm for finding the set of rectangles that has one side on x-axis and each of the other sides either contains a point of a staircase or is unbounded. Our subroutine at first computes the rectangle anchored by p, q and r_m (if any). Then it uses Orlowski's algorithm to compute the remaining rectangles of C'. It also computes the number of red points inside each rectangle it scans by making a query and returns the rectangle that contains the maximum number of red points. The following theorem is due to Goswami *et al.* [24].

Theorem 4. *[24] For any set of n points, using $O(n^2)$ preprocessing time and space, one can create a data structure that can handle rectangular (of any arbitrary orientation) query in $O(\log n)$ time.*

We note, that given two sorted lists of blue points in non-decreasing order of x and y coordinates, respectively, by Theorem 4 the subroutine runs in $O(|C'| \log n)$ time. Hence we have the following lemma.

Lemma 5. *Using $O(n^2 + m \log m)$ preprocessing time and $O(n^2 + m)$ space a maximum red rectangle of C_{pq} can be computed in $O(|C_{pq}| \log n)$ time.*

Note that this subroutine can be trivially used for finding a maximum red rectangle in C by calling it for each $p \in B$ and $q \in R \cup B$, and by returning a maximum red rectangle among all such choices. But, as we need to compute the sorted lists in every rotated plane (defined by p and q) we need $O(m^2(n + m) \log m)$ time in total just for sorting. Thus this trivial algorithm runs in $O(m^2(n + m) \log m + n^2 + |C| \log n)$ time. In the next subsection, we improve this time complexity by using a careful observation that rescues us from the burden of sorting in every rotated plane.

2.2.2 The Improved Algorithm

The rectangles of C are oriented at angles with respect to the x-axis in the range $[0, 360)$. Moreover, there are $O(m(n+m))$ orientations (or angles) of these rectangles each defined by a point of B and a point of $R \cup B$. Note that we need the sorted lists of blue points in each such orientation. We use a novel technique for maintaining the sorted lists using some crucial observations. We note that this problem itself might be of independent interest.

Consider two points $a, b \in B$ and an angle θ. We want to find the ordering of these two points with respect to x coordinates, in the plane oriented counterclockwise at the angle θ. Let us denote this plane by P_θ. Also let B_x^θ (resp. B_y^θ) be the set of blue points in P_θ sorted in increasing order of x (resp. y) coordinates. The following lemma explains how the relative ordering of two points gets changed with changes in plane orientation.

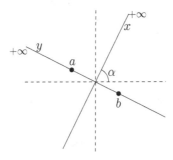

Fig. 2. The plane generated by rotation of axes by an angle α where $x_a = x_b$.

Lemma 6. *For any two points $a, b \in B$, there exists an angle $\phi < 180$ such that $x_a = x_b$ in P_ϕ and $P_{\phi+180}$, and exactly one of the following is true,*

(i) $x_a < x_b$ in P_θ for $\theta \in [0, \phi) \cup (\phi + 180, 360)$ and $x_a > x_b$ in P_θ for $\theta \in (\phi, \phi + 180)$

(ii) $x_a > x_b$ in P_θ for $\theta \in [0, \phi) \cup (\phi + 180, 360)$ and $x_a < x_b$ in P_θ for $\theta \in (\phi, \phi + 180)$

Proof. Denote the line segment connecting a and b (w.r.t P_0) by l. We translate a, b in a way so that the midpoint of l is now at the origin. Note that translation does not change the ordering of the points. Let f_a (resp. f_b) be the function such that $f_a(\theta)$ (resp. $f_b(\theta)$) is the x-coordinate value of a (resp. b) in P_θ. Now consider a continuous rotation process of the axes (or the plane) in counterclockwise direction with respect to the origin, keeping the points a, b fixed. If $f_a(0)$ is equal to $f_b(0)$, then a and b are on the y-axis in P_0. Let $\alpha = 0$ in this case. Otherwise, a and b are on the opposite sides of the y-axis. In this case, as we rotate the axes the value of $|f_a(\theta) - f_b(\theta)|$ becomes zero at some angle $\theta = \alpha$ when both a and b lie on the y-axis (see Fig. 2). Note that if $f_a(0) < f_b(0)$, $f_a(\theta) < f_b(\theta)$ for all $\theta \in [0, \alpha)$, as a and b do not change their sides with respect to the y-axis during this rotation process. Similarly, if $f_a(0) > f_b(0)$, $f_a(\theta) > f_b(\theta)$ for all $\theta \in [0, \alpha)$. Now in both of the cases ($f_a(0) = f_b(0)$ or $f_a(0) \neq f_b(0)$), after a slight rotation a and b will be on the opposite sides of the y-axis. As we rotate further the value of $|f_a(\theta) - f_b(\theta)|$ again becomes zero at some angle $\theta = \beta$ when again both a and b lie on the y-axis. In the first case, either $f_a(\theta) > f_b(\theta)$ for all $\theta \in (\alpha, \beta)$, or $f_a(\theta) < f_b(\theta)$ for all $\theta \in (\alpha, \beta)$. In the second case, if a was on the left (resp. right) of the y-axis in P_θ for $\theta \in [0, \alpha)$, now a is on the right (resp. left) side of the y-axis. Thus if $f_a(\theta) < f_b(\theta)$ for $\theta \in [0, \alpha)$, $f_a(\theta) > f_b(\theta)$ for all $\theta \in (\alpha, \beta)$. Similarly, if $f_a(\theta) > f_b(\theta)$ for $\theta \in [0, \alpha)$, $f_a(\theta) < f_b(\theta)$ for all $\theta \in (\alpha, \beta)$. As we rotate past β in both of the cases, a and b will again be on the opposite sides of the y-axis. If $f_a(\theta) < f_b(\theta)$ for $\theta \in (\alpha, \beta)$, $f_a(\theta) > f_b(\theta)$ for all $\theta \in (\beta, 360)$. Similarly, if $f_a(\theta) > f_b(\theta)$ for $\theta \in (\alpha, \beta)$, $f_a(\theta) < f_b(\theta)$ for all $\theta \in (\beta, 360)$.

We let $\phi = \alpha$. Note that $f_a(\theta)$ can be equal to $f_b(\theta)$ for $\theta \in [0, 360)$ only when y-axis is aligned with the line through a and b. Also note that during a complete $360°$ rotation of the axes this can happen exactly twice. Moreover, the difference between the corresponding angles should be 180, i.e. $\beta = \phi + 180$. This completes the proof of the lemma. □

Similarly, we get the following lemma for ordering of two points with respect to y-coordinates.

Lemma 7. *For any two points $a, b \in B$, there exists an angle $\phi < 180$ such that $y_a = y_b$ in P_ϕ and $P_{\phi+180}$, and exactly one of the following is true,*

(i) $y_a < y_b$ in P_θ for $\theta \in [0, \phi) \cup (\phi + 180, 360)$ and $y_a > y_b$ in P_θ, for $\theta \in (\phi, \phi + 180)$

(ii) $y_a > y_b$ in P_θ for $\theta \in [0, \phi) \cup (\phi + 180, 360)$ and $y_a < y_b$ in P_θ, for $\theta \in (\phi, \phi + 180)$

We refer to the angles ϕ and $\phi+180$ in Lemmas 6 and 7 as *critical angles* with respect to a and b. Let E_1 (resp. E_2) be the set of all critical angles corresponding to pairs of blue points with respect to x (resp. y) coordinates. An element of E_1 (resp. E_2) will be referred to as an event point of first (resp. second) type. Let A be the set of angles (or orientations) defined by pairs of points (p, q) such that $p \in B$ and $q \in R \cup B$. We refer to an element of A as an event point of third type. We construct the set $E = E_1 \cup E_2 \cup A$ of event points. Also we add the angle 0 to E as an event point of both first and second type.

Lemmas 6 and 7 hint the outline of an algorithm. We sort the list E of event points in increasing order of their angles. We process the event points in this order. To start with we construct the set B_x^0 (resp. B_y^0) at the very first event point which is 0. In case the algorithm encounters an event point of the first (resp. second) type it swaps the corresponding two points in the current sorted list B_x^ϕ (resp. B_y^ϕ), where $\phi \in [0, 360)$. At an event point of the third type it makes a call to the subroutine of Sect. 2.2.1. Lastly, a rectangle that contains the maximum red points over all orientations is returned as the solution.

The correctness of this algorithm depends on the correctness of maintaining the sorted lists in all the orientations. As the sorted order with respect to the x (resp. y) coordinates do not change in between two consecutive event points of the first (resp. second) type, it is sufficient to prove the following lemma.

Lemma 8. *At each event point ϕ of the first (resp. second) type, the correct sorted list B_x^ϕ (resp. B_y^ϕ) is maintained.*

Proof. We prove this lemma for the event points of the first type. The proof for the event points of the second type is similar. We use induction argument on the event points. In the base case for $\phi = 0$ we correctly compute the set B_x^0. Now assume that for $\alpha \in E_1$ and any $\phi \leq \alpha$ such that $\phi \in E_1$, the correct sorted list B_x^ϕ is maintained. Let β be the successor of α in E_1. Let a and b be the two points corresponding to β. By Lemma 6, the ordering of x_a and x_b should be different in P_ψ and P_θ for $\psi \in (\alpha, \beta)$ and $\theta \in (\beta, \beta + 180)$. Thus our algorithm rightly swaps a and b in B_x^α to get B_x^β. The location of any other point remains same. Now if there is any point c in B_x^α in between a and b, then the order of a and c (resp. b and c) also gets changed and we end up computing a wrong list. Thus it is sufficient to prove the following claim.

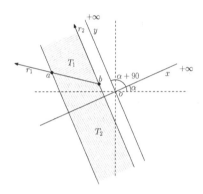

Fig. 3. The sets T_1 and T_2. T_1 is contained in the wedge defined by r_1 and r_2.

Claim 9. *a and b must be consecutive in B_x^α.*

Proof. We consider the case where $\beta < 180$. The other case is symmetric. Now consider the plane P_α. Without loss of generality assume that $x_a \leq x_b$ in this plane. Suppose there is a point c in B_x^α in between a and b. If $x_a = x_b$, then $x_a = x_c$. But this violates the general position assumption. Thus we consider the case where $x_a < x_b$. Let l_a (resp. l_b) be the line passing through a (resp. b) which is parallel to the y-axis. Let T be the strip between the two lines l_a and l_b, i.e. T contains all the points p such that $x_a \leq x_p \leq x_b$. Then c must be contained in T. Note that the line segment that connects a and b divides T into two sets T_1 and T_2, above and below $\overline{ab}$ respectively (see Fig. 3). Suppose c is in T_1. Consider two rays r_1 and r_2 both originated at b that pass through the point a and $(x_b, y_b + 1)$, respectively. Note that r_1 and r_2 are oriented at angles $\beta + 90°$ and $\alpha + 90°$, respectively. Also the wedge defined by r_1 and r_2 with angle less than $180°$ contains T_1 (see Fig. 3). Thus as c is in T_1 there is a critical angle θ with respect to b and c such that $\alpha \leq \theta < \beta$. If $\alpha < \theta < \beta$, we get a contradiction to the assumption that β is the next event point in E_1. If θ is equal to α, $x_c = x_b$. As $c \in T_1$, $y_c > y_b$. Thus in any plane P_θ with $\theta \in (\alpha, \beta)$, $x_b < x_c$. Hence if c appears before b in B_x^α, we get a contradiction to the assumption that B_x^α is the correct sorted list. In case c is in T_2, one can get similar contradiction. Hence the claim follows. □

Now consider the time complexity of our algorithm. Sorting of $O(m^2 + mn)$ event points takes $O((m^2 + mn) \log(m + n))$ time. Handling of any event point of first (resp. second) type takes constant time. By Lemma 5, handling of all the event points of third type takes in total $O(|C| \log n)$ time. As $|C| = \Omega(m^2 + mn)$ we get the following theorem.

Theorem 10. *MPP can be solved in $O(|C| \log(m+n)+n^2)$ time with $O(n^2+m^2)$ space required.*

3   Maximum Coloring Problem

Recall, that we are given the collection $Q = \{(a_1, b_1), \ldots, (a_n, b_n)\}$ of pairs of points and $P = \cup_{1 \le i \le n} \{a_i, b_i\}$. A coloring of the points in P is valid if for each pair of points in Q, exactly one point is colored by red and the other is colored by blue. For any such valid coloring C and any half plane h, we denote the number of red points and the number of blue points in h by $h_r(C)$ and $h_b(C)$ (see Fig. 4). For any valid coloring C, $\eta(C) = \max |h_r(C)|$, where the maximum is taken over all halfplanes h for which $|h_b(C)| = 0$.

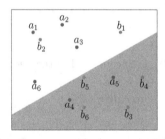

Fig. 4. An example of a valid coloring. For the halfplane h shown in grey, $h_r(C) = 4$ and $h_b(C) = 2$. (Color figure online)

In the Maximum Coloring Problem (MaxCol), the objective is to find a coloring C^* which maximizes $\eta()$ over all valid colorings. Next we show that a maximum valid coloring can be found by solving the following problem.

Problem 1. *Given a collection $Q = \{(a_1, b_1), \ldots, (a_n, b_n)\}$ of pair of points find a halfplane that contains at most one point from each pair and contains maximum number of points from P where $P = \cup_{1 \le i \le n} \{a_i, b_i\}$.*

The following lemma shows that it is sufficient to solve Problem 1 to get a solution for MaxCol.

Lemma 11. *Suppose there exist an algorithm $\mathcal{A}$ which finds an optimal solution to Problem 1 in time $T(n)$. Then MaxCol can be solved in $T(n) + O(n)$ time.*

Proof. Let the Algorithm $\mathcal{A}$ outputs the halfplane h^* and $|h^* \cap P| = k$. We prove the claim that given h^*, we can find a valid coloring C^*, where $\eta(C^*) = k$, which maximizes $\eta()$ over all valid colorings.

We define C^* as follows. Now for each pair (a_i, b_i) such that $|h^* \cap \{a_i, b_i\}| = 1$, we color the point in $h^* \cap \{a_i, b_i\}$ by red and the point in $h^* \setminus \{a_i, b_i\}$ by blue. For each pair (a_i, b_i) such that $|h^* \cap \{a_i, b_i\}| = 0$, we arbitrarily color one point by red and the other point by blue. Given h^* such a coloring can be found in $O(n)$ time.

Clearly this coloring is a valid coloring. Next we show that this is maximum as well. Suppose C^* is not maximum and there exists a valid coloring C such that $\eta(C^*) < \eta(C)$. Let h_C be the halfplane corresponding to C that contains only red points and $|P \cap h_C| = \eta(C)$. But then h_C is a half plane such that h_C contains at most one point from each pair in Q and $|P \cap h_C| > |P \cap h^*|$ which contradicts the optimality of h^*.   □

Instead of directly solving Problem 1, we consider its dual. We use the standard point/line duality that preserves the above/below relationship between points and lines. Let $\mathbb{L} = \{(l_1, \overline{l_1}), \ldots, (l_n, \overline{l_n})\}$ be the collection of the corresponding n pairs of lines in the dual space and let L be the set of those $2n$ lines. Also let A be the set of points p such that for each pair in $\mathbb{L}$, at most one line of the pair lie below p. For any $p \in A$, let $\nu(p)$ denote the number of lines in L that lie below p. Similarly, let B be the set of points p such that for each pair in $\mathbb{L}$, at most one line of the pair lie above p. For any $p \in B$, let $\nu(p)$ denote the number of lines in L that lie above p. Using duality, we get the following problem.

Problem 2. *Given a collection $\mathbb{L}$ of n pairs of lines in the plane, find a point in $A \cup B$ that maximizes the function $\nu()$.*

Here we describe how to find a point in A that maximizes $\nu()$. The point in B which maximizes $\nu()$ can be found similarly. More specifically we solve the following decision problem.

Problem 3. *Given a collection $\mathbb{L}$ of n pairs of lines in the plane and an integer k, does there exist a point p in A such that $\nu(p) = k$.*

If one can solve this decision version in time $T'(n)$, then Problem 2 can be solved in $O(T'(n) \log n)$ time by using a simple binary search on the values of k. Observe, that for an "YES" instance of Problem 3, the point p must be on the k-level in the arrangement of those $2n$ lines (the k-level of an arrangement of a set of lines is the polygonal chain formed by the edges that have exactly k other lines strictly below them). Thus it is sufficient to compute the k-level and decide if there is a point p on it such that $p \in A$. Our approach is the following.

Let Γ_k be the k-level. We traverse the vertices of Γ_k in sorted order of their x-coordinates. Throughout the traversal we maintain a list $Arr[]$ of size n, where $Arr[j]$ denotes the number of lines from $\{l_j, \overline{l_j}\}$ that are currently below the k-level. Note that the value of $Arr[j]$ can be 0, 1 or 2. We also maintain an integer n_b which denotes the number of pairs of lines currently below Γ_k. In other words n_b is the number of 2's in $Arr[]$. We update $Arr[]$ and n_b at each vertex of Γ_k. If a line belonged to a pair $\{l_j, \overline{l_j}\}$ leaves Γ_k, we reduce the value of $Arr[j]$ by one. Moreover, if the value of $Arr[j]$ was 2 before, we reduce n_b by one. Similarly, if a line belonged to a pair $\{l_j, \overline{l_j}\}$ enters Γ_k, we increase the value of $Arr[j]$ by one. If $Arr[j]$ becomes 2, we also increase n_b by one. At any point during the traversal if n_b becomes zero, we report yes. Otherwise we report no at the end.

Chan [13] designed an algorithm that computes the k-level in an arrangement of n lines in the plane in $O(n \log b + b^{1+\epsilon})$ time. From the result of Dey [20] we know $b = O(n(k+1)^{\frac{1}{3}})$. Thus k-level can be computed in $O(n \log n + nk^{\frac{1}{3}})$ time. Hence we have the following lemma.

Lemma 12. *Problem 3 can be solved in $O(n \log n + n^{1+\epsilon} k^{\frac{1+\epsilon}{3}})$ time.*

Hence by using a binary search on the values of k we have the following result.

Theorem 13. *Problems 1 and 2 can be solved in $O(n^{\frac{4}{3}+\epsilon} \log n)$ time.*

Acknowledgements. We would like to thank an anonymous reviewer of an earlier version of this paper for suggestions that has helped us improve the running time of the algorithm for MaxCol.

References

1. Aggarwal, A., Suri, S.: Fast algorithms for computing the largest empty rectangle. In: SoCG, Waterloo, Canada, pp. 278–290 (1987)
2. Arkin, E.M., Banik, A., Carmi, P., Citovsky, G., Katz, M.J., Mitchell, J.S.B., Simakov, M.: Conflict-free covering. In: CCCG, Kingston, Ontario, Canada, 10–12 August 2015
3. Arkin, E.M., Banik, A., Carmi, P., Citovsky, G., Katz, M.J., Mitchell, J.S.B., Simakov, M.: Choice is hard. In: Elbassioni, K., Makino, K. (eds.) ISAAC 2015. LNCS, vol. 9472, pp. 318–328. Springer, Heidelberg (2015). doi:10.1007/978-3-662-48971-0_28
4. Arkin, E.M., Díaz-Báñez, J.M., Hurtado, F., Kumar, P., Mitchell, J.S.B., Palop, B., Pérez-Lantero, P., Saumell, M., Silveira, R.I.: Bichromatic 2-center of pairs of points. Comput. Geom. 48(2), 94–107 (2015)
5. Armaselu, B., Daescu, O.: Maximum area rectangle separating red and blue points. In: CCCG 2016, British Columbia, Canada, 3–5 August 2016, pp. 244–251 (2016)
6. Aronov, B., Har-Peled, S.: On approximating the depth and related problems. SIAM J. Comput. 38(3), 899–921 (2008)
7. Backer, J., Keil, J.M.: The mono- and bichromatic empty rectangle and square problems in all dimensions. In: López-Ortiz, A. (ed.) LATIN 2010. LNCS, vol. 6034, pp. 14–25. Springer, Heidelberg (2010). doi:10.1007/978-3-642-12200-2_3
8. Backer, J., Mark Keil, J.: The bichromatic square and rectangle problems. Technical report 2009–01, University of Saskatchewan (2009)
9. Bar-Noy, A., Cheilaris, P., Smorodinsky, S.: Deterministic conflict-free coloring for intervals: from offline to online. ACM Trans. Algorithms 4(4), 44 (2008)
10. Biniaz, A., Bose, P., Maheshwari, A., Smid, M.: Plane bichromatic trees of low degree. In: Mäkinen, V., Puglisi, S.J., Salmela, L. (eds.) IWOCA 2016. LNCS, vol. 9843, pp. 68–80. Springer, Heidelberg (2016). doi:10.1007/978-3-319-44543-4_6

11. Biniaz, A., Maheshwari, A., Nandy, S.C., Smid, M.: An optimal algorithm for plane matchings in multipartite geometric graphs. In: Dehne, F., Sack, J.-R., Stege, U. (eds.) WADS 2015. LNCS, vol. 9214, pp. 66–78. Springer, Heidelberg (2015). doi:10. 1007/978-3-319-21840-3_6

12. Bitner, S., Cheung, Y.K., Daescu, O.: Minimum separating circle for bichromatic points in the plane. In: ISVD 2010, Quebec, Canada, June 28–30, 2010, pp. 50–55 (2010)

13. Chan, T.M.: Output-sensitive results on convex hulls, extreme points, and related problems. In: SOCG, Vancouver, B.C., Canada, 5–12 June 1995, pp. 10–19 (1995)

14. Chaudhuri, J., Nandy, S.C., Das, S.: Largest empty rectangle among a point set. J. Algorithms **46**(1), 54–78 (2003)

15. Chazelle, B., (Scot) Drysdale III, R.L., Lee, D.T.: Computing the largest empty rectangle. SIAM J. Comput. **15**(1), 300–315 (1986)

16. Cheilaris, P., Gargano, L., Rescigno, A.A., Smorodinsky, S.: Strong conflict-free coloring for intervals. Algorithmica **70**(4), 732–749 (2014)

17. Chen, K., Fiat, A., Kaplan, H., Levy, M., Matousek, J., Mossel, E., Pach, J., Sharir, M., Smorodinsky, S., Wagner, U., Welzl, E.: Online conflict-free coloring for intervals. SIAM J. Comput. **36**(5), 1342–1359 (2007)

18. Cortés, C., Díaz-Báñez, J.M., Pérez-Lantero, P., Seara, C., Urrutia, J., Ventura, I.: Bichromatic separability with two boxes: a general approach. J. Algorithms **64**(2–3), 79–88 (2009)

19. Cristianini, N., Shawe-Taylor, J.: An Introduction to Support Vector Machines: And Other Kernel-based Learning Methods. Cambridge University Press, New York (2000)

20. Dey, T.K.: Improved bounds on planar k-sets and k-levels. In: FOCS, Miami Beach, Florida, USA, 19–22 October 1997, pp. 156–161 (1997)

21. Duda, R.O., Hart, P.E., Stork, D.G.: Pattern Classification. Wiley, New York (2000)

22. Eckstein, J., Hammer, P.L., Liu, Y., Nediak, M., Simeone, B.: The maximum box problem and its application to data analysis. Comp. Opt. Appl. **23**(3), 285–298 (2002)

23. Even, G., Lotker, Z., Ron, D., Smorodinsky, S.: Conflict-free colorings of simple geometric regions with applications to frequency assignment in cellular networks. SIAM J. Comput. **33**(1), 94–136 (2003)

24. Goswami, P.P., Das, S., Nandy, S.C.: Triangular range counting query in 2d and its application in finding k nearest neighbors of a line segment. Comput. Geom. **29**(3), 163–175 (2004)

25. Kaneko, A., Kano, M.: Discrete geometry on red and blue points in the plane a survey. In: Aronov, B., Basu, S., Pach, J., Sharir, M. (eds.) Discrete and Computational Geometry, Algorithms and Combinatorics, vol. 25, pp. 551–570. Springer, Heidelberg (2003)

26. Katz, M.J., Lev-Tov, N., Morgenstern, G.: Conflict-free coloring of points on a line with respect to a set of intervals. Comput. Geom. **45**(9), 508–514 (2012)

27. Liu, Y., Nediak, M.: Planar case of the maximum box and related problems. In: CCCG 2003, Halifax, Canada, 11–13 August 2003, pp. 14–18 (2003)

28. Naamad, A., Lee, D.T., Hsu, W.-L.: On the maximum empty rectangle problem. Discret. Appl. Math. **8**(3), 267–277 (1984)

29. Orlowski, M.: A new algorithm for the largest empty rectangle problem. Algorithmica **5**(1), 65–73 (1990)

Voronoi Diagram for Convex Polygonal Sites with Convex Polygon-Offset Distance Function

Gill Barequet[1] and Minati De[2(✉)]

[1] Department of Computer Science, The Technion—Israel Institute of Technology, Haifa, Israel
barequet@cs.technion.ac.il
[2] Department of Computer Science and Automation, Indian Institute of Science, Bangalore, India
minati@csa.iisc.ernet.in

Abstract. The concept of convex polygon-offset distance function was introduced in 2001 by Barequet, Dickerson, and Goodrich. Using this notion of point-to-point distance, they showed how to compute the corresponding nearest- and farthest-site Voronoi diagram for a set of points. In this paper we generalize the polygon-offset distance function to be from a point to any convex object with respect to an m-sided convex polygon, and study the nearest- and farthest-site Voronoi diagrams for sets of line segments and convex polygons. We show that the combinatorial complexity of the nearest-site Voronoi diagram of n disjoint line segments is $O(nm)$, which is asymptotically equal to that of the Voronoi diagram of n point sites with respect to the same distance function. In addition, we generalize this result to the Voronoi diagram of disjoint convex polygonal sites. We show that the combinatorial complexity of the nearest-site Voronoi diagram of n convex polygonal sites, each having at most k sides, is $O(n(m + k))$. Finally, we show that the corresponding farthest-site Voronoi diagram is a tree-like structure with the same combinatorial complexity.

Keywords: Polygon-offset distance function · Nearest-site Voroni diagram · Farthest-site Voroni diagram · Convex polygonal sites · Line segments

1 Introduction

The Voronoi diagram is a very powerful tool which is widely used in diverse fields like epidemiology, ecology, computational chemistry, robot motion planning, and architecture, to name just a few. Being studied from as early as in the 17th century by Descartes [7], the topic has a vast literature. Many variations of Voronoi diagrams, differing by dimension, sites, measure of distance, etc., have been studied. To unify the concept, at least in two dimensions, Klein [11] introduced the notion of an *abstract* Voronoi diagram.

M. De—Supported by DST-INSPIRE Faculty Grant (DST-IFA14-ENG-75).

© Springer International Publishing AG 2017
D. Gaur and N.S. Narayanaswamy (Eds.): CALDAM 2017, LNCS 10156, pp. 24–36, 2017.
DOI: 10.1007/978-3-319-53007-9_3

Chew and Drysdale [6] proposed a Voronoi diagram for point sites with respect to convex distance function (also known as Minkowski functionals [2,9]), which is based on the notion of scaling of a convex polygon.

Barequet et al. [3] introduced the concept of convex polygon-offset distance function between two points. This function captures the distance according to the notion of shrinking or expanding a convex polygon along its medial axis. In the cited paper, the authors pointed out that polygon-offset distance functions are more natural than convex distance functions for many applications, e.g., manufacturing processes of physical three-dimensional objects, because a convex distance function varies, for the same underlying polygon, if different center (reference) points are used, which is not the case with polygon-offset distance function whose center is determined by the polygon. However, if the underlying polygon is not regular, the respective polygon-offset distance function does not satisfy the triangle inequality. In the same work, the authors also presented an algorithm for computing the Voronoi diagram for points with respect to this distance function.

In the current paper, we generalize the convex polygon-offset distance function to measure distances from a point to a convex object. This distance measure is useful, for example, to model the distribution of pollution expanded from an industrial zone to a residential area. Assume that an industrial zone $\mathcal{P}$ should be placed such that the pollution from $\mathcal{P}$ takes maximum time to reach any part of the residential area. Here, the residential area is described as a polygon set $\mathcal{S}$ and the (polygon-offset) distance is measured with respect to the industrial zone $\mathcal{P}$. The Voronoi vertices of the nearest-site Voronoi diagram are the candidates for the placement of the industrial zone. The vertex with the maximum distance to a site is the sought solution.

1.1 Related Work

The Voronoi diagram of point sites has been studied extensively in the literature. In the Euclidean metric, the combinatorial complexity of both the nearest- and farthest-site Voronoi diagram is $O(n)$, where n is the number of point sites [8] or disjoint line segments [1,16] in the diagram. These diagrams can be constructed in optimal $O(n \log n)$ time. For a set of k disjoint convex polygonal sites with total complexity n, the combinatorial complexity of the nearest-site Voronoi diagram in the Euclidean metric is also $O(n)$ [10,13,17]. McAllister et al. [14] presented an algorithm to compute a compact representation of the diagram in $O(k \log n)$ time using only $O(k)$ space. This compact representation can be used to answer closest-site queries in $O(\log n)$ time. Recently, Cheong et al. [5] showed that in the Euclidean metric, the combinatorial complexity of the farthest-site counterpart is also $O(n)$, and the diagram can be computed in $O(n \log^3 n)$ time. On a related note, Bohler et al. [4] introduced recently the notion of abstract higher-order Voronoi diagram and studied its combinatorial complexity.

With respect to the polygon-offset distance function, the combinatorial complexity of both nearest- and farthest-site Voronoi diagrams of a set of n points is $O(nm)$ [3], where m is the complexity of the underlying convex polygon. Both diagrams can be computed in expected $O(n \log n \log^2 m + m)$ time. Neither the combinatorial complexity nor the computation algorithm were investigated so far for sites which are disjoint line segments or convex polygons, with respect to the polygon-offset distance function.

1.2   Our Contribution

In this paper, we generalize the definition of the polygon-offset distance function $D_{\mathcal{P}}(p, S)$ to be from a point p to an object S (either a line segment or a convex polygon, rather than just to another point), where $\mathcal{P}$ is, as before, an m-sided convex polygon. Next, we study the properties and complexities of both nearest- and farthest-site Voronoi diagrams of disjoint line segments and convex polygons with respect to this distance function.

We show that the combinatorial complexity of the nearest-site Voronoi diagram of n disjoint line segments is $O(nm)$, which is asymptotically the same as that of the nearest-site Voronoi diagram of n point sites. Next, we prove that the combinatorial complexity of the nearest-site Voronoi diagram of n polygonal sites, each having at most k sides, is $O(n(m + k))$. Finally, we show that the farthest-site Voronoi diagrams for both line segments and convex polygons are tree-like structures; they have the same combinatorial complexity as their nearest-site counterpart.

1.3   Organization

First, we give some relevant definitions and preliminaries in Sect. 2. Then, the nearest-site Voronoi diagrams of line segments and convex polygons are studied in Sect. 3. The farthest-site Voronoi diagram is studied in Sect. 4. We end in Sect. 5 with some concluding remarks.

2   Preliminaries

We begin with illustrating the process of *offsetting* a convex polygon, following closely the description of Barequet et al. [3]. Given a convex polygon $\mathcal{P}$, described by the intersection of m closed half-planes $\{H_i\}$, an offset copy of $\mathcal{P}$, denoted as $O_{\mathcal{P}, \varepsilon}$, is defined as the intersection of the closed half-planes $\{H_i(\varepsilon)\}$, where $H_i(\varepsilon)$ is the half-plane parallel to H_i with bounding line translated by ε. Depending on whether the value of ε is positive or negative, the translation is done outward or inward of P. See Fig. 3(a) for an illustration. The value of $\varepsilon_0 < 0$ for which the $O_{\mathcal{P}, \varepsilon}$, degenerates into a point c (or a line segment s) is the *radius* of $\mathcal{P}$, and the point c (or any point on s) is referred to as the *center* of $\mathcal{P}$.

Using the above concept, Barequet et al. [3] defined the polygon-offset distance function $D_{\mathcal{P}}$ between two points as follows.

Definition 1 (Point to point distance [3]**).** *Let* z_1 *and* z_2 *be two points in* $\mathbb{R}^2$ *and* $O_{\mathcal{P},\varepsilon}$ *be an offset of* $\mathcal{P}$ *such that a translated copy of* $O_{\mathcal{P},\varepsilon}$, *centered at* z_1, *contains* z_2 *on its boundary. The offset distance is defined as* $D_{\mathcal{P}}(z_1, z_2) = \frac{\varepsilon + |\varepsilon_0|}{|\varepsilon_0|} = \frac{\varepsilon}{|\varepsilon_0|} + 1.$

Note that this distance function is not a metric since it is not symmetric. We generalize the offset distance function $D_{\mathcal{P}}$ to measure distance from any point z to any object o in $\mathbb{R}^2$ in a natural way, as follows.

Definition 2 (Point to object distance). *Let* z *be any point, and let* o *be any object in* $\mathbb{R}^2$. *The offset distance* $D_{\mathcal{P}}(z, o)$ *is defined as* $D_{\mathcal{P}}(z, o) = \min_{z' \in o} D_{\mathcal{P}}(z, z')$.

2.1 Properties of $D_{\mathcal{P}}$:

The following properties [3] also hold for the generalized distance function $D_{\mathcal{P}}$.

Property 1 [3, Sect. 3.1, Property 1]. The distance function $D_{\mathcal{P}}$ induces a Euclidean topology in the plane. In other words, each small neighborhood of a point contains an L_2-neighborhood of it, and vice versa.

Property 2 [3, Sect. 3.1, Property 2]. The distance between every pair of points is invariant under translation.

Property 3 [3, Theorem 6]. The distance function $D_{\mathcal{P}}$ is complete, and for each pair of points $z_1, z_3 \in \mathbb{R}^2$, there exists a point $z_2 \notin \{z_1, z_3\}$ such that $D_{\mathcal{P}}(z_1, z_2) + D_{\mathcal{P}}(z_2, z_3) = D_{\mathcal{P}}(z_1, z_3)$.

Property 4 [3, Theorem 7]. For each pair of points $z_1, z_2 \in \mathbb{R}^2$, there exists a point $z_3 \neq z_2\}$ such that $D_{\mathcal{P}}(z_1, z_2) + D_{\mathcal{P}}(z_2, z_3) = D_{\mathcal{P}}(z_1, z_3)$.

It is easy to observe the following property of $D_{\mathcal{P}}$.

Observation 1. *Let* z *be a point and* o *be a convex object. Then, the function* $D_{\mathcal{P}}(z, o)$ *increases monotonically when* z *moves along a ray originating at a point on the boundary of* o *and never crossing it again.*

2.2 Definitions of Voronoi Diagram

Under the convex polygon offset distance function, the bisector of two points (as defined originally [3]) can be 2-dimensional instead of 1-dimensional (see Fig. 1 for an illustration). This makes the Voronoi diagram of points unnecessarily complicated. To make it simple, as is also defined by Klein and Woods [12], we redefine the bisector and Voronoi diagram with respect to the offset distance function $D_{\mathcal{P}}$ as follows.

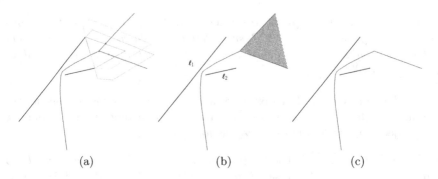

(a)   (b)   (c)

Fig. 1. (a) Three different positions of the offset polygons from where both line segments are equidistant; (b) The bisector of two line segments according to the definition used in [3]; and (c) The bisector according to our definition.

Let z be a point, and $\Pi = \{\sigma_i\}$ a set of objects in the plane. In order to avoid 2-dimensional bisectors between two objects in Π, we define the index of the objects as the "tie breaker" for the relation '$\prec$' between distances from z to the sites. That is,

$$D_{\mathcal{P}}(z, \sigma_i) \prec D_{\mathcal{P}}(z, \sigma_j)$$

if $D_{\mathcal{P}}(z, \sigma_i) < D_{\mathcal{P}}(z, \sigma_j)$ or, in case $D_{\mathcal{P}}(z, \sigma_i) = D_{\mathcal{P}}(z, \sigma_j)$, if $i < j$.[1] Note that the relation '$\prec$' does not allow equality if $i \neq j$. Therefore, the definition below uses the *closure* of portions of the plane in order to have proper boundaries between the regions of the diagram.

Definition 3 (Nearest-site Voronoi diagram). *Let $\Pi = \{\sigma_1, \sigma_2, \ldots, \sigma_n\}$ be a set of n sites in $\mathbb{R}^2$. For any $\sigma_i, \sigma_j \in \Pi$, we define the region of σ_i with respect to σ_j as $NV_{\mathcal{P}}^{\sigma_j}(\sigma_i) = \{z \in \mathbb{R}^2 | D_{\mathcal{P}}(z, \sigma_i) \prec D_{\mathcal{P}}(z, \sigma_j)\}$. The bisecting curve $B_{\mathcal{P}}(\sigma_i, \sigma_j)$ is defined as $\overline{NV_{\mathcal{P}}^{\sigma_j}(\sigma_i)} \cap \overline{NV_{\mathcal{P}}^{\sigma_i}(\sigma_j)}$, where $\overline{X}$ is the closure of X. The region of a site σ_i in the Voronoi diagram of Π is defined as $NV_{\mathcal{P}}(\sigma_i) = \{z \in \mathbb{R}^2 | D_{\mathcal{P}}(z, \sigma_i) \prec D_{\mathcal{P}}(z, \sigma_j) \forall j \neq i\}$. The nearest-site Voronoi diagram is the union of the regions: $NVD_{\mathcal{P}}(\Pi) = \bigcup_i NV_{\mathcal{P}}(\sigma_i)$.*

In other words, the diagram $NVD_{\mathcal{P}}(\Pi)$ is a partition of the plane, such that if a point $p \in \mathbb{R}^2$ has more than one closest site, then it belongs to the region of the site with the smallest index. The bisectors between regions are defined by taking the closures of the open regions. The farthest-site Voronoi diagram is defined analogously.

---

[1] A disadvantage of this approach is that relabeling of the input sites will change the diagram. One can adopt the rule of Klein and Wood [12], who break ties by the lexicographic order of the input points, but with such a solution, the Voronoi diagram will not be invariant under rotation of the plane.

3   Nearest-Site Voronoi Diagram

3.1   Line Segments

Let $\mathscr{S} = \{s_1, s_2, \ldots, s_n\}$ be a set of n line segments in $\mathbb{R}^2$. First, we study the combinatorial complexity of the nearest-site Voronoi diagram $\mathrm{NVD}_\mathcal{P}(\mathscr{S})$ with respect to $D_\mathcal{P}$, where $\mathcal{P}$ is an m-sided convex polygon. We use the abstract Voronoi diagram paradigm of Klein and Wood [12].

Theorem 2   [12, Theorem 4.6]. *If $D_\mathcal{P}$ satisfies Property 1, Property 3, and Property 4, then all Voronoi regions are simply connected.*

As a result, we have the following.

Lemma 3. *Every Voronoi region $NV_\mathcal{P}(s_i)$ in $NVD_\mathcal{P}(\mathscr{S})$ is simply connected.*

Let s_1, s_2 be two line segments in the plane, and $B(s_1, s_2)$ and $B_\mathcal{P}(s_1, s_2)$ be the bisectors of s_1 and s_2 with respect to the Euclidean distance and the polygon-offset distance $D_\mathcal{P}$, respectively. In the sequel, we will use the term *polyline* to denote a piecewise-simple curve all of whose elements are described by low-degree polynomials, like line segments and parabolic arcs. It is well known, then, that $B(s_1, s_2)$ is a polyline. Let N_t be the normal to the bisector $B(s_1, s_2)$ at a point t on the bisector. Now, observe the following.

Observation 4. *If we move a point x from t along N_t towards s_1, then $D_\mathcal{P}(x, s_2)$ increases monotonically and $D_\mathcal{P}(x, s_1)$ behaves like a convex function. At the minimum of $D_\mathcal{P}(x, s_1)$, we have that $D_\mathcal{P}(x, s_1) \leq D_\mathcal{P}(x, s_2)$ (see Fig. 3(b) for an illustration). A similar claim holds (with the roles of s_1 and s_2 exchanged) when we move x from t in the opposite direction along N_t (towards s_2).*

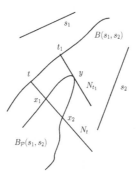

Fig. 2. Illustration of Lemma 5.

Now, we prove the following.

Lemma 5. *For any two line segments s_1 and s_2, the curve $B_\mathcal{P}(s_1, s_2)$ is monotone with respect to $B(s_1, s_2)$.*

Proof. We need to show that for any point t on the Euclidean bisector $B(s_1, s_2)$, the normal N_t to the bisector $B(s_1, s_2)$ at t intersects $B_\mathcal{P}(s_1, s_2)$ at most once. Assume to the contrary that there exist on N_t two points x_1 and x_2, such that $\mathcal{D}_\mathcal{P}(x_1, s_1) = \mathcal{D}_\mathcal{P}(x_1, s_2)$ and $\mathcal{D}_\mathcal{P}(x_2, s_1) = \mathcal{D}_\mathcal{P}(x_2, s_2)$. Assume further, without loss of generality, that both x_1 and x_2 are closer to s_2 than to s_1 (see Fig. 2). Then, since both $B(s_1, s_2)$ and $B_\mathcal{P}(s_1, s_2)$ are continuous, there is a point t_1 on $B(s_1, s_2)$ for which the line N_{t_1} is tangent to $B_\mathcal{P}(s_1, s_2)$, say, at point y. Thus, there exist a direction along which the two functions $\mathcal{D}_\mathcal{P}(x, s_1)$ and $\mathcal{D}_\mathcal{P}(x, s_2)$ are tangent but do not cross. This can happen only where both functions are monotone increasing or monotone decreasing. However, in light of Observation 4, this behavior cannot happen near the $\mathcal{D}_\mathcal{P}$-bisector, where one function decreases and the other increases (see Fig. 3(b)), which is a contradiction. The claim follows.   □

Corollary 1. *For any two line segments s_1 and s_2, the curve $B_\mathcal{P}(s_1, s_2)$ is unbounded.*

Lemma 6

  (i) *Let s_1, s_2 be two line segments in the plane. The bisecting curve $B_\mathcal{P}(s_1, s_2)$ is a polyline with $O(m)$ arcs and line segments.*
 (ii) *Two different bisecting curves intersect $O(m)$ times.*

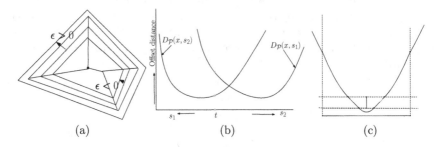

(a)   (b)   (c)

Fig. 3. (a) Different offset copies of a polygon; (b) Illustration of Observation 4; and (c) Euclidean bisector of two line segments.

Proof

  (i) Let us swipe a point t along the bisector $B_\mathcal{P}(s_1, s_2)$ from end to end, and characterize every position of t by (a) which elements (vertices or sides) of the current offset of $\mathcal{P}$ (which now touches simultaneously s_1 and s_2) touch the two segments; and (b) the location of the touching points on the two segments (either one of the endpoints, or a point internal to the segment, in which case we also distinguish between the two sides of the segment). We call this characterization the *status* of the bisector at t; see Fig. 4(a) for an illustration. During this sweep, a new basic piece of the bisector is manifested by a change in the status. In between such changes, the bisector is a line segment or a parabolic arc, depending of the elements of the status. Estimating the complexity of the bisector boils down to setting an upper bound on the total number of such changes in the status.

There are two crucial issues to observe. First, when t is swept along $B_{\mathcal{P}}(s_1, s_2)$, the offset-distance from $B_{\mathcal{P}}(s_1, s_2)$ to s_1 (or s_2) first decreases monotonically until a minimum point (or an interval of one minimum value, in the degenerate case in which the two segments are parallel and their mutual orthogonal projection is non-empty), and then it increases monotonically. This implies that $\mathcal{P}$ first shrinks continuously (a process in which sides only "disappear") and then expands (the opposite process in which sides only "appear"). During this process, the touching point moves cyclically around $B_{\mathcal{P}}$. Hence, the status can change $O(m)$ times due to changes of the touching element of $B_{\mathcal{P}}$. The fact that $B_{\mathcal{P}}$ first shrinks and then expands does not change this bound—it only implies that some elements of $\mathcal{P}$ are skipped without contributing to these changes.

Second, the touching point on each segment also rotates about it. This means that each segment contributes an additional constant number of changes to the status.

In conclusion, the total number of changes of the status during the sweep is $O(m)$.

(ii) Let $B_{\mathcal{P}}(s_1, s_2)$ and $B_{\mathcal{P}}(s_3, s_4)$ be the two bisectors (with respect to $D_{\mathcal{P}}$ of two pairs of segments s_1, s_2 and s_3, s_4. As we already know, the complexity of each one of them is $O(m)$. Every pair of basic elements of the two bisectors may intersect at most two times (as they are partial polynomials of degree 1 or 2). Suppose that we advance along the two bisectors from end to end and detect one (or two) intersection(s) between the same basic pieces of the bisectors. It is crucial to observe that if we continue further along the one of the bisectors switching to another basic piece, and discover another intersection, this intersection must lie on a basic element which is *further* along the other bisector. This follows from the monotonicity of the polygon-offset distance functions. Indeed, as seen in Fig. 1(b), one can decompose the range of the parameter t into three intervals: One in which both functions decrease, one in which one increases and the other decreases, and one in which both functions increase. Hence, the number of intersections behaves like a merge between two ordered lists, each of complexity $O(m)$: The complexity of the merged list is $O(m)$ as well. □

Theorem 7 [12, Theorem 2.5]. *Assume that a distance function $D_{\mathcal{P}}$ induces the Euclidean topology in the plane. Furthermore, assume that each bisector consists of disjoint simple curves, and curves belonging to different bisectors can intersect only finitely often within each bounded area. Finally, assume that all possible Voronoi regions are connected sets. Then $NVD_{\mathcal{P}}(\mathscr{S})$, where $\mathscr{S}$ is a set of n sites, has n faces and $O(n)$ edges and vertices.*

Property 1, Lemma 3, and Lemma 6(i) ensure that all the assumptions in Theorem 7 are satisfied. As a corollary to the theorem above, and due to Lemma 6(ii), we have the following.

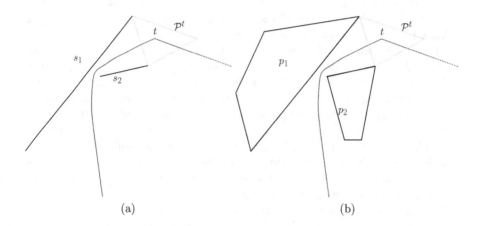

Fig. 4. Status configuration for (a) line segments, and (b) convex polygons.

Theorem 8. *For a set $\mathscr{S}$ of n line segments, the combinatorial complexity of the Voronoi diagram $NVD_{\mathcal{P}}(\mathscr{S})$ is $O(nm)$, where m is the number of sides of $\mathcal{P}$.*

Remark. The bound $O(nm)$ on the complexity of the diagram is attainable. For example, put n line segments more or less aligned horizontally and well spaced, so that most of the bisector of every pair of consecutive segments is present in the diagram, for a total complexity of $\Omega(nm)$. Hence, we conclude that the complexity of the diagram is $\Theta(nm)$ in the worst case.

3.2   Convex Polygons

We can easily extend the sites from line segments to convex polygons. Let $\mathscr{Q} = \{p_1, p_2, \ldots, p_n\}$ be a set of n convex polygonal sites, each having at most k sides, and let $\mathrm{NVD}_{\mathcal{P}}(\mathscr{Q})$ be the nearest-site Voronoi diagram of these sites with respect to the convex polygon-offset distance function $D_{\mathcal{P}}$, where $\mathcal{P}$ is an m-sided convex polygon. With similar arguments given for Lemmata 3 and 5 for the nearest-site Voronoi diagram of a set of line segments (with respect to $D_{\mathcal{P}}$), we can prove the following.

Lemma 9. *Every Voronoi region $NV_{\mathcal{P}}(p_i)$ in the nearest-site Voronoi diagram $NVD_{\mathcal{P}}(\mathscr{Q})$ is simply-connected.*

Lemma 10. *For any convex polygons p_1 and p_2, the curve $B_{\mathcal{P}}(p_1, p_2)$ is monotone with respect to $B(p_1, p_2)$.*

In the same manner, the following generalizes Lemma 6 to deal with polygonal sites.

Lemma 11

(i) *The bisecting curve $B_{\mathcal{P}}(p_1, p_2)$ of a pair of convex polygons p_1, p_2, each having at most k sides, is a polyline with $O(m + k)$ arcs and segments.*

(ii) *Two such bisecting curves intersect $O(m + k)$ times.*

The proof of the lemma above is identical to that of Lemma 6 with the following refinements.

1. The offset of $\mathcal{P}$ can touch any of the up to k corners and k sides of each of the sites.
2. When swiping the point t along the bisector, the touching points on the two sites move sequentially along their boundaries without turning back, therefore, decomposing the bisector according to the touching points contributes at most $4k$ additional pieces.
3. Except that, the proofs of the two parts of the lemmas are identical. Therefore, the complexity $O(m)$ is replaced by $O(m + k)$.

Similarly to the argument in the proof of Theorem 8, in the polygonal case Property 1, Lemma 9, and Lemma 11 also ensure that all the assumptions in Theorem 7 are satisfied, and due to Lemma 11(ii), we have the following.

Theorem 12. *For a set $\mathcal{Q}$ of n convex polygons, each having at most k sides, the combinatorial complexity of the Voronoi diagram $NVD_{\mathcal{P}}(\mathcal{Q})$ is $O(n(m+k))$, where m is the number of sides of $\mathcal{P}$.*

4   Farthest-Site Voronoi Diagram

We are given a set $\mathscr{S} = \{\sigma_1, \sigma_2, \ldots, \sigma_n\}$ of sites (line segments or convex polygons) in $\mathbb{R}^2$. In this section, following the framework of Mehlhorn et al. [15] (and generalizing the approach of Barequet et al. [3, Sect. 5]), we obtain the combinatorial complexity of the farthest-site Voronoi diagram of $\mathscr{S}$ with respect to the distance function $D_{\mathcal{P}}$, where $\mathcal{P}$ is a convex polygon with m sides.

Let σ_i, σ_j be a pair of sites in $\mathscr{S}$. As in Sect. 2, we define $\mathrm{NV}_{\mathcal{P}}^{\sigma_j}(\sigma_i)$ to contain all the points in the plane that are closer to σ_i than to σ_j with respect to $D_{\mathcal{P}}$. Let us also define the *dominant set* $M(\sigma_i, \sigma_j)$ to be equal to $\mathrm{NV}_{\mathcal{P}}^{\sigma_j}(\sigma_i)$ *except* the bisecting curve $\delta M(\sigma_i, \sigma_j) = B_{\mathcal{P}}(\sigma_i, \sigma_j)$.

Consider the family $M = \{M(\sigma_i, \sigma_j) | 1 \le i \ne j \le n\}$. The family M is a called a *dominance system* if for all $\sigma_i, \sigma_j \in S$, the following properties are satisfied:

1. $M(\sigma_i, \sigma_j)$ is open and non-empty;
2. $M(\sigma_i, \sigma_j) \cap M(\sigma_j, \sigma_i) = \emptyset$ and $\delta M(\sigma_i, \sigma_j) = \delta M(\sigma_j, \sigma_i)$; and
3. $\delta M(\sigma_i, \sigma_j)$ is homeomorphic to the open interval $(0, 1)$.

Similarly to the argument given in the cited work [3, Theorem 16], we can prove the following.

Lemma 13. *The family M is a dominance system.*

Proof. We argue why the properties of a dominance system are satisfied.

1. $M(\sigma_i, \sigma_j)$ is open because $D_{\mathcal{P}}$ is monotone and continuous.
   Let ℓ be a line which splits the plane into two parts and separating between σ_i and σ_j. Without loss of generality, assume that ℓ is vertical and that σ_i is to the left side of ℓ. In this situation, all the points which are further to the left of σ_i belong to $M(\sigma_i, \sigma_j)$. Hence, $M(\sigma_i, \sigma_j)$ is always non-empty.
2. Centering $\mathcal{P}$ at any point $r \in \mathbb{R}^2$, if we "pump" $\mathcal{P}$ up, then either it hits σ_i first, or σ_j first, or hits simultaneously σ_i and σ_j. In the first case, $r \in M(\sigma_i, \sigma_j)$; in the second case, $r \in M(\sigma_j, \sigma_i)$; and in the third case, if $D_{\mathcal{P}}(r, \sigma_i) \prec D_{\mathcal{P}}(r, \sigma_j)$ (that is, if $i < j$), then $r \in M(\sigma_i, \sigma_j)$, otherwise $r \in M(\sigma_j, \sigma_i)$. Hence, $M(\sigma_i, \sigma_j) \cap M(\sigma_j, \sigma_i) = \emptyset$. On the other hand, $\delta M(\sigma_i, \sigma_j) = B_{\mathcal{P}}(\sigma_i, \sigma_j) = \overline{\mathrm{NV}_{\mathcal{P}}^{\sigma_j}(\sigma_i)} \cap \overline{\mathrm{NV}_{\mathcal{P}}^{\sigma_i}(\sigma_j)} = B_{\mathcal{P}}(\sigma_j, \sigma_i) = \delta M(\sigma_j, \sigma_i)$.
3. $\delta M(\sigma_i, \sigma_j)$ is homeomorphic to the open interval $(0, 1)$ since $B_{\mathcal{P}}(\sigma_i, \sigma_j)$ is a polyline (see Lemma 11(i)). $\qquad\square$

A dominance system is *admissible* if it also satisfies the following properties.

4. Any two bisecting curves intersect finitely-many times.
5. For all non-empty subsets S' of S, and for every reordering of indices in S,
   (a) The (nearest neighbor) Voronoi cell of every sites $\sigma_i \in S'$ (with respect to S') is connected and has a non-empty interior.
   (b) The union of all the (nearest neighbor) Voronoi cells of all sites $\sigma_i \in S'$ (with respect to S') is the entire plane.

A dominance system which satisfies only Properties 4 and 5(b) is called *semi-admissible*.

It is easy to verify from Lemmata 9 and 11(ii) that the family M is admissible. Thus, we have the following.

Theorem 14. *The family M is admissible.*

The fact that M is semi-admissible suffices for our purposes. Consider now the family M^*, the "dual" of M, in which the dominance relation as well as the ordering of the sites are reversed. From [15], we have the following theorem.

Theorem 15. [3,15] *If M is semi-admissible, then M^* is semi-admissible too. Moreover, the farthest site Voronoi diagram that corresponds to M^* is identical to the nearest site Voronoi diagram that corresponds to M.*

As a consequence, we conclude the following.

Theorem 16. *Let $\mathcal{P}$ be a convex polygon with m sides.*

(i) *For a set of n line segments, the combinatorial complexity of the farthest-site Voronoi diagram (with respect to $D_{\mathcal{P}}$) is $O(nm)$.*

(ii) For a set of n convex polygons, each having at most k sides, the combinatorial complexity of the farthest-site Voronoi diagram (with respect to $D_{\mathcal{P}}$) is $O(n(m + k))$.

As we could show that the farthest-site Voronoi diagram can be defined as a dominance system, we could benefit from all the results of [15] on this diagram, namely, that it is a tree.

5 Conclusion

We investigate the combinatorial complexity of the nearest- and farthest-site Voronoi diagram of line segments or convex polygons under a convex polygon-offset distance function. It would be interesting to see how fast one can compute these diagrams. Another important direction for future research is to investigate higher-order Voronoi diagram in this setting.

References

1. Aurenhammer, F., Drysdale, R.L.S., Krasser, H.: Farthest line segment Voronoi diagrams. Inf. Process. Lett. **100**(6), 220–225 (2006)
2. Aurenhammer, F., Klein, R., Lee, D.-T.: Voronoi Diagrams and Delaunay Triangulations. World Scientific, Singapore (2013)
3. Barequet, G., Dickerson, M.T., Goodrich, M.T.: Voronoi diagrams for convex polygon-offset distance functions. Discret. Comput. Geom. **25**(2), 271–291 (2001)
4. Bohler, C., Cheilaris, P., Klein, R., Liu, C.-H., Papadopoulou, E., Zavershynskyi, M.: On the complexity of higher order abstract Voronoi diagrams. Comput. Geom. Theory Appl. **48**(8), 539–551 (2015)
5. Cheong, O., Everett, H., Glisse, M., Gudmundsson, J., Hornus, S., Lazard, S., Lee, M., Na, H.-S.: Farthest-polygon Voronoi diagrams. Comput. Geom. Theory Appl. **44**(4), 234–247 (2011)
6. Chew, L.P., Drysdale III, R.L.S.: Voronoi diagrams based on convex distance functions. In: O'Rourke, J. (ed.) Proceedings of the First Annual Symposium on Computational Geometry, Baltimore, Maryland, USA, 5–7 June 1985, pp. 235–244. ACM (1985)
7. Descartes, R.: Principia Philosophiae. Ludovicus Elzevirius, Amsterdam (1644)
8. Fortune, S.: A sweepline algorithm for Voronoi diagrams. Algorithmica **2**, 153–174 (1987)
9. Kelley, J.L., Namioka, I.: Linear Topological Spaces. Springer, Heidelberg (1976)
10. Kirkpatrick, D.G.: Efficient computation of continuous skeletons. In: 20th Annual Symposium on Foundations of Computer Science, San Juan, Puerto Rico, 29–31 October 1979, pp. 18–27. IEEE Computer Society (1979)
11. Klein, R.: Concrete and Abstract Voronoi Diagrams. Lecture Notes in Computer Science, vol. 400. Springer, Heidelberg (1989)
12. Klein, R., Wood, D.: Voronoi diagrams based on general metrics in the plane. In: Cori, R., Wirsing, M. (eds.) STACS 1988. LNCS, vol. 294, pp. 281–291. Springer, Heidelberg (1988). doi:10.1007/BFb0035852

13. Leven, D., Sharir, M.: Planning a purely translational motion for a convex object in two-dimensional space using generalized Voronoi diagrams. Discret. Comput. Geom. **2**, 9–31 (1987)
14. McAllister, M., Kirkpatrick, D.G., Snoeyink, J.: A compact piecewise-linear Voronoi diagram for convex sites in the plane. Discret. Comput. Geom. **15**(1), 73–105 (1996)
15. Mehlhorn, K., Meiser, S., Rasch, R.: Furthest site abstract Voronoi diagrams. Int. J. Comput. Geom. Appl. **11**(6), 583–616 (2001)
16. Papadopoulou, E., Dey, S.K.: On the farthest line-segment Voronoi diagram. Int. J. Comput. Geometry Appl. **23**(6), 443–460 (2013)
17. Yap, C.-K.: An $O(n \log n)$ algorithm for the Voronoi diagram of a set of simple curve segments. Discret. Comput. Geom. **2**, 365–393 (1987)

Optimum Gathering of Asynchronous Robots

Subhash Bhagat$^{(\boxtimes)}$ and Krishnendu Mukhopadhyaya

ACM Unit, Indian Statistical Institute, Kolkata, India
{sbhagat_r,krishnendu}@isical.ac.in

Abstract. This paper considers the problem of *gathering* a set of asynchronous robots on the two dimensional plane under the additional requirement that the maximum distance traversed by the robots should be minimized. One of the implications of this optimization criteria is the energy efficiency for the robots. The results of this paper are two folds. First, it is proved that multiplicity detection capability is not sufficient to solve the constrained gathering problem for a set of oblivious robots even when the robots are fully synchronous. The problem is then studied for the robots having $O(1)$ bits persistent memory and a distributed algorithm is proposed for the problem in this model for a set of $n \geq 5$ robots. The proposed algorithm uses only two bits of persistent memory.

Keywords: Asynchronous · Gathering · Swarm robots · Robots with lights

1 Introduction

A *swarm* of robots is a distributed system of small, autonomous, inexpensive mobile robots designed to work cooperatively to achieve some goal. They can execute some task which is beyond the capability of a single robot. The system is usually a collection of autonomous (without any centralized control), homogeneous (same capabilities), anonymous (without any identity i.e., indistinguishable) robots. They do not have any explicit communication. The implicit communications are achieved via observing the positions of other robots using their endowed sensors. They do not have any global coordinate system. However, each robot has its own local coordinate system. The directions and orientations of the axes and the unit distances of the local coordinate systems may vary from robot to robot. The robots may be oblivious (they do not remember any information from their past computations).

At any point of time, a robot is either active or inactive (idle). An active robot operates in *Look-Compute-Move* cycles. In the *Look* phase, it takes a snapshot of its surrounding to capture the locations of other robots. In the *Compute* phase, it computes a destination point using information collected in the *Look* phase. Finally, it moves towards the computed destination point in the *Move* phase. An inactive robot does not perform any action i.e., it is in sleep mode. To solve a variety of problems, robots are endowed with some additional capabilities. *Weak multiplicity detection* helps a robot to identify multiple occurrences of robots at

© Springer International Publishing AG 2017
D. Gaur and N.S. Narayanaswamy (Eds.): CALDAM 2017, LNCS 10156, pp. 37–49, 2017.
DOI: 10.1007/978-3-319-53007-9_4

a single point. Whereas, *strong multiplicity detection* enables the robots also to count the total number of robots occupying the same location. The robots may have an agreement on a common orientation (clockwise direction) i.e., *chirality*. They may have some agreement on the directions of their local coordinate axes. In memory model, robots are endowed with externally visible lights, which can assume a constant number of predefined colours, to indicate their states [8,10].

Depending on the timings of the operations and activation schedules of the robots, three types of models are used. The most general model is the *asynchronous (ASYNC or CORDA)* model [14]. The activation of the robots are arbitrary and independent of each other. The time spans of the operations are finite but unpredictable. Thus, robots may compute on some obsolete data. In the *semi-synchronous (SSYNC or ATOM)* [16] model, robots operate in rounds and a subset of robots is activated simultaneously in each round. The operations are instantaneous and hence a robot is not observed while in motion. The unpredictability lies in the activated subset of robots in each round. The most restrictive of these three is the *fully synchronous (FSYNC)* model in which all robots are activated in all rounds. We assume a fair scheduler which activates each robot infinitely often [9].

A variety of problems can be solved by designing proper coordination strategies. Fundamental geometric problems like *gathering, circle formation, arbitrary pattern formation, flocking, scattering* etc. have been studied extensively in the literature. The *gathering* problem is defined as follows: a swarm of robots, in the two dimensional plane, should coordinate their motions in such way that in finite time all of them meet at a single point which is not defined in advanced. The *constrained gathering* problem asks robots to achieve gathering by minimizing the maximum distance traversed by any robot.

1.1   Earlier Works

Considering different schedulers and different capabilities of the robots, a variety of solutions have been proposed by researchers for the *gathering* problem. The primary goal of these works has been understanding the minimal set of capabilities of the robots which enables the robots to gather at a point not known in advanced, under different scheduling models.

- *FSYNC Model:* In this model the gathering problem is solvable without any extra assumption [10].
- *SSYNC Model:* Suzuki and Yamashita [17] proved that gathering of $n = 2$ robots is impossible without any agreement on the local coordinate systems even with strong multiplicity detection. Prencipe [15] studied the problem for $n > 2$ robots and proved that there does not exist any deterministic algorithm for the gathering problem in absence of multiplicity detection and any form of agreement on the local coordinate systems [15]. Bramas and Texeuil [3] presented an algorithm to solve the problem in the presence of arbitrary number of crashed robots, when the robots are endowed only with strong multiplicity detection capability. A study of probabilistic gathering was presented by Défago et al. [9].

– *ASYNC Model:* Cieliebak et al. [6] solved the gathering problem for $n > 2$ robots, with weak multiplicity detection capability. Bhagat et al. [2] presented a fault-tolerant distributed algorithm, for $n \geq 2$ robots with agreement in one direction, which solves the problem in the presence of arbitrary number of crashed robots even if robots are opaque i.e., they obstruct the visibility of the other robots. Flocchini et al. [11] showed that gathering is possible in *limited visibility* model (the robots can see up to a limited radius around themselves) if robots have agreements in the directions and orientations of both the axes. The gathering problem for robots, represented as unit discs (*fat robots*), have also been investigated by the researchers [1,4,7]. A restricted version of the gathering problem has been studied in [5] where robots are asked to gather at any one of the predefined fixed points (known as meeting points) on the plane. Their solution also satisfies the constraint that the maximum distance traversed by any robot is minimized. To the best of our knowledge, this is the only work which considers the constrained version of the gathering problem, but for a restricted model. This paper considers the general version of the problem.

The use of externally visible lights was first suggested by Peleg [13]. Das et al. [8] investigated the characterizations of the model, in which robots are endowed with externally visible lights. In memory model, the gathering problem for $n = 2$ robots (also known as *rendezvous*), was studied under different restrictions. The studies of [18] and [8] proposed solutions to the *rendezvous* problem, using the lights for both internal memory and communication purpose. Flocchini et al. [12] investigated the possibilities of solving the *rendezvous* problem in two models: (i) the lights are used only to remember internal states and (ii) light are used for communication purposes.

1.2   Our Contribution

In this paper, we study the constrained gathering problem for a set of autonomous robots. The contribution of this paper is in two parts. While the gathering problem is solvable for $n > 2$ asynchronous robots with weak multiplicity detection only, it is shown that even in the $FSYNC$ model, multiplicity detection capability is not sufficient to solve the constrained gathering problem for oblivious robots. A distributed algorithm is then presented to solve the problem in finite time for a set of $n \geq 5$ asynchronous, oblivious robots under the assumptions that robots are endowed with only two bits of memory. We do not make any extra assumption like agreements in coordinate systems, unit distance and chirality, rigidity of movements. In spite of these weak assumptions, we have showed that the constrained gathering problem is solvable for asynchronous robots using only four colours. Our solution also provides collision free movements for the robots. To the best of our knowledge, this paper is the first attempt to study the constrained gathering problem, in general, for asynchronous robots. One may view this constrained version of gathering problem as a solution to energy efficiency.

2   General Model and Definitions

The robots are autonomous, homogeneous, anonymous, asynchronous in nature. They are considered as points in the two dimensional plane and they can move freely on the plane. We consider the $ASYNC$ model ($CORDA$). Each robot has its own local coordinate system (Cartesian coordinate system) having origin at its current position. The directions and orientations of the axes and unit distances of the local coordinate systems may differ. They do not have any common chirality. All the measurements are done with respect to the local coordinate systems of the robots. We assume that initially all the robots occupy distinct points. The visibility range of the robots is unlimited. The movements of the robots are non-rigid i.e., a robot may stop before reaching its destination point and start a fresh computational cycle. However, there exists a constant $\delta > 0$ such that it moves at least a distance $minimum\{\delta, d\}$ towards its destination point where d is the distance of its destination point from its current position. It assures finite time reachability of the robots to their respective destinations. The value of δ is not known to the robots. Since the timing of operations by the robots are unpredictable, a robot may be observed by other robots in the system while it is in motion and the computation of robot may be done on some obsolete data.

- **Configuration of the Robots:** Let $\mathcal{R} = \{r_1, r_2, \ldots, r_n\}$ denote the set of n robots. A robot configuration is denoted by the multi set $\mathcal{R}(t) = \{r_1(t), \ldots, r_n(t)\}$, where $r_i(t)$ is the position of the robot r_i at time t. Let $\tilde{\mathcal{R}}$ denote the set of all such robot configurations. We assume that the initial configuration $\mathcal{R}(t_0)$ does not contain any multiplicity point (a point having multiple robots on it).
- **Smallest Enclosing Circle:** Let $SEC(\mathcal{R}(t))$ denote the smallest enclosing circle of the points in $\mathcal{R}(t)$ and $\mathcal{O}_t$ denote its centre. Let $C_{out}(t)$ denote the set of robot positions on the circumference of $SEC(\mathcal{R}(t))$ and $C_{int}(t)$ the set of robot positions lying within $SEC(\mathcal{R}(t))$. When there is no ambiguity, we use $SEC(t)$ instead of $SEC(\mathcal{R}(t))$.
- Let $\overline{ab}$ denote the closed line segment joining two points a and b (including the end points a and b) and (a, b) the open line segment joining the points a and b (excluding the two end points a and b). By $|a, b|$, we denote the distance between the points a and b. For two sets A and B, by $A \backslash B$ we denote the set difference of A and B. The angle between two given line segments is considered as the angle which is less than or equal to π.

We use the following terms, as defined in [3], to describe our algorithm:

- **View of a Robot:** For a robot $r_i \in \mathcal{R}$, the view $\mathcal{V}(r_i(t))$ of r_i is defined as the set of polar coordinates of the points in $\mathcal{R}(t)$ where the polar coordinate system of r_i is defined as follows: (i) the centre of the coordinate system is $r_i(t)$ and (ii) the point $(1, 0)$ is $\mathcal{O}_t$ if $r_i(t) \neq \mathcal{O}_t$, otherwise it is any point $r_k(t) \neq r_i(t) \in \mathcal{R}(t)$ that maximizes $\mathcal{V}(r_k(t))$. The orientation of the polar coordinate system should maximize $\mathcal{V}(r_i(t))$. The view of each robot is defined uniquely. While comparing the views of two robots, lexicographic sorting is used.

- **Rotational Symmetry:** An equivalence relation $\sim$ is defined on $\mathcal{R}(t)$ as follows: $\forall r_i(t), r_j(t) \in \mathcal{R}(t)$, $r_i(t) \sim r_j(t)$ iff $\mathcal{V}(r_i(t)) = \mathcal{V}(r_j(t))$ with same orientation. Let $sym(\mathcal{R}(t))$ denote the cardinality of the largest equivalence class defined by $\sim$. The set $\mathcal{R}(t)$ is said to have rotational symmetry if $sym(\mathcal{R}(t)) > 1$.

- **Successor:** Given a robot configuration $\mathcal{R}(t)$ and a fixed point $c \in \mathbb{R}^2$, the clockwise successor of a point $r_i(t) \in \mathcal{R}(t)$ around c, denoted by $S(r_i(t), c)$, is the point $r_j(t) \in \mathcal{R}(t)$ defined as follows: if $(c, r_i(t))$ contains at least one point of $\mathcal{R}(t)$, then $r_j(t)$ is the point in $\mathcal{R}(t) \cap (c, r_i(t))$ which minimizes $|r_i(t), r_j(t)|$. Otherwise, $r_j(t)$ is the point in clockwise direction such that $\angle(r_i(t), c, r_j(t))$ contains no other point of $\mathcal{R}(t)$ and $|c, r_j(t)|$ is maximized. The counter-clockwise successor of r_i can be defined analogously. The kth *clockwise successor* of $r_i(t)$ around c, denoted by $S^k(r_i(t), c)$, is defined by the recursive relation: *for* $k > 1$, $S^k(r_i(t), c) = S(S^{k-1}(r_i(t), c))$, $S^1(r_i(t), c) = S(r_i(t), c)$ and $S^0(r_i(t), c) = r_i(t)$.

- **String of Angles:** Let $SA(r_i(t), c)$ denote the string of angles $\alpha_1(t), \alpha_2(t), \ldots, \alpha_m(t)$ where $m = n - mult(c)$, $mult(c)$ is the number of robots at the point c and $\alpha_i(t) = \angle(S^{i-1}(r_i(t)), c, S^i(r_i(t)))$. The length of $SA(r_i(t), c)$ is $|SA(r_i(t), c)| = m$. The string $SA(r_i(t), c)$ is *k-periodic* if there exists a constant $1 \leq k \leq m$ such that $SA(r_i(t), c) = X^k$, where X is a sub-string of $SA(r_i(t), c)$. The *periodicity* of $SA(r_i(t), c)$, denoted by $per(SA(r_i(t), c))$, is the largest value of k for which $SA(r_i(t), c)$ is *k-periodic*.

- **Regularity:** A robot configuration $\mathcal{R}(t)$ is said to be *regular* if $\exists$ a point $c \in \mathbb{R}^2$ and an integer m such that $per(SA(r_i(t), c)) = m > 1$ and the *regularity* of $\mathcal{R}(t)$ is denoted by $reg(\mathcal{R}(t)) = m$. The point c is called the centre of regularity.

- **Quasi Regularity:** A robot configuration $\mathcal{R}(t)$ is said to be *quasi regular* or *Q-regular* iff $\exists$ a configuration $\mathcal{B}(t)$ and a point $c \in \mathbb{R}^2$ such that $reg(\mathcal{B}(t)) > 1$, c is the centre of regularity of $\mathcal{B}(t)$ and $p \in \mathcal{R}(t) \backslash \mathcal{B}(t)$, $p = c$. In other words, $\mathcal{B}(t)$ can be obtained from $\mathcal{R}(t)$ by moving the robot positions located at c, if any, along particular half lines starting at c (including c). If $\mathcal{R}(t)$ is Q-regular, then the point c is called as the centre of Q-regularity. By c_q, we denote the centre of Q-regularity. The quasi-regularity of $\mathcal{R}(t)$ is denoted $qreg(\mathcal{R}(t)) = reg(\mathcal{B}(t))$. If $\mathcal{R}(t)$ is not quasi-regular, then $qreg(\mathcal{R}(t)) = 1$.

Fact 1. *Let $\mathcal{R}(t_0)$ be an initial robot configuration. Then the centre $\mathcal{O}_{t_0}$ of $SEC(t_0)$ is the unique point which minimizes the maximum distance from any point in $\mathcal{R}(t_0)$ to it.*

For a set of points $\mathcal{A}$, the *Weber point* of $\mathcal{A}$ is a point c which minimizes the sum of distances of all the points in $\mathcal{A}$ to it. It is known that the Weber point is not computable, in general, for a set of more than four points.

Fact 2. *Weber point of a non-linear configuration is unique and this point remains invariant under the straight movements of the robots towards it.*

Fact 3. *If a configuration has multiple lines of symmetry, then it is Q-regular. However, the converse is not true.*

Fact 4. *The centre of Q-regularity of a non-linear configuration coincides with its unique Weber point and it is computable in finite time.*

Fact 5. *For a set A of $n \geq 3$ points, there exists a subset $B \subseteq A$ such that $|B| \leq 3$ and the smallest enclosing circles of A and B are same.*

We state the following theorem without proof:

Theorem 1. *The constrained gathering problem for a set of oblivious robots is deterministically unsolvable even with strong multiplicity detection capability under the FSYNC model.*

3   Gathering Algorithm with Persistent Memory

By Fact 1, for an initial configuration $\mathcal{R}(t_0)$, the centre $\mathcal{O}_{t_0}$ of the circle $SEC(\mathcal{R}(t_0))$ is the only candidate for the gathering point which satisfies the optimization criteria. But this point does not remain invariant under the movements of the robots. Thus, strategies are designed in such a way that even when configuration is changed, the robots can identify the point $\mathcal{O}_{t_0}$. We assume that each robot has two bits of persistent memory. This is implemented by endowing the robots with externally visible lights. These lights can assume a finite number of colours, each colour indicates a different state. The colours do not change automatically and they are persistent. The lights are used in two different ways: one way is to remember the robot's own state and the other way is to broadcast its current state i.e., for both internal memory and communication purpose [12]. A robot can identify the colours of the lights of all the robots in the system even when multiple robots occupy same position. It may be noted that strong multiplicity detecting follows directly from it; the light model provides some additional powers to the robots also. Except for the colour of its light, robots are oblivious i.e., they do not remember any other information of its previous computational cycles. Our algorithm assumes total four colours i.e., two bits memory.

3.1   States of the Robots

The robots use colours for their external lights to indicate their current states. This set of colours is denoted by $\mathcal{X}$. Let $\mathcal{R}(t_0)$ be an initial configuration. Following are the list of states and their corresponding colours: *yellow* indicates that it is an inactive robot or it has not changed its colour yet, *red* indicates that it has found $\mathcal{R}(t_0)$ as a Q-regular configuration in which the centre of $SEC(\mathcal{R}(t_0))$ coincides with centre of Q-regularity i.e., $\mathcal{O}_{t_0} = c_q$, *green* indicates that it has found either $\mathcal{R}(t_0)$ as not Q-regular or as Q-regular with $\mathcal{O}_{t_0} \neq c_q$ and initially the robot was not at $\mathcal{O}_{t_0}$ and *blue* indicates that it has found either

$\mathcal{R}(t_0)$ as not Q-regular or as Q-regular with $\mathcal{O}_{t_0} \neq c_q$ and it is at $\mathcal{O}_{t_0}$. Thus, $\mathcal{X} = \{yellow, red, green, blue\}$. Let $s_i(t)$ denote the colour of the light for the robot r_i at time t.

3.2 Configurations

Let $\mathcal{R}_c(t)$ denote the set of all tuples $(r_i(t), s_i(t))$ for all $r_i(t) \in \mathcal{R}(t)$ and $s_i(t) \in \mathcal{X}$. The set of all such $\mathcal{R}_c(t)$ is denoted by $\widetilde{\mathcal{R}}_c$ where $\mathcal{R}(t) \in \widetilde{\mathcal{R}}$. We partition $\widetilde{\mathcal{R}}_c$ into the following classes:

- **Central ($\mathcal{CL}$):** A robot configuration $\mathcal{R}_c(t)$ belongs to this class if it satisfies exactly one of the following: (i) $\mathcal{R}_c(t)$ contains a multiplicity point p_m such that at least one robot at p_m has *red* light or (ii) $\mathcal{R}_c(t)$ contains a point such that at least one robot at this point has *blue* light. A configuration in this class is called a central configuration and the multiplicity point or the point with blue robot is called the central point.
- **Q*-regular ($\mathcal{QR}_0$):** A robot configuration $\mathcal{R}_c(t)$, containing no multiplicity point, is in this class if (i) $\mathcal{R}(t)$ is Q-regular with $\mathcal{O}_t = c_q$ and all the robots have *yellow* light or (ii) contains at least one tuple $(r_i(t), s_i(t))$ such that the value of $s_i(t)$ is *red*.
- **Non Q*-regular ($\mathcal{NQ}$):** This class contains all the configurations $\mathcal{R}(t) \in \widetilde{\mathcal{R}}_c \backslash (\mathcal{CL} \cup \mathcal{QR}_0)$. The configurations which are asymmetric or have exactly one line of symmetry belong to this class.

3.3 Algorithm *MoveToDestination*()

Let r_i be a robot which has a destination point p_x. The robot r_i follows following steps to reach the point p_x:

- If $(r_i(t), p_x) \cap \mathcal{R}(t) = \emptyset$ i.e., there is no other robot position in between $r_i(t)$ and p_x on the line segment $\overline{r_i(t)p_x}$, the robot r_i moves directly to p_x along $\overline{r_i(t)p_x}$.
- If $(r_i(t), p_x) \cap \mathcal{R}(t) \neq \emptyset$ i.e., there is at least one robot position in between $r_i(t)$ and p_x on the line segment $\overline{r_i(t)p_x}$, the robot r_i waits until it finds a free corridor straight to p_x.

3.4 Algorithm *GatheringLight*()

We assume that (i) the initial robot configuration $\mathcal{R}(t_0)$ does not contain any multiplicity point (ii) all the robots initially have *yellow* lights and (iii) $n \geq 5$. An initial configuration $\mathcal{R}_c(t_0)$ belongs to either $\mathcal{QR}_0$ or $\mathcal{NQ}$. The basic idea is to convert the initial configuration, within finite time, into a central one i.e., one in $\mathcal{CL}$ in which $\mathcal{O}_{t_0}$ is the central point. During the conversion phase, the movements of the robots are coordinated in such way that the initial $SEC(t_0)$ does not change until a central configuration is created. Once a central configuration is created, the point $\mathcal{O}_{t_0}$ remains recognizable by the robots even if the initial $SEC(t_0)$ changes. The movements of the robots are designed to satisfy the constraint of the problem. Let r_i be an arbitrary active robot in $\mathcal{R}$, at time t. Robot r_i takes one of the following actions, depending on the configuration and the position of the robot:

- **Case-1 $\mathcal{R}(t) \in \mathcal{QR}_0$:** In this case, all the active robots have c_q as their destination point. Robot r_i finds that either all robots have *yellow* lights or at least one robot has *red* light. If $r_i(t) = c_q$, the robot r_i does not move. Otherwise, it does one of the following: (i) if there is a *yellow* robot at c_q, the robot r_i waits (ii) otherwise, it sets c_q as its destination point. In all these sub-cases, if the light of r_i is *yellow*, it also changes the colour to *red*.

- **Case-2 $\mathcal{R}(t) \in \mathcal{NQ}$:** The robot r_i finds either all robots with *yellow* light or at least one robot with *green* light and it acts according to the following:

  - **Case-2.1 $C_{int}(t) \neq \emptyset$:** If $r_i(t) \in C_{out}(t)$, the robot r_i does nothing. Otherwise, it does one of the following: (i) if $r_i(t) = \mathcal{O}_t$, the robot r_i does not move and it turns its light *blue*. (ii) if $r_i(t) \neq \mathcal{O}_t$, the robots r_i sets $\mathcal{O}_t$ as its destination and it changes its colour to *green*.

  - **Case-2.2 $C_{int}(t) = \emptyset$:** In this case all the robots lie on the circumference of $SEC(t)$.

    * **Case-2.2.1 $\mathcal{R}(t)$ is not Q-regular:** Fact 3 implies that the robot positions in $\mathcal{R}(t)$ have at most one line of symmetry.

      · **Case-2.2.1.1 $\mathcal{R}(t)$ does not have any line of symmetry:** The robot positions in $\mathcal{R}(t)$ are orderable in this case [4]. Consider an ordering $\mathcal{G}$ (the ordering algorithm is same for all robots) of the points in $\mathcal{R}(t)$. Select the robot position $r_u(t) \in \mathcal{R}(t)$ such that the smallest enclosing circle of $\mathcal{R}(t) \backslash \{r_u(t)\}$ is same as $SEC(\mathcal{R}(t))$ and $r_u(t)$ has the highest order in $\mathcal{G}$ among all the points satisfying this property. Since $n \geq 5$, such a point exists by Fact 5. If $r_i(t) = r_u(t)$, then it moves towards $\mathcal{O}_t$ and turns its light *green*. Otherwise, it does nothing.

      · **Case-2.2.1.2 $\mathcal{R}(t)$ has exactly one line of symmetry $\mathcal{L}$:** There are two possibilities: (i) the line $\mathcal{L}$ passes through at least one robot position in $\mathcal{R}(t)$ or (ii) there are four robot positions, say $H_1 = \{r_{v_1}(t), r_{v_2}(t), r_{v_3}(t), r_{v_4}(t)\}$, belonging to $C_{out}(t)$ which are closest to $\mathcal{L}$. If $r_i(t) \in H_1$ or $\mathcal{L}$ does not pass through $r_i(t)$, the robot r_i does nothing. Otherwise, r_i has $\mathcal{O}_t$ as its destination point and it changes its colour to *green*.

    * **Case-2.2.2 $\mathcal{R}(t)$ Q-regular with $\mathcal{O}_t \neq c_q$:** There are at most two distinct robot positions in C_{out} which are farthest from c_q. Let $\mathcal{U}$ denote the set of these points. $\mathcal{L}_z$ is the ray defined as follows (i) if $|\mathcal{U}| = 1$ and $r_l(t) \in \mathcal{U}$, then $\mathcal{L}_z$ is the ray from $r_l(t)$, which passes through $\mathcal{O}_t$ (ii) otherwise, it is the ray from the middle point of the two robot positions in $\mathcal{U}$, which passes through $\mathcal{O}_t$. Let $\mathcal{L}_z$ intersect the circumference of $SEC(t)$ at p_z. $\mathcal{W}$ is the set defined as follows: (i) if p_z contains a robot position, $\mathcal{W}$ is the singleton set containing this

point (ii) otherwise, $\mathcal{W} = \{r_j(t), r_k(t)\}$ where $r_j(t), r_k(t) \in C_{out}(t)$ and they lie on two different sides of $\mathcal{L}_z$. If $r_i(t) \neq \mathcal{U} \cup \mathcal{W}$, it sets $\mathcal{O}_t$ as its destination point and turns its light *green*. Otherwise, the robot r_i does nothing. Note that $|\mathcal{U} \cup \mathcal{W}|$ is at most 4 and the robot positions in $\mathcal{U} \cup \mathcal{W}$ keep $SEC(t)$ intact. Since $n \geq 5$, at least one robot moves inside $SEC(t)$.

- **Case-3 $\mathcal{R}(t) \in \mathcal{CL}$:** An initial configuration $\mathcal{R}_c(t_0)$ does not belong to this class. A configuration in this class is generated by the movements of the robots as described in case-1 and case-2. The destination point of each robot is $\mathcal{O}_{t_0}$. First, consider the case when $\mathcal{R}_c(t)$ contains a multiplicity point at $\mathcal{O}_{t_0}$, with all robots at this point having *red* colour. If $r_i(t) = \mathcal{O}_{t_0}$, the robot r_i does not move. Otherwise, it sets $\mathcal{O}_{t_0}$ as its destination point. In both the cases, if the colour of r_i is *yellow*, it changes its colour to *red*. Since robots can distinguish colours of all the robots occupying the same position, they can easily identify the point $\mathcal{O}_{t_0}$. Now, let $\mathcal{R}_c(t)$ contain a *blue* robot at the point $\mathcal{O}_{t_0}$. Same strategies are followed. In this case a robot does not change its colour.

In all of the above cases, a robot moves to its destination point according to algorithm $MoveToDestination()$ described in Sect. 3.3. Following list shows the transactions between different states of the robots: $\{yellow\} \xrightarrow{\mathcal{R}(t) \in \mathcal{QR}_0} \{red\}$, $\{yellow\} \xrightarrow{\mathcal{R}(t) \in \mathcal{NQ} \wedge C_{int}(t) \neq \emptyset \wedge \mathcal{O}_t = r_i(t)} \{blue\}$, $\{yellow\} \xrightarrow{\mathcal{R}(t) \in \mathcal{NQ} \wedge \mathcal{O}_t \neq r_i(t)} green$, $\{green\} \xrightarrow{\mathcal{R}(t) \in \mathcal{NQ} \wedge C_{int}(t) \neq \emptyset \wedge \mathcal{O}_t = r_i(t)} \{blue\}$.

3.5   Correctness of *GatheringLight()*

In this section, it is proved that *GatheringLight()* solves the constrained gathering problem.

Lemma 1. *Algorithm MoveToDestination() guarantees collision free movements for the robots during the whole execution of algorithm GatheringLight().*

Proof. During the whole execution of algorithm *GatheringLight()*, the destination point for each robot in the system is $\mathcal{O}_t$. Let r_i be robot which wants to move to the point $\mathcal{O}_t$. The robot r_i starts moving towards $\mathcal{O}_t$ only when it finds a free corridor straight to this point. Otherwise, it waits, until all the robots in $(r_i(t), \mathcal{O}_t)$ reach their respective destinations. Thus, algorithm *MoveToDestination()* guarantees a collision free movement for the robot r_i.   □

Lemma 2. *During the whole execution of algorithm GatheringLight(), $\nexists$ a time t such that $s_i(t) = red$ and $s_j(t) = green \vee blue$ for any two robots $r_i, r_j \in \mathcal{R}$.*

Proof. Let $\mathcal{R}(t_0)$ be an initial robot configuration. Initially all the robots have *yellow* lights.

- **Case-1 $\mathcal{R}_c(t_0) \in \mathcal{QR}_0$:** Since $\mathcal{O}_{t_0} = c_q$, when a robot wakes up, it either finds that all robots are *yellow* or at least one robot has *red* light. In both the cases, it turns its light *red* and it never changes its colour again during the whole execution of the algorithm. Thus through out the whole execution of the algorithm, each robot has any one of the colours from the set $\mathcal{F}_1 = \{yellow, red\}$ for its light.
- **Case-2 $\mathcal{R}_c(t_0) \in \mathcal{NQ}$:** The robots which wake up first, find that $\mathcal{R}(t_0) \in \mathcal{NQ}$ and all the robots have *yellow* lights. First, they turn their lights *green* or *blue*, depending upon their positions and then execute *move* phase. The robots which wake up after this, find that either $\mathcal{R}_c(t)$ has at least one robot with *green* light or it contains unique point occupied by at least one *blue* robot. They change their colours to *green* or to *blue*. Hence, in this case, the set of colours of consumed by the robots is $\mathcal{F}_2 = \{yellow, green, blue\}$.
- **Case-3 $\mathcal{R}_c(t) \in \mathcal{CL}$:** This case is applicable for a robot configuration $\mathcal{R}_c(t)$ where $t > t_o$. From case-1 and case-2, it is clear that $\mathcal{R}_c(t)$ does not contain two robots; one with *red* colour and other with *blue* colour. When a robot finds a multiplicity point with *red* robots, it changes its colour to *red*. Otherwise, when it finds a robot with *blue* colour, it does not change its colour. This implies at time $t' > t$, the system does not have two robots; one with *red* light and other with *blue* light. Hence the lemma holds. □

Lemma 3. *Algorithm GatheringLight() converts any initial configuration $\mathcal{R}(t_0)$ with more than 4 distinct robot positions, to a central configuration in finite time.*

Proof. Let $\mathcal{R}_c(t_0) \notin \mathcal{CL}$ be an initial robot configuration with more than 4 distinct robot positions. The algorithm *GatheringLight()* maintains one of the two invariants (i) if the initial configuration is Q-regular with $\mathcal{O}_{t_0} = c_q$, it remains Q-regular until a multiplicity point with *red* robots is created (ii) otherwise, the smallest enclosing circle $SEC(t_0)$ of $\mathcal{R}(t_0)$ remains the same until at least one robot reaches $\mathcal{O}_{t_0}$ and changes its colour to *blue*. Thus each active robot has exactly one desired destination point $\mathcal{O}_{t_0}$.

- **Case-1 $\mathcal{R}(t_0) \in \mathcal{QR}_0$:** Here, $\mathcal{O}_{t_0} = c_q$ and the robots move towards c_q. By Fact 2, the point c_q, remains invariant under the straight movements of the robots towards it. The robots which are moving towards c_q, have *red* colour lights to indicate the state of the initial configuration. By Lemma 2, within finite time at least two *red* robots occupy the point $\mathcal{O}_{t_0} = c_q$. This converts c_q as a multiplicity point with all *red* robots. The uniqueness of the multiplicity point follows from Lemma 1. Thus, within finite time, we would have a central configuration.
- **Case-2 $\mathcal{R}(t_0) \in \mathcal{NQ}$:** The robots which discover this case first, turn their lights *green* or *blue* so that all other robots, whenever they wake up, can have this information about the initial configuration. The robots have any one of the following scenarios:
  - **Case-2.1 $C_{int}(t_0) \neq \emptyset$:** In this case, the robots in $C_{int}(t_0)$ move towards $\mathcal{O}_{t_0}$. Since the robots in $C_{out}(t_0)$ do not move, the circle $SEC(t_0)$ and

hence $\mathcal{O}_{t_0}$, remain invariant. Within finite time at least one robot reaches $\mathcal{O}_{t_0}$ and turns its light *blue*, which converts the configuration into a central one. Since a robot at $\mathcal{O}_{t_0}$ decides to turn its colour to *blue* when it finds no robot with *blue* light in the system, the point with *blue* robots is unique.

- **Case-2.2 $C_{int}(t_0) = \emptyset$:** The objective, in this case, is to move at least one robot inside $SEC(t_0)$ so that $|C_{int}(t)|$ becomes at least 1. The selection of this robot depends on the symmetry of $\mathcal{R}(t_0)$.

  * **Case-2.2.1 $\mathcal{R}(t_0)$ is not Q-regular:** There are two possibilities:
    · **Case-2.2.1.1 $\mathcal{R}(t_0)$ does not have any line of symmetry:** The robot positions in $\mathcal{R}(t_0)$ are orderable. Exactly one robot can be selected deterministically to move inside $SEC(t_0)$ such that $SEC(t_0)$ remains invariant.
    · **Case-2.2.1.2 $\mathcal{R}(t_0)$ has exactly one line of symmetry $\mathcal{L}$:** Since $n \geq 5$ and at most four robots retain their positions on the circumference of $SEC(t_0)$ to keep it intact, there is at least one robot which is eligible to move inside $SEC(t_0)$. Whenever one such robot moves inside $SEC(t_0)$, we are done.
  * **Case-2.2.2 $\mathcal{R}(t_0)$ is Q-regular with $\mathcal{Q}_{t_0} \neq c_q$:** Same arguments as in the case-2.2.1.2 above, works for this case also.

  Within finite time, $|C_{int}(t)|$ will be at least 1 and case-2.1 shall be applicable. This implies that the initial configuration will be converted into a central one within finite time.

Hence, the lemma is true.   □

Lemma 4. *Algorithm GatheringLight() solves the constrained gathering problem in the ASYNC model for $n \geq 5$ robots in finite time.*

Proof. By Lemma 3, an initial configuration $\mathcal{R}_c(t_0)$ can be converted into a central configuration in finite time if the number of distinct robot positions in the configuration is more than 4. Now, let $\mathcal{R}(t) \in \mathcal{CL}, \forall\, t > t_0$. By Lemma 2, exactly one of the following is the case: (i) $\mathcal{R}(t)$ contains a unique multiplicity point with *red* robots or (ii) $\mathcal{R}(t)$ contains a unique point with *blue* robots. In both the cases, the special point is created at $\mathcal{O}_{t_0}$. Thus, a robot not at $\mathcal{O}_t$ can easily identify the location of $\mathcal{O}_{t_0}$ even when the circle $SEC(t_0)$ has changed. All the robots move towards $\mathcal{O}_{t_0}$. They follow algorithm $MoveToDestination()$ to reach their destination point. Algorithm $MoveToDestination()$ guarantees that (i) the movements of the robots are collision free, by Lemma 1 and (ii) the movements of the robots satisfy the constraint of the problem (since robots move along the straight lines to the destination points). Thus, algorithm $GatheringLight()$ achieves the required goal. It is easy to see that the constrained gathering problem is not solvable, in general, for $n \leq 4$ robots with lights.   □

Theorem 2. *The constrained gathering problem is solvable in the ASYNC model for $n \geq 5$ robots when robots are endowed with externally visible lights with 4 different colours.*

4   Conclusion

This paper presents a distributed algorithm to solve the constrained version of gathering problem for asynchronous robots, when robots are endowed with externally visible lights using only 4 colours ($O(1)$ bit of memory) for $n \geq 5$ robots. It is also proved that the problem is not solvable solely with multiplicity detection under non-rigid motion even for fully synchronous robots. One of the future directions is to study the problem in the presence of faulty robots.

References

1. Agathangelou, C., Georgiou, C., Mavronicolas, M.: A distributed algorithm for gathering many fat mobile robots in the plane. In: Proceedings of ACM Symposium on Principles of Distributed Computing (PODC), pp. 250–259 (2013)
2. Bhagat, S., Chaudhuri, S.G., Mukhopadhyaya, K.: Fault-tolerant gathering of asynchronous oblivious mobile robots under one-axis agreement. J. Discret. Algorithms **36**, 50–62 (2016)
3. Bramas, Q., Tixeuil, S.: Wait-free gathering without chirality. In: Scheideler, C. (ed.) Structural Information and Communication Complexity. LNCS, vol. 9439, pp. 313–327. Springer, Heidelberg (2015). doi:10.1007/978-3-319-25258-2_22
4. Chaudhuri, S.G., Mukhopadhyaya, K.: Leader election and gathering for asynchronous fat robots without common chirality. J. Discret. Algorithms **33**, 171–192 (2015)
5. Cicerone, S., Di Stefano, G., Navarra, A.: Minmax-distance gathering on given meeting points. In: Paschos, V.T., Widmayer, P. (eds.) CIAC 2015. LNCS, vol. 9079, pp. 127–139. Springer, Heidelberg (2015). doi:10.1007/978-3-319-18173-8_9
6. Cieliebak, M., Flocchini, P., Prencipe, G., Santoro, N.: Distributed computing by mobile robots: gathering. SIAM J. Comput. **41**(4), 829–879 (2012)
7. Czyzowicz, J., Gasieniec, L., Pelc, A.: Gathering few fat mobile robots in the plane. Theor. Comput. Sci. **410**(6), 481–499 (2009)
8. Das, S., Flocchini, P., Prencipe, G., Santoro, N., Yamashita, M.: The power of lights: synchronizing asynchronous robots using visible bits. In: Proceedings of IEEE 32nd International Conference on Distributed Computing Systems (ICDCS), pp. 506–515 (2012)
9. Défago, X., Gradinariu, M., Messika, S., Raipin-Parvédy, P.: Fault-tolerant and self-stabilizing mobile robots gathering. In: Dolev, S. (ed.) DISC 2006. LNCS, vol. 4167, pp. 46–60. Springer, Heidelberg (2006). doi:10.1007/11864219_4
10. Flocchini, P., Prencipe, G., Santoro, N.: Distributed computing by oblivious mobile robots. In: Synthesis Lectures on Distributed Computing Theory. Morgan & Claypool Publishers (2012)
11. Flocchini, P., Prencipe, G., Santoro, N., Widmayer, P.: Gathering of asynchronous robots with limited visibility. Theor. Comput. Sci. **337**(1–3), 147–168 (2005)
12. Flocchini, P., Santoro, N., Viglietta, G., Yamashita, M.: Rendezvous of two robots with constant memory. In: Moscibroda, T., Rescigno, A.A. (eds.) SIROCCO 2013. LNCS, vol. 8179, pp. 189–200. Springer, Heidelberg (2013). doi:10.1007/978-3-319-03578-9_16
13. Peleg, D.: Distributed coordination algorithms for mobile robot swarms: new directions and challenges. In: Pal, A., Kshemkalyani, A.D., Kumar, R., Gupta, A. (eds.) IWDC 2005. LNCS, vol. 3741, pp. 1–12. Springer, Heidelberg (2005). doi:10.1007/11603771_1

14. Prencipe, G.: *Instantaneous actions* vs. *full asynchronicity*: controlling and coordinating a Sset of autonomous mobile robots. In: Restivo, A., Della Rocca, S.R., Roversi, L. (eds.) ICTCS 2001. LNCS, vol. 2202, pp. 154–171. Springer, Heidelberg (2001). doi:10.1007/3-540-45446-2_10
15. Prencipe, G.: Impossibility of gathering by a set of autonomous mobile robots. Theor. Comput. Sci. **384**(2–3), 222–231 (2007)
16. Suzuki, I., Yamashita, M.: Formation and agreement problems for anonymous mobile robots. In: Proceedings of 31st Annual Conference on Communication, Control and Computing, pp. 93–102 (1993)
17. Suzuki, I., Yamashita, M.: Distributed anonymous mobile robots: formation of geometric patterns. SIAM J. Comput. **28**, 1347–1363 (1999)
18. Viglietta, G.: Rendezvous of two robots with visible bits. In: Flocchini, P., Gao, J., Kranakis, E., Meyer auf der Heide, F. (eds.) ALGOSENSORS 2013. LNCS, vol. 8243, pp. 291–306. Springer, Heidelberg (2014). doi:10.1007/978-3-642-45346-5_21

Improved Bounds for Poset Sorting in the Forbidden-Comparison Regime

Arindam Biswas[1], Varunkumar Jayapaul[2]([✉]), and Venkatesh Raman[1]

[1] The Institute of Mathematical Sciences, HBNI, Chennai 600113, India
{barindam,vraman}@imsc.res.in
[2] Chennai Mathematical Institute, Chennai 603103, India
varunkumarj@cmi.ac.in

Abstract. We study the classical problem of sorting when comparison between certain pair of elements are forbidden. Along with the set of elements V, the input to our problem is an undirected graph $G(V, E)$, whose edges represent the pairs that can be directly compared in constant time. We call this the *comparison graph*. It is also possible that the set of elements forms a partial-order, and not a total-order in which case, the sorting problem is the problem of determining all possible relations in the partial order, i.e. determining the (transitive) orientations of the edges of the graph.

If q is the number of edges missing in the graph, we first give a sorting algorithm that takes $O\left((q + n)(\lg(n^2/q))\right)$ comparisons improving on the recent upper bound of $O\left((q + n)\lg n\right)$. We also show the first lower bound by giving a graph and an orientation by an adversary where $\Omega\left(q + n\lg n\right)$ comparisons are necessary. Then, we give an $O\left(n\lg n\right)$ algorithm (independent of q) when the comparison graph is from a special class of graphs like chordal or comparability graphs. Finally, we make some remarks regarding the complexity of sorting with forbidden comparisons when the elements form a total order.

Keywords: Poset · Sort · Topological · Chordal · Comparability · Forbidden · Query · Complexity

1 Introduction

Comparison based sorting algorithms is one of the most studied areas in computer science. We continue the investigation recently initiated by Banerjee and Richards [1] on sorting when certain pairs of elements are forbidden to be compared.

A forbidden pair is a pair of elements x and y which cannot be directly compared to each other. This does not necessarily imply that the two elements are mutually incomparable, i.e. there may exist some element z, such that $x > z$ and $z > y$, thereby implying $x > y$. The input is an undirected graph $G(V, E)$, where V is the set of elements and E represents the allowed comparisons/edge queries (the two terms are used interchangeably). We call G as the comparison

© Springer International Publishing AG 2017
D. Gaur and N.S. Narayanaswamy (Eds.): CALDAM 2017, LNCS 10156, pp. 50–59, 2017.
DOI: 10.1007/978-3-319-53007-9_5

graph, and we assume that the algorithm is given this undirected graph (so it doesn't have to spend time to determine whether a given pair is comparable or not).

The comparison graph G has an underlying directed acyclic orientation realizing a poset P_G maintained by an adversary. The goal of the problem is to determine the orientation of all the edges by probing the adversary for as few edges as possible (and using transitivity). We call this problem loosely as sorting the poset underlying the graph G or simply sorting the graph G. The number of queries made to the adversary is defined as the query complexity.

Poset sorting has been well studied for width bounded posets in [2]. It is known that if a poset P has width (the size of the largest anti-chain) at most w, then the information theoretic bound for the query complexity is $\Omega\left((w + \lg n)n\right)$. A query optimal algorithm n for width bounded posets whose total complexity is $O\left(nw^2 \lg(n/w)\right)$ is presented in the same paper.

Let q be the number of forbidden pairs in the given graph and let w be the width of the poset P_G realized by the vertices of the graph. The parameters q and w are known to be related, since $q \geq \#$ of incomparable pairs in $P_G \geq \binom{w}{2}$. Hence, $w = O\left(\sqrt{q}\right)$, although $\sqrt{q}$ can be substantially larger than w. Take, for example, a total order with the comparison graph as simply a path on n vertices. The width of the graph is 1, as there is no anti-chain of length more than 1, however the graph is missing $\binom{n}{2} - n + 1$ edges. Banerjee and Richards [1] used q as a measure how difficult it is to sort G. When q is 0, there is no forbidden pair, and the standard comparison based sorting algorithms can determine the poset using $O\left(n \lg n\right)$ edge queries. At the other extreme, when $|E|$, the number of edges present in the graph is small, one can simply probe every edge in the graph and sort much faster. A good structure on the input graph can also help to sort much faster, regardless of how large q is, as we show later in this paper.

Banerjee and Richards [1] showed a $O\left((q + n) \lg n\right)$ query bound for sorting a graph with q missing edges. We modify the algorithm resulting in a query complexity of $O\left((q + n) \lg(n^2/q)\right)$. For $q = \Theta(n^2)$, this bound is better than the bound of Banerjee and Richards. We also show the first lower bound of $\Omega\left(q + n \lg n\right)$ by exhibiting a comparison graph with q forbidden pairs and an orientation of the edges that requires $\Omega\left(q + n \lg n\right)$.

Then, we investigate the problem for a couple of special classes of comparison graphs. A graph is *chordal* if every cycle of length at least 4 in the graph has a chord (an edge joining two non-adjacent vertices of the cycle). We show that if the comparison graph is a chordal graph, then we can sort the graph using $O\left(n \lg n\right)$ queries using the simplicial ordering of the vertices of the graph. A graph is a *comparability* graph if its edges can be oriented such that for every triple x, y, z of vertices, if there is a directed edge from x to y, and if there is a directed edge from y to z, then there is a directed edge from x to z. We call such an orientation a comparability orientation. Note that a directed acyclic orientation is not necessarily a comparability orientation. For example, if G, the underlying undirected graph is a path on three vertices a, b and c, then the orientation $a \rightarrow b \rightarrow c$ is not a comparability orientation though it is a directed acyclic orientation. The two comparability orientations for such a path are $a \rightarrow b \leftarrow c$ and $a \leftarrow b \rightarrow c$.

We show that if the comparison graph is a comparability graph and if the adversary answers the edge queries based on a comparability orientation of the graph, then we can sort the poset underlying the graph using $O(n \lg n)$ queries. A consequence of this result is the following. Suppose we have a poset represented by a directed graph such that the directed edges of the graph represent *all* the order relations between the elements of the poset (so the missing edges represent incomparable elements). And suppose we are not given the orientation, but only the underlying undirected graph. Then we can find the poset probing the adversary for $O(n \lg n)$ edge queries (the rest of the edge directions are deduced using transitivity). This may be of independent interest.

1.1 Organization of the Paper

In Sect. 2, we give an improved upper bound and the first lower bound for sorting a comparison graph based on q. In Sect. 3, we outline the $O(n \lg n)$ query algorithm to sort comparability and chordal graphs. In Sect. 4, we give some concluding remarks on sorting the graph when it is known that the elements form a total order.

1.2 Definitions and Notation

Let G be a graph. We denote by $V(G)$ the vertex set of G, and by $E(G)$ the edge set. If $|V(G)| = n$ and $|E(G)| = m$, G is said to have *order* n and *size* m. We define $|G| = |V(G)|$ and $\|G\| = |E(G)|$.

For any $v \in V(G)$, $N(v)$ denotes the set of neighbors of v, and we define $\deg(v) = |N(v)|$. For a subset $S \subseteq V(G)$, $G[S]$ denotes the subgraph of G induced by S and for $v \in S$, $N_S(v)$ denotes the neighbors of v in $G[S]$. Analogously, we define $\deg_S(v) = |N_S(v)|$. The transitive reduction of a directed acyclic graph G is a minimal subgraph H of G such that the transitive closure of H (that has all directed edges uv whenever there is a directed path from u to v in H) is G.

A tournament is a directed graph in which there is exactly one directed edge between every pair of vertices, and a (directed or undirected) Hamiltonian path is a (directed or undirected appropriately) path that visits every vertex of the graph exactly once.

We omit floors and ceilings when talking about the sizes that are integers to simplify the notation, it should be clear from the context and the asymptotic bounds are not affected by this approximation.

2 Sorting Under Forbidden Comparisons

2.1 Upper Bounds

We start by revising the algorithm of [1]. The central claim in the algorithm in [1] is the following lemma, which is not stated explicitly, but is presented as a collection of lemmas culminating in the result in Sect. 2.2 of [1].

Lemma 1. *Every graph with $q < n^2/320$ missing edges[1] has an approximate median vertex m, such that m is greater than at least $n/40$ elements and less than at least $n/40$ elements. Furthermore m can't be compared with at most $O(q/n)$ elements, and it can be found using $O(q+n)$ queries.*

Both their and our algorithms are recursive, and the main difference is that we bail out of the recursion earlier which improves the bounds. The output of the algorithm is an orientation of some of the edges of the graph which form a supergraph of the transitive reduction of (the oriented) G, i.e. the orientation of the remaining edges can be deduced using transitivity.

Algorithm 1. *Sort $G(V, E, q, depth)$*

---

1   **if** $q \geq n^2/320$ *or* $depth = \lg(\frac{n^2}{q})/\lg(\frac{40}{39})$ **then**

2   query every edge in E and output their orientations;

3   **end**

4   **else**

5   find an approximate median vertex m using Lemma 1;

6   Compare m with all its neighbors in V and output their orientations;

7       $V_l = \{v/v \in V, v < m\}$ and $V_h = \{v/v \in V, v > m\}$;

8       $V_{incomp} = \{v/v \in V, \{v, m\}$ *is not in* $E\}$;

9   Compare every vertex in V_{incomp} with all its neighbors in V and output their orientations;

10   Sort $G(V_l, E_l, q_l, depth + 1)$ and Sort $G(V_h, E_h, q_h, depth + 1)$;

11   (Here E_l is the set of edges in $G[V_l]$ and E_h is the set of edges in $G[V_h]$ and q_l is the number of missing edges in $G[V_l]$ and q_h is the number of missing edges in $G[V_h]$)

12 **end**

Theorem 1. *Sorting a comparison graph with q forbidden edges can be done in $min\{|E|, O((q+n)\lg(\frac{n^2}{q}))\}$ edge queries.*

Proof. The details of the algorithm are given in Algorithm 1 which is called with $depth = 0$. It checks if the value of q or the depth of the recursion is less than a threshold and then it breaks the problem into two disjoint problems using $O(q+n)$ queries by Lemma 1. Suppose at level l of the recursion, the sizes of the subproblems are $n_1, n_2, n_3...n_t$ and the number of missing edges in these subproblems is $q_1, q_2, q_3...q_t$ respectively. The algorithm either breaks them into smaller subproblems or queries every edge in that subproblem (with $q_i \geq n_i^2/320$ missing edges) in which case the algorithm performs at most $160q_i$ queries. The query cost incurred by the incomparable elements at any internal node of recursion tree is also $O((q_i/n_i) * n_i) = O(q_i)$. In either case, the total number of queries done at this level is at most $\sum_{i=1}^{l}(q_i + n_i) = O(q+n)$. Thus at any level of the recursion tree, the algorithm makes at most $O(q+n)$ queries.

---

[1] The constant used in the paper is 200, instead of 320. We have made the small change to factor in a calculation gap in last equation on page 7 in [1].

The algorithm is essentially the same as that of [1], except that it is forced to stop the recursion, when the depth of the recursion i is $(\lg(\frac{n^2}{q}))/\lg(\frac{40}{39})$. At this point the number of subproblems would be at most $O\left(n^2/q\right)$ and the size of each subproblem would be at most $(39/40)^i n = q/n$, since each subproblem has at most $39/40$ fraction of the vertices of its parent subproblem by Lemma 1.

Even if all these subproblems were complete graphs, the total number of edges in all these subproblems would be at most $O\left(n^2/q * (q^2/2n^2)\right) = O\left(q\right)$. At this point we just ask all the edge queries without recursing any further using $O(q)$ edge queries. The algorithm creates a recursion tree which has $O\left(\lg(\frac{n^2}{q})\right)$ levels and queries $O\left(q+n\right)$ edges at each level. Thus the total edge queries made by the algorithm is $O\left((q+n)\lg(\frac{n^2}{q})\right)$. If $|E| < (q+n)\lg(\frac{n^2}{q})$, then algorithm just asks all edge queries without optimizing in any way.   □

In the above theorem, we have just counted the query complexity and ignored the time it takes to find the queries to make. One can easily see that the rest of the running time remains as $O\left(n^2 + (\sqrt{q})^\omega\right)$ as shown in Theorem 6 of [1]. This running time is essentially required to compute the transitive closure of the directed graph to deduce the missing relations.

2.2   Lower Bounds

We now exhibit lower bounds on the number of edge queries needed to sort a graph $G = (V, E)$ in terms of $|V|$ and $q = \binom{n}{2} - |E|$, the number of missing edges. When q is large, we have the following lower bound.

Lemma 2. *There exists a graph with $q \geq n^2/4$ and an orientation such that, $\Omega(|E|)$ edge queries are needed to sort the graph.*

Proof. The graph which the adversary constructs is a complete bipartite graph with A and B as the equal sized parts. The adversary orients the edges from A to B. Here the number of missing edges, as well as the number of edges present is, roughly $n^2/4$. The adversary also orients the edges from A to B forcing the algorithm to probe every edge, as the algorithm can not deduce any of the edges using transitivity. If the algorithm fails to query an edge, the algorithm has a choice of flipping its direction.   □

For $q < n^2/4$, we have the following bound.

Theorem 2. *When $q < n^2/4$, there exists a graph and an orientation of the edges such that any algorithm has to probe $\Omega(q + n \lg n)$ queries to sort the graph.*

Proof. In this case, the graph constructed by the adversary consists first of a complete bipartite graph B with partitions X and Y of size roughly $\sqrt{q}$ each such that it has q edges and has q edges missing. Then it forms a complete graph K on the remaining $n - 2\sqrt{q}$ vertices and maintains a total order among those vertices. And it has all the edges between every vertex of K and every vertex of B. If a query comes between a vertex b in the bipartite graph B and a vertex c in the complete graph K, the adversary directs the edge from b to c. If the edge query is between two vertices inside the complete graph, the adversary answers consistent with the total order it maintains. If the edge query is between two elements inside the bipartite graph, the adversary always directs the edges from set X to set Y.

The number of edge queries required to sort the complete graph K would be $\Omega(n \lg n)$ and the number of edge queries required to sort the bipartite graph B is atleast $\Omega(q)$ from Lemma 2, which gives a lower bound of $\Omega(q + n \lg n)$ edge queries.   □

3 Sorting Posets with Special Comparison Graphs

In general, the number of edge queries needed to sort a poset can depend on both its size (n) and the number of forbidden pairs (q). However, when the comparison graph accompanying the poset has some additional structure, the number of edge queries needed can be at most $O(n \lg n)$, and more importantly becomes independent of q.

In this section, we show that when the comparison graph is *chordal* or *transitively orientable* (i.e. it is a comparability graph), the graph can be sorted by making $O(n \log n)$ edge queries. In fact, both algorithms presented below output linear extensions (i.e. topological orderings) of the input posets. A topological ordering $v_1, v_2...v_n$ of its vertices is an ordering where if (v_i, v_j) is a directed edge in the graph, then $i < j$. It is well-known that every directed acyclic graph has a topological sort.

Topologically sorting a general directed acyclic graph $G = (V, E)$ needs $\Omega(|V| + |E|)$ running time, while the algorithms in this section make $O(n \lg n)$ queries. This is due to the fact that the algorithms exploit the additional information about the input poset which the comparison graph provides and more importantly, we only care about the query complexity, and not bother about finding the orientation of every edge. One can deduce the edge directions from the topological sort of the vertices using transitivity, but they are not counted in the query complexity. Hence we only worry about finding a topological sort of the vertices.

3.1 The General Idea

The algorithms in the next two subsections have the following general outline.

1. Pick an appropriate (constant-size) subposet of the input and topologically sort it.
2. Iteratively extend the topological ordering by inserting one element at a time using a binary search type procedure among its neighbors.

A binary insertion of a vertex v in the topological order $v_1, v_2, \ldots v_t$ of its neighbors is the process of finding whether $v \to v_1$ or $v_t \to v$ or some adjacent vertices v_i and v_{i+1}, $i < t$ such that $v_i \to v$ and $v \to v_{i+1}$. This is similar to the process of searching a value in a sorted array using binary search and can be performed using $\lceil \lg t \rceil + 1$ queries.

When the comparison graph is a complete graph, then any directed acyclic orientation of its edges has a unique directed Hamiltonian path (as with the orientation, the directed graph is a transitive tournament), and hence has a unique topological ordering. In this case, the above procedure is essentially the binary insertion sort to sort the graph.

What we show is that such a binary insertion procedure works even if the comparison graph is not a complete graph. Towards that we capture the following two simple observations.

Lemma 3. *If G is a directed acyclic orientation of a complete graph, then it has a unique topological order.*

Lemma 4. *Let $G(V, E)$ be a directed acyclic graph, and let $S \subseteq V$ be such that $G[S]$, the induced subgraph on S, is a tournament on s vertices. Let $v_1, v_2, \ldots v_s$ be the unique topological order of vertices of S. In any topological order of the vertices of V, the elements of S appear in the unique topological order within S.*

Proof. Let $u, v \in S$, and let u appear before v in one topological order, then (u, v) is directed from u to v ((u, v) edge is there as $G[S]$ is a tournament). Then v cannot appear before u in any other topological order by the definition of topological order. $\square$

3.2   Chordal Comparison Graphs

Chordal graphs [3] form a well-studied class of graphs as they can be recognized in linear time [6], and several problems that are hard on other classes of graphs such as graph coloring can be solved in polynomial time for chordal graphs [5]. An undirected graph is chordal if every cycle of length greater than three has a chord, namely an edge connecting two non-consecutive vertices of the cycle [3].

In a graph G, a vertex v is called *simplicial* if and only if the subgraph of G induced by the vertex set $\{v\} \cup N(v)$ is a complete graph. A graph G on n vertices is said to have a perfect elimination ordering (PEO) if and only if there is an ordering $v_1, v_2 \ldots v_n$ of G's vertices, such that each v_i is simplicial in the subgraph induced by the vertices $v_1, v_2 \ldots v_i$. Every chordal graph has a perfect elimination ordering which can be found in $O(|E| + |V|)$ time [5,6]. Note that no edge queries are made while finding a perfect elimination ordering.

Now we apply the idea outlined in Sect. 3.1 in the perfect elimination ordering. Suppose $v_1, v_2, \ldots, v_n$ is a perfect elimination ordering (PEO) for the graph.

We obtain a topological sort of the vertices of the graph, in the inductive order of the PEO. Let G_i be the induced graph on the first i vertices of the PEO, and suppose that we know all the orientations of the edges of G_i, and we have a topological sort of the orientation of G_i. Now we insert v_{i+1} using binary search as follows. We consider the projection of the topological order of vertices of G_i that are neighbors of v_{i+1}, and insert v_{i+1} using binary search among them. In particular, suppose we have vertices $v_p, v_q \in G_i$ such that $v_p \to v_{i+1} \to v_q$ where v_p and v_q are consecutive vertices in the topological order of G_i among neighbors of v_{i+1} (it is possible that one of v_p or v_q does not exist; v_p may not exist if v_q is the first vertex of G_i in the topological order among neighbors of v_{i+1} and v_q may not exist if v_p is the last vertex of G_i in the topological order among neighbors of v_{i+1}). Then we simply insert v_{i+1} after v_p (or before every vertex if v_p doesn't exist) in the topological order. We claim the following.

Claim. The resulting order is a topological order of the directed acyclic orientation of G_{i+1}.

Proof. We only need to worry about edges incident on v_{i+1} as for every other pair of edges, their relative order has not been changed by the insertion and hence the topological order property is satisfied by induction. By Lemma 4, v_{i+1} has been inserted properly in the unique topological order on the induced subgraph of v_{i+1} and its neighbors, as they form a clique (because of simplicial ordering property). □

Thus we have the following result.

Theorem 3. *If the comparison graph is a chordal graph, then the poset underlying the graph can be sorted using $O(n \lg n)$ edge queries.*

3.3 Comparability Comparison Graphs

An undirected graph is a *comparability graph* [3] if the edges of the graph can be oriented in a way that for every triple x, y, z of vertices, if there is a directed edge from x to y, and a directed edge from y to z, then there is a directed edge from x to z. While both comparability graphs and chordal graphs are sub-classes of the well-known class of *perfect* graphs, there is really no relation between both these classes of graphs. In particular, in comparability graphs, we have no guarantee about the existence of a simplicial ordering. However, we argue that we can incrementally insert new vertices and get the orientation of its edges by doing the variation of binary search outlined in Sect. 3.1.

The idea is the same as what we did for chordal graphs. However, here as we don't have simplicial ordering, we simply start with an arbitrary ordering of the vertices and maintain the topological sort of the subgraph induced by the initial set of vertices and incrementally insert the new vertex. Let G_i be the induced graph on the first i vertices ($i \geq 1$) in the arbitrary order, and suppose that we know all the orientations of the edges of G_i, and we have a topological sort of the orientation of G_i. Now we insert v_{i+1} using binary search as before. We consider the projection of the topological order of vertices of G_i among the neighbors of v_{i+1}, and insert v_{i+1} using binary search among them. In particular, suppose we have vertices $v_p, v_q \in G_i$ such that $v_p \to v_{i+1} \to v_q$ where v_p and v_q

are consecutive vertices in the topological order of G_i among neighbors of v_{i+1}. (Here again, it is possible that one of v_p or v_q does not exist; v_p may not exist if v_q is the first vertex of G_i in the topological order among neighbors of v_{i+1} and v_q may not exist if v_p is the last vertex of G_i in the topological order among neighbors of v_{i+1}.) Then we simply insert v_{i+1} after v_p (or before every vertex if v_p doesn't exist) in the topological order. We claim the following.

Claim. The resulting order is a topological order of the directed acyclic orientation of G_{i+1}.

Proof. Here again we only need to worry about edges incident on v_{i+1} as for every other pair of edges, their relative order has not been changed by the insertion and hence the topological order property is satisfied by induction. Suppose that there exists a vertex $u \in G_i$ that comes before v_{i+1} in the resulting topological order and we have the edge v_{i+1} to u. Then $u \neq v_p$ (as $v_p \to v_{i+1}$) and u has to come before v_p in the order (as v_p immediately preceded v_{i+1} in the order). But as $v_p \to v_{i+1} \to u$, there must be an edge from $v_p \to u$ as the orientation is a comparability orientation, which is a contradiction to the fact that we have a topological order of vertices of G_i (as u comes before v_p in the topological order of vertices of G_i). The case when there is a vertex u that comes after v_{i+1} in the topological order, is similar to the case discussed. $\square$

Thus we have,

Theorem 4. *If the comparison graph is a comparability graph, and if the adversary answers the queries according to a comparability orientation, then the poset underlying the graph can be sorted using* $O(n \lg n)$ *edge queries.*

Note that our proof crucially used the fact that adversary answers the query according to a comparability orientation, and this is not just the artefact of the proof. Otherwise we may not be able to sort using $O(n \lg n)$ queries. For example, a complete bipartite graph is a comparability graph, but if we are told that the orientation of the adversary is a comparability orientation, then only two orientations are possible, and we can sort in just one query. However we have shown an $\Omega(n^2)$ lower bound in Lemma 2 if the adversary is free to choose any directed acyclic orientation. While this may appear as a restriction, another way to view the theorem is as follows, which can be of independent interest.

Corollary 1. *Suppose we have a poset represented by a directed graph whose directed edges represent all the underlying relations between elements of the poset. Suppose that we are given only the underlying undirected graph, and there is an adversary that answers queries according to the partial order. Then we can determine the relations of the poset using* $O(n \lg n)$ *queries to the adversary.*

4   Concluding Remarks

We have given an improved upper bound of $O\left((q+n)\lg(n^2/q)\right)$ and the first lower bound of $\Omega(q + n \lg n)$ for sorting an undirected graph on n vertices and q

missing edges. There is still a gap between the upper and lower bound, narrowing this gap is an interesting open problem.

We gave algorithms which make $O(n \lg n)$ edge queries when the input comparison graph is a comparability or a chordal graph, an interesting open problem is to find the largest class of graphs for which an $O(n \lg n)$ query algorithm is possible.

Finally, the problem is wide open when we know that there is a total order underlying the vertices of the comparison graph. For example, in that case, the complete bipartite graph can not be oriented as shown in Lemma 2 and hence we do not know of any lower bound other than $\Omega(n \lg n)$ regardless of the number of missing edges. In particular, sorting the graph when the graph is a complete bipartite graph is the famous *nuts and bolts* problem and can be sorted using $O(n \lg n)$ queries [4]. We conjecture that sorting any undirected graph whose vertices represent an underlying total order, can be done in $O(n \lg n)$ queries. In other words, if we know that the directed acyclic orientation of the comparison graph has a directed hamiltonian path, then we conjecture that we can find the path using $O(n \lg n)$ queries.

In fact, for total-orders in the forbidden pairs model, we do not even know how to find an element of any given rank (or the median element) in $O(n)$ queries, while this is possible for the smallest or the largest element.

References

1. Banerjee, I., Richards, D.: Sorting under forbidden comparisons. In: 15th Scandinavian Symposium and Workshops on Algorithm Theory (SWAT 2016). Leibniz International Proceedings in Informatics (LIPIcs), vol. 53, pp. 22:1–22:13. Schloss Dagstuhl-Leibniz-Zentrum fuer Informatik (2016)
2. Daskalakis, C., Karp, R.M., Mossel, E., Riesenfeld, S.J., Verbin, E.: Sorting and selection in posets. SIAM J. Comput. **40**(3), 597–622 (2011)
3. Golumbic, M.C., Rheinboldt, W.: Algorithmic Graph Theory and Perfect Graphs. Computer Science and Applied Mathematics. Elsevier Science, Amsterdam (2014)
4. Komlós, J., Ma, Y., Szemerédi, E.: Matching nuts and bolts in o(n log n) time. SIAM J. Discret. Math. **11**(3), 347–372 (1998)
5. Rose, D.J., Tarjan, R.E., Lueker, G.S.: Algorithmic aspects of vertex elimination on graphs. SIAM J. Comput. **5**(2), 266–283 (1976)
6. Tarjan, R.E., Yannakakis, M.: Simple linear-time algorithms to test chordality of graphs, test acyclicity of hypergraphs, and selectively reduce acyclic hypergraphs. SIAM J. Comput. **13**(3), 566–579 (1984)

Positional Dominance: Concepts and Algorithms

Ulrik Brandes, Moritz Heine, Julian Müller$^{(\boxtimes)}$, and Mark Ortmann

Computer and Information Science, University of Konstanz, Konstanz, Germany
Julian.Mueller@uni-konstanz.de

Abstract. Centrality indices assign values to the vertices of a graph such that vertices with higher values are considered more central. Triggered by a recent result on the preservation of the vicinal preorder in rankings obtained from common centrality indices, we review and extend notions of domination among vertices. These may serve as building blocks for new concepts of centrality that extend more directly, and more coherently, to more general types of data such as multilayer networks. We also give efficient algorithms to construct the associated partial rankings.

1 Introduction

One of the core concepts of network analysis is the identification of central vertices [13]. The most commonly applied centrality indices measure, e.g., the number of vertices a vertex can communicate with directly (*degree*), the expenses of a vertex to reach each other vertex in the network (*closeness* [21]), and the control over communication of others in the network (*betweenness* [6]).

While all centrality indices assign numerical values to each vertex in the graph, one is typically only interested in the derived ranking. Although well established centrality indices differ substantially in their definition, the rankings they induce all coincide on the vicinal preorder. In the *vicinal preorder* [5], a vertex $w \in V$ dominates another vertex $v \in V$, i.e. $v \leq w$, if and only if $N(v) \subseteq N[w]$ where $N(u)$ is the neighborhood of vertex u in the graph and $N[u] = N(u) \cup \{u\}$. This implies that it is possible to construct a partial ranking of the vertices by simply comparing their neighborhoods, and this ranking is preserved by any centrality index [22].

The vicinal preorder, or *neighborhood inclusion*, is itself the union of two other preorders: (i) the *dominance preorder* where $v \leq_a w \iff N[v] \subseteq N[w]$ and (ii) the *structural preorder* where $v \leq_n w \iff N(v) \subseteq N(w)$. Furthermore, it is an instantiation of *positional dominance* [1], a generic concept that allows for valued relationships and the expression of levels of homogeneity, i.e., admissible substitutions of vertices in the comparison of neighborhoods. Positional dominance provides a building block on which concepts of centrality can not only be generalized more easily, but also more coherently, to more complex kinds of data. While we are motivated by the implications of variant preorders for centrality, we are especially interested in their computational complexity here.

We gratefully acknowledge financial support from Deutsche Forschungsgemeinschaft (DFG) under grants Br 2158/6-1 and Br 2158/11-1.

© Springer International Publishing AG 2017
D. Gaur and N.S. Narayanaswamy (Eds.): CALDAM 2017, LNCS 10156, pp. 60–71, 2017.
DOI: 10.1007/978-3-319-53007-9_6

Contribution. We present efficient algorithms for instances of positional dominance. Our main contribution is an algorithm with $\mathcal{O}(nm \log \log \Delta(G))$ running time for the homogeneous case with weights on both edges and vertices. This is an improvement over the straightforward approach with an $\mathcal{O}(nm\Delta(G)^{3/2})$ time bound. In addition we give lower bounds for worst-case running times by constructing families of graphs with large output size, i.e., dense preorders. Although we consider simple undirected graphs, our results can be adapted for weighted, directed, and graphs with a given bipartition (*two-mode graphs*).

Note, however, that we assume throughout this paper that our input graphs do not contain isolated vertices because these are dominated by every other vertex in the graph (or no other vertex in the dominance preorder), so that their relationships can be checked in constant time and are best represented implicitly.

2   Preliminaries

For the most part, we consider simple undirected graphs $G = (V, E)$, where both vertices and edges may carry weights $\omega : V \cup E \to \mathbb{R}$. For edges $\{v, w\} \in E$, we use shorthand notation $\omega(v, w) = \omega(\{v, w\})$. Weights can be thought of as non-negative reals for convenience but any ordered range of values will do. Following the usual convention, we denote the number of vertices and edges by $n = n(G) = |V|$ and $m = m(G) = |E|$. We write $H \subseteq G$ if H is a subgraph of G, and $G[W]$ for the subgraph induced by $W \subseteq V$.

The (open) *neighborhood* of a vertex $v \in V$ is defined as $N(v) = \{w : \{v, w\} \in E\}$ and the *closed neighborhood* as $N[v] = N(v) \cup \{v\}$. We assume that $N(v) \neq \emptyset$, $v \in V$, throughout the paper. The *degree* of $v \in V$ is $\deg(v) = |N(v)|$, and since $2m = \sum_{v \in V} \deg(v)$ the *average degree* is $\langle \deg \rangle = \frac{2m}{n}$. Let $\Delta(G) = \max\{\deg(v) : v \in V\}$ denote the *maximum degree* of a graph.

The *arboricity* $\alpha(G)$ of a graph G is the minimum number of forests needed to cover its edges. Arboricity is an indicator of sparseness as it is closely related to the average degree in a densest subgraph via $\alpha(G) = \max_{H \subseteq G} \{\lceil \frac{m(H)}{n(H)-1} \rceil\}$ [16].

A binary relation $R \subseteq (V \times V)$ is called a *preorder* if it is reflexive and transitive. Since a total preorder gives a ranking, we may refer to a preorder also as a *partial ranking*. If a (partial) ranking is antisymmetric, it is a (partial) order.

3   Dominance

The dominance preorder is a restriction of the more general vicinal preorder. A vertex $w \in V$ (*vertex*) *dominates* a vertex $v \in V$, $v \leq_a w$, if $N[v] \subseteq N[w]$. The subscript indicates that a relation w.r.t. the dominance preorder can only exist for pairs of adjacent vertices. Any two vertices that dominate each other are also referred to as *true twins*, because they are adjacent and have exactly the same neighborhood. Consequently, each equivalence class of the dominance relation $\leq_a$ induces a clique.

Algorithm 1: Dominance Preorder

---

input : simple undirected graph $G = (V, E)$
output : partial ranking $\leq_a$ on V (dominance)

initialize $\leq_a$ with $v \leq_a v$ for all $v \in V$;
for $\{v, w\} \in E$ **do** $\deg(v, w) \leftarrow 0$
for $v_i = v_1, \ldots, v_n$ *where* $\deg(v_1) \geq \ldots \geq \deg(v_n)$ **do**
    mark all $w \in N^+(v_i)$ with v_i;
    for $w \in N^+(v_i)$ **do**
        for $u \in N^+(w)$ **do**
            if u *is marked with* v_i **then**
                foreach $\{i, j\} \in \binom{\{v_i, w, u\}}{2}$ **do** increment $\deg(i, j)$
    if $\deg(v_i, w) = \deg(v_i) - 1$ **then** add $v_i \leq_a w$
    if $\deg(v_i, w) = \deg(w) - 1$ **then** add $w \leq_a v_i$

To construct the dominance preorder, we extend the concept of neighborhood to edges $\{v, w\} \in E$ via $N(v, w) = N(v) \cap N(w)$, and denote $\deg(v, w) = |N(v, w)| < \min\{\deg(v), \deg(w)\}$. Since $\deg(v, w) = \deg(v) - 1$ means every neighbor of v other than w itself is also a neighbor of w, it implies $v \leq_a w$. Thus, the dominance preorder can be determined using any algorithm that counts the number of triangles an edge is part of.

Algorithm 1 is based on an efficient realization [17] of the triangle listing algorithm of Chiba and Nishizeki [2]. Given any ordering of the vertices, here specifically from higher to lower degrees, we let $N^+(v)$ denote the number of adjacent vertices that appear after v in the ordering, i.e., all edges are oriented from earlier to later respective to the ordering.

Theorem 1. *Algorithm 1 determines the dominance preorder of a simple undirected graph in time $\mathcal{O}(\alpha(G)m)$.*

Proof. The vertex ordering ensures that for each edge, the neighbors of the vertex with smaller degree are inspected. Following Chiba and Nishizeki's reasoning for their algorithm K3 [2], the claimed runtime is a consequence of the inequality

$$\sum_{\{u,v\}\in E} \min\{\deg(u), \deg(v)\} \leq 2\alpha(G)m$$

□

As the arboricity of a graph can be as large as n, Algorithm 1's running time is far from being linear in the size of the input and output, since clearly the size of the dominance preorder is bounded from above by m.

However, although in the worst case the arboricity is linear in n, it is often small in social networks [4] and can, in fact, be even smaller than the average degree [2]. We show next that there is no simple relationship between these two graph invariants because arboricity is determined by the densest subgraph.

Theorem 2. *There is a family of graphs G_n, $n > 3$, with $\langle \deg \rangle \in \mathcal{O}(1)$ and $\alpha(G_n) \in \Omega(n^{1/2})$.*

Proof. Let G be a graph with $n = k^2$ vertices with $k \in \mathbb{Z}$ consisting of a $\sqrt{n}$-clique where each vertex of the clique except for one has additionally $\sqrt{n}$ pendants. Consequently G has $m < 3n$ edges and therefore $\langle \deg \rangle \in O(1)$. However its arborocity is $\alpha(G) \geq \lceil \frac{\sqrt{n}}{2} \rceil$ [2]. □

4 Structural Equivalence and Neighborhood Inclusion

The dominance preorder requires that dominating vertices are adjacent. This is a severe restriction, as all non-adjacent pairs are necessarily incomparable. A natural extension of the dominance preorder that softens this requirement is the *vicinal preorder* [5]. In the vicinal preorder, a vertex $w \in V$ *dominates* a vertex $v \in V$, $v \leq w$, if $N(v) \subseteq N[w]$. Another way to look at the vicinal preorder is that it is the union of the dominance preorder and the following.

In the *structural preorder* a vertex $w \in V$ dominates a vertex $v \in V$, $v \leq_n w$, if $N(v) \subseteq N(w)$. Analogously to the dominance preorder the subscript indicates that this relation can only exist between non-adjacent pairs of vertices. The resulting equivalence classes induce independent sets, and vertices that dominate each other are also known as *structurally equivalent* [12], or *false twins*. As the vicinal preorder is the union of dominance and structural preorder, each equivalence class induces either a clique or an independent set. The graphs for which the vicinal preorder is complete are known as *threshold graphs* [15].

Computing the set of false twins as well as recognizing threshold graphs can be done in $\mathcal{O}(m)$ time [8,11,18]. Moreover, constructing the dominance preorder using Algorithm 1 and counting cycles of length 4 (the problem underlying the structural and thus the vicinal preorder) is possible in time $\mathcal{O}(\alpha(G)m)$ [2]. However, these algorithms cannot be adapted for our purposes without increasing their running time. This is due to the fact that the sizes of structural and vicinal preorder are not bounded by $\mathcal{O}(\alpha(G)m)$, as we will show now using a concept closely related to the structural and vicinal preorder: *Subset partial orders.*

Given a family of subsets of a domain, the subset partial order represents all the subset inclusions between the subsets. Expressing our problem in terms of subset partial orders, the domain is the vertex set, the subsets are the neighborhoods of the vertices and the preorders correspond to the subset partial order.

Yellin and Jutla [23] constructed subset partial orders of size $\Theta(m^2/\log^2 m)$, which was later shown to be a tight upper bound [19]. The example below adapts Yellin and Jutla's construction to graphs. It shows the $\Omega(m^2/\log^2 m)$ lower bound for the structural and vicinal preorder, and demonstrates that we cannot hope to construct both preorders in time $\mathcal{O}(\alpha(G)m)$ like the dominance preorder.

Theorem 3. *There exists a family of graphs for which even the transitive reductions of both vicinal and structural preorder have size $\Theta(m^2/\log^2 m)$ and thus $\omega(\alpha(G)m)$.*

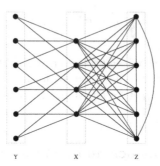

Y   X   Z

Fig. 1. Graph G_4 as produced by the construction in the proof of Theorem 3.

Proof. Let k denote an even positive integer. Let $X = \{x_1, \ldots, x_k\}$, $Y = \{y_1, \ldots, y_{\binom{k}{k/2}}\}$ and $Z = \{z_0, \ldots, z_{\binom{k}{k/2}-1}\}$ be disjoint sets. Let $A_1, \ldots, A_{\binom{k}{k/2}}$ be the subsets of X of size $k/2$. We construct the graph G_k as follows: The vertex set of G_k is given by $V = X \uplus Y \uplus Z$. Each vertex y_i is adjacent to all vertices in A_i. Finally, each vertex z_i is adjacent to all vertices in X, to vertex $z_{i-1 \mod \binom{k}{k/2}}$ and to $z_{i+1 \mod \binom{k}{k/2}}$; i.e., the vertices in Z form a cycle. Figure 1 exemplifies the construction for $k = 4$.

First, the graph has $m = \frac{k}{2}\binom{k}{k/2} + (k+1)\binom{k}{k/2} \in \Theta(k\binom{k}{k/2})$ edges. Second, note that all vertices in Z dominate all vertices in Y in the vicinal and structural preorder, but any two other vertices are incomparable. Thus, both preorders have size $\binom{k}{k/2}^2$, and since the preorders do not contain any transitive relationships, this is also the size of the transitive reductions. Finally, the graph has arboricity at most $k + 2$, as we can cover all edges by k stars centered at the vertices in X and two paths that cover the edges of the cycle within Z.

Using Stirling's formula, we obtain $\binom{k}{k/2} \in \Theta(2^k/\sqrt{k})$ and thus $k \in \Theta(\log m)$. Putting it all together, the transitive reductions of the preorders have size $\binom{k}{k/2}^2 \in \Theta(m^2/\log^2 m) \subseteq \Omega(\alpha(G)m^2/\log^3 m) \subseteq \omega(\alpha(G)m)$. $\qquad\square$

Several algorithms have been developed that compute the subset partial order in $\mathcal{O}(m^2/\log m)$ randomized or worst-case time [19,20,23]. However, these algorithms require substantive book keeping [19], cannot be generalized to weighted edges [20] or use complex data structures [23].

Algorithm 2 adapts a simple subset partial order algorithm introduced by Pritchard [20] to the vicinal preorder, and it can also be straightforwardly modified to determine the structural preorder instead. As this algorithm can also be used to count all cycles of length 4 in a graph, it can also be viewed as an adaption of Chiba and Nishizeki's algorithm C4 [2].

Theorem 4. *Algorithm 2 determines the vicinal preorder of a simple undirected graph in time $\mathcal{O}(\Delta(G)m)$ with space linear in the size of the input.*

Proof. For each non-isolated vertex $v \in V$, the algorithm marks neighbors $u \in N(v)$ and vertices $w \in N(u) \setminus \{v\}$ at distance two. For a marked vertex w, $t[w]$

Algorithm 2: Vicinal Preorder

---

input   : graph $G = (V, E)$
output : partial ranking $\leq$ on V (neighborhood-inclusion)

initialize $\leq$ with $v \leq v$ for all $v \in V$;
for $v \in V$ **do**
    for $u \in N(v)$ **do**
        for $w \in N[u] \setminus \{v\}$ **do**   // use $N(u)$ to determine $\leq_n$
            if w *not marked with* v **then**
                mark w with v;
                $t[w] \leftarrow 0$;
            increment $t[w]$;
            if $t[w] = \deg(v)$ **then** add $v \leq w$

---

holds the number of times it was encountered from a neighbor u of v (plus one for the neighbors themselves). This counter reaches $\deg(v)$ if and only if all (other) neighbors of v are also neighbors of w.

As for the running time, note that the first two loops yield $\sum_{v \in V} \deg(v) \in \mathcal{O}(m)$ iterations in total. During each iteration, the inner loop is executed $\deg(u) \leq \Delta(G)$ times.   □

We note that Pritchard also gave an optimized variant of the algorithm that runs in $\mathcal{O}(\min\{m^2/\log n, \Delta(G)m\})$ time [20]. This can be a substantial improvement on sparse graphs with $o(n \log n)$ edges, and the algorithm can also be faster on graphs with $o(\log n)$ high-degree vertices. However, there appears to be no simple generalization of this optimization to weighted edges.

4.1   A Heuristic Based on Modular Decomposition

Closely related to the problem of computing the preorders is modular decomposition. The modular decomposition of a graph can be computed in $\mathcal{O}(m)$ time [7], and it lends itself to a heuristic approach to compute the dominance, structural and vicinal preorders on unweighted graphs that we will describe now.

In modular decomposition, a module defines a subset $M \subseteq V$ such that all vertices in M have exactly the same neighborhood in $V \setminus M$. A module is strong if there is no other module overlapping it. The modular decomposition tree $MD(G)$ represents the inclusion structure of strong modules in the graph. The representative graph $R(M)$ of the module M is the quotient graph $G[M]/P$, where P is the partition of M given by the child modules of M in $MD(G)$; in other words, it is the graph obtained from the subgraph $G[M]$ by contracting all child modules into single vertices. There are three types of strong modules: In a series module, $R(M)$ is a complete graph; in a parallel module, $R(M)$ is an empty graph; otherwise, the module is prime.

The heuristic computes the preorders by walking the modular decomposition tree $MD(G)$. Suppose we are currently at module M in the walk of $MD(G)$ and consider two vertices $v, w \in V$ that are contained in different child modules

C_v, C_w of M in $\mathrm{MD}(G)$. The heuristic decides the relation between v and w in the dominance and structural preorders as summarized in Table 1. The vicinal preorder arises by combining the cases for the other two preorders.

Table 1. Modular decomposition heuristic: ordering of v and w based on the type of the containing module M and the distinct children $C_v \ni v$, $C_w \ni w$ of M in $\mathrm{MD}(G)$

M	Dominance preorder	Structural preorder
Series	$N[v] \subseteq N[w] \iff C_w = \{w\}$	$N(v) \nsubseteq N(w)$
Parallel	$N[v] \nsubseteq N[w]$	$N(v) \subseteq N(w) \iff C_v = \{v\}$
Prime	$N[v] \subseteq N[w] \iff$ 1. $N_{R(M)}[C_v] \subseteq N_{R(M)}[C_w]$ 2. $C_w = \{w\}$ or $\quad C_w$ is series with child $\{w\}$	$N(v) \subseteq N(w) \iff$ 1. $N_{R(M)}(C_v) \subseteq N_{R(M)}(C_w)$ 2. $C_v = \{v\}$ or $\quad C_v$ is parallel with child $\{v\}$

The bottleneck in this heuristic is condition 1 for prime modules. While we can construct the representative graphs in $\mathcal{O}(m)$ time [10], we also need algorithms that compute the preorders on them. Using Algorithms 1 and 2, the heuristic computes the dominance preorder in time $\mathcal{O}(m + \alpha'(G)m') \subseteq \mathcal{O}(\alpha(G)m)$ and the structural and vicinal preorders in time $\mathcal{O}(m + \Delta'(G)m' + |\leq|) \subseteq \mathcal{O}(\Delta(G)m)$, where $\alpha'(G)$ and $\Delta'(G)$ denote the maximum arboricity and the maximum degree of a representative prime graph in G, m' is the total number of edges in representative prime graphs in G, and $|\leq|$ refers to the size of the output. This heuristic can thus significantly improve runtime on decomposable graphs.

5 Positional Dominance

The notions of dominance considered so far all require adjacency with identical neighbors. Common centrality indices, on the other hand, are typically invariant under automorphisms. In degree centrality, for instance, it is sufficient to have more neighbors, no matter which. Positional dominance [1] generalizes such assumptions by allowing comparison of neighbors using sets of admissible permutations. We here consider the case in which any neighbor with at least the same vertex weight and at least an equally strong relationship may serve as a replacement.

A vertex w dominates vertex v w.r.t. positional dominance,

$$v \leq w \quad \text{if} \quad \begin{cases} \text{there ex. } \pi : V \to V \text{ such that } \omega(u) \leq \omega(\pi(u)) \text{ and} \\ \omega((v,u)) \leq \omega((w, \pi(u))) \quad \forall (v,u) \in E : \omega((v,u)) \neq 0. \end{cases}$$

Restricting π to the identity permutation or transpositions we can derive the structural and dominance preorders, therefore positional dominance is a generalization of the previous notions. Note that diagonal entries and the dyads $(v,w), (w,v)$ may be treated specially when comparing v and w in the positional dominance approach, however we will not address this topic in more detail here.

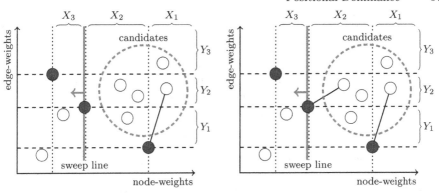

Fig. 2. Sweep line approach: blue/white vertices are neighbors of v/w, and white vertices are partitioned into bins X_i and Y_j along dotted and dashed lines, respectively. Vertices connected by a black line have been matched. (left) potential candidates for the matching of the vertex under the sweep line; (right) matching an unmatched vertex in the non-empty bin Y_k (here Y_2) that contains the vertices with smallest edge-weight greater than or equal to the one of the vertex under the sweep line. (Color figure online)

The straightforward approach to decide whether $v \leq w$ or not consists of two phases.[1] In the first phase for each vertex $u \in N(w)$ the subset of vertices in $N(v)$ that is dominated by u is computed, which requires $\mathcal{O}(k + \deg(w) \log \deg(v))$ time [14] with k denoting the total number of dominance pairs. Based on these dominance relations, it is tested in the second phase if there exists a mapping π such that $v \leq w$. Finding this mapping is equivalent to finding a perfect matching in the bipartite graph induced by the dominance relations and can be done in $\mathcal{O}(\sqrt{\deg(v) + \deg(w)}k)$ [9]. Consequently computing the positional dominance preorder using this approach has a complexity of $\mathcal{O}(nm\Delta(G)^{3/2})$.

In the following we will show that this problem can be solved more efficiently using a greedy sweep line approach, cf. Algorithm 3 and Fig. 2. The basic idea of this approach is to process all neighbors v_i of vertex v in decreasing order of their vertex-weights. The algorithm maintains bins $\mathcal{Y} = (Y_1, \ldots, Y_{\deg(v)})$ that partition the set of unmatched candidates u with vertex-weight $\omega(u) \geq \omega(v_i)$. Here, a bin Y_j contains all those unmatched candidates u whose associated edge-weights lie between $\omega(v, v_{\sigma(j)}) \leq \omega(w, u) < \omega(v, v_{\sigma(j+1)})$, where σ is a permutation such that neighbors $v_{\sigma(1)}, \ldots, v_{\sigma(\deg(v))}$ of v are sorted in increasing order of their associated edge-weights. Hence, the bins $Y_{\sigma^{-1}(i)}, \ldots, Y_{\deg(v)}$ contain all unmatched candidates with edge-weight at least $\omega(v, v_i)$. To find a vertex to match with v_i, we thus identify a non-empty bin Y_k with $k \geq \sigma^{-1}(i)$. If there are several such bins, we choose the one with smallest index $k \geq \sigma^{-1}(i)$. We then match v_i with an arbitrary vertex in bin Y_k. This greedy matching is correct since v's remaining unprocessed neighbors all have smaller vertex-weights than all the vertices in the bins $\mathcal{Y}$ due to the way we process the neighbors of v. Matching v_i with a candidate from the non-empty bin Y_k with minimal index $k \geq \sigma^{-1}(i)$,

---

[1] In the following we assume that $\deg(v) \leq \deg(w)$, since otherwise w cannot dominate v w.r.t. positional dominance.

Algorithm 3: Positional Dominance Test $v \leq w$

---

input   : graph $G = (V, E; \omega : V \cup E \rightarrow \mathbb{R})$ and $v, w \in V$

data   : bins $\mathcal{X} = (X_1, \ldots, X_{\deg(v)}), \mathcal{Y} = (Y_1, \ldots, Y_{\deg(v)})$,

vertex array of bin indices $b[u]$, list of indices of non-empty bins T

output : boolean indicating whether $v \leq w$, or not

---

if $\deg(w) < \deg(v)$ **then return** *FALSE*

let $N(v) = \langle v_1, \ldots, v_{\deg(v)} \rangle$ s.t. $\omega(v_1) \geq \cdots \geq \omega(v_{\deg(v)})$;

let σ be permutation s.t. $\omega(v, v_{\sigma(1)}) \leq \cdots \leq \omega(v, v_{\sigma(\deg(v))})$;

partition $N(w)$ into $X_i = \{u \in N(w) : \omega(v_i) \leq \omega(u) < \omega(v_{i-1})\}$ where $\omega(v_0) := \infty$;

for $u \in N(w)$ **do**

    if $\omega(v, v_{\sigma(1)}) \leq \omega(w, u)$ **then** $b[u] \leftarrow \max\{k : \omega(v, v_{\sigma(k)}) \leq \omega(w, u)\}$

$T, Y_1, \ldots, Y_m \leftarrow \emptyset$;

for $i = 1, \ldots, \deg(v)$ **do**

    for $u \in X_i$ **do**

        if $\omega(v, v_{\sigma(1)}) \leq \omega(w, u)$ **then**

            if $Y_{b[u]} = \emptyset$ **then** $T \leftarrow T \cup \{b[u]\}$

            $Y_{b[u]} \leftarrow Y_{b[u]} \cup \{u\}$

    if $k < \sigma^{-1}(i)$ for all $k \in T$ **then return** *FALSE*

    $k \leftarrow \min\{\ell \in T : \ell \geq \sigma^{-1}(i)\}$;

    remove some vertex u from Y_k // match v_i and some vertex $u \in Y_k$

    if $Y_k = \emptyset$ **then** $T \leftarrow T \setminus \{k\}$

return *TRUE*

---

consequently, retains those candidates in the bins that have the highest potential to also dominate the remaining neighbors of v respective their edge-weights.

Before we actually prove the correctness and running time of this algorithm, we first show an invariant of the outer for loop. Assume that vertex v is dominated by vertex w w.r.t. positional dominance via the permutation $\pi' : V \rightarrow V$. This means that Algorithm 3 would succeed if it matched according to the permutation π', since neighbor $\pi'(v_i)$ of w would always be available for matching in some bin while processing v_i. Now let $\mathcal{Y} = (Y_1, \ldots, Y_{\deg(v)})$ be the actual bins produced by the algorithm, and $\mathcal{Y}' = (Y_1', \ldots, Y_{\deg(v)}')$ be the bins that would be produced by the algorithm if it matched according to the permutation π' instead. Observe that at any point during the execution of the algorithm, we have $\sum_{j=1}^{\deg(v)} |Y_j| = \sum_{j=1}^{\deg(v)} |Y_j'|$, and T and T' always contain the indices of all non-empty bins $\mathcal{Y}$ and $\mathcal{Y}'$, respectively. We say that $\mathcal{Y}$ *covers* $\mathcal{Y}'$ if and only if for all $k \in \{1, \ldots, \deg(v)\}$ we have $\sum_{j=k}^{\deg(v)} |Y_j| \geq \sum_{j=k}^{\deg(v)} |Y_j'|$.

Lemma 1. $\mathcal{Y}$ *covers* $\mathcal{Y}'$ *at the end of each successful iteration performed by the outer for loop.*

Proof. We prove the loop invariant by induction on the number of loop iterations performed by the outer for loop.

– Basis: Before the first iteration of the outer for loop, all bins are empty, so $\mathcal{Y}$ trivially covers $\mathcal{Y}'$.

– Inductive step: By induction, $\mathcal{Y}$ covers $\mathcal{Y}'$ at the end of the $(i-1)$-th iteration. During the i-th iteration, the inner for loop adds the same vertices to the bins in $\mathcal{Y}$ and $\mathcal{Y}'$, therefore at the end of this loop $\mathcal{Y}$ still covers $\mathcal{Y}'$.

Finally, Algorithm 3 tries to match the current neighbor v_i of v with some neighbor of w. Since w dominates v via permutation π', $\omega(w, \pi'(v_i))$ is at least as large as $\omega(v, v_i)$, and since $\pi'(v_i)$ thus must be matchable, $\pi'(v_i)$ must be in some bin Y'_k. Additionally, $\mathcal{Y}$ still covers $\mathcal{Y}'$ after the inner for loop, so there is a non-empty bin Y_ℓ with $\sigma^{-1}(i) \leq \ell$ minimal. Algorithm 3 will pick a vertex from Y_ℓ and match it with v_i. If $\ell \leq k$, then the invariant still holds trivially after the matching. If $\ell > k$, observe that $\sum_{j=\ell}^{\deg(v)} |Y_j| = \sum_{j=k}^{\deg(v)} |Y_j| \geq \sum_{j=k}^{\deg(v)} |Y'_j| \geq 1 + \sum_{j=k+1}^{\deg(v)} |Y'_j|$ before the removal, as bins $Y_k, \ldots, Y_{\ell-1}$ are empty. Thus, $\mathcal{Y}$ still covers $\mathcal{Y}'$ after removing vertices from Y_ℓ and Y'_k in both cases. $\qquad\square$

Theorem 5. *Algorithm 3 decides for a given pair of vertices v, w with $\deg(v) \leq \deg(w)$ and weights ω in $\mathcal{O}(\deg(w) \log \deg(v))$ time whether $v \leq w$ or not.*

Proof. We will start by proving the correctness of the algorithm and thereafter show the correctness of the claimed running time.

Correctness: Assume for now that w dominates v w.r.t. positional dominance via permutation π'. Suppose that Algorithm 3 already succeeded in performing $0 \leq i - 1 < \deg(v)$ iterations of the outer for loop, and let $\mathcal{Y}$ and $\mathcal{Y}'$ denote the respective bins at the beginning of the i-th iteration. During the i-th iteration, Algorithm 3 tries to match neighbor v_i of v with some neighbor of w. In π', v_i has already been matched with $\pi'(v_i) \in N(w)$. First, assume that $\pi'(v_i)$ was already in a bin Y'_k at the beginning of the i-th iteration; i.e., it was already a candidate in previous iterations. Since $\mathcal{Y}$ covers $\mathcal{Y}'$ (Lemma 1), there is a non-empty bin Y_ℓ with $\ell \geq k$. Since $\ell \geq k \geq \sigma^{-1}(i)$, we have $\omega(w, u) \geq \omega(v, v_i)$ for any vertex $u \in Y_\ell$, so the algorithm finds a vertex to match with v_i in the i-th iteration. Next, suppose that $\pi'(v_i)$ was newly added to the bins $\mathcal{Y}'$ in the inner for loop of the i-th iteration. All newly added vertices are inherently unmatched in the algorithm and thus can be matched with v_i. Therefore, the algorithm will also succeed in performing the i-th iteration, and hence, correctly decide that $v \leq w$.

Conversely, assume the algorithm succeeds. Let π be the permutation computed during the run of the algorithm. The algorithm ensures that any neighbor v_i of v is only matched with a single neighbor $\pi(v_i)$ of w if $\omega(\pi(v_i)) \geq \omega(v_i)$, as $\pi(v_i)$ will not be added to the bins $\mathcal{Y}$ otherwise, and $\omega(w, \pi(v_i)) \geq \omega(v, v_i)$, as otherwise $\pi(v_i)$ cannot be extracted from a bin Y_k, $k \geq \sigma^{-1}(i)$. Thus, w dominates v w.r.t. positional dominance via permutation π. $\qquad\square$

Time complexity: Sorting the neighbors of v (lines 2–3) and (pre-)binning the neighbors of w (lines 4–6) can be done in $\mathcal{O}(\deg(w) \log \deg(v))$ time. It remains to be shown that the work done by the outer for loop does not exceed this complexity. The cost of the outer for loop is composed of (i) adding neighbors of w to a bin Y_k, (ii) adding indices of non-empty bins to T (iii) testing if T contains a bin Y_k with $k \geq \sigma^{-1}(i)$ and (iv) possibly removing the index of a bin from T if that bin becomes empty again. Since each neighbor of w is sorted into a bin Y_k at most once, (i) costs in total $\mathcal{O}(\deg(w))$ time. Furthermore, there are at most $\deg(v)$ deletions and tests in T, so steps (ii), (iii) and (iv) are performed at most $\mathcal{O}(\deg(v))$ times. When T is implemented by a specialiced data structure on a bounded integer domain like a van Emde Boas tree [3], (ii), (iii) and (iv) cost in total $\mathcal{O}(\deg(v) \log \log \deg(v))$ time. Thus, the total running time of the outer for loop of Algorithm 3 is in $\mathcal{O}(\deg(w) + \deg(v) \log \log \deg(v))$.   □

To compute the positional dominance preorder, we can sort the neighborhoods of all vertices in advance in $\mathcal{O}(m \log \Delta(G))$ time. Then we no longer need to sort the neighborhoods (lines 2–3) and (pre-)binning (lines 4–6) is possible in $\mathcal{O}(\deg(w))$ time in Algorithm 3. That means that on pre-sorted neighborhoods, the runtime is dominated by the outer for loop, which requires $\mathcal{O}(\deg(w) + \deg(v) \log \log \deg(v))$ time. Hence, we can compute the positional dominance preorder of a graph G with weights ω in time $\mathcal{O}(nm \log \log \Delta(G))$, since

$$\sum_{\substack{v,w \in V \\ \deg(v) \le \deg(w)}} (\deg(w) + \deg(v) \log \log \deg(v)) \le 2n \left(\sum_{v \in V} \deg(v) \right) \log \log \Delta(G)$$

$$= 4nm \log \log \Delta(G).$$

6   Conclusion

We studied various notions of dominance which can serve as potential building blocks for the generalization of the concept of centrality. Using a greedy sweep line approach, cf. Algorithm 3, we were able to show that positional dominance can be computed in $\mathcal{O}(nm \log \log \Delta(G))$ time compared to $\mathcal{O}(nm\Delta(G)^{3/2})$ required by a straight-forward algorithm to solve this problem. For this problem, we see the greatest potential for further runtime improvements in avoiding some of the pairwise comparisons between vertices.

Restricting positional dominance to the identity permutation, i.e., assuming heterogeneity, translates into the structural preorder, which is a restriction of the vicinal preorder and a variant of (vertex) dominance. With Algorithm 1 we presented an algorithm running in $\mathcal{O}(a(G)m)$ to compute the dominance preorder. The running time may be far from linear in the size of input and output, for social networks, however, where the arboricity is often negligible [4], it is acceptable, not least since we are not aware of any faster solution to this problem. While the main challenge in the computation of the structural preorder lies in finding cycles of length four we proved that a running time of $\mathcal{O}(a(G)m)$, which is the running time of an efficient algorithm to solve this problem [2], is not achievable as the size of the preorder can already be much larger. As a result of this finding we proposed with Algorithm 2 a procedure to compute the structural as well as vicinal preorder in time $\mathcal{O}(\Delta(G)m)$. For computing dominance, structural and vicinal preorder, we additionally presented a heuristic that can yield substantial speed-ups on unweighted graphs that are decomposable through modular decomposition.

References

1. Brandes, U.: Network positions. Methodol. Innov. **9**, 2059799116630650 (2016)
2. Chiba, N., Nishizeki, T.: Arboricity and subgraph listing algorithms. SIAM J. Comput. **14**(1), 210–223 (1985)
3. van Emde Boas, P.: Preserving order in a forest in less than logarithmic time and linear space. Inf. Process. Lett. **6**(3), 80–82 (1977)
4. Eppstein, D., Spiro, E.S.: The h-index of a graph and its application to dynamic subgraph statistics. J. Graph Algorithms Appl. **16**(2), 543–567 (2012)
5. Foldes, S., Hammer, P.L.: The Dilworth number of a graph. Ann. Discret. Math. **2**, 211–219 (1978)
6. Freeman, L.C.: A set of measures of centrality based on betweenness. Sociometry **40**(1), 35–41 (1977)
7. Habib, M., Paul, C.: A survey of the algorithmic aspects of modular decomposition. Comput. Sci. Rev. **4**(1), 41–59 (2010)
8. Heggernes, P., Kratsch, D.: Linear-time certifying recognition algorithms and forbidden induced subgraphs. Nordic J. Comput. **14**(1–2), 87–108 (2007)
9. Hopcroft, J.E., Karp, R.M.: An $n^{5/2}$ algorithm for maximum matchings in bipartite graphs. SIAM J. Comput. **2**(4), 225–231 (1973)
10. Lagraa, S., Seba, H.: An efficient exact algorithm for triangle listing in large graphs. Data Min. Knowl. Disc. **30**(5), 1350–1369 (2016)
11. Lerner, J.: Role assignments. In: Brandes, U., Erlebach, T. (eds.) Network Analysis. LNCS, vol. 3418, pp. 216–252. Springer, Heidelberg (2005). doi:10.1007/978-3-540-31955-9_9
12. Lorrain, F., White, H.C.: Structural equivalence of individuals in social networks. J. Math. Soc. **1**(1), 49–80 (1971)
13. Lü, L., Chen, D., Ren, X.L., Zhang, Q.M., Zhang, Y.C., Zhou, T.: Vital nodes identification in complex networks. Phys. Rep. **650**, 1–63 (2016)
14. Lueker, G.S.: A data structure for orthogonal range queries. In: 19th Annual Symposium on Foundations of Computer Science, Ann Arbor, Michigan, USA, 16–18 October 1978, pp. 28–34 (1978)
15. Mahadev, N.V., Peled, U.N.: Threshold Graphs and Related Topics, Annals of Discrete Mathematics, vol. 56. Elsevier, Amsterdam (1995)
16. Nash-Williams, C.S.J.A.: Decomposition of finite graphs into forests. J. Lond. Math. Soc. **39**(1), 12 (1964)
17. Ortmann, M., Brandes, U.: Triangle listing algorithms: back from the diversion. In: Proceedings of the 16th Workshop on Algorithm Engineering and Experiments (ALENEX 2014), pp. 1–8 (2014)
18. Paige, R., Tarjan, R.E.: Three partition refinement algorithms. SIAM J. Comput. **16**(6), 973–989 (1987)
19. Pritchard, P.: On computing the subset graph of a collection of sets. J. Algorithms **33**(2), 187–203 (1999)
20. Pritchard, P.: A simple sub-quadratic algorithm for computing the subset partial order. Inf. Process. Lett. **56**(6), 337–341 (1995)
21. Sabidussi, G.: The centrality index of a graph. Psychometrika **31**(4), 581–603 (1966)
22. Schoch, D., Brandes, U.: Re-conceptualizing centrality in social networks. Eur. J. Appl. Math. **27**(6), 971–985 (2016)
23. Yellin, D.M., Jutla, C.S.: Finding extremal sets in less than quadratic time. Inf. Process. Lett. **48**(1), 29–34 (1993)

Accurate Low-Space Approximation of Metric k-Median for Insertion-Only Streams

Vladimir Braverman[1], Harry Lang[2],
and Keith Levin[1(✉)]

[1] Department of Computer Science, Johns Hopkins University,
Baltimore, MD, USA
klevin@jhu.edu
[2] Department of Mathematics, Johns Hopkins University,
Baltimore, MD, USA

Abstract. We present a low-constant approximation for metric k-median on an insertion-only stream of n points using $O(\epsilon^{-3}k \log n)$ space. In particular, we present a streaming $(O(\epsilon^{-3}k \log n), 2 + \epsilon)$-bicriterion solution that reports cluster weights. It is well-known that running an offline algorithm on this bicriterion solution yields a $(17.66 + \epsilon)$-approximation.

Previously, there have been two lines of research that trade off between space and accuracy in the streaming k-median problem. To date, the best-known (k, ϵ)-coreset construction requires $O(\epsilon^{-2}k \log^4 n)$ space [8], while the best-known $O(k \log n)$-space algorithm provides only a $(O(k \log n), 1063)$-bicriterion [3]. Our work narrows this gap significantly, matching the best-known space while significantly improving the accuracy from 1063 to $2 + \epsilon$. We also provide a matching lower bound, showing that any polylog(n)-space streaming algorithm that maintains an (α, β)-bicriterion must have $\beta \geq 2$.

Our technique breaks the stream into segments defined by jumps in the optimal clustering cost, which increases monotonically as the stream progresses. By a storing an accurate summary of recent segments and a lower-space summary of older segments, our algorithm maintains a $(O(\epsilon^{-3}k \log n), 2 + \epsilon)$-bicriterion solution for the entire input.

Keywords: Streaming algorithms · k-median · Clustering

1 Introduction

In metric k-median clustering over insertion-only streams, we are sequentially given n points from a metric space and attempt to return a set of k centers that

---

V. Braverman—This material is based upon work supported in part by the National Science Foundation under Grants IIS-1447639 and CCF-1650041.

H. Lang—This research is supported by the Franco-American Fulbright Commission. The author thanks INRIA (l'Institut national de recherche en informatique et en automatique) for hosting him during the writing of this paper.

D. Gaur and N.S. Narayanaswamy (Eds.): CALDAM 2017, LNCS 10156, pp. 72–82, 2017.
DOI: 10.1007/978-3-319-53007-9_7

approximately minimize the sum of the distances of each point to its nearest center. We present an improved algorithm for this problem, maintaining the best-known space bound while drastically improving the approximation-ratio.

Streaming clustering has a long history since the work of Guha, Meyerson, Mishra, Motwani and O'Callaghan [11]. There have been two main classes of algorithms that have polylogarithmic space complexity and solve the streaming version of metric k-median. The first class contains facility-based algorithms, starting with the first polylogarithmic solution for the streaming k-median by Charikar, O'Callaghan, and Panigrahy [6]. These methods build upon the connection between the k-median problem and the facility location problem, using the online algorithm of Meyerson [14] as a subroutine. Facility-based algorithms achieve low storage (currently $O(k \log n)$-space due to [3]), but suffer from an extremely large approximation ratio. The best-space algorithm in this class provides a $(O(k \log n), 1063)$-bicriterion (see Sect. 7.2 for a calculation of this constant). The second class contains coreset-based algorithms, such as the works of [1,7,8,12,13]. By coreset-based, we refer to algorithms that can return a (k, ϵ)-coreset for any $\epsilon > 0$. A (k, ϵ)-coreset of a set A is a set B such that the cost of clustering A and B with any set of centers differ by at most a factor of $(1 \pm \epsilon)$. These algorithms achieve an arbitrarily low approximation ratio, but yet require significantly more storage, the lowest being a $O(\epsilon^{-2} k \log^4 n)$-space coreset due to [8]. In this line of research, an offline coreset construction is provided, which is then transformed into a streaming construction using the merge-and-reduce technique of [2] from 1980. Merge-and-reduce multiplies the space-bound by a factor of $\Omega(\log^3 n)$, and although other methods have been found for the Euclidean case [4,9], this remains the only technique available for coresets in general metric spaces. Without overcoming this 35-year-old barrier, coreset-based algorithms cannot match the space-bounds of facility-based algorithms.

The two classes of algorithms suggest a possible trade-off between the favorable space-bounds of facility-based methods and the favorable approximation ratio of coreset-based methods. A natural question is if it is possible to design an algorithm that performs well in both space and approximation ratio. For Euclidean space, this question was answered in the affirmative by [15]. We now answer this in the affirmative for general metric spaces, using a technique entirely different from that of [15]. Our algorithm achieves a low-approximation ratio using $O(\epsilon^{-3} k \log n)$-space and maintaining a $(O(\epsilon^{-3} k \log n), 2 + \epsilon)$-bicriterion.

In 2009, an important result by Guha [10] was a facility-based $(34 + \epsilon)$-approximation using $O(\epsilon^{-3} \log \frac{1}{\epsilon} k \log^2 n)$-space. This straddles the above-mentioned space-accuracy trade-off by offering a low-constant (although not as low as offered by coreset-based algorithms) as well as low-space (although not as low as the $O(k \log n)$ offered by facility-based algorithms). In comparison, our algorithm offers both lower space and a lower approximation ratio than [10]. It is a well-known result [3,6,10,11] that running an offline γ-approximation on a (α, β)-bicriterion solution yields a $(\beta + 2\gamma(1 + \beta))$-approximation. With our $(O(\epsilon^{-3} k \log n), 2 + \epsilon)$-bicriterion, running the offline 2.61-approximation of [5] yields a $(17.66 + \epsilon)$-approximation. In relation to [10], this is a 48% reduction in the approximation factor and the space requirement is improved from $O(k \log^2 n)$ to $O(k \log n)$. Additionally, we show in the Appendix (see Sect. 7.1) that no polylog(n)-space algorithm can improve upon our approximation ratio.

2 Our Contribution

We present an algorithm that maintains a $(k, 2 + \epsilon)$-bicriterion and uses $O(\epsilon^{-3} k \log n)$ space. Our algorithm works in three layers. The first layer is a black-box $O(1)$-approximation; a single instance simply runs in the background while the higher layers save information from it. The second layer (in Sect. 4) maintains a prefix A that contributes at most an ϵ-small portion to OPT of the stream; this layer only requires the space needed to store the output of the first layer at two previous moments. The third layer (in Sect. 5) runs the facility location algorithm of [14], and at any moment only four instances of facility location are required to run in order to maintain the $1 - \frac{1}{n}$ probability guarantee.

Our techniques differ from previous facility-based algorithms in crucial ways. Like the previous works of [3,6], our algorithm operates in phases. However, the techniques used in these algorithms cause additional costs to be compounded during each phase. In contrast, our algorithm manages the stream so that we only incur cost during the two most recent phases. We avoid additional costs by maintaining a prefix A of the stream S such that $\text{OPT}(A, k) \leq \epsilon \text{OPT}(S)$ and such that we have an $O(1)$-approximate estimate of $\text{OPT}(S)$ before processing the suffix $S \setminus A$.

Of course, it is impossible to have an $O(1)$-approximate estimate of $\text{OPT}(S)$ before processing the suffix. However, Algorithm 1 allows to pretend that we have such an estimate. The fundamental idea is to always maintain the "next prefix" A'. If we ever detect that $\text{OPT}(S)$ may have surpassed the upper bound of our estimate, then we replace the prefix A with A' and update the estimate accordingly.

Given an $O(1)$-approximate estimate of $\text{OPT}(S)$, we can contruct a good approximation of $S \setminus A$ (see Sect. 3). Because $\text{OPT}(A, k) \leq \epsilon \text{OPT}(S)$, even a poor approximation on the prefix is sufficient. Combining both these pieces, we are able to maintain a low-constant solution over the stream.

3 Definitions

Our algorithm works for weighted sets of integral weight. For the bounds, let n be the total weight of the stream (the sum of the weights of each point in the stream). In fact, if n is not known in advance, a polynomial upper-bound will suffice. Note that n is assumed to be known in [3,6,15], so this does not add any additional restrictions. Let $(\mathcal{X}, d)$ be a metric space.

Definition 1 (Cost Function). *Given sets $A, C \subset \mathcal{X}$, the function $Cost(A, C)$ gives the cost of clustering A with center set C. Explicitly, $Cost(A, C) = \sum_{a \in A} \min_{c \in C} d(a, c)$.*

Definition 2 (Optimum Cost). *The value* $\mathrm{OPT}(A, B, k)$ *is the lowest possible cost of clustering A with k centers from B. Explicitly,* $\mathrm{OPT}(A, B, k) = \min_{C \in B^k} Cost(A, C)$. *As shorthand,* $\mathrm{OPT}(A, k) = \mathrm{OPT}(A, \mathcal{X}, k)$ *where $\mathcal{X}$ is the entire metric space.*

Definition 3 (Connect Function). *Let A, B be multisets of equal weight. $Connect(A, B)$ is the minimum connection cost over all possible bijective maps t from A to B, where the connection cost of t is defined as $\sum_{a \in A} d(a, t(a))$.*

Definition 4 (Bicriterion). *An (α, β)-bicriterion approximation of the k-median clustering of A is a set B such that $\mathrm{COST}(A, B) \leq \alpha\,\mathrm{OPT}(A, k)$ and $|B| \leq \beta k$.*

We will make use of the following observation in Sect. 4.

Observation 1. *For any set C and equally weighted multisets A and B, $Cost(A, C) \leq Connect(A, B) + Cost(B, C)$.*

Proof. Let g be the optimal map from B to C. Let t be the optimal bijective map from A to B. Then by the triangle inequality, for every $a \in A$, $d(a, g(t(a))) \leq d(a, t(a)) + d(t(a), g(t(a)))$. Let h be the optimal map from A to C. The result followed by summing over all $a \in A$ and then noting that $d(a, h(a)) \leq d(a, g(t(a)))$.

The following observation is used in Sect. 6.

Observation 2. *If $Connect(A_1, B_1) \leq v_1$ and $Connect(A_2, B_2) \leq v_2$, then $Connect(A_1 \cup A_2, B_1 \cup B_2) \leq v_1 + v_2$.*

Proof. Let t_i be the optimal bijective map from A_i to B_i. Then consider $g(a) = t_i(a)$ if $a \in A_i$. Although g may not be the optimal bijective map from $A_1 \cup A_2$ to $B_1 \cup B_2$, it yields an upper bound.

4   Phase Manager

Over an insertion-only stream S, the algorithm of [3] maintains a multiset Q such that $Connect(S, Q) \leq \alpha\,\mathrm{OPT}(S, \mathcal{X}, k)$ from some constant α. We refer to this algorithm as **PLS** (which is the name of the algorithm in [6] that provides a similar guarantee). It constructs Q through a technique that connects points in S to other points in S and weights them accordingly. Therefore, we can make a simple modification to additionally maintain a value q such that $Connect(S, Q) \leq q$. For a section P of the stream, we denote the multiset Q by **PLS**(P) and we denote the value q by $q(P)$.

By running an offline γ-approximation for k-median on Q, we obtain a θ-approximation on the original stream where $\theta = 2\gamma(1 + \alpha)$. This is a standard result, and the reader is referred to [6] for details.

We denote the first N points of the stream by $[1, N]$. Our algorithm requires a monotonically increasing function $f([1, N])$ such that $\mathrm{OPT}([1, N], \mathcal{X}, k) \leq$

$f([1, N]) \leq \theta \ \text{OPT}([1, N], \mathcal{X}, k)$. We compute this function as follows. We define f' to be the sum of $q([1, N])$ and the cost of clustering $\mathbf{PLS}([1, N])$ with its γ-approximation. By Observation 1, f' satisfies the desired inequalities, but f' may not be monotonically increasing because the γ-approximation may decrease at times. We define f recursively as $f([1, N]) = \max\{f'([1, N], f([1, N-1]))\}$. Updating f requires $O(1)$ time and space because it is computed by taking the maximum of two already stored values. Now f is guaranteed to be monotonically increasing, and moreover is still satisfies the desired inequalities because $\text{OPT}([1, N], \mathcal{X}, k)$ is monotonically increasing.

Our algorithm relies on maintaining a partition of the stream into three segments. After processing S_N (the first N points from the stream), we write the elements of the filtration as A_N and B_N such that we have $\emptyset \subset A_N \subset B_N \subset S_N$. Here both A_N and B_N are prefixes of the stream, meaning that they are equal to $[1, m]$ for some $1 \leq m \leq N$. The following two loop invariants will be maintained, where $\beta = \alpha\theta/\epsilon$.

1. $f(A_N) \leq \beta^{-1} f(B_N)$
2. $f(B_N) > \beta^{-1} f(S_N)$

At the beginning of the stream, it will be necessary to establish the two loop invariants. We do this by letting B_m be the first $k+1$ distinct points and letting A_m be empty. Even if $k+1$ distinct points do not arrive until m is much greater than $k+1$, it is not difficult to see that this initialization procedure can be performed in $O(k \log n)$ memory.

Having established the loop invariants, Algorithm 1 maintains these invariants with a single instance of $\mathbf{PLS}$. When a point arrives, it simply updates $\mathbf{PLS}$ and (if necessary) redefines the filtration to satisfy the invariants. Note that Algorithm 1 does not store any information besides the state of $\mathbf{PLS}$ for each element of the current filtration, resulting in memory requirement equal to that of $\mathbf{PLS}$.

Algorithm 1. Update Process, upon arrival of point p_N

1: Update $\mathbf{PLS}$ with p_N and compute $f(S_N)$
2: **if** $f(S_N) \geq \beta f(B_{N-1})$ **then**
3:      $A_N \leftarrow B_{N-1}$
4:      $B_N \leftarrow S_N$
5: **else**
6:      $A_N \leftarrow A_{N-1}$
7:      $B_N \leftarrow B_{N-1}$

Theorem 1. *Using $O(k \log n)$ memory, Algorithm 1 maintains a filtration $\emptyset \subset A_N \subset B_N \subset S_N$ such that $f(A_N) \leq \beta^{-1} f(B_N)$ and $f(B_N) > \beta^{-1} f(S_N)$.*

Proof. If the condition on Line 2 is not satisfied, then this implies that both invariants continue to hold. If the condition on Line 2 is satisfied, then the second invariant has been violated and must be re-established on Lines 3–4. We recursively assume that both invariants held for the filtration of S_{N-1}. The first invariant reads $f(A_N) \leq \beta^{-1} f(B_N)$ which is equivalent to $f(B_{N-1}) \leq \beta^{-1} f(S_{N-1})$; this is guaranteed to hold since the second invariant was violated. The second invariant reads $f(B_N) > \beta^{-1} f(S_N)$. Since on Line 4 we have $B_N = S_N$, this clearly holds for $\beta > 1$.

Algorithm 1 guarantees that when a phase change occurs, $\mathrm{OPT}(S, k)$ will remain within a constant multiplicative range before the next phase change. We now prove that this is the case.

Lemma 1. *Algorithm 1 guarantees that $f(B_N)/\theta \leq \mathrm{OPT}(S_N, k) < \beta f(B_N)$.*

Proof. The second inequality follows from the second loop invariant of Algorithm 1 and noting that $\mathrm{OPT}(S_N, k) \leq f(S_N)$. The first inequality follows from the approximation guarantee of f and monotonicity.

In the next two sections, we will use the guarantees of Algorithm 1 to construct a $(O(\epsilon^{-3} k \log n), 2 + \epsilon)$-bicriterion. Other subroutines will observe (but not influence) Algorithm 1 and store two sets: **PLS**(A_N) and **PLS**(B_N).

5 Facility Manager

In this section, we present Algorithm 3 that will run in parallel with Algorithm 1. We will use a modified version of the online facility location algorithm of [14] as a subroutine. The main result of this section is Theorem 3 stating that Algorithm 3 maintains a weighted set Q_N such that $Connect(S_N \setminus A_N, Q_N) \leq (3 + \epsilon) \mathrm{OPT}(S_N, k)$ with high probability.

We recall the **OFL** Algorithm 2 with facility cost κ as used in [6]. We maintain a weighted set of facilities Φ, and denote $d(p, \Phi) = \min_{\phi \in \Phi} d(p, \phi)$, with $d(p, \emptyset) = \infty$ by convention. Upon receiving a point p, we open a weight $w(p)$ facility there with probability $w(p)d(p, \Phi)/\kappa$; otherwise we connect it to the nearest facility, incrementing that facilities weight by $w(p)$ and paying service cost $w(p)d(p, \Phi)$.

Algorithm 2. OFL(facility cost κ)

1: $ServiceCost \leftarrow 0$
2: $FacilityCount \leftarrow 0$
3: $\Phi \leftarrow \emptyset$
      Update Process, upon receiving point p_N:
4: **if** a probability $\min(1, w(p_N)d(p_N, \Phi)/\kappa)$ event occurs **then**
5:    Open a facility at p_N with weight $w(p_N)$
6:    $FacilityCount \leftarrow FacilityCount + 1$
7: **else**
8:    Increment weight of a nearest facility to p_N by $w(p_N)$
9:    $ServiceCost \leftarrow ServiceCost + w(p_N)d(p_N, \Phi)$

The following theorem follows from a tuning of parameters based on Theorem 3.1 of [3]. The original statement was for $\epsilon = 1$, so we include a sketch of how we modify their proof.

Theorem 2. *If* **OFL** *is run on a weighted set* A *of weight at most* n *using facility cost* $\frac{L}{k(1+\log n)}$ *where* $L \leq \epsilon\,\mathrm{OPT}(A, k)$, *then with probability at least* $1 - \frac{1}{n}$ *the service cost is at most* $(2 + 7\epsilon)\,\mathrm{OPT}(A, k)$ *and at most* $7\epsilon^{-1}k(1 + \log n)\frac{\mathrm{OPT}(A,k)}{L}$ *facilities are opened.*

Proof (Proof Sketch). Consider an optimal center c that services the set $S \subset A$. Let Σ be the total service cost of assigning S to c. For $j \geq 0$, define regions S_j such that $|S_j| = \epsilon|S|/(1 + \epsilon)^j$ and each point in S_j is not farther from c than any point point in S_{j+1}. Then $\cup_{j > j'} S_j$ consists of at most a single point for $j' = \log_{1+\epsilon}(n/2) \leq 2\epsilon^{-1}\log n$ (whenever $\epsilon \leq 1/2$). As in the proof of [3], the service cost of all points after a facility is opened in a region is deterministically at most $(\frac{\epsilon}{1-\epsilon} + (1 + \epsilon))\Sigma$. This follows by applying Markov's inequality to show the cost of connecting the nearest $\epsilon|S|$ points is at most $\frac{\epsilon}{1-\epsilon}\Sigma$.

As for before a facility opens, it is shown in [3] that the probability of having total service cost over x regions of at least y before a facility opens is at most $e^{x-y\frac{e-1}{e}}$. Here we now set $x = 2\epsilon^{-1}k(1 + \log n)$ and $y = 2\frac{e}{e-1}\epsilon^{-1}k(1 + \log n)$ to yield the result.

Algorithm 3 maintains a set of **OFL** instances, where n is the weight of the stream. After each phase change, it begins running $d + 1$ instances of **OFL** with facility cost set to $\epsilon f(B_N)/\theta$. Run this instance until the end of the phase, and then increase the service cost and duplicate the instance $d + 1$ times. At any moment, provide Q_N as the weighted set of facilities of the instance running in the bucket of the current phase with minimal service cost.

We will refer to an instance running in "bucket t". This is to avoid confusion because there will be instances running during phase t in buckets t and $t + 1$. We discard buckets $t - 1$ and earlier.

We now present the main theorem of this section.

Theorem 3. *With probability at least* $1 - n^{-d}$, *where* d *is a chosen parameter, Algorithm 3 maintains a weighted set* Q_N *such that* $\mathrm{Connect}(S_N \setminus A_N, Q_N) \leq (2 + 7\epsilon)(1 + \epsilon)\,\mathrm{OPT}(S_N, k)$. *The storage requirement is* $O(d\epsilon^{-3}k \log n)$.

Proof. The space bound is deterministic and follows easily from Algorithm 3. This is because Line 9 guarantees that we have at most $7\alpha\theta^2\epsilon^{-3}k(1 + \log n)$ facilities per instance. We store at most $d + 1$ instances in bucket $t + 1$ and at most $d + 1$ instances in bucket t, resulting in an overall storage of at most $14(d + 1)\alpha\theta^2\epsilon^{-3}k(1 + \log n)$ facilities.

We now prove that at least one instance in bucket t remained active throughout phase $t - 1$ (by not opening too many facilities and thus terminating on line 9). Consider the instances in bucket t, which were first begun as a batch of $d + 1$ instances at the beginning of phase $t - 1$. Since Algorithm 1 shifts $A_N \leftarrow B_{N-1}$ at a phase change, these instances were started with facility cost

Algorithm 3. Update Process, upon receiving point p_N

1: **if** p_N causes phase t to begin **then**
2:     Terminate all instances in bucket $t-1$
3:     Force all instances in bucket t to open p_N as a facility
4:     $\Phi_1 \leftarrow$ facilities of a bucket t instance with minimal service cost
5:     $\kappa \leftarrow \epsilon f(B_N)/\theta k(1+\log n)$
6:     Initialize $d+1$ instances of **OFL**(κ) in bucket $t+1$
7: **else**
8:     Update all running instances of **OFL** with point p_N
9:     Terminate instances with facility-count above $7\alpha\theta^2\epsilon^{-3}k(1+\log n)$
10: **if** bucket t contains a running instance **then**
11:     $Q_N \leftarrow$ facilities of a bucket t instance with minimal service cost
12: **else**
13:     $\Phi_2 \leftarrow$ facilities of a bucket $t+1$ instance with minimal service cost
14:     $Q_N \leftarrow \Phi_1 \cup \Phi_2$

$\epsilon f(A_N)/(\theta k(1+\log n))$ and ran on the segment $B_N \setminus A_N$. We let B'_N denote B_N without the final point that caused the transition to phase t, and we apply Theorem 2 to B'_N. By Theorem 1 we have $\mathrm{OPT}(B'_N,k)/(\epsilon f(A_N)/\theta) < \beta\theta\epsilon^{-1} = \alpha\theta^2\epsilon^{-2}$. With this bound, Theorem 2 guarantees with probability $1 - n^{-d-1}$ that at least one of the $d+1$ instances will run on B'_N with at most $7\alpha\theta^2\epsilon^{-2}k(1+\log n)$ facilities. Since the number of facilities is monotonically increasing during run-time, this implies the same bound on the number of facilities when running **OFL** on the segment $B'_N \setminus A_N$. Therefore with probability $1 - n^{-d-1}$ at least one instance survives to the beginning of phase t by not being terminated on Line 9. At the beginning of phase t, we apply the same analysis to $S_N \setminus B_N$ (without the need to remove the final point, since the phase has not ended) and arrive at the same probabilistic bound on the number of facilities for instances in bucket $t+1$.

Let $c = 3 + 7\epsilon$, let L_t (and similarly L_{t-1}) be $k(1+\log n)\kappa$ where κ is the facility cost used for instances in bucket t. We break into two cases and analyze each seperately. In the first case, suppose $\mathrm{OPT}(B'_N,k) \geq \epsilon\,\mathrm{OPT}(S_N,k)$. We repeat the previous analysis with Theorem 2 of running **OFL** on the segment $S_N \setminus A_N$ instead of $B'_N \setminus A_N$. Since $\mathrm{OPT}(S_N,k)/(\epsilon f(A_N)/\theta) \leq \epsilon^{-2}\,\mathrm{OPT}(B'_N,k)/(f(A_N)/\theta) < \alpha\theta^2\epsilon^{-3}$ and $L_t = \epsilon f(A_N)/\theta \leq \epsilon\,\mathrm{OPT}(A_N,k) < \epsilon\,\mathrm{OPT}(S_N,k)$, Theorem 2 gives a high-probability guarantee that an instance in bucket t has opened at most $7\alpha\theta^2\epsilon^{-3}k(1+\log n)$ facilities with service cost at most $c\,\mathrm{OPT}(S_N,k)$, and we are done. In the second case, suppose $\mathrm{OPT}(B'_N,k) < \epsilon\,\mathrm{OPT}(S_N,k)$. If there is an active instance in bucket t, then the connection cost is at most $c\,\mathrm{OPT}(S_N,k)$. If there are no active instances in bucket t, we return $\Phi_1 \cup \Phi_2$. We apply Theorem 2 to $B'_N \setminus A_N$ to show $Connect(B'_N \setminus A_N, \Phi_1) \leq c\,\mathrm{OPT}(B'_N,k)$. Line 3 implies $Connect(B'_N \setminus A_N, \Phi_1) = Connect(B_N \setminus A_N, \Phi_1)$, and therefore $Connect(B_N \setminus A_N, \Phi_1) < c\,\mathrm{OPT}(B'_N,k)$. $L_{t+1} = \epsilon f(B_N)/\theta \leq \epsilon\,\mathrm{OPT}(B_N,k) \leq \epsilon\,\mathrm{OPT}(S_N,k)$, and we apply the theorem again to $S_N \setminus B_N$ to show $Connect(S_N \setminus B_N, \Phi_2) \leq c\,\mathrm{OPT}(S_N,k)$. Then by Observation 2, we bound connection costs as $Connect(S_N \setminus A_N, \Phi_1 \cup \Phi_2) \leq c\,\mathrm{OPT}(B'_N, k) + c\,\mathrm{OPT}(S_N, k) < c(1 + \epsilon)\,\mathrm{OPT}(S_N, k)$.

The above proof holds for a single phase. There is at least one point per phase, implying that there are at most n phases. Thus the $1 - n^{-d-1}$ probability guarantee for each phase becomes a $1 - n^{-d}$ probability guarantee over the stream.

From now on, we refer to the result of Theorem 3 as guaranteeing connection cost $(2 + \epsilon)\,\mathrm{OPT}(S_N, k)$ instead of $(2 + 7\epsilon)(1 + \epsilon)\,\mathrm{OPT}(S_N, k)$. This follows by selecting $\epsilon' = \epsilon/7$.

6   Combining Both Algorithms

Algorithm 1 provides a weighted set $\mathbf{PLS}(A_N)$ such that $Connect(A_N, \mathbf{PLS}(A_N)) \leq \alpha\,\mathrm{OPT}(A_N, k)$. Moreover, the algorithm guarantees that $f(A_N) \leq \beta^{-1} f(S_N)$, where $\beta = \alpha\theta/\epsilon$, and therefore $\mathrm{OPT}(A_N, k) \leq f(A_N) \leq \beta^{-1} f(S_N) \leq \beta^{-1}\theta\ \mathrm{OPT}(S_N, k) \leq \epsilon\alpha^{-1}\,\mathrm{OPT}(S_N, k)$. Together, this implies that $Connect(A_N, \mathbf{PLS}(A_N)) \leq \epsilon\,\mathrm{OPT}(S_N, k)$. Algorithm 3 provides us with a weighted set Q_N such that $Connect(S_N \setminus A_N, Q_N) \leq (2 + \epsilon)\,\mathrm{OPT}(S_N, k)$. Define Σ_N as the union $\mathbf{PLS}(A_N) \cup Q_N$. This is the desired bicriterion since, by Observation 2, $Connect(S_N, \Sigma_N) \leq (2 + 2\epsilon)\,\mathrm{OPT}(S_N, k)$.

Our algorithm maintains Σ_N using $O(\epsilon^{-3} k \log n)$ space. Moreover, as shown above, Σ_N is a $(O(\epsilon^{-3} k \log n), 2 + \epsilon)$-bicriterion for the stream S_N.

7   Appendix

7.1   Lower Bound

The following lower bound relies on the fact that the algorithm, if it uses sub-linear space, must forget most of the input points. Missing a critical input point can prevent anything better than a 2-approximation existing among the points remaining in storage.

Theorem 4. *For the metric k-median problem, no polylog(n)-space streaming algorithm can (with constant probability) maintain a (α, β)-bicriterion for $\beta < 2$.*

Proof. Consider a specific algorithm. Let $S(n)$ be the space-complexity of this algorithm, measured in the number of points able to be stored. Suppose that $S(n) \in o(n)$. Fix a value of n. Define $R(n) = \lceil \sqrt{n S(n)} \rceil$ and note that $R(n) \in o(n)$ We will construct an input for the 1-median case, and then show it can be modified for k-median. Let the input begin with $(p_1, \ldots, p_{R(n)})$ where $d(p_i, p_j) = 1 \forall i \neq j$ where $j \in \{1, \ldots, R(n)\}$. Thus the first $R(n)$ points are indistinguishable, so even for a non-deterministic algorithm there must exist a deterministic $c \in \{1, \ldots, R(n)\}$ such that after the algorithm passes the first $R(n)$ points, c is stored in memory with probability at most $S(n)/R(n) \leq \sqrt{S(n)/n} \in o(1)$. The entire input is then $(p_1, \ldots, p_{R(n)}, q_{R(n)+1}, \ldots, q_n)$ where $d(p_c, q_i) = 1 \forall i \in$

$\{R(n)+1, \ldots, n\}$ and all other distances are given by the shortest path. Without p_c stored as a potential center, the next best clustering (using one of the first $R(n)$ points as the center) yields a cost of $R(n) - \alpha + 2(n - R(n))$ while the optimum (with p_c as the center) is $n - 1$. Since α is a lower-bound on the storage requirement of the algorithm (it must at least store the bicriterion is provides), in the limit $n \to \infty$ this input has a cost-ratio approaching 2 with probability approaching 1. To extend to k-median, use the above input with size n/k and duplicate it k times (where each duplicate is at least distance 2 from any other). The value of c may be different for each of the k pieces, but there must exist a deterministic $(c_1, \ldots, c_k)$ such that the above argument extends.

7.2 Computing the Constant of Previous Algorithms

In this section, we compute the constant of the approximation-algorithm in [6] which the authors leave unspecified. The lower-space algorithm of [3] has an even larger constant due to the high-probability guarantee on the facility location lemma of a $(3 + \frac{2e}{e-1})$-approximation instead of a 4-approximation.

In Sect. 2 of [6], the connection cost of the maintained PLS set is seen to be $\alpha = 4(1 + 4(\gamma + \beta))$. The constants γ and β are freely selected subject to the restraint that $\gamma + 4(1 + 4(\gamma + \beta)) \leq \gamma\beta$. By minimizing the function $\alpha(\gamma, \beta)$ subject to this constraint, we obtain a lower bound on the approximation-ratio of these algorithms. Since the function $\alpha(\gamma, \beta)$ has no critical points, its minimum must occur on the boundary of the constraint equation. Using Lagrange multipliers to minimize $\gamma + \beta$ subject to $\gamma + 4(1 + 4(\gamma + \beta)) = \gamma\beta$, we find $\gamma = 16 + \sqrt{276}$ and $\beta = \gamma + 1$, setting the final approximation ratio to over 1063.5.

References

1. Bădoiu, M., Har-Peled, S., Indyk, P.: Approximate clustering via core-sets. In: Proceedings of the Thiry-Fourth Annual ACM Symposium on Theory of Computing, STOC 2002, pp. 250–257. ACM, New York (2002)
2. Bentley, J.L., Saxe, J.B.: Decomposable searching problems I. Static-to-dynamic transformation. J. Algorithms 1(4), 301–358 (1980)
3. Braverman, V., Meyerson, A., Ostrovsky, R., Roytman, A., Shindler, M., Tagiku, B.: Streaming k-means on well-clusterable data. In: Proceedings of the Twenty-Second Annual ACM-SIAM Symposium on Discrete Algorithms, SODA 2011, pp. 26–40. SIAM (2011)
4. Bury, M., Schwiegelshohn, C.: Random projections for k-means: maintaining core-sets beyond merge & reduce. CoRR, abs/1504.01584 (2015)
5. Byrka, J., Pensyl, T., Rybicki, B., Srinivasan, A., Trinh, K.: An improved approximation for k-median, and positive correlation in budgeted optimization. In: Proceedings of the Twenty-Sixth Annual ACM-SIAM Symposium on Discrete Algorithms, SODA 2015, pp. 737–756. SIAM (2015)
6. Charikar, M., O'Callaghan, L., Panigrahy, R.: Better streaming algorithms for clustering problems. In: Proceedings of the Thirty-Fifth Annual ACM Symposium on Theory of Computing, STOC 2003, pp. 30–39. ACM, New York (2003)

7. Chen, K.: On coresets for k-median and k-means clustering in metric and euclidean spaces and their applications. SIAM J. Comput. **39**(3), 923–947 (2009)
8. Feldman, D., Langberg, M.: A unified framework for approximating and clustering data. In: Proceedings of the Forty-Third Annual ACM Symposium on Theory of Computing, STOC 2011, pp. 569–578. ACM, New York (2011)
9. Fichtenberger, H., Gillé, M., Schmidt, M., Schwiegelshohn, C., Sohler, C.: BICO: BIRCH meets coresets for k-means clustering. In: Bodlaender, H.L., Italiano, G.F. (eds.) ESA 2013. LNCS, vol. 8125, pp. 481–492. Springer, Heidelberg (2013). doi:10.1007/978-3-642-40450-4_41
10. Guha, S.: Tight results for clustering and summarizing data streams. In: Proceedings of the 12th International Conference on Database Theory, ICDT 2009, pp. 268–275. ACM, New York (2009)
11. Guha, S., Meyerson, A., Mishra, N., Motwani, R., O'Callaghan, L.: Clustering data streams: theory and practice. IEEE Trans. Knowl. Data Eng. **15**(3), 515–528 (2003)
12. Har-Peled, S., Kushal, A.: Smaller coresets for k-median and k-means clustering. Discrete Comput. Geom. **37**(1), 3–19 (2007)
13. Har-Peled, S., Mazumdar, S.: Coresets for k-means and k-median clustering and their applications. In: STOC 2004, pp. 291–300 (2004)
14. Meyerson, A.: Online facility location. In: Proceedings of the 42nd IEEE Symposium on Foundations of Computer Science, FOCS 2001, p. 426. IEEE Computer Society, Washington, DC (2001)
15. Shindler, M., Wong, A., Meyerson, A.W.: Fast and accurate k-means for large datasets. In: Shawe-Taylor, J., Zemel, R., Bartlett, P., Pereira, F., Weinberger, K. (eds.) Advances in Neural Information Processing Systems 24, pp. 2375–2383. Curran Associates Inc., Red Hook (2011)

Querying Relational Event Graphs
Using Colored Range Searching Data Structures

Farah Chanchary$^{(\boxtimes)}$, Anil Maheshwari, and Michiel Smid

School of Computer Science, Carleton University, Ottawa, ON K1S 5B6, Canada
farah.chanchary@carleton.ca, {anil,michiel}@scs.carleton.ca

Abstract. We present a general approach for analyzing structural parameters of a relational event graph within arbitrary query time intervals using colored range query data structures. Relational event graphs generally represent social network datasets, where each graph edge carries a timestamp. We provide data structures based on colored range searching to efficiently compute several graph parameters (e.g., density, neighborhood overlap, h-index).

Keywords: Colored range searching · Relational event graph · Social network analysis · Timestamp

1 Introduction

A *relational event (RE) graph* $G = (V, E)$ is defined to be an undirected graph with set of vertices V and a set of edges (or relational events) $E = \{e_k | 1 \leq k \leq m\}$ between pairs of vertices. We assume that each edge has a unique *timestamp*. We denote the timestamp of an edge $e_k \in E$ by $t(e_k)$. Without loss of generality, we assume that $t(e_1) < t(e_2) < \cdots < t(e_m)$. Given a relational event graph G, for a pair of integers $1 \leq i \leq j \leq m$, we define the *graph slice* $G_{i,j} = (V', E' = \{e_i \cup e_{i+1} \cup \cdots \cup e_j\})$, where V' is the set of vertices incident on edges of E'. In this paper, for a query time interval $[i, j]$, where $1 \leq i \leq j \leq m$, we are interested in answering questions about various graph parameters on the graph slice $G_{i,j}$.

A social network can naturally be represented by an RE graph, where each vertex of the graph represents an entity of the social network, and edges represent communication events between pair of entities occurred at some specific time. The RE graph model was first proposed by Bannister et al. [1]. They presented data structures to find the number of connected components, number of components containing cycles, number of vertices with some predetermined degree and number of reachable vertices on a time-increasing path within a query time window. Later, Chanchary and Maheshwari [6] presented data structures to solve subgraph counting problems for triangles, quadrangles and complete subgraphs in RE graphs. In this paper, we present a general approach to construct data

---

This research work was supported by NSERC and OGS.

D. Gaur and N.S. Narayanaswamy (Eds.): CALDAM 2017, LNCS 10156, pp. 83–95, 2017.
DOI: 10.1007/978-3-319-53007-9_8

structures on a set of colored points in $\mathbb{R}^d$, $(d \geq 1)$, that support colored (or generalized) range queries, and efficiently count and/or report various structural parameters of the underlying RE graph G within a query pair of indices $[i, j]$.

Preliminaries: We define some structural graph parameters that we want to compute using our data structures. One of the basic indicators for measuring graph structure is the *density* of a graph. It evaluates how close the graph is to a complete graph. The density of an undirected simple graph $G = (V, E)$ is defined as $D(G) = \frac{|E|}{\binom{|V|}{2}}$ [17].

In social networks, center vertices of k-stars are considered as *hubs*. In network analysis, hubs have been extensively studied as they are the basis of many tasks, for example web search and epidemic outbreak detection [2]. A *k-star* is defined to be a complete bipartite graph $K_{1,k}$, i.e., a tree with one internal node and k leaves. The *h-index* is the largest number h such that the graph contains h vertices of degree at least h. For any graph with m edges, $h = O(\sqrt{m})$ [1].

Embeddedness of an edge (u, v), denoted as $emb(u, v)$, in a network is the number of common neighbors the two endpoints u and v have, i.e., $emb(u, v) = |N(u) \cap N(v)|$ [9]. Embeddedness of an edge (u, v) in a network represents the trustworthiness of its neighbors, and the confidence level in the integrity of the transactions that take place between two vertices u and v [9]. A *local bridge* in a graph G is an edge whose endpoints have no common neighbour. *Neighborhood overlap* of an edge (u, v), denoted as $NOver(u, v)$, is the ratio of the number of vertices who are neighbors of both u and v, and the number of vertices who are neighbors of only one of them [9].

$$NOver(u, v) = \frac{emb(u, v)}{|N(u) \cup N(v)| - emb(u, v) - 2} \tag{1}$$

In the denominator, u or v are not counted as the neighbor of one another. Neighborhood overlap of an edge represents the strength (in terms of connectivity) of that edge in its neighborhood. The neighborhood overlap of an entire graph G is defined as the average of the neighborhood overlap values of all the edges of G, i.e., $NOver(G) = \frac{1}{|E|} \sum_{k=1}^{|E|} NOver(e_k)$ [16]. Suppose $G = (V = A \cup B, E)$ is a bipartite graph. Neighborhood overlap of a pair of vertices $u, v \in A$ of G is defined as the following ratio [9].

$$NOver(u, v) = \frac{emb(u, v)}{|N(u) \cup N(v)| - emb(u, v)} \tag{2}$$

In social network analysis, this bipartite graph is known as the *affiliation network*. An affiliation network represents how entities of a network (i.e., vertices in A) are affiliated with other groups or activities (i.e., vertices in B).

Modularity of a network is a measure that quantifies the quality of a given network division into disjoint partitions or communities [17]. Many real world complex networks, e.g., the world wide web, biological or social networks, demonstrate some forms of community structures that are important for various topological studies. The *modularity of a vertex pair* (u, v) is defined as $\frac{d(u) \times d(v)}{2|E|}$, where $d(u)$ is the degree of vertex u [17].

Analysis of the *diffusion phenomena* in graphs is a very widely studied research area in many communities including computer science, social sciences and epidemiology. Although the basic model of diffusion shares common properties across various disciplines, it is highly context dependent [3,5,10]. In particular, the spread of information in any social network requires some influence from a set of designated agents (vertices) and depends on the connectivity of the network. We want to solve the problem of counting and reporting influenced vertices in an RE graph slice using the following model.

Suppose, $G = (V, E)$ is an RE graph with n vertices and m edges with a fixed set of influential vertices $V' \subseteq V$. Let $f : V \to N$ be a function assigning thresholds values to vertices such that, (a) $f(v) = 0$ if v is an influential vertex; (b) $1 \le f(v) \le d(v)$ otherwise, where $d(v)$ is the degree of $v \in V$. A vertex can be influenced only if it is on a path of influence. A simpler model for counting influential vertices has been presented in [1].

Definition 1. *For a pair of vertices u and v, and a positive integer r, a path $\pi = (u = v_1, v_2, v_3, \cdots, v_k, v_{k+1} = v)$ is a path of influence with parameter r, if the following holds:*

1. u is an influential vertex and v is a non-influential vertex
2. $v_2, v_3, \cdots, v_k$ are influenced vertices
3. $t(v_i, v_{i+1}) < t(v_{i+1}, v_{i+2})$
4. $t(v_k, v_{k+1}) - t(v_1, v_2) \le r$.

Definition 2. *A vertex v is influenced with respect to r if either v is an influential vertex, i.e., $v \in V'$ or v is a non influential vertex and there are at least $f(v)$ edge-disjoint paths of influence with parameter r by which v can be reached from some influential vertices.*

We assume an influential vertex u can influence other vertices $v_1, v_2, \cdots, v_{k+1}$ on a path of influence π such that the time difference $t(v_k, v_{k+1}) - t(v_1, v_2) \le r$, where $r > 0$. After r rounds, v_i's influence becomes inactive on other vertices on this particular path of influence. This is known as the degradation of influence in social networks. A similar model has been proposed by Gargano et al. [10] for non-temporal graphs.

New Results: Let G be a relational event graph consisting of m edges and n vertices, and let $q = [i, j]$ be an arbitrary query time interval, where $1 \le i \le j \le m$. Our data structures can efficiently answer queries on computing the graph density, number of vertices, degrees of all vertices, k-stars, h-index, embeddedness of edges, average neighborhood overlap (for general and bipartite graphs) and the number of influenced vertices in the graph slice $G_{i,j}$. The contributions of this paper are summarized in Table 1. Some of the queries mentioned above are obtained by reducing the problem to a new colored range query problem stated as follows: Given a set of n colored points on the real line and a fixed threshold value k, we can build a data structure in $O(n \log n)$ time using $O(n \log n)$ space that can report those colors that have at least k elements within any query interval $q = [a, b]$ in $O(\log^2 n + w)$ time, where w is the number of reported colors (Theorem 1).

Table 1. Summary of results.

Problems	Preprocessing time	Query time	
$|V_{i,j}|$	$O(m+n)$	$O(\log n)$	(Theorem 2)
$D(G_{i,j})$	$O(m+n)$	$O(\log n)$	(Theorem 2)
$d(v_k \in V_{i,j})$	$O(m+n)$	$O(\log^2 n + w)$	(Theorem 3)
k-stars	$O(m+n)$	$O(\log^2 n + w)$	(Theorem 4)
h-index (approx.)	$O(m+n)$	$O(\log^3 n + h \log n)$	(Theorem 5)
Embeddedness	$O(a(G)m)$	$O(\log^2 n + w \log n)$	(Theorem 6)
$NOver(G_{i,j})$	$O(mn)$	$O((\log^2 n + w \log n + s)$	(Theorem 6)
$NOver(G_{i,j})$-bipartite	$O(a(G)m)$	$O(\log^2 n + w \log n + k)$	(Theorem 7)
Influenced vertices	$O(m \log n)$	$O(\log n + w)$	(Theorem 8)

2   Colored Range Searching Data Structures

Colored range searching problems are variations of the standard range searching problems, where a set of colored objects (e.g., points, lines or rectangles) is given. Typically, the color of an object represents its category. In the generic instance of the range searching problems, a set of objects S is to be preprocessed into a data structure so that given a query range q, it can efficiently answer counting or reporting queries based on the intersection of q with the elements in S. In the colored version, the query should efficiently report those colors that have at least one object intersecting the query range. For example, we have a set of points $S \in \mathbb{R}^2$ and each point is colored by one of the four different colors, i.e., $C = \{c_1, c_2, c_3, c_4\}$ (see Fig. 1). Suppose a query rectangle $q = [a, b] \times [c, d]$ is given. The standard colored range counting (reporting) query will count (report) the number of colors that have at least one point inside q. In this example, there are three colors (red, green and blue) that have points inside q, hence they will be reported. A different version namely *Type-2 counting problem* reports the number of points for each color intersected by q as a pair of values (c_k, #c_k-colored points intersected by q). Following our example, the type-2 counting query will generate

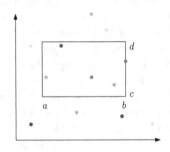

Fig. 1. Example of colored range queries. (Color figure online)

the output as $(red, 2), (green, 2), (blue, 1)$. A variety of colored range searching problems have been studied extensively, see for example [4,11,12,14].

2.1 k-Threshold Color Queries

We wish to preprocess a set S of colored points on a horizontal line so that given a fixed threshold k and a query interval $q = [a, b]$, we can quickly report the colors that have at least k points in $q \cap S$. To solve this problem, we transform each 1-dimensional point to a 3-dimensional point by applying the following chaining technique. For each color c, we sort the distinct points of this color by increasing values of their x-coordinates. For each point x_i of color c, we map x_i to $x_i' = (x_{i-1}, x_i, x_{i+k-1}) \in \mathbb{R}^3$ and assign it color c. For the leftmost point x_l, let $x_{l-1} = -\infty$ and for the rightmost point x_r, let $x_{r+1} = +\infty$. Let S' be the new set of points in $\mathbb{R}^3$. Now a given query interval $q = [a, b]$ is mapped to a new query $q' = (-\infty, a) \times [a, +\infty) \times (-\infty, b]$.

Lemma 1. *There are at least k points of color c in the query slice $q = [a, b]$ if and only if there is a point of color c in $q' = (-\infty, a) \times [a, +\infty) \times (-\infty, b]$. Moreover, the c-colored point in q' is unique, if there exists any.*

Proof. Let $x_i' = (x_{i-1}, x_i, x_{i+k-1})$ be a c-colored point in q' for some c-colored point x_i in the original set of points. Since x_i' is in q', it is clear that $x_i \geq a$ and $x_{i+k-1} \leq b$ which indicates that starting from x_i there are at least k c-colored elements in the original query interval $[a, b]$. Thus $x_i \in [a, b]$.

Assume there are at least k points of color c in q. Let x_i be the leftmost point of color c in the query interval $[a, b]$. Therefore, $a \leq x_i \leq b$ and the k-th element from x_i, i.e., $x_{i+k-1} \leq b$. Moreover, since x_i is the leftmost point in $[a, b]$ of color c, we know $x_{i-1} \in (-\infty, a)$, i.e., $x_{i-1} < a$. It follows that $x_i' = (x_{i-1}, x_i, x_{i+k-1})$ is in q'.

We prove the uniqueness of point x_i' in q' by contradiction. Suppose, there is a second c-colored point y_i' in q'. It implies that $y_i \geq a$ and $y_{i+k-1} \leq b$. Since x_i is the leftmost point in $[a, b]$, it must be the case that $y_i > x_i$ and thus $y_{i-1} \geq x_i \geq a$. Then it contradicts with the assumption that $y_i' \in q'$.   □

Now we have a set of n colored points in $\mathbb{R}^3$. We build a standard 2-dimensional range tree [8], where the first level of the tree is based on the x-coordinates of the points. At the second level of the tree, for each canonical node we build a priority search tree (PST) [15] using the y and the z-coordinates of our 3-dimensional points. There can be at most $O(\log n)$ canonical nodes on the search path of a range tree. For each canonical node, the second level PST takes linear space. So, total space requirement is $O(n \log n)$. This data structure can be built naively in $O(n \log^2 n)$ time. This time can be improved by building PSTs in linear time by bottom-up heap construction technique using points sorted on z-coordinates [8]. Therefore, total time required to built this data structure becomes $O(n \log n)$. PSTs support insertions and deletions in $O(\log n)$ time and can report w points lying inside a grounded query rectangle in $O(\log n + w)$

time. We query the first level of the range tree using $(-\infty, a)$. In the second level, we query using $[a, +\infty) \times (-\infty, b]$. Thus, given any query interval $q = [a, b]$ this data structure can report the colors that have at least k elements in q in $O(\log^2 n + w)$ time, where w is the number of reported colors.

Theorem 1. *Given a set of n colored points on the real line and a fixed threshold value k, we can build a data structure in $O(n \log n)$ time using $O(n \log n)$ space that can report those colors that have at least k elements within any query interval $q = [a, b]$ in $O(\log^2 n + w)$ time, where w is the number of reported colors.*

3   General Approach for Modeling Problems

In this paper we present the following general approach to solve problems in RE graphs. Suppose an RE graph $G = (V, E)$ is given and we want to answer queries about some structural parameters of G. To solve each problem, we will define a set of colors $C = \{c_1, c_2, \cdots, c_p\}$, where each color $c_k \in C$ is encoded as an integer in the range $[1, p]$ for some integer p. We process each edge $e = (u, v)$ of G according to the timestamps of the edges in increasing order and scan either each adjacent edge of e or each neighboring vertex of both u and v. Depending on the problem in hand, our algorithm associates C either to the set of vertices V (i.e., $|C| = n$) or to the set of edges E (i.e., $|C| = m$). When each vertex $v_k \in V$ is associated with a color c_k, the algorithm scans each neighbor N_i of v_k, where $1 \leq i \leq d(v_k)$, generates a c_k colored point p, and assigns the timestamp $t(v_k, N_i)$ as p's coordinate value (see Fig. 2(a)). Similarly, each edge $e_k \in E$ is associated with a color c_k, the algorithm scans each adjacent edge M_i of e_k, and generates a c_k colored point p with the timestamp of M_i assigned as p's coordinate value (see Fig. 3(b)). This general approach will be extended when we report influenced vertices or preprocess bipartite graphs. We explain these in later sections.

Now, we summarize the components of our general approach as follows. To solve any problem with our model, we need to specify the following components.

1. The set of colors C and the corresponding graph element (i.e., vertices, edges);
2. The representation of points in $\mathbb{R}^d$, where $d \geq 1$;
3. Appropriate colored range searching data structure to answer queries.

4   Algorithms for Solving Queries

4.1   Counting Vertices, Density, and Degrees of Vertices

Given an RE graph $G = (V, E)$ and a query time slice $[i, j]$, we want to find the number of vertices $|V_{i,j}|$, the density $D(G_{i,j})$, and report all the vertices with non-zero degrees in $V_{i,j}$.

Counting Number of Vertices $|V_{i,j}|$: We use a set $C = \{c_1, c_2, \cdots, c_n\}$ of n colors, where each c_k is associated with a vertex v_k, for $1 \leq k \leq n$. For each vertex

$v_k \in V$, we maintain a linked list $adj[v_k]$, where each node of the list contains its neighboring vertex N_i, where $1 \leq i \leq d(v_k)$, and the timestamp of the edge between them, i.e., $t(v_k, N_i)$. We associate color c_k with all timestamps stored in $adj[v_k]$, for all $1 \leq k \leq n$. Now we have exactly $2|E|$ colored points on the real line (see Fig. 2 for an example). Let this set of colored points be P. Using 1-dimensional colored range counting data structure on P with query interval $q = [i, j]$, we can find all distinct colors intersected by q. Each of these distinct colors represents a vertex having adjacent edges in $G_{i,j}$. Thus, we can report $|V_{i,j}|$ in $O(\log m) = O(\log n)$ time (by [12], Theorem 3.2) using a data structure of size $O(m \log n)$.

Fig. 2. (a) A vertex v_k and its three neighbors N_1, N_2 and N_3. Since the timestamp $t(v_k, N_1)$ of the edge (v_k, N_1) is 2, (v_k, N_1) is shown as a c_k-colored point on the line with x-coordinate value $= 2$. (b) The set of all c_k colored points on the line. (Color figure online)

Computing Density $D(G_{i,j})$: This step requires computing the value of $|E_{i,j}|$ in addition to $|V_{i,j}|$. For any graph slice $G_{i,j}$, the value of $|E_{i,j}|$ can be computed in constant time since $|E_{i,j}| = j - i + 1$. We summarize these results as follows.

Theorem 2. *Given an RE graph G with m edges and n vertices, and a query time slice $[i, j]$, the problem of computing $|V_{i,j}|$ and the density $D(G_{i,j})$ of the graph slice can be reduced to 1-dimensional colored range queries in linear time. Queries can be answered in $O(\log n)$ time using $O(m \log n)$ space.*

Reporting Degrees of Vertices $d(v \in G_{i,j})$: Now we want to report number of neighboring edges of each vertex in $G_{i,j}$. So, we construct a 1-dimensional type-2 color counting data structure on P. This query will report all vertices that have non-zero degrees in $G_{i,j}$ in $O(\log n + w)$ time, where w is the number of reported vertices.

Theorem 3. *Given an RE graph G with m edges and n vertices, and a query time slice $[i, j]$, the problem of computing degrees of all vertices $v_k \in V_{i,j}$ can be reduced to 1-dimensional type-2 colored range queries in linear time. Queries can be answered in $O(\log n + w)$ time, where w is the number of reported vertices.*

4.2   Counting k-Stars and h-Index

Given an RE graph G with m edges and n vertices, a fixed threshold k and a query time slice $[i, j]$, we want to count all k-stars in $G_{i,j}$. We use a similar model as described in the previous section (see Fig. 2). Each edge adjacent to a vertex v_k will have the color c_k. Thus we again have a set of colored points on the real line. By applying k-threshold color queries (Theorem 1) with query interval $q = [i, j]$ we can report all vertices that have at least k neighbors in $G_{i,j}$, hence are at the center of k-stars in $G_{i,j}$.

Theorem 4. *Given an RE graph G with m edges and n vertices, a query time slice $[i, j]$, and a fixed threshold value k, the problem of finding all k-stars in $G_{i,j}$ can be reduced to k-threshold color queries in linear time. Queries can be answered in $O(\log^2 n + w)$ time, where w is the number of k-stars.*

Given an RE graph G with m edges and n vertices, and a query time slice $[i, j]$, we want to compute the h-index of $(G_{i,j})$. We use the same model and build the parameterized k-threshold color counting data structures to answer this query. A k-threshold color counting data structure reports the number of vertices (w) that have at least k neighbors in $G_{i,j}$. Therefore, if the number of reported vertices $w \geq k$, then the h-index of $G_{i,j}$ is at least k. Now we perform a set of decision problems to compute the h-index$(G_{i,j})$, i.e., *'Are there at least k vertices of degree k in $G_{i,j}$?'*

We query the k-threshold color counting data structure with parameter $k = 1$ and compare the number of colors (w) reported by this query with the value of k. We perform the following steps for $k = 1, (1+\epsilon), (1+\epsilon)^2, \cdots$ for up to $\log_{1+\epsilon} n$ blocks, where ϵ is an arbitrarily small positive constant:

1. If $w = k$, we return k as the h-index of $G_{i,j}$.
2. If $w > k$, we search using the data structure with the next value of k in the sequence.
3. Otherwise, if $k' < w < k$, where k' was the previous block size, we return w as the approximation to h-index.

In case we do not find the exact answer for h, we report an approximate h-index$(G_{i,j})$ with approximation ratio of $(1 + \epsilon)$. Total preprocessing time required to build this data structure will be $O(n \log^2 n)$. The following theorem summarizes this result.

Theorem 5. *Given an RE graph G with m edges and n vertices, and a query time slice $[i, j]$, the problem of computing the h-index of $G_{i,j}$ can be reduced to the parameterized k-threshold color counting problem in linear time. Queries can be answered in $O(\log^3 n + h \log n)$ time with an approximation ratio of $(1 + \epsilon)$, where ϵ is a small positive constant and h is the h-index of $G_{i,j}$.*

4.3 Computing Neighborhood Overlap and Embeddedness

We notice, from the definition of neighborhood overlap of an edge (u, v) that the numerator term represents the embeddedness of the edge (u, v) (i.e., the number of triangles containing the edge (u, v)) and the denominator is the number of edges adjacent to edge (u, v) minus the common edges (triangles). So, to answer the main query we need to solve two subproblems; i.e., for each edge $e_k \in G_{i,j}$, where $i \leq k \leq j$, we count (a) the number of edges adjacent to e_k, and (b) the number of triangles containing e_k. We describe below the preprocessing steps for these subproblems.

Counting the Number of Adjacent Edges: We use a set of m colors, $C = \{c_1, c_2, \cdots, c_m\}$, where each color c_k will be associated with an edge e_k, for $1 \leq k \leq m$. Now, for each edge e_k we associate each adjacent edge of e_k with a color $c_k \in C$. Suppose, e_k has a set of adjacent edges $A(e_k) = \{e', e'', \cdots\}$. We create a c_k-colored point on the real line for each edge in $A(e_k)$ with the value of its timestamp as its coordinate (see Fig. 3). The total number of colored points is $\sum_{e=(u,v)} (d(u) + d(v)) = \sum_u d(u)^2 \leq (n-1) \sum_u d(u) = 2m(n-1) = O(mn)$. Using 1-dimensional type-2 counting query data structure (Theorem 1.4, [11]) with query interval $[i, j]$ the number of edges adjacent to each $e_k \in E_{i,j}$ can be found in $O(\log mn + s)$ time using $O(mn)$ space, where s is the total number of edges in $E_{i,j}$ having some neighboring edges.

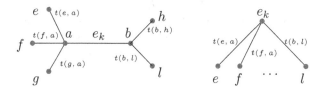

Fig. 3. (a) Adjacent vertices of an edge e_k, and (b) Colored points from edge e_k. (Color figure online)

Counting the Number of Triangles: We observe that we can preprocess G in $O(a(G)m)$ time to count the number of triangles in $G_{i,j}$ ([6], Theorem 2), where $a(G)$ is the arboricity of G defined as the minimum number of edge-disjoint spanning forests into which G can be partitioned [13]. For a general connected graph G, $a(G) = O(\sqrt{m})$ [7]. Algorithm-1 from [6] represents each triangle of G as a point $p = (high, low) \in \mathbb{R}^2$, where $high$ (low) is the maximum (minimum) timestamp of the participating edges of that triangle. Now we modify the algorithm so that each point p in the original algorithm representing a triangle will now have three copies of itself and each copy will be associated with the color of each participating edge of the triangle. There will be exactly $3\mathcal{K}$ colored points in $\mathbb{R}^2$, where $\mathcal{K}$ is the number of triangles in G. Using 2-dimensional type-2 range counting query data structure ([4], Theorem 6.2) with query rectangle $[i, j] \times [i, j]$, the number of triangles containing edges e_k, where ($i \leq k \leq j$), can be found in $O(\log^2 \mathcal{K} + w \log \mathcal{K})$ time, where w is the number of output edges. The maximum number of triangles in any graph G can be $O(n^3)$. So the query time can be re-defined as $O(\log^2 n + w \log n)$. From this result, we can also report (i) total number of local bridges $\#LB(G_{i,j}) = (j - i - w + 1)$ and (ii) value of embeddedness of any edge in $G_{i,j}$.

Analysis: For part (a), there are overall $O(mn)$ points representing all edge adjacencies in G. For part (b), identifying all triangles in G requires $O(a(G)m)$ preprocessing time. Thus, the total time required to reduce the problem of computing neighborhood overlap to colored range queries takes $O(a(G)m + mn) = O(mn)$ time. Total query time is $O(\log^2 n + w \log n) + O(\log mn + s) = O(\log^2 n + w \log n + s)$. The following theorem summarizes the results.

Theorem 6. *Given an RE graph G with m edges and n vertices, and a query time slice $[i, j]$, the problem of computing the average neighborhood overlap of $G_{i,j}$ can be reduced to the colored range counting in $O(mn)$ time. $NOver(G_{i,j})$ can be computed in $O(\log^2 n + w \log n + s)$ time, where w is the number of edges in $G_{i,j}$ with positive embeddedness and s is the number of edges having some neighboring edges in $E_{i,j}$.*

Similarly, for bipartite RE graphs we can obtain the following results.

Theorem 7. *Given a bipartite RE graph G with m edges and n vertices, and a query time slice $[i, j]$, the problem of computing the average neighborhood overlap of a bipartite graph slice $G_{i,j}$ can be reduced to type-2 colored range searching problem in $O(a(G)m)$ time. $NOver(G_{i,j})$ can be computed in $O(\log^2 n + w \log n + k)$ time, where w is the total number of vertex pairs having some common neighbors in $G_{i,j}$ and k is the total number of vertices having some neighbors in $G_{i,j}$.*

4.4   Reporting Influenced Vertices

Given an RE graph $G = (V, E)$ with a fixed set of influential vertices $V' \subseteq V$, a positive integer r, and a query time slice $[i, j]$, we want to find the total number of influenced vertices in $G_{i,j}$. We associate each vertex $v_k \in V$ with a color $c_k \in C = \{c_1, c_2, \cdots, c_n\}$. Each point representing an influenced vertex v_k will be colored with c_k.

Preprocessing Step: Recall Definitions 1 and 2. For each vertex $v \in V$ having thresholds $f(v) \geq 1$, we associate it with a queue $Q[v]$ of size $f(v)$, initially empty. Queues are maintained by balanced binary search trees, so that the minimum element of the queue can be retrieved when required. In addition, we maintain $\alpha(v)$, for all $v \in V$, where $\alpha(v)$ is the largest timestamp of an edge e such that e is adjacent to an influential vertex v' and there is a path of influence from v' to v. Initially $\alpha(v)$ is set to zero, for all $v \in V$. We process each edge $e_k = (u_k, v_k)$ according to the sequence of their timestamps, i.e., $t(e_1) < t(e_2) < \cdots < t(e_m)$, and set $\alpha(u_k)$ and $\alpha(v_k)$. We explain the process as follows. At time k, let $e_k = (u_k, v_k)$.

Case 1: Both u_k and v_k are influential vertices. Then u_k and v_k are already influenced. We set both $\alpha(u_k)$ and $\alpha(v_k)$ to k. We also store two points (k, k) of colors c_{u_k} and c_{v_k} respectively.

Case 2: u_k is an influential vertex and v_k is a non-influential vertex. We store a point (k, k) of color c_{u_k} and set $\alpha(v_k)$ to k.

Case 3: u_k is already an influenced non-influential vertex and $\alpha(u_k) = l$ has been set. If $k - l \leq r$, u_l influences v_k. If $\alpha(v_k) < \alpha(u_k)$ and we set $\alpha(v_k) = \alpha(u_k) = l$. Otherwise, we keep $\alpha(v_k)$ unchanged.

Each time we set $\alpha(v_k) = l$ for some vertex v_k, we add l to $Q[v_k]$. In case $Q[v_k]$ is full after inserting l, we remove the smallest element s from the queue, and store (k, s) as a point in $\mathbb{R}^2$ with color c_{v_k}. The c_{v_k}-colored point (k, s) implies that v_k is influenced by $f(v_k)$ paths of influence in the graph slice $G_{s,k}$. Vertices v_s and v_l are respectively the first and the last influential vertex that influence v_k in $G_{s,k}$. See Fig. 4 for an example.

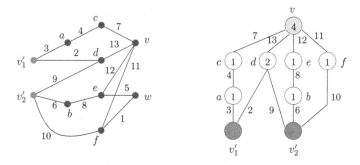

Fig. 4. (a) An RE graph with two influential vertices v_1' and v_2', (b) An influenced vertex v with $r = 3$ and $f(v) = 3$. The influence threshold $f(v)$ is mentioned inside the circle of each vertex v. (Color figure online)

Preprocessing Analysis: All initializations take linear time. There will be n queues, one for each vertex and each queue can have size at most equal to the degree of the vertex $d(v)$. Maintaining a queue for each vertex using balanced binary search tree requires $O(d(v) \log d(v))$ time, and both insertion and removal of a queue element requires $O(\log d(v))$ time. We visit each edge exactly once to set the values of α and to perform operations on queues. Thus, the total preprocessing time will be $\sum_v O(d(v) \log d(v)) = O(m \log n)$.

Now, we can reduce this problem to the colored range searching problem. In the worst case, every vertex can be influenced by each of its neighboring vertices. Thus, we have a set of at most $O(n^2)$ colored points in plane. Now our problem of reporting all influenced vertices within $G_{i,j}$ reduces to the problem of reporting the number of distinct colors in plane. By using 2-dimensional range reporting data structure with query rectangle $[0, j] \times [i, \infty]$, we can report all influenced vertices in $G_{i,j}$ in $O(\log n + w)$ time, where w is the number of influenced vertices ([11], Theorem 1.12).

Theorem 8. *Given an RE graph G with m edges and n vertices, and a query time slice $[i, j]$, the problem of reporting the total number of influenced vertices can be reduced to 2-dimensional colored range reporting problem in $O(m \log n)$ time. Queries can be answered in $O(\log n + w)$ time, where w is the number of influenced vertices in $G_{i,j}$.*

5   Conclusion

In this paper, we have presented a general approach for solving queries regarding various structural parameters of relational event graphs in an arbitrary query time slice using colored range query data structures. Our approach models the reachability relationships between vertices and edges of a given RE graph by transforming them into colored points in $\mathbb{R}^d$, where $d \geq 1$. Subsequently, we reduce original problems into colored range searching problems to efficiently answer the queries.

References

1. Bannister, M.J., DuBois, C., Eppstein, D., Smyth, P.: Windows into relational events: data structures for contiguous subsequences of edges. In: Proceedings of the 24th ACM-SIAM SODA, pp. 856–864. SIAM (2013)
2. Berlingerio, M., Coscia, M., Giannotti, F., Monreale, A., Pedreschi, D.: The pursuit of hubbiness: analysis of hubs in large multidimensional networks. J. Comput. Sci. **2**(3), 223–237 (2011)
3. Bettencourt, L.M., Cintrón-Arias, A., Kaiser, D.I., Castillo-Chávez, C.: The power of a good idea: quantitative modeling of the spread of ideas from epidemiological models. Phys. A: Stat. Mech. Appl. **364**, 513–536 (2006)
4. Bozanis, P., Kitsios, N., Makris, C., Tsakalidis, A.: New upper bounds for generalized intersection searching problems. In: Fülöp, Z., Gécseg, F. (eds.) ICALP 1995. LNCS, vol. 944, pp. 464–474. Springer, Heidelberg (1995). doi:10.1007/3-540-60084-1_97
5. Centola, D., Macy, M.: Complex contagions and the weakness of long ties. Am. J. Sociol. **113**(3), 702–734 (2007)
6. Chanchary, F., Maheshwari, A.: Counting subgraphs in relational event graphs. In: Kaykobad, M., Petreschi, R. (eds.) WALCOM 2016. LNCS, vol. 9627, pp. 194–206. Springer, Heidelberg (2016). doi:10.1007/978-3-319-30139-6_16
7. Chiba, N., Nishizeki, T.: Arboricity and subgraph listing algorithms. SIAM J. Comput. **14**(1), 210–223 (1985)
8. De Berg, M., Van Kreveld, M., Overmars, M., Schwarzkopf, O.C.: Computational Geometry. Springer, Heidelberg (2000)
9. Easley, D., Kleinberg, J.: Networks, Crowds, and Markets: Reasoning About a Highly Connected World. Cambridge University Press, Cambridge (2010)
10. Gargano, L., Hell, P., Peters, J.G., Vaccaro, U.: Influence diffusion in social networks under time window constraints. Theor. Comput. Sci. **584**, 53–66 (2015)
11. Gupta, P., Janardan, R., Rahul, S., Smid, M.: Computational geometry: generalized (or colored) intersection searching. In: Handbook of Data Structures and Applications, 2nd edn. (in press)

12. Gupta, P., Janardan, R., Smid, M.: Further results on generalized intersection searching problems: counting, reporting, and dynamization. J. Algorithms **19**(2), 282–317 (1995)
13. Harary, F.: Graph Theory. Addison-Wesley, Reading (1969)
14. Janardan, R., Lopez, M.: Generalized intersection searching problems. Int. J. Comput. Geom. Appl. **3**(01), 39–69 (1993)
15. McCreight, E.M.: Priority search trees. SIAM J. Comput. **14**(2), 257–276 (1985)
16. Meghanathan, N.: A greedy algorithm for neighborhood overlap-based community detection. Algorithms **9**(1), 8 (2016)
17. Santoro, N., Quattrociocchi, W., Flocchini, P., Casteigts, A., Amblard, F.: Time-varying graphs and social network analysis: temporal indicators and metrics. In: 3rd AISB SNAMAS, pp. 32–38 (2011)

Axiomatic Characterization of the Interval Function of a Bipartite Graph

Manoj Changat[1(✉)], Ferdoos Hossein Nezhad[1], and Narayanan Narayanan[2]

[1] Department of Futures Studies, University of Kerala, Thiruvananthapuram, India
mchangat@gmail.com, ferdows.h.n@gmail.com
[2] Department of Mathematics, IIT Madras, Chennai, India
narayana@gmail.com

Abstract. The axiomatic approach with the interval function and induced path transit function of a connected graph is an interesting topic in metric and related graph theory. In this paper, we introduce a new axiom:

(bp) for any $x, y, z \in V$, $R(x,y) = \{x,y\} \Rightarrow y \in R(x,z)$ or $x \in R(y,z)$.

We study axiom (bp) on the interval function and the induced path transit function of a connected, simple and finite graph. We present axiomatic characterizations of the interval function of bipartite graphs and complete bipartite graphs. Further, we present an axiomatic characterization of the induced path transit function of a tree or a 4-cycle.

Keywords: Interval function · Induced path transit function · Bipartite graph · Axiomatic characterization

1 Introduction

Transit functions on discrete structures were introduced by Mulder [14] to generalize mainly the concept of betweenness in an axiomatic way. Betweenness has been extensively studied on connected simple graphs with the three basic transit functions like interval function, induced path function and all-paths function as models. Our concern here is also on these three standard path functions on finite connected simple graphs. A transit function is an abstract notion of an interval. A *transit function* R defined on a non empty set V is a function $R\colon V \times V \to 2^V$ satisfying the three axioms

(t1) $x \in R(x,y)$ for all $x, y \in V$,
(t2) $R(x,y) = R(y,x)$ for all $x, y \in V$,
(t3) $R(x,x) = \{x\}$ for all $x \in V$.

Basically, a transit function on a simple connected undirected graph G describes how we can get from vertex u to vertex v: via vertices in $R(u,v)$. An element $x \in R(u,v)$, can be considered as *"between"* the points u and v and hence the set $R(u,v)$ consisting of the set of all elements between u and v abstracts the notion

D. Gaur and N.S. Narayanaswamy (Eds.): CALDAM 2017, LNCS 10156, pp. 96–106, 2017.
DOI: 10.1007/978-3-319-53007-9_9

of an interval on set V. Transit functions defined on vertex sets of a connected graph are extensively studied. For example, on betweenness [3,5,6,12,13]; on intervals [1,2,5,8,13,15–19] and on convexity [2,4,11–13].

The underlying graph G_R of a transit function R is the graph with vertex set V, where two distinct vertices u and v are joined by an edge if and only if $R(u,v) = \{u,v\}$. We note here that, much of the axiomatic studies on transit functions have captured attention due to the presence of simple first order axioms that are defined on $R(u,v)$. It is important that the axioms must be first order definable. On the other hand, for ease of handling they are not always stated in FO syntax. In this paper, we search for axioms that can be defined on an arbitrary transit function and try to obtain an axiomatic characterization of the interval function and the induced path function of a connected graph using other standard axioms already studied.

A u,v-shortest path in a connected graph G is a u,v-path in G containing the minimum number of edges. The length of a shortest u,v-path P (that is, the number of edges in P) is the standard distance in G.

The interval function of a connected graph G is defined as

$$I(u,v) = \{w \in V \colon w \text{ lies on a shortest } u,v - path\}.$$

The induced paths naturally generalizes shortest path in the sense that every shortest path is an induced path and replacing shortest paths by induced paths in the previous definition, we obtain the induced path transit function, which is defined as

$$J(u,v) = \{w \in V \colon w \text{ lies on an induced } u,v - path\}.$$

Replacing shortest paths or induced paths by any path in the graph, we obtain the all-paths transit function of G, defined as

$$A(u,v) = \{w \in V \colon w \text{ lies on some } u,v - path\}.$$

The three transit functions, the interval function, induced path function and the all-paths function of a connected graph G is denoted respectively as I_G, J_G and A_G. From the definition of I and J of G, it follows that $I(u,v) = \{u,v\}$ and $J(u,v) = \{u,v\}$ if and only if uv is an edge. Hence the underlying graph G_I of I and G_J of J are both isomorphic to G. But this is not the case with the all-paths transit function $A(u,v)$.

The following non-trivial betweenness axioms were considered by Mulder in [14].

(b1) $x \in R(u,v)$, $x \neq v \Rightarrow v \notin R(u,x)$,
(b2) $x \in R(u,v) \Rightarrow R(x,v) \subseteq R(u,v)$,

It is proved in [1,5] that if R satisfies axioms (b1) and (b2), then the underlying graph G_R of R is connected and both the axioms (b1) and (b2) are necessary for the connectedness of G_R.

The first systematic study on the interval function is by Mulder [13]. The notion of interval might be part of folklore. Sholander studied median semi-lattices using segments [20,21]. Axiomatic characterization of interval function of arbitrary connected graphs has been first given by Nebeský [16–18] and finally Mulder and Nebeský in [15] gave an optimal characterization. Recently, the axiomatic characterization of the interval function of a connected graph is extended to that of a disconnected graph by modifying the $(t1)$ and $(t3)$ axioms and including an additional axiom [9].

The axiomatic characterization of the interval function of trees was presented by Sholander in [20], but he gave only a partial proof for this characterization. Later on a completion of this proof is obtained by Chvátal et al. in [10]. Recently in 2015, characterizations of the interval function of trees using three different sets of axioms weaker than those in [10,20] is presented in [1]. Moreover, axiomatic characterization of the interval function of block graphs, median graphs, modular graphs and geodetic graphs are respectively shown in [1,13,15,16]. Recently, the axiomatic characterization of the interval function of claw and paw-free graphs have been obtained by Changat et al. [8].

Induced path transit function is also studied by several authors [3,5,6,12,19]. One of the interesting result on this function is due to Nebeský [19] who proved that there does not exist a characterization of the induced path function J of a connected graph using first order transit axioms alone. Mulder and Morgana [12] gave a characterization of induced path transit function of a connected graph which satisfies betweenness axioms $(b1)$ and $(b2)$. Changat and Mathew [3] gave a characterization of J-monotone graphs using forbidden induced subgraphs. Moreover, Changat et al. [5] characterized axiomatically the induced path transit function of HHP-free and HHD-free graphs. Changat et al. [7] characterized graphs for which the induced path transit function satisfies the Pasch axiom and recently that of the claw and paw-free graphs in [8].

Among the graph classes whose axiomatic characterization of the interval functions already mentioned, the class of trees, median graphs and modular graphs are examples of bipartite graphs. But an axiomatic characterization of the interval function of an arbitrary bipartite graph is not attempted so far. In this paper, we try towards this goal by introducing the following axiom:

(bp) for any $x, y, z \in V$, $R(x, y) = \{x, y\} \Rightarrow y \in R(x, z)$ or $x \in R(y, z)$.

We name the axiom as (bp), because this is the key axiom involved in the axiomatic characterization of the interval function of a bipartite graph. We prove that the interval function of a connected graph G satisfy the axiom (bp) if and only if G is a bipartite graph and further, the induced path transit function satisfies the axiom (bp) if and only if G is triangle-free. This is natural, since the induced path interval $J(u, v)$ contains the geodetic interval $I(u, v)$, for every pair of vertices u, v in a connected graph.

In Sect. 2, we discuss the interval function I and in Sect. 3 the induced path transit function J. But first, we observe that the axiom (bp) is satisfied by the all-paths transit function of every graph. The all-paths transit function is the coarsest path transit function and an axiomatic characterization of all-paths transit function is presented in [2].

Proposition 1. *The all-paths transit function A of a connected graph G, satisfies axiom (bp) on every graph.*

Proof. Let $x, y, z \in V$ be such that $A(x, y) = \{x, y\}$. That is, xy is a block, that is, xy is a cut-edge in G. Hence $G - xy$ is a disconnected graph and consists of two components, one containing the vertex x and the other containing y. If z lies in the component containing y, then $A(x, y) \subseteq A(x, z)$ and hence $y \in A(x, z)$. Similarly, if z lies in the component containing x, then $x \in A(y, z)$. Hence the proposition follows. □

2   Interval Function of Bipartite Graphs

We note that $I(u, v) = \{x \in V | d(u, x) + d(x, v) = d(u, v)\}$, where $d(u, v) = d_G(u, v)$ denotes the shortest path distance between u and v in a connected graph G. In Proposition 1.1.2 of [13] the first few properties of the interval function of a connected graph were presented. We state this fundamental proposition in the following.

Proposition 2 [13]**.** Let G be a connected graph with interval function I. Then for any u and v in V,

   (i) $u, v \in I(u, v)$
  (ii) $I(u, v) = I(v, u)$
 (iii) If $x \in I(u, v)$, then $I(u, x) \subseteq I(u, v)$
 (iv) If $x \in I(u, v)$, then $I(u, x) \cap I(x, v) = \{x\}$
  (v) If $x \in I(u, v)$ and $y \in I(u, x)$, then $x \in I(y, v)$

The properties (i) and (ii) are precisely the transit axioms $(t1)$ and $(t2)$. The third, fourth and fifth properties are respectively the betweenness axioms $(b2)$, $(b4)$ and $(b3)$.

In this section, we first prove that the interval function I of a graph G satisfies the axiom (bp) if and only if G is a bipartite graph. Then we give an axiomatic characterization of the interval function of a bipartite graph.

Theorem 1. *The interval function I on a connected graph G, satisfies axiom (bp) if and only if G is a bipartite graph.*

Proof. We first show that if G is non-bipartite, then we can find a triplet that violates axiom (bp). Since G is not bipartite, consider a shortest odd cycle C of length (say) $2t + 1$. Since every shortest odd cycle is isometric, for any edge xy in C, there exists a unique vertex $z \in C$ equidistant from both x and y. That is, $d_G(x, z) = d_G(y, z) = t$. Therefore, $y \notin I(x, z) \wedge x \notin I(y, z)$. Hence I does not satisfy axiom (bp), a contradiction. Thus, G must be bipartite.

Conversely, assume that G is a bipartite graph with bipartition X, Y of the vertex set. Consider some edge xy with $x \in X$ and $y \in Y$, without loss of generality, let $z \in X$ be any other vertex.

Let s, t be the lengths of x, z-shortest path and y, z-shortest path respectively. Since xy is an edge, notice that either $s = t - 1$ or $s = t + 1$, without loss of generality, assume that $s = t - 1$.

Since the vertices of all y, z-shortest paths (of length t) belong to $I(y, z)$, any x, z-path of length $s = t - 1$ together with the edge xy forms a y, z-shortest path of length t containing x. Thus $x \in I(y, z)$. The remaining case is analogous.

Thus, for any edge xy (whenever $I(x, y) = \{x, y\}$), we have either $x \in I(y, z)$ or $y \in I(x, z)$ for every z. Thus the axiom (bp) holds true for I.   □

In [15], Mulder and Nebeský used a set of five axioms also known as *classical axioms* and two supplementary axioms. The five classical axioms are precisely $(t1)$, $(t2)$, the betweenness axiom $(b2)$, and the following two axioms.

$(b3)$ if $x \in R(u, v)$ and $y \in R(u, x)$,then $x \in R(y, v)$ for all u, v, x, y.
$(b4)$ if $x \in R(u, v)$, then $R(u, x) \cap R(x, v) = \{x\}$ for all u, v, x,

It is clear that axioms $(t1)$ and $(b4)$ imply axiom $(t3)$. Hence any function satisfying the five classical axioms is a transit function. Furthermore, the following implications are immediate. Axioms $(t1)$, $(t2)$ and $(b4)$ imply $(b1)$, see [1]. Axioms $(t1)$, $(t2)$, $(t3)$ and $(b3)$ imply axioms $(b4)$, see [15]. Hence if R is a transit function satisfying axiom $(b3)$ on V then R satisfies axioms $(b4)$ and $(b1)$ on V.

Below, we state the two supplementary axioms from [15] and restate the main theorem characterizing the interval function of arbitrary connected graphs.

$(s1)$ If $R(u, \bar{u}) = \{u, \bar{u}\}$, $R(v, \bar{v}) = \{v, \bar{v}\}$, $u \in R(\bar{u}, \bar{v})$ and $\bar{u}, \bar{v} \in R(u, v)$, then $v \in R(\bar{u}, \bar{v})$ for all $u, \bar{u}, v, \bar{v}$.
$(s2)$ If $R(u, \bar{u}) = \{u, \bar{u}\}$, $R(v, \bar{v}) = \{v, \bar{v}\}$, $\bar{u} \in R(u, v)$, $\bar{v} \notin R(u, v)$, and $v \notin R(\bar{u}, \bar{v})$, then $\bar{u} \in R(u, \bar{v})$ for all $u, \bar{u}, v, \bar{v}$.

Theorem 2 [15]. *Let $R: V \times V \to 2^V$ be a function on V, satisfying the axioms $(t1)$, $(t2)$, $(b2)$, $(b3)$, $(b4)$ with the underlying graph G, and let I be the interval function of G. The following statements are equivalent:*

(a) $R = I$
(b) R satisfies axioms $(s1)$ and $(s2)$.

In the following, we prove that any transit function R satisfying axioms $(b1)$, $(b2)$ and (bp) also satisfies axioms $(s1)$ and $(s2)$.

Lemma 1. *Let R be a transit function satisfying axioms $(b1)$, $(b2)$ and (bp) on V. Then R satisfies axioms $(s1)$ and $(s2)$.*

Proof. First we show that R satisfies $(s1)$. Suppose R does not satisfy $(s1)$. Then there exist $u, \bar{u}, v, \bar{v} \in V$ such that $R(u, \bar{u}) = \{u, \bar{u}\}$, $R(v, \bar{v}) = \{v, \bar{v}\}$, $u \in R(\bar{u}, \bar{v})$, $\bar{u}, \bar{v} \in R(u, v)$ and $v \notin R(\bar{u}, \bar{v})$. Since $\bar{u}, \bar{v} \in R(u, v)$, by $(b1)$ we have $u \notin R(\bar{u}, v)$. As $R(v, \bar{v}) = \{v, \bar{v}\}$, by (bp), either $\bar{v} \in R(v, \bar{u})$ or $v \in R(\bar{v}, \bar{u})$. By assumption, $v \notin R(\bar{u}, \bar{v})$, and hence, $\bar{v} \in R(v, \bar{u})$. Therefore, by $(b2)$, $R(\bar{u}, \bar{v}) \subseteq$

$R(v, \bar{u})$ and since $u \in R(\bar{u}, \bar{v})$, $u \in R(v, \bar{u})$. This is a contradiction since $u \notin R(\bar{u}, v)$. Thus, R satisfies axiom $(s1)$.

Next we prove that R satisfies axiom $(s2)$. Suppose not. Then there exist $u, \bar{u}, v, \bar{v} \in V$ such that $R(u, \bar{u}) = \{u, \bar{u}\}$, $R(v, \bar{v}) = \{v, \bar{v}\}$, $\bar{u} \in R(u, v)$, $\bar{v} \notin R(u, v)$, $v \notin R(\bar{u}, \bar{v})$ and $\bar{u} \notin R(u, \bar{v})$. Since $R(u, \bar{u}) = \{u, \bar{u}\}$, by (bp), $\bar{u} \in R(u, \bar{v})$ or $u \in R(\bar{v}, \bar{u})$. Thus $u \in R(\bar{v}, \bar{u})$ because $\bar{u} \notin R(u, \bar{v})$. Furthermore, $R(v, \bar{v}) = \{v, \bar{v}\}$ together with axiom (bp) implies that $v \in R(u, \bar{v})$ or $\bar{v} \in R(v, u)$. As before $v \in R(u, \bar{v})$, as $\bar{v} \notin R(v, u)$. Since $u \in R(\bar{v}, \bar{u})$, by $(b2)$, $R(u, \bar{v}) \subseteq R(\bar{v}, \bar{u})$. Thus, $v \in R(\bar{v}, \bar{u})$, a contradiction to the assumption that $v \notin R(\bar{v}, \bar{u})$. Therefore R satisfies axiom $(s2)$. $\square$

We have the following theorem characterizing the interval function of bipartite graphs.

Theorem 3. *Let R be a transit function on a finite set V and G_R be the underlying graph of R, then R satisfies axioms $(b3)$, $(b2)$ and (bp) if and only if G_R is a bipartite graph and $I_{G_R} = R$.*

Proof. Let R be a transit function on V satisfying axioms $(b3)$, $(b2)$ and (bp). Since for a transit function R, $(b3) \Rightarrow (b4) \Rightarrow (b1)$, see [1,15], R satisfies $(b4)$ and $(b1)$. By Lemma 1 R satisfies $(s1)$ and $(s2)$. Therefore, by Theorem 2, I_{G_R} coincides with R. Hence, G_R is isomorphic to G and by Theorem 1, it follows that G_R is bipartite. Conversely suppose R is a transit function such that G_R is bipartite and $I_{G_R} = R$. From Proposition 2 of Mulder [13], it follows that the interval function R satisfies axioms $(b2), (b3)$. From Theorem 1, R satisfies (bp). $\square$

The following example shows that axioms $(b1), (b2), (bp)$ do not imply $(b3)$ and $(b4)$.

Example 1. $(b1), (b2), (bp) \not\Rightarrow (b3)$ and $(b4)$.
Let $V = \{a, b, c, d\}$ and define R as follows: $R(a, b) = \{a, b, c, d\}$, $R(a, c) = \{a, c, d\}$ and $R(b, c) = \{b, c, d\}$ and for any other distinct pair $x, y \in V$, $R(x, y) = \{x, y\}$ and $R(x, x) = \{x\} \forall x$ and $R(x, y) = R(y, x)$ for $x, y \in V$. It is easy to see that R is a transit function and R satisfies axioms $(b1), (b2), (bp)$, but R does not satisfy $(b4)$, since $c \in R(a, b)$, but $R(a, c) \cap R(b, c) = \{c, d\} \neq \{c\}$. Thus, R also does not satisfy $(b3)$, because for a transit function R, $(b3) \Rightarrow (b4)$.

The following examples show that axioms $(b3)$, $(b2)$ and (bp) are independent. Example 1 shows that $(b2), (bp) \not\Rightarrow (b3)$.

Example 2. $(b2), (b3) \not\Rightarrow (bp)$.
Let G be a non-bipartite connected graph and let I be its interval function. Since I is the interval function, from Proposition 2 it follows that I satisfies $(b2)$ and $(b3)$. Moreover, since G is a non-bipartite graph from Theorem 1, it follows that I does not satisfy (bp).

Example 3. $(b3), (bp) \not\Rightarrow (b2)$.

Let $V = \{a, b, c, d, e\}$ and define R as follows, $R(a, b) = \{a, b, c\}$, $R(a, c) = \{a, c, d\}$, $R(a, d) = \{a, d, e\}$, $R(a, e) = \{a, e, b\}$, $R(b, d) = \{b, d, c, e\}$, $R(c, e) = \{c, e, d, b\}$, and for any other distinct pair $x, y \in V$, $R(x, y) = \{x, y\}$ and for any $x \in V$, $R(x, x) = \{x\}$. It is easy to see that R satisfies axioms $(t1), (t2), (t3), (b3)$ and (bp) on V, but does not satisfy $(b2)$, since $c \in R(a, b)$ and $R(a, c) \not\subseteq R(a, b)$.

Next, we define a new axiom (cbp). Then using axioms (bp) and (cbp) we give a characterization of the interval function of complete bipartite graphs.

(cbp) For distinct $x, y, z \in V$, if $z \in R(x, y)$, then $R(x, z) = \{x, z\}$ and $R(z, y) = \{z, y\}$.

Remark 1. Let R be a transit function satisfying axiom (cbp). Then R satisfies axioms $(b1), (b2), (b3)$ and $(b4)$.

Hence we have the following theorem characterizing the interval function of complete bipartite graphs.

Theorem 4. *Let R be a transit function, then R satisfies axioms (bp) and (cbp) if and only if G_R is a complete bipartite graph and $I_{G_R} = R$.*

Proof. Let R be a transit function satisfying axioms (bp) and (cbp). By Remark 1, it follows that R satisfies $(b3), (b4), (b2)$ and $(b1)$. Hence by Theorems 2 and 1, G_R is bipartite and $I_{G_R} = R$. Let X, Y be the bipartition of G_R. Assume that there exist $x \in X$ and $y \in Y$, such that $R(x, y) \neq \{x, y\}$. Using axiom (cbp), for every $z \in R(x, y)$, $R(x, z) = \{x, z\}$ and $R(y, z) = \{y, z\}$. Since $(b1)$ and $(b2)$ holds, there must exist at least one such z. Now as G_R is bipartite, we have $z \notin X$ and $z \notin Y$. This is a contradiction since $X \cup Y = V(G_R)$.

To prove the converse it suffices to prove that R satisfies (cbp) by Theorem 3. Since G_R is complete bipartite (with bipartition X, Y), and $R = I$, we have the following:

If $u, v \in X$, $R(u, v) = Y \cup \{u, v\}$. If $u, v \in Y$, $R(u, v) = X \cup \{u, v\}$ and $R(u, v) = \{u, v\}$ if $u \in X, v \in Y$. Let $z \in R(u, v), z \neq u, z \neq v$. Then $u, v \in X$ or $u, v \in Y$. Without loss of generality, let $u, v \in X$. Then $z \in Y$ and hence, $R(u, z) = \{u, z\}$ and $R(v, z) = \{v, z\}$.   □

The following examples show that (bp) and (cbp) are independent.

Example 4. $(cbp) \not\Rightarrow (bp)$.

Let G be a triangle (K_3) with vertex set $V = \{a, b, c\}$ and let I be its interval function. Note that I satisfies axioms $(t1), (t2), (t3), (b2), (b3)$, but I does not satisfy axiom (bp), since $I(a, b) = \{a, b\}$, but $b \notin I(a, c)$ and $a \notin I(b, c)$.

Example 5. $(bp) \not\Rightarrow (cbp)$.

Let G be a 6-cycle on $v_1 \ldots v_6$ and I, its interval function. By Proposition 1 of Mulder [13] and Theorem 1, I satisfies $(t1), (t2), (t3)$ and (bp). Now, I does not satisfy (cbp), because $I(v_2, v_5) = V$ and $v_3 \in I(v_2, v_5)$ but $I(v_3, v_5) \neq \{v_3, v_5\}$.

3   Induced Path Transit Function

In this section, we use the axiom (bp) on the induced path transit function J of a connected graph G, to present a characterization of triangle-free graphs. We have the following theorem.

Theorem 5. *The induced path transit function J of a connected graph G satisfies axiom (bp) if and only if G is triangle-free.*

Proof. Let J be the induced path transit function of a connected graph G. First assume that G contains a triangle. Then, the triplet of vertices of the triangle violates axiom (bp). Conversely suppose G is a triangle-free graph. Assume that G does not satisfy (bp). Then there exists a vertex x and an edge yz such that $z \notin J(x,y)$ and $y \notin J(x,y)$. Let P be a shortest x,y-path. Since $z \notin J(x,y)$, it follows that y is not on P. So P followed by the edge yz is an x,z-path. Since $y \notin J(x,z)$ this path cannot be induced, so that there must be a chord between z and a vertex on P distinct from y. Let zy' be such a chord with y' on P closest to y. Since G is triangle-free, y' is not adjacent to y. But now the subpath P' of P between x and y' followed by the edges $y'z$ and zy is an induced x,y-path containing z, so that $z \in J(x,y)$. This contradicts the choice of vertex x and edge yz.   □

Changat et al. in [5] showed that for several special classes of graphs (for e.g., HHP-free graphs), there exist axiomatic characterizations of the induced path transit function using axioms $(b1)$, $(b2)$ and the axioms $(j1)$, $(j2)$ and $(j3)$ given below.

$(j1)$ $w \in R(u,v)$, $w \notin \{u,v\} \Rightarrow$ there exists $u_1 \in R(u,w) \backslash R(v,w)$, $v_1 \in R(v,w) \backslash R(u,w)$, such that $R(u_1,w) = \{u_1,w\}$, $R(v_1,w) = \{v_1,w\}$ and $w \in R(u_1,v_1)$.

$(j2)$ $R(u,x) = \{u,x\}$, $R(x,v) = \{x,v\}$, $u \neq v$, $R(u,v) \neq \{u,v\} \Rightarrow x \in R(u,v)$.

$(j3)$ $x \in R(u,y)$, $y \in R(x,v)$, $x \neq y$, $u \neq v$ and $R(u,v) \neq \{u,v\} \Rightarrow x \in R(u,v)$.

A graph is said to be HHP-free if it does not contain a hole, a house or a P-graph (see Fig. 1) as an induced subgraph. A *hole* is an induced cycle of length at least five and a *house* is a 5-cycle having one chord. We quote some results from Changat et al. [5].

Theorem 6 [5]. Let R be a transit function on a non-empty finite set V satisfying the axioms $(b1)$, $(b2)$, $(j1)$, $(j2)$ and $(j3)$. Then the underlying graph G_R of R is HHP-free and moreover $J_{G_R} = R$.

Theorem 7 [5]. Let G be a connected HHP-free graph, and let J be the induced path function of J. Then J satisfies the axioms $(b1)$, $(b2)$, $(j1)$, $(j2)$ and $(j3)$.

We show that a transit function R satisfying axioms $(b2)$ and $(j1)$ also satisfies axiom $(b1)$. Therefore we can omit $(b1)$ in the characterization of induced path transit function of HHP-free graphs in Theorem 6.

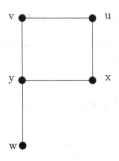

Fig. 1. P-graph

Lemma 2. *Let R be a transit function satisfying axioms $(b2)$ and $(j1)$. Then R satisfies axiom $(b1)$ and G_R is connected.*

Proof. By $(b2)$, for any $c \in R(a,b)$, we have $R(b,c) \subseteq R(a,b)$. Suppose that R does not satisfy $(b1)$. Then, there exists distinct a, b, c such that $c \in R(a,b)$ and $b \in R(a,c)$. By $(j1)$, there must exist $x \in R(b,c) \setminus R(a,b)$. This is not possible since $R(b,c) \subseteq R(a,b)$. Therefore R satisfies $(b1)$. Together with $(b2)$ connectedness of G_R follows. $\qquad\qquad\qquad\qquad\qquad\qquad\qquad\qquad\qquad\qquad\qquad \Box$

Hence by Lemma 2 we can rewrite Theorems 6 and 7 as follows.

Theorem 8. *Let R be a transit function on a non-empty finite set V satisfying the axioms $(b2)$, $(j1)$, $(j2)$ and $(j3)$. Then the underlying graph G_R of R is HHP-free and moreover $J_{G_R} = R$.*

Theorem 9. *Let G be a connected HHP-free graph, and let J be the induced path function of J. Then J satisfies the axioms $(b2)$, $(j1)$, $(j2)$ and $(j3)$.*

Next, we present examples to show that for an arbitrary transit function R, $(j1), (j2)$ and $(j3)$ form an independent set of axioms.

Example 6. $(b2), (j2), (j3) \nRightarrow (j1)$.
Let $V = \{a, b, c, d\}$ and define R on V as follows: $R(a,b) = V = R(b,c), R(a,c) = \{a, c, d\}$, for any other distinct pair $x, y \in V$, $R(x,y) = \{x, y\}$. For any $x \in V$, $R(x,x) = \{x\}$ and $R(x,y) = R(y,x)$ for all x, y. It is easy to see that R is a transit function and satisfies axioms $(b2), (j2), (j3)$. But R does not satisfy axiom $(j1)$ since $c \in R(a,b)$ and $\nexists x \in R(a,c) \setminus R(b,c)$ such that $R(a,x) = \{a, x\}$.

Example 7. $(b2), (j1), (j3) \nRightarrow (j2)$.
Let $V = \{a, b, c, d\}$ and define R on V as follows: $R(a,c) = \{a, b, c\}$ and for any other distinct pair $x, y \in V$, $R(x,y) = \{x, y\}$ and for any $x \in V$, $R(x,x) = \{x\}$. It is easy to see that R is a transit function and satisfies axioms $(b2), (j1), (j3)$. But R does not satisfy axiom $(j2)$ since $R(a,d) = \{a, d\}$ and $R(d,c) = \{d, c\}$ and $R(a,c) \neq \{a, c\}$ but $d \notin R(a,c)$.

Example 8. $(b2), (j1), (j2) \not\Rightarrow (j3)$.

Let $V = \{a, b, c, d\}$ and define R on V as follows: $R(a, b) = R(c, d) = V$ and for any other distinct pair $x, y \in V$, $R(x, y) = \{x, y\}$ and for any $x \in V$, $R(x, x) = \{x\}$. Notice that R is a transit function and satisfies axioms $(b2), (j1), (j2)$. But R does not satisfy axiom $(j3)$, since $c \in R(a, b)$ and $b \in R(c, d)$, but $c \notin R(a, d)$.

We prove that the axiom (bp) implies axiom $(j2)$.

Proposition 3. *Let R be a transit function satisfying axiom (bp), then R satisfies $(j2)$.*

Proof. Let $R(a, b) = \{a, b\}$ and $R(b, c) = \{b, c\}$. By (bp), either $a \in R(b, c)$ or $b \in R(a, c)$. Since $a \notin R(b, c)$, it follows that $b \in R(a, c)$. Therefore R satisfies $(j2)$. □

It is easy to see that, Example 6 also satisfies $(b1)$ and (bp) and hence, $(b1), (b2), (bp) \not\Rightarrow (j1)$ and moreover Example 6 shows that $(bp), (b2), (j3) \not\Rightarrow (j1)$. Example 7 shows that $(j1), (j3), (b2) \not\Rightarrow (bp)$ and Example 8 shows that $(j1), (b2), (bp) \not\Rightarrow (j3)$.

Finally, we have the following theorem.

Theorem 10. *Let R be a transit function on non empty finite set V then G_R is either C_4 or tree if and only if R satisfies axioms $(b2), (bp), (j1)$ and $(j3)$ and $J_{G_R} = R$.*

Proof. Let R satisfy axioms $(b2), (bp), (j1)$ and $(j3)$. Then, by Lemma 2 and Proposition 3, R satisfies axioms $(b1)$ and $(j2)$ and G_R is connected. By Theorem 8, $J_{G_R} = R$ and $G_R \simeq G$. From Theorem 9, it follows that G_R is HHP-free and by Theorem 5, G_R is triangle-free. Therefore G_R is either C_4 or a tree. □

Concluding Remarks

From Examples 6, 7, 8, the axioms $(j1), (j2), (j3)$ are necessary for the characterization of the induced path transit function in Theorem 6. But, so far we do not have an example to show that $(b2)$ is independent from the axioms $(j1), (j2)$ and $(j3)$. So, we leave it as a problem. That is, whether the axiom $(b2)$ is necessary for the characterization of the induced path transit function in Theorem 6. Further, do not have an example to show that $(b2)$ is independent from the axioms $(j1), (bp)$ and $(j3)$.

Acknowledgments. This research work is supported by NBHM-DAE, Govt. of India under grant No. 2/48(9)/2014/NBHM(R.P)/R & D-II/4364 DATED 17TH NOV, 2014.

References

1. Balakrishnan, K., Changat, M., Lakshmikuttyamma, A.K., Mathews, J., Mulder, H.M., Narasimha-Shenoi, P.G., Narayanan, N.: Axiomatic characterization of the interval function of a block graph. Discret. Math. **338**, 885–894 (2015)

2. Changat, M., Klavžar, S., Mulder, H.M.: The all-paths transit function of a graph. Czech. Math. J. **51**(126), 439–448 (2001)
3. Changat, M., Mathew, J.: Induced path transit function, monotone and Peano axioms. Discret. Math. **286**(3), 185–194 (2004)
4. Changat, M., Mulder, H.M., Sierksma, G.: Convexities related to path properties on graphs. Discret. Math. **290**(23), 117–131 (2005)
5. Changat, M., Mathews, J., Mulder, H.M.: The induced path function, monotonicity and betweenness. Discret. Appl. Math. **158**(5), 426–433 (2010)
6. Changat, M., Lakshmikuttyamma, A.K., Mathews, J., Peterin, I., Narasimha-Shenoi, P.G., Seethakuttyamma, G., Špacapan, S.: A forbiddensubgraph characterization of some graph classes using betweenness axioms. Discret. Math. **313**, 951–958 (2013)
7. Changat, M., Peterin, I., Ramachandran, A., Tepeh, A.: The induced path transit function and the Pasch axiom. Bull. Malays. Math. Sci. Soc. **39**, 1–12 (2015)
8. Changat, M., Hossein Nezhad, F., Narayanan, N.: Axiomatic characterization of claw and paw-free graphs using graph transit functions. In: Govindarajan, S., Maheshwari, A. (eds.) CALDAM 2016. LNCS, vol. 9602, pp. 115–125. Springer, Heidelberg (2016). doi:10.1007/978-3-319-29221-2_10
9. Changat, M., Hossein Nezhad, F., Mulder, H.M., Narayanan, N.: A note on the interval function of a disconnected graph, Discussiones Mathematicae Graph Theory (2016, accepted)
10. Chvátal, V., Rautenbach, D., Schäfer, P.M.: Finite sholander trees, trees, and their betweenness. Discret. Math. **311**(20), 2143–2147 (2011)
11. Duchet, P.: Convex sets in graphs II minimal path convexity. J. Combin. Theory Ser. B **44**, 307–316 (1988)
12. Morgana, M.A., Mulder, H.M.: The induced path convexity, betweenness and svelte graphs. Discret. Math. **254**, 349–370 (2002)
13. Mulder, H.M.: The interval function of a graph. MC Tract **132**, 1–191 (1980). Mathematisch Centrum, Amsterdam
14. Mulder, H.M.: Transit functions on graphs (and posets). In: Changat, M., Klavžar, S., Mulder, H.M., Vijayakumar, A. (eds.) Convexity in Discrete Structures. Lecture Notes Series, pp. 117–130. Ramanujan Mathematical Society, Mysore (2008)
15. Mulder, H.M., Nebeský, L.: Axiomatic characterization of the interval function of a graph. Eur. J. Combin. **30**, 1172–1185 (2009)
16. Nebeský, L.: A characterization of the interval function of a connected graph. Czech. Math. J. **44**, 173–178 (1994)
17. Nebeský, L.: Characterizing the interval function of a connected graph. Math. Bohem. **123**(2), 137–144 (1998)
18. Nebeský, L.: Characterization of the interval function of a (finite or infinite) connected graph. Czech. Math. J. **51**, 635–642 (2001)
19. Nebeský, L.: The induced paths in a connected graph and a ternary relation determined by them. Math. Bohem. **127**, 397–408 (2002)
20. Sholander, M.: Trees, lattices, order, and betweenness. Proc. Am. Math. Soc. **3**(3), 369–381 (1952)
21. Sholander, M.: Medians and betweenness. Proc. Am. Math. Soc. **5**(5), 801–807 (1954)

Analysis of 2-Opt Heuristic for the Winner Determination Problem Under the Chamberlin-Courant System

Ante Ćustić, Ehsan Iranmanesh$^{(\boxtimes)}$, and Ramesh Krishnamurti

Simon Fraser University, Burnaby, BC V5A 1S6, Canada
{acustic,iranmanesh,ramesh}@sfu.ca

Abstract. Winner determination problem under Chamberlin-Courant system deals with the problem of selecting a fixed-size assembly from a set of candidates that minimizes the sum of misrepresentation values. This system does not restrict the candidates to have a minimum number of votes to be selected. The problem is known to be NP-hard. In this paper, we consider domination analysis of a 2-Opt heuristic for this problem. We show that the 2-Opt heuristic produces solutions no worse than the average solution in polynomial time. We also show that the domination number of the 2-Opt heuristic is at least $\binom{m-1}{k-1}k^{n-1}$ for n voters and m candidates.

Keywords: Domination analysis · Proportional representation · Chamberlin-courant system · 2-Opt heuristic

1 Introduction

Proportional representation voting arises in voting procedures which elect multiple winners in an election. Such procedures can be used to elect any representative body comprising more than one individual, such as a committee, a council, or a legislative assembly. Informally, in this voting procedure, the size of the subset of candidates elected as winners is proportional to the number of voters that prefer them. Thus, special interest groups gain representation in proportion to their size [1]. Brams and Fishburn [2,3] first proposed proportional representation as a voting procedure. Choosing a set of candidates that minimizes the sum of the misrepresentation values of voters was proposed by Monroe [4]. In this system, for each voter, the 'misrepresentation' of the candidate (in the chosen subset) that best represents the voter is counted. The objective is to minimize the sum of the misrepresentation count over all the voters. This rule selects as winners a subset of candidates that minimizes the sum of the misrepresentation values of the voters. The misrepresentation values are obtained from the profiles of the

E. Iranmanesh and R. Krishnamurti—Supported by NSERC Discovery Grant, Canada.

© Springer International Publishing AG 2017
D. Gaur and N.S. Narayanaswamy (Eds.): CALDAM 2017, LNCS 10156, pp. 107–117, 2017.
DOI: 10.1007/978-3-319-53007-9_10

voters. Each voter ranks all the candidates in her profile. If μ_{ij} is the rank of a candidate j in voter i's profile, then the misrepresentation of that candidate for the voter is defined to be $\mu_{ij} - 1$ (the candidate with rank 1 has a misrepresentation of 0). Monroe also recommended approval balloting, a procedure similar to approval voting, where each voter approves as many alternatives as she wishes. In this paper, we study a restricted version of Monroe's voting procedure, called the Chamberlin-Courant rule, proposed by Chamberlin and Courant [5]. While Monroe's rule imposes upper and lower limits on the number of votes assigned to a chosen candidate, the Chamberlin-Courant system imposes no such limits. We call these versions of the problem the 'minisum' versions. Procaccia [6] show that the minisum versions for both these voting procedures are computationally intractable in general. They also provide a polynomial-time algorithm for the problem, where the degree of the polynomial is proportional to the size of the assembly. Betzler et al. [7] propose variants of the problem where the objective is to minimize the maximal misrepresentation. The 'minimax' versions of the problem continue to be NP-Hard for both the Monroe rule and the Chamberlin-Courant rule. Betzler et al. also provide fixed-parameter tractability results for both the minisum and the minimax versions under the Monroe system and the Chamberlin-Courant system. They also provide polynomial algorithms for the problem when there is an ordering of candidates such that the ranking in each voter's profile is single-peaked. Skowron et al. [8] consider the complexity of the problem under the single-crossing assumption. In this restricted version of the problem, there is an ordering of voters such that for every pair of candidates a and b there is a voter x in the ordering such that all voters to the left of x prefer candidate a and all voters to the right of x prefer candidate b (voter x either prefers candidate a or candidate b). Skowron et al. provide a polynomial algorithm for this version under the Chamberlin-Courant system, and show its intractability under the Monroe system. Yu et al. [9] consider the complexity of the problem for elections that are single-peaked on a tree. A profile is single-peaked on a tree if we can construct a tree whose vertices are the candidates in the election, and for any path in this tree, the ordering of candidates with respect to that path has the single-peaked property. Yu et al. show that the minimax version of the Chamberlin-Courant system is polynomially solvable for any tree and any misrepresentation function under this assumption. For the minisum version, their algorithm runs in polynomial time when the tree has a small number of leaves. Clearwater et al. [10] consider the single-crossing on a tree assumption and show that both minimax and minisum versions of the Chamberlin-Courant system are polynomially solvable under this assumption. A profile is single-crossing on a tree if we can construct a tree whose vertices are the candidates in the election, and for any path in this tree, the ordering of candidates with respect to that path has the single-crossing property. Skowron et al. [11] show that no constant-factor approximation algorithms exist for the minisum version under both the Monroe system and the Chamberlin-Courant system. In work more related to our work, Skowron et al. [12] provide experimental results for several simple and fast heuristics that obtain near optimal solutions.

Here we consider the minisum version of the Winner Determination Problem under Chamberlin-Courant system (WDPCC). We are given a set of n voters ($\mathcal{V}$), and m candidates ($\mathcal{C}$). Each voter ranks candidates, giving them a distinct number between 1 and m (μ_{ij} is the ranking of voter i for candidate j, and the misrepresentation value of voter i for candidate j is $r_{ij} = \mu_{ij} - 1$). Let $\mathcal{S}_k = \{S \subseteq \mathcal{C} : |S| = k\}$. For each $S \in \mathcal{S}_k$, we can define an assignment function, $\mathcal{A}_S$, that assigns each voter to a candidate in S ($\mathcal{A}_S : \mathcal{V} \to S$). Under assignment function $\mathcal{A}_S$, the misrepresentation value of voter i is $r_{i\mathcal{A}_S(i)}$, and the sum of misrepresentation values is $\sum_{i \in \mathcal{V}} r_{i\mathcal{A}_S(i)}$. The objective under WDPCC is to select $S \in \mathcal{S}_k$ and $\mathcal{A}_S$ that minimizes the sum of misrepresentation values.

We design a heuristic for WDPCC. Different metrics in the literature are used to measure the quality of a heuristic. This includes worst case analysis [13], domination ratio [14,15], domination number [14–18], and comparison to average solution value [17–19].

In this paper we focus on comparison to average solution value and domination number. Let $\mathbf{x}^{\Phi} \in \mathcal{F}(I)$ be a solution provided by algorithm Φ on a given instance I of an optimization problem ($\mathcal{F}(I)$ denotes the set of feasible solutions for the instance I). Let $\mathcal{G}^{\Phi}(I) = \{\mathbf{x} \in \mathcal{F}(I) : f(\mathbf{x}) \geq f(\mathbf{x}^{\Phi})\}$ where $f(\mathbf{x})$ denotes the objective function value of $\mathbf{x}$. Furthermore, let $\mathcal{I}$ be the collection of all instances of the problem with a fixed instance size. Then

$$\inf_{I \in \mathcal{I}} \left| \mathcal{G}^{\Phi}(I) \right| \quad \text{and} \quad \inf_{I \in \mathcal{I}} \frac{|\mathcal{G}^{\Phi}(I)|}{|\mathcal{F}(I)|},$$

are called *domination number* and *domination ratio* of Φ, respectively [14,15].

We first compute the average solution value of the WDPCC. We show that any solution to WDPCC with an objective function value less than the average value dominates $\binom{m-1}{k-1}k^{n-1}$ solutions (n is the number of voters, m the number of candidates, and k the number of candidates that needs to be selected). We then design a simple heuristic to obtain a solution better than the average in polynomial time. Hence, we present a heuristic with domination number no worse than $\binom{m-1}{k-1}k^{n-1}$, and domination ratio no worse than $\frac{1}{m}$.

2   Integer Linear Programming Formulation for WDPCC

The first integer programming formulation for this system under both the Monroe rule and the Chamberlin-Courant rule was given by Potthof and Brams [20]. Each voter provides a ranking of the candidates in set $\mathcal{C}$ in her profile, i.e. assigns a distinct integer from 1 to m ($|\mathcal{C}| = m$) to each of the candidates. We let μ_{ij} denote the rank voter i assigns to candidate j, and $r_{ij} = \mu_{ij} - 1$ denote the misrepresentation of voter i by candidate j.

There are two sets of decision variables in the formulation: x_{ij} and y_i, where $x_{ij} = 1$ if candidate j is assigned to voter i ($x_{ij} = 0$ otherwise), and $y_i = 1$ if candidate i is chosen ($y_i = 0$ otherwise).

$$\text{Minimize} \sum_{i \in \mathcal{V}} \sum_{j \in \mathcal{C}} r_{ij} x_{ij} \qquad (1)$$

$$\text{s.t.} \sum_{j \in \mathcal{C}} y_j = k \qquad (2)$$

$$\sum_{j \in \mathcal{C}} x_{ij} = 1 \qquad \forall i \in \mathcal{V} \qquad (3)$$

$$y_j - x_{ij} \geq 0 \qquad \forall i \in \mathcal{V}, \forall j \in \mathcal{C} \qquad (4)$$

$$x_{ij} \in \{0,1\} \qquad \forall i \in \mathcal{V}, \forall j \in \mathcal{C} \qquad (5)$$

$$y_j \in \{0,1\} \qquad \forall j \in \mathcal{C} \qquad (6)$$

The objective function (1) minimizes the total misrepresentation value. Constraint (2) ensures that a subset of size k is selected from the set of candidates. Constraint (3) ensures that each voter is assigned to only one candidate in the selected subset. Constraint (4) ensures that each voter is only assigned to a selected candidate. Constraints (5), and (6) are the integrality constraints.

Let $\mathcal{F}$ denote the set of all feasible solutions of the WDPCC, i.e., the set of all pairs $(\mathbf{x}, \mathbf{y})$ where $\mathbf{x} = [x_{ij}]_{n \times m}$ and $\mathbf{y} = [y_i]_m$ satisfy constraints (2) to (6). Furthermore, let $f(\mathbf{x}, \mathbf{y})$ denote the objective function value of $(\mathbf{x}, \mathbf{y}) \in \mathcal{F}$. For every fixed $\bar{\mathbf{y}}$ that determines the subset of k candidates from $\mathcal{C}$, there are k^n solutions $(\mathbf{x}, \bar{\mathbf{y}})$ in $\mathcal{F}$. Therefore, it is clear that $|\mathcal{F}| = \binom{m}{k} k^n$. In the rest of this paper, we will interchangeably refer to $\mathbf{y}$ both as a vector of variables y_j and as a corresponding subset of k candidates. The correct meaning will be clear from the context.

3   2-Opt Heuristic

In this section, we present a simple 2-Opt heuristic for the winner determination problem under the Chamberlin-Courant system. Let $v : \binom{\mathcal{C}}{k} \to \mathbb{N}$ be a function that evaluates the objective function of the best solution with a given k-subset of candidates. That is, $v(\bar{\mathbf{y}}) = \min\{f(\mathbf{x}, \bar{\mathbf{y}}) : (\mathbf{x}, \bar{\mathbf{y}}) \in \mathcal{F}\}$. Function v counts the misrepresentation of the candidate (in the input subset) that best represents each voter. Note that, given a $\mathbf{y}$, $v(\mathbf{y})$ can be calculated in linear time. Algorithm 1 shows the pseudo code of the heuristic.

The heuristic starts with a randomly generated subset of size k, $\mathbf{y}_{best}$, iteratively selects a pair of candidates $i \in \mathbf{y}_{best}$, $j \in \mathcal{C} \setminus \mathbf{y}_{best}$, and checks if swapping i with j leads to an improved objective function value. The algorithm stops only when no improvement is possible by swapping i with j, for every possible pair of candidates $i \in \mathbf{y}_{best}$, $j \in \mathcal{C} \setminus \mathbf{y}_{best}$.

3.1   Computing the Average Solution Value

Given a set of voters $\mathcal{V}$, a set of candidates $\mathcal{C}$ and a misrepresentation matrix $\mathbf{r}$, let $A(\mathcal{V}, \mathcal{C}, \mathbf{r})$ denote the average objective function value among all feasible solutions $\mathcal{F}$. That is,

---

Algorithm 1. 2-Opt heuristic for the winner determination problem under the Chamberlin Courant System

---

1: **Input:** misrepresentation matrix $\mathbf{r} = [r_{ij}]_{n \times m}$ and number of desired candidates $k \leq m$
2: **Output:** a subset of k candidates
3: Let $\mathbf{y}_{best}$ be a randomly generated subset of size k of candidates;
4: Set $best\_solution\_value = v(\mathbf{y}_{best})$;
5: **while** TRUE **do**
6:     **for** every pair (i, j) such that $i \in \mathbf{y}_{best} \wedge j \in \mathcal{C} \setminus \mathbf{y}_{best}$ **do**
7:         Set $\mathbf{y}' = \mathbf{y}_{best} \setminus \{i\} \cup \{j\}$;
8:         **if** $v(\mathbf{y}') < best\_solution\_value$ **then**
9:             $\mathbf{y}_{best} = \mathbf{y}'$
10:            $best\_solution\_value = v(\mathbf{y}')$
11:            BREAK;
12:        **end if**
13:    **end for**
14:    **if** every possible pair (i, j) is checked at step 6, and there was no improvement **then**
15:        BREAK;
16:    **end if**
17: **end while**
18: **return** $\mathbf{y}_{best}$;

$$A(\mathcal{V}, \mathcal{C}, \mathbf{r}) = \frac{\sum_{(\mathbf{x}, \mathbf{y}) \in \mathcal{F}} f(\mathbf{x}, \mathbf{y})}{|\mathcal{F}|}.$$

Theorem 1. *The average objective function value $A(\mathcal{V}, \mathcal{C}, \mathbf{r})$ for the WDPCC is equal to* $\frac{\sum_{i \in \mathcal{V}} \sum_{j \in \mathcal{C}} r_{ij}}{m} = \frac{n(m-1)}{2}$.

Proof

$$Z := \sum_{i \in \mathcal{V}} \sum_{j \in \mathcal{C}} r_{ij} x_{ij}$$

$$\mathbb{E}[Z] = \mathbb{E}\left[\sum_{i \in \mathcal{V}} \sum_{j \in \mathcal{C}} r_{ij} x_{ij}\right] = \sum_{i \in \mathcal{V}} \sum_{j \in \mathcal{C}} r_{ij} \mathbb{E}[x_{ij}]$$

$$\mathbb{E}[x_{ij}] = \Pr[x_{ij} = 1] = \Pr[y_j = 1 \text{ and } i \text{ is assigned to } j] = \Pr[y_j = 1] \times \Pr[x_{ij} = 1 | y_j = 1]$$

$\Pr[y_j = 1]$ is given by

$$\Pr[y_j = 1] = \frac{\binom{m-1}{k-1}}{\binom{m}{k}} = \frac{k}{m}$$

and the conditional probability $Pr[x_{ij} = 1 | y_j = 1]$ is given by

$$Pr[x_{ij} = 1 | y_j = 1] = \frac{1}{k}.$$

Thus, $\Pr[y_j = 1$ and voter i is assigned to $j]$ is given by

$$\Pr[y_j = 1 \text{ and voter } i \text{ is assigned to } j] = \frac{k}{m} \cdot \frac{1}{k} = \frac{1}{m}$$

The average objective function value is given by $A(\mathcal{V}, \mathcal{C}, \mathbf{r}) = \frac{\sum_{i \in \mathcal{V}} \sum_{j \in \mathcal{C}} r_{ij}}{m} = \frac{nm(m-1)}{2m} = \frac{n(m-1)}{2}$, since $\{r_{ij} : j = 1, 2, \ldots, m\} = \{0, 1, \ldots, m-1\}$ for every $i \in \mathcal{V}$. $\qquad\square$

3.2   Computing the Domination Number

In this section we calculate the number of feasible solutions with objective function value no better than the average objective function value $A(\mathcal{V}, \mathcal{C}, \mathbf{r})$. We make use of the following result.

Theorem 2 (Baranyai's theorem [21]**).** *If k divides m, then the complete k-uniform hypergraph on m vertices decomposes into 1-factors, where a 1-factor is a set of m/k pairwise disjoint k-sets.*

Using set theory terminology, Baranyai's theorem can be stated as follows. Let S be a set and k a positive integer that divides $|S|$. Then the set of all k-subsets of S can be split into groups, such that each group is a partition of S.

Lemma 1. *For the WDPCC where k divides m, there are at least $\binom{m-1}{k-1} k^{n-1}$ feasible solutions with objective function value greater than or equal to the average objective function value $A(\mathcal{V}, \mathcal{C}, \mathbf{r})$.*

Proof. Recall that for every feasible solution $(\mathbf{x}, \mathbf{y}) \in \mathcal{F}$, $\mathbf{y}$ determines a subset of k candidates, and $\mathbf{x}$ assigns a candidate determined by $\mathbf{y}$ to each of n voters. In the rest of this proof, we will identify every such vector $\mathbf{y}$ with the subsets of candidates it determines.

Let $\{\mathcal{Y}_1, \mathcal{Y}_2, \ldots, \mathcal{Y}_{\binom{m-1}{k-1}}\}$ be a partition of all k-subsets of candidates, such that each $\mathcal{Y}_i = \{\mathbf{y}_1^i, \mathbf{y}_2^i, \ldots, \mathbf{y}_{\frac{m}{k}}^i\}$ is a partition of candidates $\mathcal{C}$, $i = 1, 2, \ldots, \binom{m-1}{k-1}$. Since k divides m, from Theorem 2 it follows that such a partition of all k-subsets exists. For example, in the case when $m = 4$ and $k = 2$, such a partition is given by

$$\Big\{ \{\{1,2\}, \{3,4\}\}, \ \{\{1,3\}, \{2,4\}\}, \ \{\{1,4\}, \{2,3\}\} \Big\}.$$

Note that a partition $\{\mathcal{Y}_1, \mathcal{Y}_2, \ldots, \mathcal{Y}_{\binom{m-1}{k-1}}\}$ also induces a partition $\{\mathcal{F}_1, \mathcal{F}_2, \ldots, \mathcal{F}_{\binom{m-1}{k-1}}\}$ of $\mathcal{F}$, where $\mathcal{F}_i = \{(\mathbf{x}, \mathbf{y}) \in \mathcal{F} : \mathbf{y} \in \mathcal{Y}_i\}$, $i = 1, 2, \ldots, \binom{m-1}{k-1}$.

Now we fix some i, and show how $\mathcal{F}_i$ can be partitioned into small classes of feasible solutions so that the average objective function value in each such class is equal to $A(\mathcal{V}, \mathcal{C}, \mathbf{r})$.

Before we formally present our construction, we illustrate it with an example: let $n = 2$, $m = 6$, $k = 3$ and $\mathcal{Y}_i = \{\mathbf{y}_1^i = \{1, 2, 3\}, \mathbf{y}_2^i = \{4, 5, 6\}\}$. Now consider some solution s^1 from $\mathcal{F}_i$. Let $s^1 = (\mathbf{x}^1, \mathbf{y})$ be such that $\mathbf{y} = \mathbf{y}_1^i$ and $x_{11}^1 = x_{22}^1 = 1$. We create two additional solutions $s^2 = (\mathbf{x}^2, \mathbf{y}_1^i)$ and $s^3 = (\mathbf{x}^3, \mathbf{y}_1^i)$ from s^1, by circularly shifting assignments for each voter to the right by one and two candidates, respectively. That is, $x_{12}^2 = x_{23}^2 = 1$ and $x_{13}^3 = x_{21}^3 = 1$. Next we create three other solutions $s^4 = (\mathbf{x}^4, \mathbf{y}_2^i)$, $s^5 = (\mathbf{x}^5, \mathbf{y}_2^i)$, $s^6 = (\mathbf{x}^6, \mathbf{y}_2^i)$ for which the set of candidates will be $\mathbf{y}_2^i$. Furthermore, s^4, s^5, and s^6, will be straightforward copies of s^1, s^2 and s^3, obtained by identifying candidates 1 with 4, 2 with 5, and 3 with 6. That is, $x_{14}^4 = x_{25}^4 = 1$, $x_{15}^5 = x_{26}^5 = 1$ and $x_{16}^6 = x_{24}^6 = 1$. Solutions $s^1, s^2, \ldots, s^6$, are depicted in Fig. 1. Note that for every $s \in \mathcal{V}$, $t \in \mathcal{C}$, there is exactly one solution $s^j \in \{s^1, \ldots, s^6\}$ for which $x_{st}^j = 1$. Hence $\sum_{j=1}^6 f(s^j) = \sum_{s \in \mathcal{V}} \sum_{t \in \mathcal{C}} r_{st}$. Moreover, the average objective function value among $s^1, \ldots, s^6$, is $\sum_{j=1}^6 f(s^j) = \sum_{s \in \mathcal{V}} \sum_{t \in \mathcal{C}} r_{st}/6 = A(\mathcal{V}, \mathcal{C}, \mathbf{r})$. Hence at least one of them is no better than the average. Note that a class of solutions with this property would be obtained if we started with any other starting solution s^1.

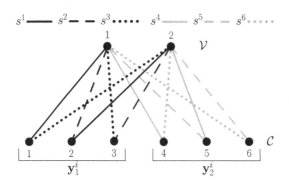

Fig. 1. Example of an equivalence class defined by $\sim_i$, for $\mathcal{Y}_i = \{\{1, 2, 3\}, \{4, 5, 6\}\}$ and π_i defined by: $\pi_i(j) = j$ for $j \le 3$, $\pi_i(j) = j - 3$ for $j > 3$.

Now we formally present the construction described above. For a fixed i, choose some function $\pi_i \colon \mathcal{C} \to \{1, 2, \ldots, k\}$ such that $\{\pi_i(c) \colon c \in \mathbf{y}_j^i\} = \{1, 2, \ldots, k\}$ for all $j = 1, 2, \ldots, \frac{m}{k}$. In other words, function π_i gives some arbitrary ordering of elements in $\mathbf{y}_j^i$, for each $j = 1, 2, \ldots, \frac{m}{k}$. (In the example of Fig. 1, π_i was given by $\pi_i(j) = j$ if $j \le 3$, $\pi_i(j) = j - 3$ otherwise.) Now we define a relation $\sim_i$ on the elements of $\mathcal{F}_i$ as follows: $(\mathbf{x}', \mathbf{y}_r^i) \sim_i (\mathbf{x}'', \mathbf{y}_s^i)$ if there exists "shift" $a \in \{0, 1, \ldots, k - 1\}$ such that for all $j = 1, 2, \ldots, n$, $x_{jp}' = 1$ and $x_{jq}'' = 1$ imply that $\pi_i(p) \equiv \pi_i(q) + a \pmod{k}$. Note that $\sim_i$ is an equivalence relation on F_i. Moreover, if S is an equivalence class defined by $\sim_i$, then it has exactly m elements, and $\sum_{(\mathbf{x}, \mathbf{y}) \in S} f(\mathbf{x}, \mathbf{y}) = \sum_{i \in \mathcal{V}} \sum_{j \in \mathcal{C}} r_{ij} = mA(\mathcal{V}, \mathcal{C}, \mathbf{r})$. Hence, there is at least one element of S with an objective function value no better than $A(\mathcal{V}, \mathcal{C}, \mathbf{r})$. Observe that the equivalence classes defined by $\sim_i$ for all $i = 1, 2, \ldots, \binom{m-1}{k-1}$, provides us a partition of the set $\mathcal{F}$ of all feasible solutions into sets of size m, each of which contains at least one solution with an objective function value no better than the average. Hence, the total number of solutions with objective function values no better than the average is at least

$$\frac{|\mathcal{F}|}{m} = \frac{1}{m}\binom{m}{k}k^n = \frac{1}{m}\frac{m}{k}\binom{m-1}{k-1}k^n = \binom{m-1}{k-1}k^{n-1},$$

which proves the lemma.   □

Now we use Lemma 1 to get an analogous result for general k, i.e., even when k does not divide m.

Theorem 3. *For the WDPCC, there are at least $\binom{k\lfloor\frac{m}{k}\rfloor-1}{k-1}k^{n-1}$ feasible solutions with an objective function value greater than or equal to the average $A(\mathcal{V},\mathcal{C},\mathbf{r})$.*

Proof. Let I be an instance of the WDPCC where k does not divide m. Let I' be the instance obtained from I by removing $t = (m \mod k)$ candidates j with the least corresponding sum of costs $\sum_{i\in\mathcal{V}} r_{ij}$. Without loss of generality, we assume that $m - t + 1, m - t + 2, \ldots, m$ are the candidates that are removed. Let A_1 and A_2 denote the average objective function values of the solutions for the instance I and I', respectively. Then it follows that

$$A_1 = \frac{\sum_{i\in\mathcal{V}}\sum_{j\in\mathcal{C}} r_{ij}}{m} \leq \frac{\sum_{i\in\mathcal{V}}\sum_{j=1,\ldots,m-t} r_{ij}}{m-t} = A_2. \tag{7}$$

Note that every feasible solution $(\mathbf{x},\mathbf{y})$ for the instance I' can be transformed to a different feasible solution for the instance I. Namely, just extend $(\mathbf{x},\mathbf{y})$ by setting $x_{ij} = 0$ and $y_j = 0$, for all $j = m - t + 1, m - t + 2, \ldots, m$. From Lemma 1, it follows that for the instance I', there are at least $\binom{k\lfloor\frac{m}{k}\rfloor-1}{k-1}k^{n-1}$ feasible solutions with an objective function value greater than or equal to the average A_2. Then from (7), it follows that there are at least $\binom{k\lfloor\frac{m}{k}\rfloor-1}{k-1}k^{n-1}$ feasible solutions corresponding to the instance I with an objective function value no better than the average A_1, which proves the theorem.   □

Note that Lemma 1 and Theorem 3 hold true also for the more general version of the problem, when the values of $\mathbf{r}$ can be arbitrary, not only when $\{r_{ij}: j = 1, 2, \ldots, m\} = \{0, 1, \ldots, m - 1\}$ for every $i \in \mathcal{V}$.

3.3   Domination Analysis of 2-Opt

Theorem 4. *If a k-subset $\mathbf{y} = \{j_1, j_2, \ldots j_k\}$ is locally optimal for 2-Opt, then, $v(\mathbf{y}) \leq A(\mathcal{V}, \mathcal{C}, \mathbf{r})$.*

Proof. Let $\mathbf{y}$ be the output of the 2-Opt, and let $\mathbf{x}$ be the corresponding optimal assignment of voters to $\mathbf{y}$, i.e., $v(\mathbf{y}) = f(\mathbf{x}, \mathbf{y})$. Then the following inequality is true for any $\mathcal{V}_j$ ($\mathcal{V}_j$ is the set of voters assigned to the candidate j at the locally optimal solution $(\mathbf{x}, \mathbf{y})$).

$$\sum_{i \in \mathcal{V}_{j_1}} r_{ij_1} \le \sum_{i \in \mathcal{V}_{j_1}} r_{ij_1}$$

$$\sum_{i \in \mathcal{V}_{j_1}} r_{ij_1} \le \sum_{i \in \mathcal{V}_{j_1}} r_{ij_2}$$

$$\sum_{i \in \mathcal{V}_{j_1}} r_{ij_1} \le \sum_{i \in \mathcal{V}_{j_1}} r_{ij_3}$$

$$\vdots$$

$$\sum_{i \in \mathcal{V}_{j_1}} r_{ij_1} \le \sum_{i \in \mathcal{V}_{j_1}} r_{ij_m}$$

Summing the left and right hand side of the above inequalities, we get:

$$m \sum_{i \in \mathcal{V}_{j_1}} r_{ij_1} \le \sum_{p=1}^{m} \sum_{i \in \mathcal{V}_{j_1}} r_{ij_p}$$

Since the above is true for each $j_q \in \mathbf{y}$, the inequalities below follow:

$$m \sum_{i \in \mathcal{V}_{j_1}} r_{ij_1} \le \sum_{p=1}^{m} \sum_{i \in \mathcal{V}_{j_1}} r_{ij_p}$$

$$m \sum_{i \in \mathcal{V}_{j_2}} r_{ij_2} \le \sum_{p=1}^{m} \sum_{i \in \mathcal{V}_{j_2}} r_{ij_p}$$

$$m \sum_{i \in \mathcal{V}_{j_3}} r_{ij_3} \le \sum_{p=1}^{m} \sum_{i \in \mathcal{V}_{j_3}} r_{ij_p}$$

$$\vdots$$

$$m \sum_{i \in \mathcal{V}_{j_k}} r_{ij_k} \le \sum_{p=1}^{m} \sum_{i \in \mathcal{V}_{j_k}} r_{ij_p}$$

Summing the m inequalities above, we get:

$$m \left(\sum_{i \in \mathcal{V}_{j_1}} r_{ij_1} + \sum_{i \in \mathcal{V}_{j_2}} r_{ij_2} + \cdots + \sum_{i \in \mathcal{V}_{j_k}} r_{ij_k} \right) \le \sum_{j=1}^{m} \sum_{i=1}^{n} r_{ij}$$

Noting that the left hand side is m times the $f(\mathbf{x}, \mathbf{y})$, we get:

$$m \cdot f(\mathbf{x}, \mathbf{y}) = m \cdot v(\mathbf{y}) \le \sum_{j=1}^{m} \sum_{i=1}^{n} r_{ij}$$

By dividing both sides by m, we get:

$$v(\mathbf{y}) \le \frac{\sum_{j=1}^{m} \sum_{i=1}^{n} r_{ij}}{m} = A(\mathcal{V}, \mathcal{C}, \mathbf{r})$$

$\square$

The next theorem shows that the 2-Opt heuristic can obtain a solution with objective function value no worse than the average $A(\mathcal{V}, \mathcal{C}, \mathbf{r})$ in polynomial time.

Theorem 5. *For the winner determination problem under the Chamberlin-Courant system, it takes at most $\frac{n(m-1)}{2}$ iterations for the 2-Opt heuristic to reach a solution with the objective function value no worse than the average $A(\mathcal{V}, \mathcal{C}, \mathbf{r})$.*

Proof. The worst possible starting solution is when all the voters are assigned to their worst candidates. So the objective function value of the worst solution is $n(m-1)$. On the other hand, $A(\mathcal{V}, \mathcal{C}, \mathbf{r}) = \frac{n(m-1)}{2}$. Since by Theorem 4 we know that 2-Opt reaches $A(\mathcal{V}, \mathcal{C}, \mathbf{r})$, and in each iteration the objective function value decreases at least by 1, the maximal number of iterations required to reach $A(\mathcal{V}, \mathcal{C}, \mathbf{r})$ is given by

$$n(m-1) - \frac{n(m-1)}{2} = \frac{n(m-1)}{2}.$$

$\square$

4   Conclusion

In this paper, we compute the average solution value and the domination number for the winner determination problem under the Chamberlin-Courant system. We show that the 2-Opt heuristic can obtain a solution no worse than the average in polynomial time, and hence achieves the domination ratio of $\frac{1}{|\mathcal{C}|}$. It would be interesting to consider other heuristics such as 3-Opt and obtain analogous results.

Acknowledgements. We wish to thank Abraham Punnen, Binay Bhattacharya, Kamyar Khodamoradi, and Vladyslav Sokol, for helpful discussions on this problem. The authors acknowledge support from a Natural Sciences and Engineering Research Council of Canada (NSERC) Discovery Grant.

References

1. Brams, S.J.: Mathematics and Democracy - Designing Better Voting and Fair-division Procedures. Princeton University Press, Princeton (2007)
2. Brams, S.J., Fishburn, P.C.: Proportional representation in variable-size legislatures. Soc. Choice Welf. **1**, 211–229 (1984)
3. Brams, S.J., Fishburn, P.C.: Some logical defects of the single transferable vote. In: Choosing an Electoral System: Issues and Alternatives (1984)
4. Monroe, B.L.: Fully proportional representation. Am. Polit. Sci. Rev. **89**(4), 925–940 (1995)
5. Chamberlin, J.R., Courant, P.N.: Representative deliberations and representative decisions: proportional representation and the Borda rule. Am. Polit. Sci. Rev. **77**(3), 718–733 (1983)

6. Procaccia, A.D.: Computational voting theory: of the agents, by the agents, for the agents. Ph.D. thesis, Hebrew University (2008)
7. Betzler, N., Slinko, A., Uhlmann, J.: On the computation of fully proportional representation. J. Artif. Intell. Res. (JAIR) **47**, 475–519 (2013)
8. Skowron, P., Yu, L., Faliszewski, P., Elkind, E.: The complexity of fully proportional representation for single-crossing electorates. In: Vöcking, B. (ed.) SAGT 2013. LNCS, vol. 8146, pp. 1–12. Springer, Heidelberg (2013). doi:10.1007/978-3-642-41392-6_1
9. Yu, L., Chan, H., Elkind, E.: Multiwinner elections under preferences that are single-peaked on a tree. In: Proceedings of the Twenty-Third International Joint Conference on Artificial Intelligence (2013)
10. Clearwater, A., Puppe, C., Slinko, A.: The single-crossing property on a tree. CoRR abs/1410.2272 (2014)
11. Skowron, P., Faliszewski, P., Slinko, A.: Fully proportional representation as resource allocation: approximability results. In: Proceedings of the Twenty-Third International Joint Conference on Artificial Intelligence (2013)
12. Skowron, P., Faliszewski, P., Slinko, A.: Achieving fully proportional representation is easy in practice. In: Proceedings of the 2013 International Conference on Autonomous Agents and Multi-agent Systems (2013)
13. Vazirani, V.V.: Approximation Algorithms. Springer, New York (2001)
14. Alon, N., Gutin, G., Krivelevich, M.: Algorithms with large domination ratio. J. Algorithms **50**(1), 118–131 (2004)
15. Glover, F., Punnen, A.P.: The travelling salesman problem: new solvable cases and linkages with the development of approximation algorithms. J. Oper. Res. Soc. **48**(5), 502–510 (1997)
16. Zemel, E.: Measuring the quality of approximate solutions to zero-one programming problems. Math. Oper. Res. **6**(3), 319–332 (1981)
17. Punnen, A.P., Sripratak, P., Karapetyan, D.: Average value of solutions for the bipartite Boolean quadratic programs and rounding algorithms. Theoret. Comput. Sci. **565**(C), 77–89 (2015)
18. Punnen, A.P., Kabadi, S.: Domination analysis of some heuristics for the traveling salesman problem. Discret. Appl. Math. **119**(1–2), 117–128 (2002)
19. Angel, E., Zissimopoulos, V.: On the quality of local search for the quadratic assignment problem. Discret. Appl. Math. **82**(1–3), 15–25 (1998)
20. Brams, S., Potthoff, R.F.: Proportional representation: broadening the options. Working papers, C.V. Starr Center for Applied Economics, New York University (1997)
21. Baranyai, Z.: On the factorization of the complete uniform hypergraph. In: Infinite and Finite Sets: Proceedings of a Colloquium, Keszthely, June 25–July 1 1973, vol. 1, pp. 91–108 (1975). Dedicated to Paul Erdos on his 60th Birthday

On Structural Parameterizations of Graph Motif and Chromatic Number

Bireswar Das, Murali Krishna Enduri$^{(\boxtimes)}$, Neeldhara Misra, and I. Vinod Reddy

IIT Gandhinagar, Gandhinagar, India
{bireswar,endurimuralikrishna,neeldhara.m,reddy_vinod}@iitgn.ac.in

Abstract. Structural parameterizations for hard problems have proven to be a promising venture for discovering scenarios where the problem is tractable. In particular, when a problem is already known to be polynomially solvable for some class of inputs, then it is natural to parameterize by the distance of a general instance to a tractable class. In the context of graph algorithms, parameters like vertex cover, twin cover, treewidth and treedepth modulators, and distances to various special graph classes are increasingly popular choices for hard problems.

Our main focus in this work is the GRAPH MOTIF problem, which involves finding a connected induced subgraph in a vertex colored graph that respects a given palette. This problem is known to be hard in the traditional setting even for fairly structured classes of graphs. In particular, Graph Motif is known to be W[1]-hard when parameterized by the distance to split graphs, and para-NP-hard when parameterized by the distance to co-graphs (which are the class of P_4-free graphs). On the other hand, it is known to be FPT when parameterized by the distance to a clique or the distance to an independent set (or equivalently, vertex cover). Towards finding the boundary of tractability, we consider parameterizing the problem by the distance to threshold graphs, which are graphs that are both split and P_4-free. Note that this is a natural choice of an intermediate parameter in that it is larger than the parameters for which the problem is hard and smaller than the ones for which the problem is tractable. Our main contribution is an FPT algorithm for the Graph Motif problem when parameterized by the distance to threshold graphs. We also address some related structural parameterizations for CHROMATIC NUMBER. Here, we show that the problem admits a polynomial kernel when parameterized by the vertex deletion distance to a clique.

1 Introduction and Motivation

The GRAPH MOTIF problem asks if, given a vertex-colored graph G and a multiset of colors M, there exists a subset S of vertices of G such that the graph induced by G on S is connected and contains every color in the set M according to the multiplicities given by M. The problem is motivated by applications

---

M.K. Enduri–Supported by Tata Consultancy Services (TCS) research fellowship.

D. Gaur and N.S. Narayanaswamy (Eds.): CALDAM 2017, LNCS 10156, pp. 118–129, 2017.
DOI: 10.1007/978-3-319-53007-9_11

in computational biology [2,17] and is also well-studied from the theoretical point of view. Because of its fundamental importance, the problem has received substantial attention from the perspective of approximation and parameterized algorithms (we refer the reader to Sect. 2 for the relevant definitions).

The use of structural parameters is an increasingly popular approach for solving hard problems, as is witnessed by several positive results in the literature of fixed-parameter algorithms [16]. One motivation for considering a structural parameter is its relevance in practice (such as the use of nesting depth for the type-checking problem in ML). However, not all parameters that are convenient in practice turn out to be tractable for the problem under consideration. For instance, although it is well-known that graphs of small treewidth are frequently encountered in datasets that arise in computational biology, the GRAPH MOTIF problem turns out to be NP-hard even on subclasses of trees (which have constant treewidth), making it intractable for treewidth. This motivates the need for considering related, but potentially "more relaxed" parameters for the problem being studied.

One consideration for identifying such parameters is to consider situations for which the problem is already polynomially solvable. Then, it is natural to parameterize by the distance of a general instance to a tractable class. For a class of graphs $\mathcal{F}$ and an arbitrary graph G, a $\mathcal{F}$-modulator is the smallest possible vertex subset W for which $G \setminus W \in \mathcal{F}$. The size of a $\mathcal{F}$-modulator is a natural measure of closeness to $\mathcal{F}$. Parameters like vertex cover, twin cover, bounded treewidth and bounded treedepth modulators, and modulators to other structured classes are increasingly popular choices for hard problems.

Our Contributions. The GRAPH MOTIF problem is known to be hard in the traditional setting even for fairly structured classes of graphs. In particular, it is easy to infer from known results that GRAPH MOTIF is para-NP-hard when parameterized by the distance to split graphs or co-graphs (which are the class of P_4-free graphs). On the other hand, it is known to be FPT when parameterized by the distance to a clique [4] or the distance to an independent set (or equivalently, vertex cover) [15]. However, these parameters can be significantly large, and in general, it is desirable to find tractable results for parameters that are as small as possible. The work of Ganian [15], for instance, addresses this issue by considering the parameters twin cover and neighborhood diversity, both of which are, in general, smaller than vertex cover. Continuing this line of work, but in a different direction, we consider parameterizing the problem by the distance to threshold graphs.

Threshold graphs were introduced by Chvátal and Hammer in 1977 [6], originally defined as graphs that can be constructed from the one-vertex graph by repeatedly adding either an isolated vertex or a universal vertex. It turns out that threshold graphs are precisely those graphs which are both split and P_4-free. Note that the distance to threshold graphs is a natural choice of an intermediate parameter because it is larger than the parameters for which the problem is hard (as every threshold graph modulator is also a split or co-graph modulator), and smaller than the ones for which the problem is tractable (see Fig. 1) (note that every clique modulator is also a threshold graph modulator). To begin with, we show that GRAPH MOTIF is polynomially solvable on the class of threshold graphs (in fact, more generally, we show that they are polynomially solvable on co-graphs). However, co-graphs evidently do not have enough structure in the context of the corresponding modulator parameter: the GRAPH MOTIF problem turns out to be NP-hard even on instances that are one vertex away from being a co-graph. However, it turns out that the situation is more positive for threshold modulators: our main contribution here is the following FPT algorithm.

Theorem 1. *The* GRAPH MOTIF *problem is* FPT *when parameterized by the size of a modulator to threshold graphs.*

We also consider the problem of CHROMATIC NUMBER parameterized by vertex deletion distance to a clique, for which CHROMATIC NUMBER is known to be FPT. We show here that the problem also admits a polynomial kernel.

Theorem 2. *The* CHROMATIC NUMBER *problem admits a polynomial kernel when parameterized by the size of a modulator to a clique.*

Our motivation for studying CHROMATIC NUMBER parameterized by the size of a modulator to a clique is that the problem is not expected to have a polynomial kernel when parameterized by vertex cover (which is a modulator to an edgeless graph) [3]. We note that the CHROMATIC NUMBER problem is FPT for the threshold graph modulator parameter. This follows from Theorem 26 [19], which shows that the problem is FPT when parameterized by sm-width, which is a measure that is bounded for threshold graphs (and therefore also for graphs with a bounded deletion set into the class of threshold graphs). Observe that we do not expect a polynomial kernel for CHROMATIC NUMBER when parameterized by the size of a threshold graph modulator, since it is already hard with respect to vertex cover.

Related Work. The GRAPH MOTIF problem was introduced by Lacroix et al. [17], and it has many application in bioinformatics, social networks [2]. As indicated above, is intractable for many restricted classes of graphs, for example it is NP-hard on bipartite graphs with maximum degree four for motifs with two colors [10], rooted trees of height two [1], trees of maximum degree three for colorful motifs [10]. The problem is also NP-hard for split graphs [15] and for graphs of constant rank-width and clique-width. The problem has been studied intensively from the parameterized lens. Ganian showed that this problem is in FPT for various graph structural parameters including vertex cover, twin cover [15], neighborhood diversity [14], and the size of a clique modulator [4]. It is W[1]-hard when parameterized by number of colors in the motif [10].

Methodology. We give a brief overview of the main ideas in our algorithm. As is standard for structural parameterizations, we assume that we are given a threshold modulator X (this is without loss of generality, since the computation of such a modulator is in FPT). We begin by guessing the intersection of a solution with X (call this part Y), and we are now left with a Steiner-like problem, where we would like to connect the components of Y using $|M| - |Y|$ vertices from $G \setminus X$. The additional complication on top of the Steiner-like connectivity requirement is that the vertices in $M \setminus Y$ must respect what remains of the palette M having accounted for Y. It is useful to think of the part of the solution that lies in $G \setminus X$ as comprising of two parts: one that induces a threshold graph, and the other induces an independent set, and within the solution, these vertices have neighbors only in Y.

To see the nature of the problem beyond the connectivity requirement, consider the special case when Y happens to be already connected. In this situation, it turns out that we can guess the neighbor of Y in $G \setminus X$ and proceed to expand from there. We will use an appropriate universal vertex that is guaranteed to exist in the part of the solution from $G \setminus X$ that induces a threshold graph. All other vertices must be neighbors of Y that belong to the independent part of $G \setminus X$, and these vertices can be identified greedily at the very end to take care of the "left over colors" in the palette.

More generally, it turns out that the threshold structure of $G \setminus X$ can be exploited to determine if Y can be extended to a complete motif solution. We do this by classifying vertices according to their neighborhoods in Y, but because there are different ways in which the solution in $G \setminus X$ interacts with Y, this classification is a slightly intricate process and needs to be done in two phases. We defer the details of this part to Sect. 3, which is dedicated to a detailed description of this algorithm.

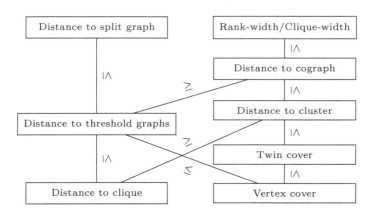

Fig. 1. Relationship between the graph parameters; $P \leqslant Q$ means there exists a function f such that for all graphs G, we have $P(G) \leqslant f(Q(G))$

2   Preliminaries

The graphs we consider in this paper are undirected and simple. For a graph $G = (V, E)$, let $V(G)$ and $E(G)$ denote the vertex and edge set of G respectively. An edge $\{x, y\} \in E(G)$ is denoted as xy for simplicity. For a subset $X \subseteq V(G)$, the graph G[X] denotes the subgraph of G induced by vertices of X. Also, we abuse notation and use $G \setminus X$ to refer to the graph obtained from G after removing the vertex set X.

For a vertex $v \in V(G)$, $N(v)$ denotes the set of vertices adjacent to v and $N[v] = N(v) \cup \{v\}$ is the closed neighborhood of v. A vertex is called *universal vertex* if it is adjacent to every other vertex of the graph. A vertex v is said to *dominate* a vertex u if $N[v] \supseteq N(u)$. For a subset of vertices $S \subseteq V$, we use $N(S)$ to denote the union of the neighborhoods of all the vertices in S.

Special Graph Classes. We now define some of the special graph classes considered in this paper. A graph is a *split graph* if its vertex set can be partitioned into a clique and an independent set. Split graphs do not contain C_4, C_5 or $2K_2$ as induced subgraphs [11]. *Co-graphs* are P_4-free graphs, that is, they do not contain any induced paths on four vertices. A graph is a *threshold graph* if it can be constructed recursively by adding an isolated vertex or a universal vertex. It is known that this class can also be characterized as the intersection of split graphs and co-graphs. Notationally, we will typically write $G = (C, I)$ when referring to a threshold graph (or a split graph), where (C, I) denotes the partition of G into a clique and an independent set, respectively. The following property of threshold graphs is easily checked.

Proposition 1. *Every connected induced subgraph of a threshold graph is also a connected threshold graph.*

It is also easy to check that every connected threshold graph has a universal vertex. The problem of checking if a graph is a threshold graph admits a linear time algorithm [20]. We have the following characterization of threshold graphs:

Theorem 3 [18]. *For a graph $G = (V, E)$, the following statements are equivalent:*

1. *G is a threshold graph.*
2. *G is a $(P_4, C_4, 2K_2)$-free graph*
3. *For any $x, y \in V(G)$ either $N(x) \subseteq N[y]$ or $N(y) \subseteq N[x]$.*

Parameterized Complexity. A parameterized problem denoted as $(I, k) \subseteq \Sigma^* \times \mathbb{N}$, where Σ is fixed alphabet and k is called the parameter. We say that the problem (I, k) is *fixed parameter tractable* with respect to parameter k if there exists an algorithm which solves the problem in time $f(k)|I|^{O(1)}$, where f is a computable function. A kernel for a parameterized problem P is an algorithm which transforms an instance (I, k) of P to an equivalent instance (I', k') in polynomial time such that $k' \leqslant k$ and $|I'| \leqslant f(k)$ for some computable function f. For a detailed survey of the methods used in parameterized complexity, we refer the reader to the texts [8,9].

3   An **FPT** Algorithm for GRAPH MOTIF

In this section, given a graph G and a fixed finite set of colors Q, use col to denote a mapping from the vertices of G to Q. For a subset X of V (or a subgraph H of G), we use col(X) (respectively, col(H)) to denote the multiset of colors that appear on the vertices of X (V(H)). We are now ready to define the GRAPH MOTIF problem:

GRAPH MOTIF
 Input: A vertex-colored graph G, and a multiset M of colors.
 Question: Does there exists a connected induced subgraph H of G
 (called a *motif*) such that the multiset of colors col(H) occur-
 ring in H is identical to M?

First we show that GRAPH MOTIF is solvable in polynomial time for co-graphs using modular decomposition [13]. Due to space restriction the proof is presented in the full version of this paper. Next we give a more direct polynomial time algorithm for the class of threshold graphs. The latter will provide a useful warmup for the next part, where we describe our FPT algorithm for the problem when parameterized by a modulator to threshold graphs.

3.1   GRAPH MOTIF for Threshold Graphs

In this Section we consider the case of threshold graphs, where we have a more direct approach that does not require the modular decomposition. The ideas here will be used in a more refined manner in the next section. Observe that any motif in a threshold graph must have a non-trivial intersection with the clique (unless it is a motif of size one, which is a case we can ignore without loss of generality). Let us call a vertex *plausible* if the color of the vertex belongs to M. Note that any motif only contains plausible vertices. For a threshold graph, it turns out that there always exists a solution which has a plausible vertex which is also universal (in the solution). The algorithm, therefore, can simply find a plausible vertex that dominates all the other plausible vertices in the clique, and then we can extend the solution arbitrarily to any subset of vertices that respects the remaining palette. Note that this works since the first vertex we pick in the solution is guaranteed to account for the connectivity requirements. We now make this approach formal.

We begin by defining the notion of an *anchor* vertex, which is a plausible vertex that dominates all the other plausible vertices.

Definition 1. *Given a graph motif instance* (G = (C, I), M) *where G is a threshold graph and M is a multiset of colors, we call a vertex* $v \in C$ *an* anchor vertex *if* $col(v) \in M$ *and* $N(u) \subseteq N[v]$ *for all* $u \in C$ *for which* $col(u) \in M$.

We now show, by a standard shifting argument, that if we have a YES instance of GRAPH MOTIF, then there exists a solution that contains an anchor vertex.

Lemma 1. *Let* (G, M) *be a* YES *instance of* GRAPH MOTIF, *where* G *is a threshold graph. Then* G *admits an anchor vertex* v. *Further, if* H *is a solution to* (G, M) *then there is a solution* H' *containing* v.

Proof. Since (G, M) is a YES-instance, it admits a motif H such that $col(H) = M$. If H contains a vertex that dominates all the plausible clique vertices, then we are done. If not, then there exists a plausible vertex w in C that is not dominated by any of the vertices in H. By property (3) in Theorem 3, we know that w dominates all the vertices in H. Since w is plausible, there is a vertex $v \in H$ for which $col(u) = col(v)$. Consider the motif H' given by $H \setminus \{v\} \cup \{w\}$. It is easily checked that H' is also a motif, since $N[w] \supseteq N(v)$. Repeating this argument until the new motif contains an anchor vertex gives us the desired claim. This procedure also shows that there is always a solution that contains an anchor vertex. □

We now turn to the algorithm, which essentially relies on identifying an anchor vertex.

Lemma 2. *The* GRAPH MOTIF *problem can be solved in linear time on threshold graphs.*

Proof. We construct motif H as follows. Find an anchor vertex v which is the vertex of highest degree that is plausible and then add v to H. Then we add a vertex from neighborhood of v to H if its color is in $M \setminus col(H)$. If, at the end of this procedure, $col(H) = M$ then H is the required motif, otherwise the algorithm returns NO. The correctness follows from Lemma 1. □

3.2 Parameterizing by the Size of a Threshold Graph Modulator

Let H be a motif for the instance (G, X, M) where $G \setminus X = (C, I)$ is a threshold graph. We use k to denote the size of the modulator, that is, $|X|$. This section is devoted to a proof of our main result, which we recall here:

Theorem 1. *The* GRAPH MOTIF *problem is* FPT *when parameterized by the size of a modulator to threshold graphs.*

We begin with some easy observations. If a motif H lies completely in the threshold graph then it can be found using Lemma 2. Otherwise, the motif H intersects X. We can guess this intersection. In particular, we try all possible subsets Y of X and try to find a motif H which intersects X exactly at Y. If $V(H) \subseteq Y \cup (N(Y) \cap I)$ then the problem reduces to solving graph motif parameterized by vertex cover [15]. Our algorithm treats this case separately as a preprocessing step, and without loss of generality we assume that $V(H) \not\subseteq Y \cup (N(Y) \cap I)$ for the rest of this section.

To develop our intuition for the algorithm, we first discuss the structure of a solution for a YES-instance of the problem. Let H be a motif for the instance (G, X, M) (where $G \setminus X = (C, I)$ is a threshold graph), and let Y denote $V(H) \cap X$. First, notice that $H \cap (G \setminus X)$ can be partitioned into H_T and H_R, where $G[H_T]$

is a connected threshold graph and $G[H_R]$ is an independent set. To make this partition unique, we consider in H_R only those vertices which have no neighbors in $C \cap H$. Since $V(H) \not\subseteq Y \cup (N(Y) \cap I)$, we also know that H_T is non-empty. Our first observation is the following:

Proposition 2. *Let H be a motif for the instance (G, X, M) where $G \setminus X = (C, I)$ is a threshold graph and let Y be $V(H) \cap X$. Then there exists a subset S of at most $k + 1$ vertices in $(G \setminus X) \cap H$ such that $G[Y \cup S]$ is connected.*

Proof. Let $Y_1, Y_2, \cdots, Y_l$ $(l \leqslant k)$ be connected components of Y. H_T be the induced connected threshold graph of $H \cap (G \setminus X)$. Without loss of generality, assume that $Y_1, Y_2, \cdots, Y_i$ are the connected components which do not have neighbors in H_T. The set $N(Y_1) \subseteq I \cap V(H)$ must contain a vertex w_1 which is a neighbor of some component Y_j, $j \in [l]$ otherwise H is not connected. Then $H[Y_1 \cup Y_j \cup \{w_1\}]$ becomes connected and reduces the number of connected components by at least one. Similarly, there exists a set $W \subseteq V(H) \cap I$ of size at most $i - 1$ such that the number of connected components in $H[Y \cup W]$ becomes at most $l - i + 1$ and each component has a neighbor in H_T. Now, for each connected component pick one neighbor in H_T. Let this set of neighbors be W'. Since H_T is a threshold graph, all vertices in W' will have common neighbor v_u in $H_T \cap C$. Therefore $H[Y \cup W \cup W' \cup \{v_u\}]$ is connected and $|W \cup W' \cup \{v_u\}| \leqslant l + 1 \leqslant k + 1$. □

For a motif H, we call the subset S given by the proposition above the *core* of H. Roughly speaking, it turns out that any subset of vertices that respects M and connects Y is the core of some motif. Therefore, the algorithm can focus on identifying such a core and then extending it to a solution. We formalize this notion now, showing that it is sufficient to identify any subset of vertices that connects Y to find a motif for the entire instance. We will then address the issue of how our algorithm can efficiently find such a subset. Since $G \setminus X = (C, I)$ is a threshold graph there exists an ordering of the clique vertices $v_1, v_2, \cdots v_{|C|}$ such that $N(v_i) \supseteq N(v_j)$ for all $i \leqslant j$. We can partition the vertex set of the clique into subsets $C_1, C_2, \ldots C_p$, such that $u, v \in C_i$ iff $N[u] = N[v]$. By the vertex ordering defined above, observe that these subsets can be ordered such that

$$N(C_1) \supsetneq N(C_2) \supsetneq \cdots \supsetneq N(C_p),$$

Consider the intersection of any motif H with $G \setminus X$. Let $i(H)$ denote the largest index for which $H \cap (C_1 \cup C_2 \cup \cdots C_{i(H)}) = \emptyset$, that is to say, we use $i(H)$ to denote the "highest layer" upto which the motif does not intersect the clique part of $G \setminus X$. We call $i(H)$ the *baseline* for H. Further, we use $D(H)$ to denote $C_1 \cup C_2 \cup \cdots C_{i(H)}$, and we let $J(H) \subseteq I \cap V(H)$ be those vertices in the independent set whose neighbors are contained entirely in $D(H)$. The following observation is easy to see.

Proposition 3. *Let H be a motif for the instance (G, X, M) where $G \setminus X = (C, I)$ is a threshold graph. Suppose the baseline of H is $i(H)$, and $D(H)$ and $J(H)$ are* *as defined above, then any vertex in* $H \cap C_{i(H)+1}$ *is a universal vertex in* H_T, *and all vertices in* $J(H)$ *are adjacent to* Y *in* H. *Also,* $H_T \subseteq (C \setminus D(H), I \setminus J(H))$ *and* $H_R \subseteq J(H)$.

The main structural lemma that we use is the following.

Lemma 3. *Let* (G, X, M) *be a* YES-*instance of* GRAPH MOTIF, *where* $G \setminus X = (C, I)$ *is a threshold graph. Suppose further that the instance admits a solution* H *with baseline* i. *Let* S *be any subset of vertices from* $G \setminus X$ *such that* $G[Y \cup S]$ *is connected,* $col(S)$ *respects the palette* $M \setminus col(Y)$, *and*

$$S \cap C_{i+1} \cap V(H) \neq \emptyset.$$

Then, $(Y \cup S)$ *can be extended to a motif for* G.

Proof. If $(Y \cup S)$ is already a motif for G respecting M, then we are done. Otherwise, let H be the solution given by the statement of the Lemma, and let v denote a vertex in $S \cap C_{i+1} \cap V(H)$. By Proposition 3, every vertex in H is adjacent to either Y or the vertex v. Let c be a color in M that is not already accounted for by $Y \cup S$. Consider the vertex u in H such that $col(u) = c$ (such a vertex must exist since H is a solution). We add this vertex to $(Y \cup S)$. Observe that the resulting subgraph is still connected, since $v \in Y \cup S$, and $G[Y \cup S]$ is connected as well. Repeating this process as long as there are colors in M that are not accounted for by the current solution, we arrive at the desired motif. □

We now turn to a discussion of how we can identify a core S as stipulated by Lemma 3. For ease of discussion, let us fix a target motif H and also fix Y as the correct guess of $H \cap X$. Let S be the core of H. A natural approach to discover S algorithmically is the following. Partition $G \setminus X$ into equivalence classes based on their neighborhood in Y. Since $|Y| \leqslant k$, we have at most 2^k classes. Since $|S| \leqslant k + 1$, we can guess the classes that the vertices of S belong to. However, it turns out that S is not completely characterized by the classes: for instance, consider the example where Y consists of two isolated vertices y_1 and y_2, and the vertex subsets $Z_1 := \{a, p\}$, $Z_2 := \{b, q\}$ are adjacent to y_1 and y_2 respectively, where $a, b \in C$ and $p, q \in I$. Note that although a and p have the same type (as do b and q), the graph induced by $Y \cup \{a, b\}$ is connected while the graph induced by $Y \cup \{p, q\}$ is not. The issue with just considering types based on neighborhoods in Y is that we fail to capture the nature of the connections within $G \setminus X$.

Fortunately these connections are simple: as we have said before, the solution vertices are either a part of a maximally connected threshold graph (recall that this is H_T), or an independent set which is not connected to the threshold part at all (which we denoted by H_R). To make the distinction of how the solution is split across these parts, will use S_T to denote $S \cap H_T$ and S_I to denote $S \cap H_R$. We will identify first the vertices from the S_I which partially connect the components in Y (using equivalence classes over the components of Y), and having fixed the components of $Y \cup S_I$ (notice we can do this even without fixing specific vertices of S_I: just the information about the types is enough), we identify vertices of S_T (again based on types). We begin with the following observation:

Proposition 4. *Let* H *be a motif for the instance* (G, X, M) *where* $G \setminus X = (C, I)$ *is a threshold graph and let* Y *be* $V(H) \cap X$. *If* S, S_I, *and* S_T *are defined as above, then all the components of* $G[Y \cup S_I]$ *have a neighbor in* S_T.

Proof. Suppose not. Then the component of $G[Y \cup S_I]$ that has no neighbors in S_T forms one side of an empty cut in $G[Y \cup S]$, which contradicts the assumption that H is connected. □

We now describe the process of forming equivalence classes in $G \setminus X$. Let $\mathcal{Y} = \{Y_1, \ldots, Y_\ell\}$ be the set of connected components of Y. For a subset $Z \subseteq [\ell]$, we say that the first-order type of a vertex v in $G \setminus X$ is Z if the set of connected components in Y that contain a neighbor of v is exactly given by $\{Y_i \mid i \in Z\}$. Note that there are at most 2^l types. For a subset $Z \subseteq [\ell]$, we use G_Z to denote the set of vertices in $G \setminus X$ whose type is Z. Also, we use $\tau(v)$ to denote the first-order type of a vertex $v \in G \setminus X$. Our next claim is that the vertices in S_I are uniquely determined by the types of the vertices that participate in S_I. The following observation is immediate from the definition of types.

Proposition 5. *Let* H *be a motif with core* S *consisting of* S_I *and* S_T. *Let* $S_I = \{u_1, \ldots, u_r\}$, *and suppose* u_i *has type* t_i. *If* $Q = \{v_1, \ldots, v_r\}$ *is any subset of* r *vertices such that* $\tau(v_i) = t_i$, *then* $G[Y \cup Q \cup S_T]$ *is also connected.*

Let $\mathcal{L} := \{t_1, \ldots, t_r\}$ be a set of first order types. We use $\mathcal{Y}[\![\mathcal{L}]\!] = \{Y_1', \ldots, Y_c'\}$ to refer to the set of connected components obtained after including in Y one vertex each of type t_i for all $1 \leqslant i \leqslant r$. We now define the second-order type of a vertex according to $\mathcal{L}$. For a subset $Z \subseteq [c]$, we say that the second-order type of a vertex v in $G \setminus X$, with respect to $\mathcal{L}$, is the set of connected components in $\mathcal{Y}[\![\mathcal{L}]\!]$ that contain a neighbor of v is exactly given by $\{Y_i' \mid i \in Z\}$. Further, we use $\tau_{\mathcal{L}}(v)$ to denote the second-order type of a vertex $v \in G \setminus X$ with respect to $\mathcal{L}$, where $\mathcal{L}$ is a set of first order types. Again, the following proposition follows from the definitions.

Proposition 6. *Let* H *be a motif with core* S *consisting of* S_I *and* S_T. *Let* $\mathcal{L}$ *be the set of types of vertices in* S_I. *Let* $S_T = \{u_1, \ldots, u_r\}$, *and suppose the second order type of* u_i *with respect to* $\mathcal{L}$ *is* t_i. *If* $Q = \{v_1, \ldots, v_r\}$ *is any subset of* r *vertices such that* $\tau_{\mathcal{L}}(v_i) = t_i$, *then* $G[Y \cup S_I \cup Q]$ *is also connected.*

We are now ready to describe our algorithm. We begin by guessing a baseline $0 \leqslant i \leqslant p$, and a vertex v from C_{i+1} to include in our solution. Then, we proceed to guess c, the size of the core of the target motif, and we also try all possible $0 \leqslant a, b \leqslant c$ such that $a + b = c$. Note that by Proposition 2, c is bounded by $(k + 1)$. We now proceed to look for a subset $S_T \subseteq N[C_{i+1} \cup \cdots \cup C_p]$, $|S_T| = a$ and a subset $S_I \subseteq N(C_1 \cup \cdots \cup C_i) \cap I$, $|S_I| = b$, such that $G[v \cup S_I \cup S_T \cup Y]$ is connected. To find the subsets S_I and S_T, we first guess a set of a first-order types $\mathcal{L}$ for the vertices of S_I and then guess a set of b second-order types with respect to $\mathcal{L}$. We sanity check our choice of types to ensure that Y is connected when considered along with one vertex of each chosen type, and the vertex v: if this is not the case, then we disregard this combination of types. It remains for the algorithm to make the choice one vertex of each guessed type to form the sets S_I and S_T—observe that this cannot be done arbitrarily because the chosen vertices have to respect the palette $M \setminus col(Y \cup \{v\})$.

In order to choose a vertex from a type, we construct a bipartite graph with one partition consisting of one vertex for each color in $M \setminus col(Y \cup \{v\})$ and the other partition contains one vertex for each selected type. There is an edge between color c_i and type t_j if there exists a vertex of color c_i in t_j. We now find a maximum matching in this bipartite graph which contains all the first and second-order types under consideration. It is easily checked that if there is no matching that saturates the types, then there is no solution that respects these types, and on the other hand, if there is a matching that saturates all types, then picking one vertex of each type according to the matching gives us the desired set S. The correctness of the algorithm follows from Propositions 5, 6 and Lemma 3. The FPT running time follows from the fact c is bounded by $k+1$ (by Proposition 2), the number of types of the first and second order are bounded by 2^k (and therefore guessing these types takes time $2^{k(a+b)} = 2^{O(k^2)}$ for a fixed choice of a and b), and that the first step where we guess of the universal vertex adds only a polynomial overhead.

4   Chromatic Number

The CHROMATIC NUMBER problem asks for the minimum number of colors needed to color the vertices of the graph such that no two adjacent vertices have the same color. This problem is FPT when parameterized by vertex cover [3], tree-width [7] and distance to clique [19]. On the other hand, is W[1]-hard when parameterized by clique-width [12] and distance to split graph [5]. Finally, it is unlikely to have a polynomial kernel when parameterized by vertex cover [3]. Adding to this complexity landscape, we give here a polynomial kernel for this problem when parameterized by distance to clique.

Theorem 2. *The CHROMATIC NUMBER problem admits a polynomial kernel when parameterized by the size of a modulator to a clique.*

Proof. Given a graph G and a set $X \subseteq V(G)$ of size k such that $C = V(G) \setminus X$ is a clique of size n. Partition the vertices of X into X' and X'', where $X' = \{x \in X / |N(x) \cap C| < n - k\}$ and $X'' = X \setminus X'$. First we show that $\chi(G) = \chi(G \setminus X')$. Each vertex x of X' has at least $k+1$ non-neighbors in C and at most $k-1$ neighbors in X therefore we can always assign x with a color of C.

Now we show that $\chi(G) = |C \setminus C'| + \chi(X'' \cup C')$, where $C' = C \setminus (\cap_{x \in X''} N(x))$. It is easy to see that $\chi(G) \leqslant |C \setminus C'| + \chi(X'' \cup C')$. For the other direction, in any optimal coloring of G each vertex of $C \setminus C'$ receives a distinct color, therefore we have $\chi(G) \geqslant |C \setminus C'| + \chi(X'' \cup C')$. Since every vertex of X'' has at least $n - k$ neighbors in C, therefore they have at least $n - k^2$ common neighbors in C which implies the size of C' is at most k^2. Therefore, the size of the reduced instance $X'' \cup C'$ is at most $k^2 + k$. □

References

1. Ambalath, A.M., Balasundaram, R., Rao H., C., Koppula, V., Misra, N., Philip, G., Ramanujan, M.S.: On the kernelization complexity of colorful motifs. In: Raman, V., Saurabh, S. (eds.) IPEC 2010. LNCS, vol. 6478, pp. 14–25. Springer, Heidelberg (2010). doi:10.1007/978-3-642-17493-3_4
2. Betzler, N., Van Bevern, R., Fellows, M.R., Komusiewicz, C., Niedermeier, R.: Parameterized algorithmics for finding connected motifs in biological networks. IEEE/ACM TCBB **8**(5), 1296–1308 (2011)
3. Bodlaender, H.L., Jansen, B.M., Kratsch, S.: Cross-composition: a new technique for kernelization lower bounds (2010). arXiv preprint arXiv:1011.4224
4. Bonnet, É., Sikora, F.: The graph motif problem parameterized by the structure of the input graph. In: LIPIcs-Leibniz International Proceedings in Informatics, vol. 43. Schloss Dagstuhl-Leibniz-Zentrum fuer Informatik (2015)
5. Cai, L.: Parameterized complexity of vertex colouring. Discret. Appl. Math. **127**(3), 415–429 (2003)
6. Chvátal, V., Hammer, P.L.: Aggregation of inequalities in integer programming. Ann. Discret. Math. **1**, 145–162 (1977)
7. Courcelle, B.: The monadic second-order logic of graphs. I. Recognizable sets of finite graphs. Inf. Comput. **85**(1), 12–75 (1990)
8. Cygan, M., Fomin, F.V., Kowalik, L., Lokshtanov, D., Marx, D., Pilipczuk, M., Pilipczuk, M., Saurabh, S.: Parameterized Algorithms, vol. 4. Springer, New York (2015)
9. Downey, R.G., Fellows, M.R.: Parameterized Complexity, vol. 3. Springer, Heidelberg (1999)
10. Fellows, M.R., Fertin, G., Hermelin, D., Vialette, S.: Sharp tractability borderlines for finding connected motifs in vertex-colored graphs. In: Arge, L., Cachin, C., Jurdziński, T., Tarlecki, A. (eds.) ICALP 2007. LNCS, vol. 4596, pp. 340–351. Springer, Heidelberg (2007). doi:10.1007/978-3-540-73420-8_31
11. Foldes, S., Hammer, P.L.: Split graphs. Institut für Ökonometrie und Operations Research, Universität Bonn (1976)
12. Fomin, F.V., Golovach, P.A., Lokshtanov, D., Saurabh, S.: Algorithmic lower bounds for problems parameterized by clique-width. In: Proceedings of 21st Annual ACM-SIAM Symposium on Discrete Algorithms, pp. 493–502. SIAM (2010)
13. Gallai, T.: Transitiv orientierbare graphen. Acta Mathematica Hungarica **18**(1), 25–66 (1967)
14. Ganian, R.: Using neighborhood diversity to solve hard problems (2012). arXiv preprint arXiv:1201.3091
15. Ganian, R.: Improving vertex cover as a graph parameter. Discret. Math. Theoret. Comput. Sci. **17**(2), 77–100 (2015)
16. Jansen, B.M., et al.: The power of data reduction: kernels for fundamental graph problems (2013)
17. Lacroix, V., Fernandes, C.G., Sagot, M.F.: Motif search in graphs: application to metabolic networks. IEEE/ACM TCBB **3**(4), 360–368 (2006)
18. Mahadev, N.V., Peled, U.N.: Threshold Graphs and Related Topics, vol. 56. Elsevier, Amsterdam (1995)
19. Sæther, S.H., Telle, J.A.: Between treewidth and clique-width. Algorithmica **75**(1), 218–253 (2016)
20. Tinhofer, G.: Bin-packing and matchings in threshold graphs. Discret. Appl. Math. **62**(1), 279–289 (1995)

On Chromatic Number of Colored Mixed Graphs

Sandip Das[1(✉)], Soumen Nandi[1], and Sagnik Sen[2]

[1] Indian Statistical Institute, Kolkata, India
sandipdas@isical.ac.in, soumen2004@gmail.com
[2] Indian Institute of Science, Bangalore, India
sen007isi@gmail.com

Abstract. An (m, n)-colored mixed graph G is a graph with its arcs having one of the m different colors and edges having one of the n different colors. A homomorphism f of an (m, n)-colored mixed graph G to an (m, n)-colored mixed graph H is a vertex mapping such that if uv is an arc (edge) of color c in G, then $f(u)f(v)$ is an arc (edge) of color c in H. The (m,n)-*colored mixed chromatic number* $\chi_{(m,n)}(G)$ of an (m, n)-colored mixed graph G is the order (number of vertices) of a smallest homomorphic image of G. This notion was introduced by Nešetřil and Raspaud (2000, J. Combin. Theory, Ser. B 80, 147–155). They showed that $\chi_{(m,n)}(G) \leq k(2m + n)^{k-1}$ where G is a acyclic k-colorable graph. We prove the tightness of this bound. We also show that the acyclic chromatic number of a graph is bounded by $k^2 + k^{2 + \lceil \log_{(2m+n)} \log_{(2m+n)} k \rceil}$ if its (m, n)-colored mixed chromatic number is at most k. Furthermore, using probabilistic method, we show that for connected graphs with maximum degree Δ its (m, n)-colored mixed chromatic number is at most $2(\Delta - 1)^{2m+n}(2m + n)^{\Delta - \min(2m+n,3)+2}$. In particular, the last result directly improves the upper bound $2\Delta^2 2^\Delta$ of oriented chromatic number of graphs with maximum degree Δ, obtained by Kostochka et al. (1997, J. Graph Theory 24, 331–340) to $2(\Delta - 1)^2 2^\Delta$. We also show that there exists a connected graph with maximum degree Δ and (m, n)-colored mixed chromatic number at least $(2m + n)^{\Delta/2}$.

Keywords: Colored mixed graphs · Acyclic chromatic number · Graphs with bounded maximum degree · Arboricity · Chromatic number

1 Introduction

An (m, n)-*colored mixed graph* $G = (V, A \cup E)$ is a graph G with set of vertices V, set of arcs A and set of edges E where each arc is colored by one of the m colors $\alpha_1, \alpha_2, \ldots, \alpha_m$ and each edge is colored by one of the n colors $\beta_1, \beta_2, \ldots, \beta_n$. We denote the number of vertices and the number of edges of the underlying graph of G by v_G and e_G, respectively. Also, we will consider only those (m, n)-colored mixed graphs for which the underlying undirected graph is simple. Nešetřil and Raspaud [6] generalized the notion of vertex coloring and chromatic number for (m, n)-colored mixed graphs by defining colored homomorphism.

© Springer International Publishing AG 2017
D. Gaur and N.S. Narayanaswamy (Eds.): CALDAM 2017, LNCS 10156, pp. 130–140, 2017.
DOI: 10.1007/978-3-319-53007-9_12

Let $G = (V_1, A_1 \cup E_1)$ and $H = (V_2, A_2 \cup E_2)$ be two (m, n)-colored mixed graphs. A colored homomorphism of G to H is a function $f : V_1 \rightarrow V_2$ satisfying

$$uv \in A_1 \Rightarrow f(u)f(v) \in A_2,$$

$$uv \in E_1 \Rightarrow f(u)f(v) \in E_2,$$

and the color of the arc or edge linking $f(u)$ and $f(v)$ is the same as the color of the arc or the edge linking u and v [6]. We write $G \rightarrow H$ whenever there exists a homomorphism of G to H (for an example see Fig. 1).

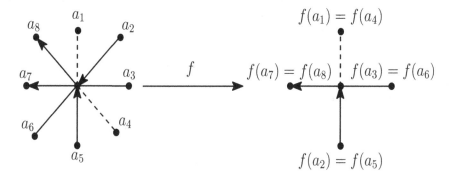

Fig. 1. An example of a colored homomorphism of $(1, 2)$-colored mixed graphs. The arcs are denoted by arrows, the edges having the first color are denoted by lines and the edges having the second color are denoted by "dashed" lines.

Given an (m, n)-colored mixed graph G let H be an (m, n)-colored mixed graph with minimum *order* (number of vertices) such that $G \rightarrow H$. Then the order of H is the (m,n)-*colored mixed chromatic number* $\chi_{(m,n)}(G)$ of G. For an undirected simple graph G, the maximum (m, n)-colored mixed chromatic number taken over all (m, n)-colored mixed graphs having underlying undirected simple graph G is denoted by $\chi_{(m,n)}(G)$. Let $\mathcal{F}$ be a family of undirected simple graphs. Then $\chi_{(m,n)}(\mathcal{F})$ is the maximum of $\chi_{(m,n)}(G)$ taken over all $G \in \mathcal{F}$.

Note that a $(0, 1)$-colored mixed graph G is nothing but an undirected simple graph while $\chi_{(0,1)}(G)$ is the ordinary chromatic number. Similarly, the study of $\chi_{(1,0)}(G)$ is the study of oriented chromatic number which is considered by several researchers in the last two decades (for details please check the recent updated survey [9]). Alon and Marshall [1] studied the homomorphism of $(0, n)$-colored mixed graphs with a particular focus on $n = 2$.

A simple graph G is *acyclic k-colorable* if we can color its vertices with k colors such that each color class induces an independent set and any two color classes induce a forest. The *acyclic chromatic number* $\chi_a(G)$ of a simple graph G is the minimum k such that G is acyclic k-colorable. Nešetřil and Raspaud [6] showed that $\chi_{(m,n)}(G) \leq k(2m+n)^{k-1}$ where G is a acyclic k-colorable graph. As planar graphs are 5-acyclic colorable due to Borodin [2], the same authors implied

$\chi_{(m,n)}(\mathcal{P}) \leq 5(2m+n)^4$ for the family $\mathcal{P}$ of planar graphs as a corollary. This result, in particular, implies $\chi_{(1,0)}(\mathcal{P}) \leq 80$ and $\chi_{(0,2)}(\mathcal{P}) \leq 80$ (independently proved before in [1,8], respectively).

Let $\mathcal{A}_k$ be the family of graphs with acyclic chromatic number at most k. Ochem [7] showed that the upper bound $\chi_{(1,0)}(\mathcal{A}_k) \leq 80$ is tight. We generalize it for all $(m,n) \neq (0,1)$ to show that the upper bound $\chi_{(m,n)}(\mathcal{A}_k) \leq k$ $(2m+n)^{k-1}$ obtained by Nešetřil and Raspaud [6] is tight. This implies that the upper bound $\chi_{(m,n)}(\mathcal{P}) \leq 5(2m+n)^4$ cannot be improved using the upper bound of $\chi_{(m,n)}(\mathcal{A}_5)$.

The arboricity $arb(G)$ of a graph G is the minimum k such that the edges of G can be decomposed into k forests. Kostochka et al. [4] showed that given a simple graph G, the acyclic chromatic number $\chi_a(G)$ of G is also bounded by a function of $\chi_{(1,0)}(G)$. We generalize this result for all $(m,n) \neq (0,1)$ by showing that for a graph G with $\chi_{(m,n)}(G) \leq k$ we have $\chi_a(G) \leq k^2 + k^{2+\lceil \log_2 \log_p k \rceil}$ where $p = 2m+n$. Our bound slightly improves the bound obtained by Kostochka et al. [4] for $(m,n) = (1,0)$. For achieving this result we first establish some relations among arboricity of a graph, (m,n)-colored mixed chromatic number and acyclic chromatic number.

Let $\mathcal{G}_\Delta$ be the family of connected graphs with maximum degree Δ. Kostochka et al. [4] proved that $2^{\Delta/2} \leq \chi_{(1,0)}(\mathcal{G}_\Delta) \leq 2\Delta^2 2^\Delta$. In fact their result was for any graph of maximum degree Δ, not necessarily connected. However, for connected graphs we improve this result in a generalized setting by proving $p^{\Delta/2} \leq \chi_{(m,n)}(\mathcal{G}_\Delta) \leq 2(\Delta-1)^p p^{\Delta-\min(p,3)+2}$ for all $(m,n) \neq (0,1)$ where $p = 2m+n$. We will use a technique introduced by Duffy [3].

2   Preliminaries

A *special 2-path* uvw of an (m,n)-colored mixed graph G is a 2-path satisfying one of the following conditions:

(i) uv and vw are edges of different colors,
(ii) uv and vw are arcs (possibly of the same color),
(iii) uv and wv are arcs of different colors,
(iv) vu and vw are arcs of different colors,
(v) exactly one of uv and vw is an edge and the other is an arc.

Observation 1. *The endpoints of a special 2-path must have different image under any homomorphism of G.*

Proof. Let uvw be a special 2-path in an (m,n)-colored mixed graph G. Let $f : G \to H$ be a colored homomorphism of G to an (m,n)-colored mixed graph H such that $f(u) = f(w)$. Then $f(u)f(v)$ and $f(w)f(v)$ will induce parallel edges in the underlying graph of H. But as we are dealing with (m,n)-colored mixed graphs with underlying simple graphs, this is not possible.   □

Let $G = (V, A \cup E)$ be an (m, n)-colored mixed graph. Let uv be an arc of G with color α_i for some $i \in \{1, 2, \ldots, m\}$. Then u is a $-\alpha_i$-neighbor of v and v is a $+\alpha_i$-neighbor of u. The set of all $+\alpha_i$-neighbors and $-\alpha_i$-neighbors of v is denoted by $N^{+\alpha_i}(v)$ and $N^{-\alpha_i}(v)$, respectively. Similarly, let uv be an edge of G with color β_i for some $i \in \{1, 2, \ldots, n\}$. Then u is a β_i-neighbor of v and the set of all β_i-neighbors of v is denoted by $N^{\beta_i}(v)$. Let $\boldsymbol{a} = (a_1, a_2, \ldots, a_j)$ be a j-vector such that $a_i \in \{\pm\alpha_1, \pm\alpha_2, \ldots, \pm\alpha_m, \beta_1, \beta_2, \ldots, \beta_n\}$ where $i \in \{1, 2, \ldots, j\}$. Let $J = (v_1, v_2, \ldots, v_j)$ be a j-tuple (without repetition) of vertices from G. Then we define the set $N^{\boldsymbol{a}}(J) = \{v \in V | v \in N^{a_i}(v_i) \text{ for all } 1 \leq i \leq j\}$. Finally, we say that G has property $Q^{t,j}_{g(j)}$ if for each j-vector $\boldsymbol{a}$ and each j-tuple J we have $|N^{\boldsymbol{a}}(J)| \geq g(j)$ where $j \in \{0, 1, \ldots, t\}$ and $g : \{0, 1, \ldots, t\} \rightarrow \{0, 1, \ldots \infty\}$ is an integral function.

Intuitively, an (m, n)-colored mixed graph G with property $Q^{t,j}_{g(j)}$ is a graph where a set of distinct j vertices for all $j \leq t$ has at least $g(j)$ common neighbors of each kind. Usually graphs with these kinds of properties are used as homomorphic target graphs.

3   On Graphs with Bounded Acyclic Chromatic Number

First we will construct examples of (m, n)-colored mixed graphs $H_k^{(m,n)}$ with acyclic chromatic number at most k and $\chi_{(m,n)}(H_k^{(m,n)}) = k(2m + n)^{k-1}$ for all $k \geq 3$ and for all $(m, n) \neq (0, 1)$. This, along with the upper bound established by Nešetřil and Raspaud [6], will imply the following result:

Theorem 1. *Let $\mathcal{A}_k$ be the family of graphs with acyclic chromatic number at most k. Then $\chi_{(m,n)}(\mathcal{A}_k) = k(2m+n)^{k-1}$ for all $k \geq 3$ and for all $(m, n) \neq (0, 1)$.*

Proof. First we will construct an (m, n)-colored mixed graph $H_k^{(m,n)}$, where $p = 2m + n \geq 2$, as follows. Let A_{k-1} be the set of all $(k-1)$-vectors. Thus, $|A_{k-1}| = p^{k-1}$.

Define B_i as a set of $(k - 1)$ vertices $B_i = \{b_1^i, b_2^i, \ldots, b_{k-1}^i\}$ for all $i \in \{1, 2, \ldots, k\}$ such that $B_r \cap B_s = \emptyset$ when $r \neq s$. The vertices of B_i's are called *bottom* vertices for each $i \in \{1, 2, \ldots, k\}$. Furthermore, let $TB_i = (b_1^i, b_2^i, \ldots, b_{k-1}^i)$ be a $(k-1)$-tuple.

After that define the set of vertices $T_i = \{t_a^i | t_a^i \in N^{\boldsymbol{a}}(TB_i) \text{ for all } \boldsymbol{a} \in A_{k-1}\}$ for all $i \in \{1, 2, \ldots, k\}$. The vertices of T_i's are called *top* vertices for each $i \in \{1, 2, \ldots, k\}$. Observe that there are p^{k-1} vertices in T_i for each $i \in \{1, 2, \ldots, k\}$.

Note that the definition of T_i already implies some colored arcs and edges between the set of vertices B_i and T_i for all $i \in \{1, 2, \ldots, k\}$. These colored arcs and edges are present in $H_k^{(m,n)}$.

As $p \geq 2$ it is possible to construct a special 2-path. Now for each pair of vertices $u \in T_i$ and $v \in T_j$ $(i \neq j)$, construct a special 2-path $uw_{uv}v$ and call these new vertices w_{uv} as *internal* vertices for all $i, j \in \{1, 2, \ldots, k\}$. This so obtained graph is $H_k^{(m,n)}$.

Now we will show that $\chi_{(m,n)}(H_k^{(m,n)}) \geq k(2m+n)^{k-1}$. Let $\boldsymbol{a} \neq \boldsymbol{a'}$ be two distinct $(k-1)$-vectors. Assume that the jth co-ordinate of $\boldsymbol{a}$ and $\boldsymbol{a'}$ is different. Then note that $t_{\boldsymbol{a}}^i b_j^i t_{\boldsymbol{a'}}^i$ is a special 2-path. Therefore, $t_{\boldsymbol{a}}^i$ and $t_{\boldsymbol{a'}}^i$ must have different homomorphic image under any homomorphism. Thus, all the vertices in T_i must have distinct homomorphic image under any homomorphism. Moreover, as a vertex of T_i is connected by a special 2-path with a vertex of T_j for all $i \neq j$, all the top vertices must have distinct homomorphic image under any homomorphism. It is easy to see that $|T_i| = p^{k-1}$ for all $i \in \{1, 2, \ldots, k\}$. Hence $\chi_{(m,n)}(H_k^{(m,n)}) \geq \sum_{i=1}^{k} |T_i| = k(2m+n)^{k-1}$.

Now we will show that $\chi_a(H_k^{(m,n)}) \leq k$. From now on, by $H_k^{(m,n)}$, we mean the underlying undirected simple graph of the (m,n)-colored mixed graph $H_k^{(m,n)}$. We will provide an acyclic coloring of this graph with $\{1, 2, \ldots, k\}$. Color all the vertices of T_i with i for all $i \in \{1, 2, \ldots, k\}$. Then color all the vertices of B_i with distinct $(k-1)$ colors from the set $\{1, 2, \ldots, k\} \setminus \{i\}$ of colors for all $i \in \{1, 2, \ldots, k\}$. Note that each internal vertex has exactly two neighbors. Color each internal vertex with a color different from its neighbors. It is easy to check that this is an acyclic coloring.

Therefore, we showed that $\chi_{(m,n)}(\mathcal{A}_k) \geq k(2m+n)^{k-1}$ while, on the other hand, Nešetřil and Raspaud [6] showed that $\chi_{(m,n)}(\mathcal{A}_k) \leq k(2m+n)^{k-1}$ for all $k \geq 3$ and for all $(m,n) \neq (0,1)$. □

Consider a complete graph K_t. For all $(m,n) \neq (0,1)$, it is possible to replace all the edges of K_t by a special 2-path to obtain an (m,n)-colored mixed graph S. Therefore, by Observation 1 we know that $\chi_{(m,n)}(S) \geq t$ whereas, it is easy to note that S has arboricity 2. Therefore, there exist graphs with arboricity 2 and arbitrarily high (m,n)-colored mixed chromatic number. Thus, the (m,n)-colored mixed chromatic number is not bounded by a function of arboricity. Though the reverse type of bound exists. Kostochka et al. [4] proved such a bound for $(m,n) = (1,0)$. We generalize their result for all $(m,n) \neq (0,1)$.

Theorem 2. *Let G be an (m,n)-colored mixed graph with $\chi_{(m,n)}(G) = k$ where $p = 2m + n \geq 2$. Then $arb(G) \leq \lceil \log_p k + k/2 \rceil$.*

Proof. Let G' be an arbitrary labeled subgraph of G consisting $v_{G'}$ vertices and $e_{G'}$ edges. We know from Nash-Williams' Theorem [5] that the arboricity $arb(G)$ of any graph G is equal to the maximum of $\lceil e_{G'}/(v_{G'} - 1) \rceil$ over all subgraphs G' of G. It is sufficient to prove that for any subgraph G' of G, $e_{G'}/(v_{G'} - 1) \leq \log_p k + k/2$. As G' is a labeled graph, so there are $p^{e_{G'}}$ different (m,n)-colored mixed graphs with underlying graph G'. As $\chi_{(m,n)}(G) = k$, there exits a homomorphism from G' to a (m,n)-colored mixed graph G_k which has the complete graph on k vertices as its underlying graph. Observe that, even though it is not a necessary condition for G_k to have the complete graph as its underlying graph, we can always add some extra edges/arcs to make G_k have that property. Note that the number of possible homomorphisms of G' to G_k is at most $k^{v_{G'}}$. For each such homomorphism of G' to G_k there are at most $p^{\binom{k}{2}}$

different (m,n)-colored mixed graphs with underlying labeled graph G' as there are $p^{\binom{k}{2}}$ choices of G_k. Therefore,

$$p^{\binom{k}{2}}.k^{v_{G'}} \geq p^{e_{G'}} \tag{1}$$

which implies

$$\log_p k \geq (e_{G'}/v_{G'}) - \binom{k}{2}/v_{G'}. \tag{2}$$

Suppose if $v_{G'} \leq k$, then $e_{G'}/(v_{G'} - 1) \leq v_{G'}/2 \leq k/2$. Now let $v_{G'} > k$. We know that $\chi_{(m,n)}(G') \leq \chi_{(m,n)}(G) = k$. So

$$\log_p k \geq \frac{e_{G'}}{v_{G'}} - \frac{k(k-1)}{2v_{G'}}$$

$$\geq \frac{e_{G'}}{(v_{G'}-1)} - \frac{e_{G'}}{v_{G'}(v_{G'}-1)} - \frac{k-1}{2}$$

$$\geq \frac{e_{G'}}{(v_{G'}-1)} - 1/2 - k/2 + 1/2$$

$$\geq \frac{e_{G'}}{(v_{G'}-1)} - k/2.$$

Therefore, $\frac{e_{G'}}{(v_{G'}-1)} \leq \log_p k + k/2$. □

We have seen that the (m,n)-colored mixed chromatic number of a graph G is bounded by a function of the acyclic chromatic number of G. Here we show that it is possible to bound the acyclic chromatic number of a graph in terms of its (m,n)-colored mixed chromatic number and arboricity. Our result is a generalization of a similar result proved for $(m,n) = (1,0)$ by Kostochka et al. [4].

Theorem 3. *Let G be an (m,n)-colored mixed graph with $arb(G) = r$ and $\chi_{(m,n)}(G) = k$ where $p = 2m + n \geq 2$. Then $\chi_a(G) \leq k^{\lceil \log_p r \rceil + 1}$.*

Proof. First we rename the following symbols: $\alpha_1 = a_0, -\alpha_1 = a_1, \alpha_2 = a_2, -\alpha_2 = a_3, \ldots, \alpha_m = a_{2m-2}, -\alpha_m = a_{2m-1}, \beta_1 = a_{2m}, \beta_2 = a_{2m+1}, \ldots, \beta_n = a_{2m+n-1}$.

Let G be a graph with $\chi_{(m,n)}(G) = k$ where $2m + n = p$. Let $v_1, v_2, \ldots, v_t$ be some ordering of the vertices of G. Now consider the (m,n)-colored mixed graph G_0 with underlying graph G such that for any $i < j$ we have $v_j \in N^{a_0}(v_i)$ whenever $v_i v_j$ is an edge of G.

Note that the edges of G can be covered by r edge disjoint forests $F_1, F_2, \ldots, F_r$ as $arb(G) = r$. Let s_i be the number i expressed with base p for all $i \in \{1, 2, \ldots, r\}$. Note that s_i can have at most $s = \lceil \log_p r \rceil$ digits.

Now we will construct a sequence of (m,n)-colored mixed graphs $G_1, G_2, \ldots, G_s$ each having underlying graph G. For a fixed $l \in \{1, 2, \ldots, s\}$ we will describe the construction of G_l. Let $i < j$ and $v_i v_j$ be an edge of G. Suppose $v_i v_j$ is an edge of the forest $F_{l'}$ for some $l' \in \{1, 2, \ldots, r\}$. Let the

l^{th} digit of $s_{l'}$ be $s_{l'}(l)$. Then G_l is constructed in such a way that we have $v_j \in N^{a_{s_{l'}(l)}}(v_i)$ in G_l.

Recall that $\chi_{m,n)}(G) \leq k$ and the underlying graph of G_l is G. Thus, for each $l \in \{1, 2, \cdots, s\}$ there exists an H_l on k vertices and a homomorphism $f_l : G_l \to H_l$. Now we claim that $f(v) = (f_0(v), f_1(v), \ldots, f_s(v))$ for each $v \in V(G)$ is an acyclic coloring of G.

For adjacent vertices u, v in G clearly we have $f(v) \neq f(u)$ as $f_0(v) \neq f_0(u)$. Let C be a cycle in G. We have to show that at least 3 colors have been used to color this cycle with respect to the coloring given by f. Note that in C there must be two incident edges uv and vw such that they belong to different forests, say, F_i and $F_{i'}$, respectively. Now suppose that C received two colors with respect to f. Then we must have $f(u) = f(w) \neq f(v)$. In particular we must have $f_0(u) = f_0(w) \neq f_0(v)$. To have that we must also have $u, w \in N^{a_i}(v)$ for some $i \in \{0, 1, \ldots, p - 1\}$ in G_0. Let s_i and $s_{i'}$ differ in their j^{th} digit. Then in G_j we have $u \in N^{a'_i}(v)$ and $w \in N^{a''_i}(v)$ for some $i' \neq i''$. Then we must have $f_j(u) \neq f_j(w)$. Therefore, we also have $f(u) \neq f(w)$. Thus, the cycle C cannot be colored with two colors under the coloring f. So f is indeed an acyclic coloring of G.   □

Thus, combining Theorems 2 and 3 we have $\chi_a(G) \leq k^{\lceil \log_p \lceil \log_p k + k/2 \rceil \rceil + 1}$ for $\chi_{(m,n)}(G) = k$ where $p = 2m + n \geq 2$. However, we managed to obtain the following bound which is better in all cases except for some small values of k, p.

Theorem 4. *Let G be an (m, n)-colored mixed graph with $\chi_{(m,n)}(G) = k \geq 4$ where $p = 2m + n \geq 2$. Then $\chi_a(G) \leq k^2 + k^{2 + \lceil \log_2 \log_p k \rceil}$.*

Proof. Let t be the maximum real number such that there exists a subgraph G' of G with $v_{G'} \geq k^2$ and $e_{G'} \geq t.v_{G'}$. Let G'' be the biggest subgraph of G with $e_{G''} > t.v_{G''}$. Thus, by maximality of t, $v_{G''} < k^2$.

Let $G_0 = G - G''$. Hence $\chi_a(G) \leq \chi_a(G_0) + k^2$. By maximality of G'', for each subgraph H of G_0, we have $e_H \leq t.v_H$.

If $t \leq \frac{v_H - 1}{2}$, then $e_H \leq (t + 1/2)(v_H - 1)$. If $t > \frac{v_H - 1}{2}$, then $\frac{v_H}{2} < t + 1/2$. So $e_H \leq \frac{(v_H - 1).v_H}{2} \leq (t + 1/2)(v_H - 1)$. Therefore, $e_H \leq (t + 1/2)(v_H - 1)$ for each subgraph H of G_0.

By Nash-Williams' Theorem [5], there exists $r = \lceil t + 1/2 \rceil$ forests $F_1, F_2, \cdots, F_r$ which covers all the edges of G_0. We know from Theorem 3, $\chi_a(G_0) \leq k^{s+1}$ where $s = \lceil \log_p r \rceil$.

Using inequality (2) we get $\log_p k \geq t - 1/2$. Therefore

$$s = \lceil \log_p (\lceil t + 1/2 \rceil) \rceil \leq \lceil \log_p (1 + \lceil \log_p k \rceil) \rceil \leq 1 + \lceil \log_p \log_p k \rceil.$$

Hence $\chi_a(G) \leq k^2 + k^{2 + \lceil \log_p \log_p k \rceil}$.   □

Our bound, when restricted to the case of $(m, n) = (1, 0)$, slightly improves the existing bound [4].

4   On Graphs with Bounded Maximum Degree

Recall that $\mathcal{G}_\Delta$ is the family of connected graphs with maximum degree Δ. It is known that $\chi_{(1,0)}(\mathcal{G}_\Delta) \leq 2\Delta^2 2^\Delta$ [4]. Here we prove that $\chi_{(m,n)}(\mathcal{G}_\Delta) \leq 2(\Delta - 1)^p.p^{(\Delta - \min(p,3)+2)} + 2$ for all $p = 2m+n \geq 2$ and $\Delta \geq 5$. Our result, restricted to the case $(m, n) = (1, 0)$, slightly improves the upper bound of Kostochka et al. [4]. However, the upper bound by Kostochka et al. [4] was proved for all graphs with maximum degree Δ, not restricted to the class of connected graphs.

Theorem 5. *Let $\mathcal{G}_\Delta$ be the family of connected graphs with maximum degree Δ. Then $p^{\Delta/2} \leq \chi_{(m,n)}(\mathcal{G}_\Delta) \leq 2(\Delta - 1)^p.p^{(\Delta - \min(p,3)+2)} + 2$ for all $p = 2m+n \geq 2$ and for all $\Delta \geq 5$.*

If every subgraph of a graph G has at least one vertex with degree at most d, then G is *d-degenerate*. Minimum such d is the *degeneracy* of G. To prove the above theorem we need the following result.

Theorem 6. *Let $\mathcal{G}'_\Delta$ be the family of connected graphs with maximum degree Δ and degeneracy $(\Delta - 1)$. Then $\chi_{(m,n)}(\mathcal{G}'_\Delta) \leq 2(\Delta - 1)^p.p^{(\Delta - \min(p,3)+2)}$ for all $p = 2m + n \geq 2$ and for all $\Delta \geq 5$.*

To prove the above theorem we need the following lemma.

Lemma 1. *There exists an (m, n)-colored complete mixed graph with property $Q_{1+(t-j)(t-2)}^{t-1,j}$ on $c = 2(t - 1)^p.p^{(t-\min(p,3)+2)}$ vertices where $p = 2m+n \geq 2$ and $t \geq 5$.*

Proof. Let C be a random (m, n)-colored mixed graph with underlying complete graph. Let u, v be two vertices of C and the events $u \in N^a(v)$ for $a \in \{\pm\alpha_1, \pm\alpha_2, \ldots, \pm\alpha_m, \beta_1, \beta_2, \ldots, \beta_n\}$ are equiprobable and independent with probability $\frac{1}{2m+n} = \frac{1}{p}$. We will show that the probability of C not having property $Q_{1+(t-j)(t-2)}^{t-1,j}$ is strictly less than 1 when $|C| = c = 2(t-1)^p.p^{(t-\min(p,3)+2)}$. Let $P(J, \boldsymbol{a})$ denote the probability of the event $|N^a(J)| < 1+(t-j)(t-2)$ where J is a j-tuple of C and $\boldsymbol{a}$ is a j-vector for some $j \in \{0, 1, \ldots, t-1\}$. Call such an event a *bad event*. Thus,

$$
\begin{aligned}
P(J, \boldsymbol{a}) &= \sum_{i=0}^{(t-j)(t-2)} \binom{c-j}{i} p^{-ij}(1 - p^{-j})^{c-i-j} \\
&< (1 - p^{-j})^c \sum_{i=0}^{(t-j)(t-2)} \frac{c^i}{i!}(1 - p^{-j})^{-i-j}p^{-ij} \\
&< 2e^{-cp^{-j}} \sum_{i=0}^{(t-j)(t-2)} c^i \\
&< e^{-cp^{-j}} c^{(t-j)(t-2)+1}.
\end{aligned}
\tag{3}
$$

Let $P(B)$ denote the probability of the occurrence of at least one bad event. To prove this lemma it is enough to show that $P(B) < 1$. Let T^j denote the set of all j-tuples and W^j denote the set of all j-vectors. Then

$$P(B) \leq \sum_{j=0}^{t-1} \sum_{J \in T^j} \sum_{a \in W^j} P(J, a) < \sum_{j=0}^{t-1} \binom{c}{j} p^j e^{-cp^{-j}} c^{(t-j)(t-2)+1}$$

$$< \sum_{j=0}^{t-1} \frac{c^j}{j!} p^j e^{-cp^{-j}} c^{(t-j)(t-2)+1}$$

$$= 2 \sum_{j=0}^{t-1} \frac{p^j}{2^j} \frac{2^{j-1}}{j!} c^j e^{-cp^{-j}} c^{(t-j)(t-2)+1}$$

$$< 2 \sum_{j=0}^{t-1} \frac{p^j}{2^j} e^{-cp^{-j}} c^{(t-j)(t-2)+1+j}.$$

(4)

It can be shown that $P(B) < 1$ and hence the result follows. □

Now we are ready to prove Theorem 6.

Proof of Theorem 6. Suppose that G is an (m, n)-colored mixed graph with maximum degree Δ and degeneracy $(\Delta - 1)$. By Lemma 1 we know that there exists an (m, n)-colored mixed graph C with property $Q_{1+(\Delta-j)(\Delta-2)}^{\Delta-1,j}$ on $2(\Delta - 1)^p . p^{(\Delta-\min(p,3)+2)}$ vertices where $p = 2m + n \geq 2$ and $\Delta \geq 5$. We will show that G admits a homomorphism to C.

As G has degeneracy $(\Delta-1)$, we can provide an ordering $v_1, v_2, \ldots, v_k$ of the vertices of G in such a way that each vertex v_j has at most $(\Delta-1)$ neighbors with lower indices. Let G_l be the (m, n)-colored mixed graph induced by the vertices $v_1, v_2, \ldots, v_l$ from G for $l \in \{1, 2, \ldots, k\}$. Now we will recursively construct a homomorphism $f : G \to C$ with the following properties:

(i) The partial mapping $f(v_1), f(v_2), \ldots, f(v_l)$ is a homomorphism of G_l to C for all $l \in \{1, 2, \ldots, k\}$.
(ii) For each $i > l$, all the neighbors of v_i with indices at most l have different images with respect to the mapping f.

Note that the base case is trivial, that is, any partial mapping $f(v_1)$ is enough. Suppose that the function f satisfies the above properties for all $j \leq t$ where $t \in \{1, 2, \ldots, k-1\}$ is fixed. Now assume that v_{t+1} has s neighbors with indices greater than $t + 1$. Then v_{t+1} has at most $(\Delta - s)$ neighbors with indices less than $t + 1$. Let A be the set of neighbors of v_{t+1} with indices greater than $t + 1$. Let B be the set of vertices with indices at most t and with at least one neighbor in A. Note that as each vertex of A is a neighbor of v_{t+1} and has at most $\Delta - 1$ neighbors with smaller indices, $|B| = (\Delta - 2)|A| = s(\Delta - 2)$. Let D be the set of possible options for $f(v_{t+1})$ such that the partial mapping is a homomorphism of G_{t+1} to C. As C has property $Q_{1+(\Delta-j)(\Delta-2)}^{\Delta-1,j}$ we have $|C| \geq 1 + s(\Delta - 1)$. So the set $D \setminus B$ is non-empty. Thus, choose any vertex from $D \setminus B$ as the image $f(v_{t+1})$. Note that this partial mapping satisfies the required conditions. □

Finally, we are ready to prove Theorem 5.

Proof of Theorem 5. First we will prove the lower bound. Let G_t be a Δ regular graph on t vertices. Thus, G_t has $\frac{t\Delta}{2}$ edges. Then we have

$$k_t = \chi_{(m,n)}(G_t) \geq \frac{p^{\Delta/2}}{p^{\binom{k_t}{2}/t}}$$

using inequality (1) (see Sect. 3). If $\chi_{(m,n)}(G_t) \geq p^{\Delta/2}$ for some t, then we are done. Otherwise, $\chi_{(m,n)}(G_t) = k_t$ is bounded. In that case, if t is sufficiently large, then $\chi_{(m,n)}(G_t) \geq p^{\Delta/2}$ as $\chi_{(m,n)}(G_t)$ is a positive integer.

Let $G = (V, A \cup E)$ be a connected (m,n)-colored mixed graph with maximum degree $\Delta \geq 5$ and $p = 2m + n \geq 2$. If G has a vertex of degree at most $(\Delta - 1)$ then it has degeneracy at most $(\Delta - 1)$. In that case by Theorem 5 we are done.

Otherwise, G is Δ regular. In that case, remove an edge uv of G to obtain the graph G'. Note that G' has maximum degree at most Δ and has degeneracy at most $(\Delta - 1)$. Therefore, by Theorem 5 there exists an (m,n)-colored complete mixed graph C on $2(\Delta - 1)^p . p^{(\Delta - \min(p,3) + 2)}$ vertices to which G' admits a homomorphism f. Let G'' be the graph obtained by deleting the vertices u and v of G'. Note that the homomorphism f restricted to G'' is a homomorphism f_{res} of G'' to C. Now include two new vertices u' and v' to C and obtain a new graph C'. Color the edges or arcs between the vertices of C and $\{u', v'\}$ in such a way so that we can extend the homomorphism f_{res} to a homomorphism f_{ext} of G to C' where $f_{ext}(u) = u'$, $f_{ext}(v) = v'$ and $f_{ext}(x) = f_{res}(x)$ for all $x \in V(G) \setminus \{u, v\}$. It is easy to note that the above mentioned process is possible.

Thus, every connected (m,n)-colored mixed graph with maximum degree Δ admits a homomorphism to C'. □

5   Conclusions

In this paper we studied the relation between (m,n)-colored mixed chromatic number of graphs and its acyclic chromatic number, arboricity and maximum degree. Considering graphs with bounded degeneracy only, there is a possibility for improvement of the upper bound of $\chi_{(m,n)}(\mathcal{G}_\Delta)$. Till now the parameter $\chi_{(m,n)}$ has been studied with a particular focus on $(m,n) = (1,0)$ and $(0,2)$ only. We think that $(m,n) = (1,1)$ is an important case which represents the family of mixed graphs should also be studied. For other families of graphs, study this parameter is an important area of further research.

Acknowledgement. The authors would like to thank the anonymous reviewer for the constructive comments towards improvement of the content, clarity and conciseness of the manuscript.

140   S. Das et al.

References

1. Alon, N., Marshall, T.H.: Homomorphisms of edge-colored graphs and Coxeter groups. J. Algebr. Combin. **8**(1), 5–13 (1998)
2. Borodin, O.V.: On acyclic colorings of planar graphs. Discret. Math. **25**(3), 211–236 (1979)
3. Duffy, C.: Homomorphisms of (j,k)-mixed graphs. Ph.D. thesis, University of Victoria/University of Bordeaux (2015)
4. Kostochka, A.V., Sopena, É., Zhu, X.: Acyclic and oriented chromatic numbers of graphs. J. Graph Theory **24**, 331–340 (1997)
5. St, C., Nash-Williams, J.A.: Decomposition of finite graphs into forests. J. Lond. Math. Soc. **1**(1), 12–12 (1964)
6. Nešetřil, J., Raspaud, A.: Colored homomorphisms of colored mixed graphs. J. Combin. Theory Ser. B **80**(1), 147–155 (2000)
7. Ochem, P.: Negative results on acyclic improper colorings. In: European Conference on Combinatorics (EuroComb 2005), pp. 357–362 (2005)
8. Raspaud, A., Sopena, É.: Good and semi-strong colorings of oriented planar graphs. Inf. Process. Lett. **51**(4), 171–174 (1994)
9. Sopena, É.: Homomorphisms and colourings of oriented graphs: an updated survey. Discret. Math. **339**(7), 1993–2005 (2016)

Optimizing Movement in Convex and Non-convex Path-Networks to Establish Connectivity

Sandip Das[1]([✉]), Ayan Nandy[1], and Swami Sarvottamananda[2]

[1] Indian Statistical Institute, Kolkata 700 108, India
sandipdas@isical.ac.in
[2] Ramakrishna Mission Vivekananda University,
Howrah 711 202, India

Abstract. We solve a *min-max movement problem* in which there are n sensors in path network in plane, where any sensor communicates only with its two immediate neighbors and only at a given maximum communication distance λ. We need to move sensors so that each sensor is in the communication range of its two neighbors, keeping the path topology intact. We present an $O(n^3)$ algorithm for min-max movement problem in a convex path-network which minimizes the maximum movement among the sensors. We also generalize our algorithm for ring, non-convex path, tethered and heterogeneous networks.

1 Introduction

There is a broad family of problems, called *movement problems*, introduced by Demaine et al. in 2005 [5]. In these problems we study the movement of multiple objects and agents, a robotic swarm, in a constrained environment. Autonomous robotic swarm movement problems deal with computing and analyzing optimal movements with respect to different criteria. These problems have practical appeal but only a few results are available which are either heuristic, approximate, hardness or inapproximation results. The earlier instances of these problems are when Corke et al. [3] studied the practical application of the movement of autonomous flying robots with limited wireless connectivity and limited mobility in the field because of energy and resource constraints. Bredin et al. [1] considered the problem of heuristically deploying or repairing a sensor network to guarantee a specified level of multi-path connectivity between all nodes and also to ensure some fault tolerance against node failure.

Euclidean 1-center and *Euclidean 1-median* problems can be seen as a *collocation* movement problem for related objective functions. There are several variants of movement problems where the objectives may differ, e.g., minimization of total movement/average movement, maximum movement, the number of moving points or objects, etc. In this paper, we consider the problem of minimization of the maximum movement, *min-max movement problem*, assuming

D. Gaur and N.S. Narayanaswamy (Eds.): CALDAM 2017, LNCS 10156, pp. 141–155, 2017.
DOI: 10.1007/978-3-319-53007-9_13

that each object has a limited local energy to move, so that all neighbors are finally within a prescribed communication distance λ.

Demaine et al. [5] studied a class of movement problems for discrete case in which the objects are constrained to move along the edges of a graph. They proved a few approximation and inapproximability results for this discrete case. It can be shown that the decision version of the min-max movement problem is NP-hard. Demaine et al. [6] also showed that if the tree-width of the final configuration is bounded then the problem can be solved by a fixed parameter algorithm.

Friggstad et al. [7] gave an 8-approximation algorithm for the minimum total movement for the *mobile facility location* problem in a graph with triangle inequality. They also showed that for min-max movement for this problem, 2-approximation is tight, unless $P = NP$. Warehelm [11] explored the problem of moving swarms of synchronous reactive robots to perform a *joint navigation/morphogenesis task* in a known world. They show that neither planning nor repairing of joint navigation task of swarms of synchronous reactive robots can be solved efficiently or correctly relative to many types of restrictions on robot and swarm architecture. Wang et al. [10] used Voronoi diagrams to study heuristic movement and deployment of sensors. A related problem in $\Re^1$ is *barrier coverage* problem, where the sensors are moved in such a way as to create a barrier which can detect any element crossing it. There is a lot of literature to solve different varieties of this problem, e.g. see [2,4,9]. Our problem can be seen as a special barrier coverage problem in which the topology of convex chain is given and we need to create a convex chain barrier using homogeneous detectors with sensing distance $\lambda/2$.

Our contributions in this paper are that we show if the configuration of final swarm architecture is known before hand as convex path, convex chain or some restricted tethered networks, then the min-max movement problem can be solved in polynomial time. We solve the min-max movement problem for the more general case of Euclidean motion in $\Re^2$ for a predetermined path, ring, tethered or heterogeneous network in $O(n^3)$ for convex versions and $O(n^4)$ for non-convex versions. Most of the related problems discussed in literature are NP-hard because final configurations are unknown [5–7]. We assume RAM model of computation with $sqrt()$ function. To the best of our knowledge, this is the first polynomial time deterministic algorithm for a non-trivial variation of the movement problems.

In Sect. 2 we present some definitions, useful concepts and ideas to solve min-max movement problem for convex path networks. In Sect. 3, we present an iterative algorithm for convex path networks. In Sect. 5, we present several generalization to ring, non-convex path, tethered and heterogeneous networks.

2   Ideas and Concepts

First we solve the problem of minimizing movement for a convex path. We define the problem for this set up in the following discussion. We define and solve the general version of the problem in Sect. 5.

Let there be n points in $\Re^2$ given as a path $G = (V, E)$ which form a part of convex polygon. Let λ be any target link length. We wish to move all points to new positions such that the distances of any point to its adjacent neighbors in G are less than or equal to λ. Such configurations of vertices are called *feasible*. We call those feasible configurations *optimal*, for which maximum displacement among vertices is minimized among all feasible configurations. We denote the maximum displacement of any point in an optimal configurations as Δ. The problem of *minimizing maximum movement* is to compute this Δ. See Fig. 1. In the paper, the vertices of G are moving points.   .

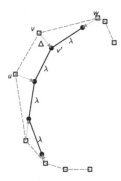   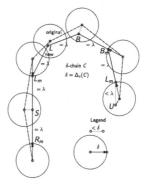

Fig. 1. Initial and final positions of the vertices of G. u is the *left vertex* of v and w is the *right vertex* of v, v' is the new optimal position of v for target link length λ and Δ is the optimal displacement.

Fig. 2. Meaning of labels R_m, S, L_m, L, B and B_m in a δ-chain C, and label U for a vertex $\notin C$. There is also an initial degeneracy.

We assume that the input satisfies a general position assumption specific to our algorithm. We will define the conditions of our general position assumption later during the discussion of algorithm in Sect. 4. However, we can take care of the degenerate cases easily by breaking the ties arbitrarily. We define the clockwise direction as *right* and the counter clockwise direction as *left* in this paper. Let $uv = d(u, v)$ be the Euclidean distance between points u and v.

In the course of our algorithm, we label each vertex as $U, L, R, B, L_m, R_m, B_m$ or S, to allow us to compute an optimal configuration. These labels indicate the displacement of vertices and the relations to their neighbors. We compute the displacements of vertices by solving a set of simultaneous equations. These simultaneous equations may be grouped for subsequences of vertices so that solutions of the groups are independent of each other. These groups of vertices are called δ-*chains*. However, the grouping is determined only during the execution of the algorithm presented in Sect. 3.3. Each δ-chain C has a corresponding maximum displacement denoted by $\Delta_c(C)$. Consequently, $\Delta = \max_C \{\Delta_c(C)\}$.

To solve the problem we propose an incremental algorithm. Every stage of the algorithm involves computing G_{k+1} from G_k, where G_k consisting of leftmost k vertices of G. We take G_k and add vertex, v_{k+1}, on the right of G_k. If $v_k v_{k+1}$ is larger than λ then we need to displace v_{k+1}. Conceptually, we continuously decrease $v_k v_{k+1}$ to target link length, λ, calculating the positions for each $v \in V$, such that the displacement of v_{k+1} is less than $\Delta_c(C)$, where C is the last δ-chain to which v_{k+1} is joined. In each stage, there will be several iterations in which the labels of vertices of the rightmost δ-chain of G_k change. Each label gives us specific equations to solve and constraints to satisfy. In every iteration we change labels of only a small subset of vertices. At the end of a stage, when $v_k v_{k+1} \le \lambda$, we have already successfully computed G_{k+1} and then we repeat the steps. Initial G_1 consists of only one vertex which is an optimal configuration without any displacement.

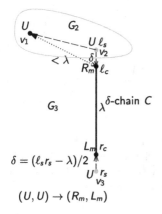

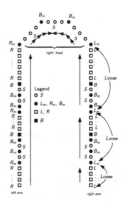

Fig. 3. In G_3 the labels of $v_2 = \ell$ and $v_3 = r$ are modified: $(U, U) \rightarrow (L_m, R_m)$.

Fig. 4. Example structure of a δ-chain (not geometrically accurate). Arrows denote the movement.

Intuitively the meaning of labels is as follows. U are the vertices which are not displaced (*unmoved*). L, R and B are vertices whose movements are constrained only by *left*, *right*, and *both* neighbors, respectively. L_m, R_m and B_m vertices are like L, R and B vertices respectively, but additionally, they are displaced by $\Delta_c(C)$, where C is the δ-chain of the vertices. The vertices labeled S are *special*; they are collinear with and equidistant to their neighbor vertices and continue to move so that they remain equidistant and collinear. See Fig. 2.

The way we label and relabel the vertices of graph G_k, the structure of a δ-chain C as a regular expression of labels emerges as follows

$$\overbrace{((S^* B_m)^* S^* B + \epsilon)(R(U + \epsilon))^*}^{\text{left arm}} (R_m S^* (B_m S^*)^* B R^* (U + \epsilon))^* \overbrace{R_m (S^* B_m S^*)^* L_m}^{\text{tight head}}$$

$$\underbrace{((U + \epsilon)L^* (B + R)(S^* B_m)^* S^* L_m)^* ((U + \epsilon)L)^* (BS^* (B_m S^*)^* + \epsilon)}_{\text{right arm}}$$

We call the middle portion of δ-chain that constraints $\Delta_c(C)$ as *tight head*, which consists of $R_m(S^*B_mS^*)^*L_m$, and the left and right subsequences of δ-chain as *left arm* and *right arm* respectively. See Fig. 4. The tight head of a δ-chain is uniquely determined as a consequence of algorithmic labeling.

The incremental algorithm works because of following generalized theorem. The proof is omitted because of lack of space.

Theorem 1. *If G_k minimizes the maximum movement then G_{k+1} also minimizes the maximum movement.*

3 Algorithm

We describe the algorithm to compute optimal positions for a given λ. Let the vertices V of G be $v_1, v_2, \ldots, v_n$ in order. Initially, G_1 consists of vertex v_1. At any stage, G_k consisting of leftmost k vertices of G and corresponding edges in an optimal configuration. See Fig. 3 for an example of G_3. Let us denote the distance $v_k v_{k+1}$ as d.

At every stage of the algorithm we add v_{k+1} to the G_k and compute G_{k+1} such that it is an optimal configuration. During computation of G_{k+1} existing labels of G_k may change and the vertices may move. Only the last δ-chain of G_{k+1}, either existing, newly created, or joined, will be affected by the movement. The procedure of moving and labeling vertices is discussed in the Subsects. 3.1, 3.2 and 3.3 below. In a few cases, when we relabel a vertex, we may also need to relabel one of its neighbors, if relative movement of the neighbor makes the configuration infeasible.

Initially, v_{k+1} vertex will be treated as U, L or L_m, depending on whether $v_k v_{k+1}$ is less than λ, between λ and $\lambda + \Delta_c(C)$ or greater than $\lambda + \Delta_c(C)$ respectively, where C is the last δ-chain to which v_{k+1} is joined or is a new δ-chain. When $v_k v_{k+1} > \lambda + \Delta_c(C)$, we conceptually decrease $v_k v_{k+1}$, denoted by d, continuously, which results in increasing $\Delta_c(C)$ and a possible change in configuration. We need to calculate the transition points of link length $v_k v_{k+1}$ when there is a change in topology of configuration, that is, when we need to relabel the vertices of last δ-chain of G_{k+1}, either existing, newly created or joined. An important property of the δ-chains is as follows.

Lemma 1. *Any δ-chain at any step of the algorithm is convex.*

Intuitively, the vertices will be relabeled if their positions, their displacements, or their collinearity conditions change such that the defining properties of the labels discussed earlier are dissatisfied.

If v_{k+1} is labeled L_m, then in every iteration, we decrease the $v_k v_{k+1}$ to a new threshold, compute the immediate next threshold where the labels change, relabel the vertices and continue till $v_k v_{k+1} \leq \lambda$, in which case, we solve for $v_k v_{k+1} = \lambda$. If v_{k+1} is labeled U or L then G_{k+1} is already optimal. In the Sect. 4, we show how we calculate the next threshold from the current configuration. First we show in the next section how an optimal configuration can be calculated from a valid labeling.

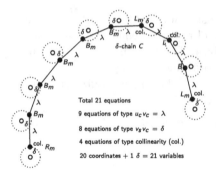

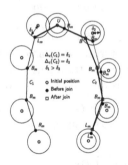

Fig. 5. Example of equations for a δ-chain C.

Fig. 6. Repositioning and relabeling of δ-chain C_2 when it is joined with δ-chain C_1

3.1 Simultaneous Equations for Solving Positions in Any δ-chain C

Let us assume that we have to determine configuration of a δ-chain C at any stage of the algorithm. Let there be m points in the δ-chain C. In a feasible configuration of G_k for any k, all link lengths are less than or equal to λ. In our algorithm, however, the link lengths will be exactly equal to λ except those links which involve U vertices. This is a consequence of the proposed algorithm. We observe that if we begin with G as a convex δ-chain, the whole graph remains convex even after displacements. This can be proved by mathematical induction on k.

To compute the configuration of the δ-chain C deterministically, we need to determine $2m$ variables, the coordinates of m vertices of C. We do this by fixing adequate number of equations. See Fig. 5. Also, we use an extra variable $\delta = \Delta_c(C)$ to simplify the set of equations. Thus we have $2m + 1$ variables to determine: $2m$ variables for m points and one variable for δ. We shall also mark a pair of equations for each vertex, when we fix $2m + 1$ equations for δ-chain C. There will be an extra unmarked equation in each δ-chain. The two equations marked for a vertex depending on its label are as follows: U: no equations needed to solve, L or R: one collinearity equation and one link-length equation, B: two link-length equations, L_m or R_m: one displacement equation and one collinearity equation, B_m: one displacement and one link-length equation, and, S: one collinearity and one link-length equation. If in any iteration of our algorithm we are able to ensure the above as well as an extra unmarked equation, then we get the right number of equations to solve.

In the following discussion, let v_s denote the initial position of $v \in V$, for example, we use $l_s, r_s, u_s, v_{i,s}$ for $l, r, u, v_i \in V$. Let v_c denote the variable for current position of $v \in V$, for example, we use $l_c, r_c, u_c, v_{i,c}$ for $l, r, u, v_i \in V$. Let uv denote the Euclidean distance between any two points $u, v \in V$. We use the notation (ℓ_1, ℓ_2) for labels of left and right vertices of a link, where ℓ_1 and ℓ_2 are one of the labels $U, L, R, B, L_m, R_m, B_m,$ or S and ℓ_1 is the label of the left vertex and ℓ_2 is the label of the right vertex. We use notation $\ell \rightarrow \ell'$

for relabeling a vertex from label ℓ to label ℓ' and notation $(\ell_1, \ell_2) \rightarrow (\ell'_1, \ell'_2)$ for relabeling two endpoints of a link from labels (ℓ_1, ℓ_2) to labels (ℓ'_1, ℓ'_2). As a shorthand notation we denote a group of cases of labels as a comma separated list of labels. Also when we say u, v and w are *collinear* in this paper, we also mean that v is in between u and w. This is important, otherwise we may get multiple solutions.

Also the link distance $v_{k,c}v_{k+1,c}$ will be equal to d instead of λ if such an equation occurs in the following discussion. The method by which the set of equations is fixed is summarized below in three broad cases, (1) a new δ-chain is created, (2) an existing δ-chain is modified, and (3) a δ-chain is deleted when two δ-chains are joined. We discuss the three cases below.

A New δ-chain is Created: A new δ-chain C is created when last vertex of G_k is a U vertex and the link length $v_k v_{k+1}$ is larger than λ. The δ-chain C is a single link containing v_k and v_{k+1}. Let ℓ stand for v_k and r for v_{k+1}. The relabeling for the corresponding link is $(U, U) \rightarrow (R_m, L_m)$ and we write five equations for five unknowns: (1) $\ell_s \ell_c = \delta$, marked for ℓ, (2) points ℓ_s, ℓ_c, and r_c are collinear, marked for ℓ, (3) $r_s r_c = \delta$, marked for r, (4) points ℓ_c, r_c and r_s are collinear, marked for r, and, (5) $\ell_c r_c = d$, unmarked for C.

An Existing δ-chain is Modified: There are following cases to be considered when any vertex in a δ-chain is relabeled. We discuss each case and specify what equations are replaced/added. Let the vertex whose label is changed be v, its left neighbor be u and its right neighbor be w.

Case $U \rightarrow R$: This happens when (1) a new vertex v immediately on the left of a δ-chain C gets added to C, resulting in a longer δ-chain, or (2) the right vertex of a U vertex in a δ-chain starts moving away. This implies that w is either labeled R, B or R_m. In this case we have two additional equations marked for v: vertices v_c, v_s and w_c are collinear, and $v_c w_c = \lambda$.

Case $L \rightarrow B$: Vertex v was earlier moving towards left and now the right vertex w is moving away relative to v, which implies w is either relabeled $L \rightarrow L_m$ or $B \rightarrow B_m$. In this case the two equations marked for v are replaced by the two equations: $u_c v_c = \lambda$, and $v_c w_c = \lambda$. *Case $R \rightarrow B$* is mirror of the case $L \rightarrow B$.

Case $R \rightarrow R_m$: Vertex v's displacement is going to be more than $\delta = \Delta_c(C)$. In this case the right vertex w needs to be relabeled too. It will be either $R \rightarrow B$ or $R_m \rightarrow B_m$. The two equations marked for v will be replaced by the equations: $v_s v_c = \delta$, and, vertices v_s, v_c and w_c are collinear. *Case $L \rightarrow L_m$* is mirror of the case $R \rightarrow R_m$.

Case $L_m \rightarrow B_m$: The right vertex w is moving away relative to v. Then w is either relabeled $L \rightarrow L_m$ or $B \rightarrow B_m$. In this case two equations marked for v will be replaced by equations: $v_s v_c = \delta$ and $v_c w_c = \lambda$. *Case $R_m \rightarrow B_m$* is mirror of the case $L_m \rightarrow B_m$.

Case $B \rightarrow B_m$: Vertex v's displacement is going to be more than $\Delta_c(C)$. In this case either the left vertex u is labeled U, L or L_m or the right vertex w

is labeled U, R or R_m, and depending on the case, either u will be relabeled $U \to R$, $L \to B$ or $L_m \to B_m$, or w will be relabeled either $U \to L$, $R \to B$ or $R_m \to B_m$, respectively. In this case two equations marked for v are replaced by equations: $v_s v_c = \delta$ and either $v_c w_c = \lambda$ (when u is relabeled) or $u_c v_c = \lambda$ (when w is relabeled).

Case $B \to S$: Either u is labeled L or L_m, or, w is labeled R or R_m, and, depending on the case, we relabel u as B or B_m, or, w as B or B_m, respectively. Then the equations for v will be replaced by equations: vertices u_c, v_c and w_c are collinear and either $v_c w_c = \lambda$ (when u is relabeled) or $u_c v_c = \lambda$ (when w is relabeled) respectively.

Case $B_m \to S$: Vertex v becomes collinear with u and w. In this case there are two equations corresponding to v. One is $v_s v_c = \delta$ and other is either $v_c w_c = \lambda$ or $u_c v_c = \lambda$. We shall replace the first equation by the equation which is then marked for v: vertices u_c, v_c and w_c are collinear.

Case $B \to R$: This case can happen when there is relatively more movement of w to the right so that v_c becomes collinear to v_s and w_c. If we relabel v from $B \to R$ then u must be relabeled $R_m \to R$, $B_m \to B$ or $S \to B$ if v is in the left arm. If v is in the right arm then this case can only happen if u, v and w make an acute angle. This can only happen a constant number of times and we deal with these separately. The equations marked for v will be a collinearity equation and a link-length equals λ as before. *Case $B_m \to R_m$* is same except for vertices with maximal displacement in the δ-chain C.

Cases $R_m \to R$ or $B_m \to B$ or $S \to B$: These cases happen when w is relabeled $B \to R$ or $B_m \to R_m$.

Two Neighbor δ-chains are Joined: Let C be the left δ-chain and C' be the right δ-chain that need to be joined. This happens when the right δ-chain C' is moving away from the left δ-chain C, and the intermediate distance will exceed λ. Let the rightmost vertex of C be u and the leftmost vertex of C' be v. The label of u is either L or L_m and the label of v is either R or R_m.

Case $\Delta_c(C) \leq \Delta_c(C')$: If $\Delta_c(C) = \Delta_c(C')$, and simultaneously if across the gap, the left vertex is an L_m vertex and the right vertex is an R_m vertex of tight head, then we relabel both of them B_m vertices. We will need to remove one of the δ variables. The two collinearity equations for u and v are also removed and we add the equation: $u_c v_c = \lambda$. Since the tight portions of both the δ-chains have one extra equation not corresponding to any vertex, we can modify the correspondences so that only one extra equation does not correspond to any vertex by reassigning correspondences in a consistent way that confirms to the labeling discussed above.

Since we are removing one variable, two equations and adding one equation, the number of variables and equations balance out.

Otherwise, in the case either $\Delta_c(C) < \Delta_c(C')$, or the rightmost vertex of C is not L_m vertex of tight head (which means right arm of C exists), or the leftmost vertex of C' is not R_m vertex of tight head (which means left arm of C' exists), we do the following.

If across the gap, the right vertex is an R vertex or the left vertex is an L vertex then we make it a B vertex and do not join the δ-chains. If we have a U vertex as the rightmost vertex of C then we cut it from C and add it to C' and relabel it R. If the B vertex of segment $(S^*B_m)^*S^*B$ in C' in the right or segment $BS^*(B_mS^*)^*$ in C in the left across the gap becomes S or B_m vertex then we make next R or L vertex, respectively, to the segment a B vertex. A B vertex allows us freedom to move. If there is a choice we relabel the right δ-chain C'.

If the B vertex of segment $(S^*B_m)^*S^*B$ in the right or segment $BS^*(B_mS^*)^*$ in the left across the gap becomes S or B_m vertex and there is R_m in C' and L_m vertex in C next to the segment on the left and right respectively, or if across the gap we have L_m in C and R_m in C', then we need to reposition and relabel a segment of C. We relabel $L_m \rightarrow B_m, S \rightarrow S$ in δ-chain C increasing their displacement to $\Delta_c(C')$ at the same time till we get to the first B, R, or $B, R, B_m \rightarrow L_m$. Then we cut the δ-chain C at that point to include B or R or exclude L_m. The B vertex might need to be relabeled R or U, and R vertex might be relabeled U if the respective conditions are satisfied. If the L_m vertex belongs to the tight head, or there is no B vertex in the tight head, then we include the tight head in the left arm of C' and then reposition and relabel the rest of δ-chain C according to the algorithm given in Subsect. 3.2.

Case $\Delta_c(C) > \Delta_c(C')$: In this case we try to reposition the whole δ-chain C' to conform to C. Since $\Delta_c(C) > \Delta_c(C')$, the vertices in δ-chain C' have now freedom to move further to a larger displacement $\Delta_c(C)$. But we cannot simply relabel the L_m, R_m and B_m vertices to L, R and B and hope for consistency, because we will then be removing several equations of the type $vv_c = \delta$ but not replacing them with new equations.

We shall reposition and relabel the vertices of δ-chain C' one by one starting from the leftmost vertex v according to the algorithm given in Subsect. 3.2. The basic idea of repositioning is to give some of the vertices on the left of C' a leftward direction movement so that the movement of the right δ-chain C' is consistent with the displacement $\Delta_c(C)$ of left δ-chain C.

By the way we add and relabel vertices we always have sufficient and necessary number of equations. Thus we have following theorem.

Theorem 2. *In k-th stage of the algorithm, in G_k, if there are m points displaced from their initial positions in a δ-chain C then there are $2m+1$ simultaneous independent equations and $2m+1$ variables with $v_k v_{k+1} = d$ as a parameter.*

Lemma 2. *All link lengths in a δ-chain at any step of the algorithm are equal to λ except those corresponding to U and R vertices in the right arm, and, U and L vertices in the left arm.*

Proof. Proof is by mathematical induction on the number of iterative stages.   □

3.2   Algorithm to Reposition and Relabeling of Vertices

Let us assume that δ-chains C and C' are being joined, C is to the left, and that $\Delta_c(C) \geq \Delta_c(C')$ (without the loss of generality). Let $\Delta_c(C) = \delta$.

Let there be m vertices in the δ-chain C'. Let the vertices of C' in order from left to right be v_i, $i = 1 \ldots m$. If $v_{1,s}$ is at a link-length less than λ from last vertex of C then we label it as U otherwise, we label it L. It will not be labeled L_m because $\Delta_c(C) \geq \Delta_c(C')$. Now, assume that we have already labeled $v_1, v_2, \ldots, v_{i-1}$ and we wish to label v_i. First we tentatively label v_i first as U or L depending on the interlink distance. In case of L, we then calculate the displacements of v_i. If the calculated displacement of v_i is less than δ, then we keep the label of v_i as L and proceed. If the calculated displacement of v_i exceeds δ, then we relabel v_i as L_m and v_{i-1} as (1) $U \rightarrow R$, or $U \rightarrow B$ if link-length of v_{i-1} from v_{i-2} exceeds λ, (2) $L \rightarrow B$, or (3) $L_m \rightarrow B_m$. In some cases v_{i-1} will have to be labeled from $U, L, L_m \rightarrow S$, if we are not able to displace v_i less than δ. Then we look for a vertex on the left one by one, till we get a vertex which will not have to be relabeled S but is either labeled R, B, B_m or relabeled $U \rightarrow R$, $U \rightarrow B$, $L \rightarrow B$, or $L_m \rightarrow B_m$. If this last vertex changes its position then we need to continue checking and relabeling backwards till we get a vertex which does not change position or label. This will be a feasible configuration for $v_1, v_2, \ldots, v_i$. Then we proceed to next i and repeat the steps. See Fig. 6. In the worst case this may take $O(n^2)$ steps.

Lemma 3. *The displacements of vertices after repositioning and relabeling of the joined δ-chain are less than or equal to δ.*

3.3   Algorithm to Relabel Vertices in Each Iteration

If last vertex of G_k is part of a δ-chain C, then we assume that $v_{k,c}v_{k+1,s} \geq \Delta_c(C) + \lambda$. Otherwise, if $v_{k,c}v_{k+1,s} < \Delta_c(C) + \lambda$, we label v_{k+1} as either L or U, if the link-length is greater than or less than λ respectively. If last vertex of G_k is not part of any δ-chain and is labeled U then we assume that $v_{k,s}v_{k+1,s} \geq \lambda$ and a new δ-chain is created (see Fig. 3). Otherwise, we label v_{k+1} as U.

In each iteration of the algorithm we have a current threshold of $d = v_{k,c}v_{k+1,c}$ and calculate the next threshold of d such that either a new δ-chain is created, any exiting δ-chain is modified, or two δ-chains are joined at the rightmost end. The threshold is such that one and only one of these events happen. If d is less than the current threshold and more than the next threshold, the positions of vertices of V can be readily computed using the simultaneous equations mentioned in the previous subsection with d as the additional parameter (Subsect. 3.1 and Sect. 4).

We calculate the next threshold value of d by giving a set of additional constraints to the above set of simultaneous equations for (1) modification of any δ-chain, (2) for a new δ-chain, or (3) the pair of neighboring δ-chains at the rightmost end. We observe that the S vertices in the structure of δ-chain can be dealt by replacing the link length by multiple of λ's and removing the

S vertices altogether. So, without loss of generality, we assume in the following discussion that no S vertices are present. Assume C is the leftmost δ-chain and contains v_{k+1}.

Case 1a (C is extended): For last δ-chain C, such that the labels of the leftmost endpoint and its left neighbor are (U, R), or (U, R_m) we will have the additional constraint as $u_s v_c \leq \lambda$. If the corresponding constraint becomes tight then following transitions will happen depending on the case: $(U, R) \rightarrow (R, R)$, or $(U, R_m) \rightarrow (R, R_m)$.

Case 1b (C is modified): For every edge uv in δ-chain C such that labels of the endpoints are (R, R) or (R, R_m) we will have the additional constraint as $u_s u_c \leq \delta$ for (R, R) or (R, R_m) where δ is the variable corresponding to $\Delta_c(C)$. If the corresponding constraint becomes tight then following transitions will happen depending on the case: $(R, R) \rightarrow (R_m, B)$, $(R, R_m) \rightarrow (R_m, B_m)$.

Case 1c (C is modified): For every vertex v in last δ-chain C with label B we will have the additional constraint as $v_s v_c \leq \delta$ if the corresponding constraint becomes tight then following transitions will happen depending on the case: $(L, B) \rightarrow (B, B_m)$, where v is right vertex, $(L_m, B) \rightarrow (B_m, B_m)$, where v is right vertex, $(B, R) \rightarrow (B_m, B)$, where v is left vertex, $(B, R_m) \rightarrow (B_m, B_m)$, where v is left vertex.

Case 1d (C is modified): For each vertex v in C with label B or B_m, we will have a constraint that u_c, v_c and w_c are clockwise convex. If the corresponding constraint becomes tight then we relabel v as S.

There are a few special cases when (1) a B vertex becomes R, or, (2) a B_m vertex becomes R_m. This case can happen only a constant number of times. The constraint for this case is that v_s, v_c and w_c are counter-clockwise convex. We deal with these cases separately. There is also a special case when $R \rightarrow R_m$ in the right arm, when we need to break the δ-chain.

Case 2 — Two leftmost δ-chains C' and C are joined: For leftmost δ-chains C' and C which are neighbors there is a chance that they may be joined when C moves away to the right. Let the rightmost vertex of C' is u and the leftmost vertex of C is v. We will have additional equation as $uv \leq \lambda$ for both the set of simultaneous equations for C' and C with renaming of δ variable as δ' and δ. If the corresponding constraint becomes tight then we reposition and relabel the δ-chain which have smaller maximum displacement.

We show how the set of simultaneous equations can be solved along with a set of constraints in (Sect. 4) such that at the next threshold one and only one of the constraints become satisfied.

4 Algorithm to Solve Simultaneous Equations with Constraints Computationally with and Without Using Jacobian

There are a few methods to solve a set of quadratic simultaneous equations. We can solve it using Jacobian method [8]. Jacobian method improves the solution iteratively using the formula $\mathbf{x}_{n+1} = \mathbf{x}_n - \mathbf{J}^{-1}\mathbf{f}(\mathbf{x}_n)$ to solve $\mathbf{f}(\mathbf{x}) = \mathbf{0}$.

Anyhow, we can do better by reducing the equations to a single function f of single variable δ, so that the Jacobian method reduces to Newton-Raphson method. Consequently we have the following lemma.

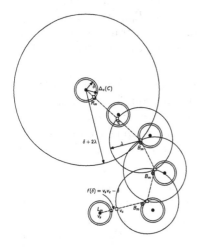

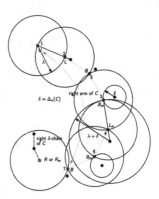

Fig. 7. Solving simultaneous equations on tight head portion of a δ-chain.

Fig. 8. Solving simultaneous equations on left or right arm of a δ-chain, with an optional appendage. $\Delta_c(C)$ is already determined from tight head. The order in which the points are determined is shown by numbers on them. The δ-chain has a right arm with a right appendage.

Lemma 4. *The simultaneous equations for any δ-chain can be solved in $O(n)$ time computationally.*

Proof. Let the δ-chain be C. Also assume without loss of generality that it is not the rightmost δ-chain. The rightmost δ-chain will have d as an extra variable and we need to calculate the next threshold using this extra variable. Otherwise, the method is exactly same. We also assume that there are B_m vertices in the tight head of δ-chain, otherwise the solution is straight forward.

First we compute $\Delta_c(C)$ using the tight head portion of the δ-chain. We can either start from left or right. Let us start from left. We take initial value of δ as the value of $\Delta_c(C)$ of the previous threshold. With the current value of δ we find the location of first B_m vertex. If there are k intermediate S vertices, then it is a point at a distance δ from its original position, and a distance $\delta + (k+1)\lambda$ from the original position of the leftmost R_m vertex. After the position of first B_m vertex is determined we can determine the location of leftmost R_m vertex and the intermediate S vertices. Next we determine the position of next undetermined B_m vertex. It will be a point at a distance δ from its original position and $(k'+1)\lambda$ from the last determined B_m vertex, where k' is the number of intermediate S vertices. When we arrive at the rightmost L_m vertices it may not be at a distance of $\lambda + (k''+1)\delta$ from the last B_m vertex, where k'' is the number of intermediate

S vertices from last B_m vertex to R_m vertex. This is because the value of δ may not be equal to $\Delta_c(C)$. The signed difference between the actual distance and $\lambda + (k'' + 1)\delta$ is denoted by $f(\delta)$ which we need to make zero. To do this, we repeat the steps with $\delta + h$ instead of δ where h, $h \ll \delta$, is infinitesimally small. $(f(\delta + h) - f(\delta))/h$ will give us an approximation of the derivative of f. We can then use any Newton-Raphson like methods to solve for $\Delta_c(C)$. See Fig. 7.

Once $\Delta_c(C)$ is determined we can easily determine the positions in left and right arm of the δ-chain C according to the labels. See Fig. 8. Also observe that each iteration of Newton-Raphson is $O(n)$. The convergence of Newton Raphson method is $O(1)$ steps however huge it may be, so the position in a δ-chain can be determined from its labels in $O(n)$.   □

At the last iteration of algorithm when $d = \lambda$ the method described in the proof can be used without change. In other iterations, the usual method can not be directly used due to presence of inequality constraints with accompanying minimizing of d. Moreover, instead of solving set of simultaneous equation repeatedly several times with only a single inequality constraint one at a time, we modify the Newton-Raphson method to take care of all inequality constraints together as discussed below.

Since we have many constraints (which are strict inequalities) and one free variable d (which is to be minimized), our task is to decrease d till one and only one of the constraints become tight, which gives us the next threshold of d. We iterate d using another Newton-Raphson method, computing d_n in n-th iteration. At any iteration three things can happen: Either all constraints are inequalities, one constraint is equality, or some of the constraints are dissatisfied. If all constraints are inequalities, we compute the difference Δd_n, following Newton-Raphson method, such that at least one constraint becomes equality. We will need to decrease d by the amount given by rate of change. If there are some constraints that are dissatisfied then we compute Δd_n so that all dissatisfied constraints are satisfied. In this case, we will need to increase d. We repeat the Newton-Raphson method iterations, till we get the solution of required precision.

As mentioned earlier, we can check the conditions of degeneracy in the algorithm while solving the simultaneous equations. The *condition of degeneracy* occurs if more than one inequality become strict equalities simultaneously. We assume the general case where this does not happen. Even otherwise, we can take care of degeneracies by breaking ties arbitrarily.

Consequently we have the following theorem.

Theorem 3. *The algorithm described in this section computes Δ correctly and the time complexity of the algorithm in $O(n^3 + n^2 T(n))$ where $T(n)$ is the time complexity of solving simultaneous equations with constraints.*

5   Generalisations

5.1   Algorithm for Ring Network

Suppose we are given the vertices in a convex ring network $G = (V, E)$ and we need to compute the optimal movement of sensors for connectivity. The only change of the algorithm is at the stage when we add the last vertex, which is also the first vertex of the δ-chain. If the last link is less than λ then we do not need to do anything, and we get the solution. However, if the last distance, $d = v_{n,c}v_{1,c}$, is larger than λ then we conceptually decrease d continuously, jumping through thresholds of d and solving for optimal configuration for each threshold, till d is less than or equal to λ. Then we solve for $d = \lambda$ using the relevant equations and an additional condition as discussed in this section. If we have more than two δ-chains at the last stage the solution can be managed. Even in the case of single δ-chain if there is left/right arm in the δ-chain we can do away with usual relabeling/joining.

The only complex case arises is when the whole ring network becomes a tight head and we need to join the last link. The problem is that when we add the corresponding length d in the set of simultaneous equations, we need to remove two equations, but add a single equation. In that case we have an extra degree of freedom. The equations then normaly will have infinite number of solutions, assuming they are consistent and non-degenerate. We need to minimize δ in the simultaneous equations among these infinite number of solutions. The way to minimize δ is adding a further constraint that the configuration cannot be rotated either left or right without increasing δ. This translates to the condition $\partial\delta/\partial\theta = 0$, where θ is angle of rotation for any arbitrary vertex from some arbitrary interior point inside the ring, the center of minimum enclosing circle will be good choice as it will always be in the interior irrespective of how small λ gets. We use the usual Jacobian method, to get the value of θ for which $\partial\delta/\partial\theta = 0$.

5.2 Algorithm for Non-convex Path Networks

Suppose we are given a path which is not convex where we need to minimize the movement. As noted earlier Whenever $L, R, L_m, R_m \to^* B, B_m \to^* L, R, L_m, R_m$ transmission occurs, an acute angle in the path is removed. Thus we can use the same algorithm except that the complexity of the algorithm will increase by a linear factor. So the algorithm for non-convex paths has complexity $O(n^4)$.

5.3 Algorithm for Heterogeneous and Tethered Networks

If we are given similar problems as discussed above except that target link-length are not uniform and the target network is a tethered network where nodes are tethered and sensors are located on edges as paths, then for this kind of heterogeneous and tethered network too, we can solve the problem of min-max movement.

References

1. Bredin, J., Demaine, E.D., Hajiaghayi, M.T., Rus, D.: Deploying sensor networks with guaranteed fault tolerance. IEEE/ACM Trans. Netw. **18**(1), 216–228 (2010)
2. Chen, D.Z., Gu, Y., Li, J., Wang, H.: Algorithms on minimizing the maximum sensor movement for barrier coverage of a linear domain. Discret. Comput. Geom. **50**(2), 374–408 (2013)

3. Corke, P.I., Hrabar, S., Peterson, R.A., Rus, D., Saripalli, S., Sukhatme, G.S.: Autonomous deployment and repair of a sensor network using an unmanned aerial vehicle. In: Proceedings of the 2004 IEEE International Conference on Robotics and Automation, pp. 3602–3608 (2004)
4. Czyzowicz, J., Kranakis, E., Krizanc, D., Lambadaris, I., Narayanan, L., Opatrny, J., Stacho, L., Urrutia, J., Yazdani, M.: On minimizing the maximum sensor movement for barrier coverage of a line segment. In: Ruiz, P.M., Garcia-Luna-Aceves, J.J. (eds.) ADHOC-NOW 2009. LNCS, vol. 5793, pp. 194–212. Springer, Heidelberg (2009). doi:10.1007/978-3-642-04383-3_15
5. Demaine, E.D., Hajiaghayi, M.T., Mahini, H., Sayedi-Roshkhar, A.S., Gharan, S.O., Zadimoghaddam, M.: Minimizing movement. ACM Trans. Algorithms 5(3), 30 (2009)
6. Demaine, E.D., Hajiaghayi, M.T., Marx, D.: Minimizing movement: fixed-parameter tractability. ACM Trans. Algorithms 11(2), 14:1–14:29 (2014)
7. Friggstad, Z., Salavatipour, M.R.: Minimizing movement in mobile facility location problems. In: 49th Annual IEEE Symposium on Foundations of Computer Science, pp. 357–366 (2008)
8. Jacobi, C.G.J.: Ueber eine neue auflsungsart der bei der methode der kleinsten quadraten vorkommenden lineare gleichungen. Astr. Nachr. 22(523), 297–306 (1845)
9. Kumar, S., Lai, T., Arora, A.: Barrier coverage with wireless sensors. Wireless Netw. 13(6), 817–834 (2007)
10. Wang, G., Cao, G., La Porta, T.F.: Movement-assisted sensor deployment. IEEE Trans. Mob. Comput. 5(6), 640–652 (2006)
11. Wareham, T.: Exploring algorithmic options for the efficient design and reconfiguration of reactive robot swarms. In: Proceedings of BICT 2015, pp. 295–302 (2016)

On Colouring Point Visibility Graphs

Ajit Arvind Diwan and Bodhayan Roy[⊠]

Department of Computer Science and Engineering, Indian Institute of Technology
Bombay, Mumbai 400076, India
{aad,broy}@cse.iitb.ac.in

Abstract. In this paper we show that the problem of deciding whether
the visibility graph of a point set is 5-colourable, is NP-complete. We
give an example of a point visibility graph that has chromatic number 6
while its clique number is only 4.

Keywords: Point set · Point visibility graph · Chromatic number ·
Graph colouring · NP-complete

1 Introduction

The visibility graph is a fundamental structure studied in the field of computational geometry and geometric graph theory [4,8]. Some of the early applications of visibility graphs included computing Euclidean shortest paths in the presence of obstacles [12] and decomposing two-dimensional shapes into clusters [17]. Here, we consider problems concerning the colouring of visibility graphs.

Let P be a set of points $\{p_1, p_2, \ldots, p_n\}$ in the plane. Two points p_i and p_j of P are said to be *mutually visible* if there is no third point p_k on the line segment joining p_i and p_j. Otherwise, p_i and p_j are said to be mutually *invisible*. The *point visibility graph* (denoted as PVG) $G(V, E)$ of P is defined as follows. The set V of vertices contains a vertex v_i for every point p_i in P. The set E contains an edge (v_i, v_j) if and only if the corresponding points p_i and p_j are mutually visible [9,11]. Point visibility graphs have been studied in the contexts of construction [2,6], recognition [1,9,10,16], partitioning [5], connectivity [14], chromatic number and clique number [3,11,15].

A graph is said to be *k-colourable* if each vertex of the graph can be assigned a colour, so that no two adjacent vertices are assigned the same colour, and the total number of distinct colours assigned to the vertices is at most k. Kára et al. characterized PVGs that are 2-colourable and 3-colourable [11]. It was not known whether the chromatic number of a PVG can be found in polynomial time. In Sect. 2 we show that the problem of deciding whether a PVG is k-colourable, for $k \geq 5$, is NP-complete.

Kára et al. also asked whether there is a function f such that for every point visibility graph G, $\chi(G) \leq f(\omega(G))$ [11]. They presented a family of PVGs that have their chromatic number lower bounded by an exponential function of their clique number. Their question was answered by Pfender, showing that

© Springer International Publishing AG 2017
D. Gaur and N.S. Narayanaswamy (Eds.): CALDAM 2017, LNCS 10156, pp. 156–165, 2017.
DOI: 10.1007/978-3-319-53007-9_14

for a PVG with $\omega(G) = 6$, $\chi(G)$ can be arbitrarily large [15]. However, it is not known whether the chromatic number of a PVG is bounded, if its clique number is only 4 or 5. In another related paper, Cibulka et al. showed that PVGs of point sets S such that there is no convex pentagon with vertices in S and no other point of S lying in the pentagon, might have arbitrarily large clique numbers [3]. In this direction, Kára et al. showed that there is a PVG G with $\omega(G) = 4$ and $\chi(G) = 5$ [11]. In Sect. 3.1 we provide some examples of PVGs whose clique number is 4 but chromatic number is 5. These examples give some forbidden induced subgraphs in 4-colourable PVGs. We do not know if they form a complete list. A complete list of such graphs would lead to a characterization of 4-colourable PVGs and possibly a polynomial time algorithm for recognizing them. In Sect. 3.2 we construct a graph G' with $\omega(G') = 4$ and $\chi(G') = 6$.

2   5-Colouring Point Visibility Graphs

In this section we prove that deciding whether a PVG with a given embedding is 5-colourable, is NP-hard. We provide a reduction of 3-SAT to the PVG 5-colouring problem. We use the reduction of 3-SAT to the 3-colouring problem of general graphs.

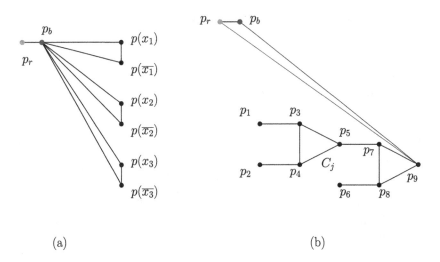

(a)   (b)

Fig. 1. (a) A variable gadget. (b) A clause gadget. (Color figure online)

2.1   3-Colouring a General Graph

For convenience, we first briefly describe the reduction for the 3-colouring of general graphs, considering the graph as an embedding ξ of points and line-segments in the plane [7]. Consider a 3-SAT formula θ with variables $x_1, x_2, \ldots, x_n$ and clauses $C_1, C_2, \ldots, C_m$. Suppose the corresponding graph is to be coloured with

$$\theta = (x_1 \vee \overline{x_2} \vee x_3) \wedge (\overline{x_1} \vee x_2 \vee \overline{x_4}) \wedge (x_2 \vee x_3 \vee x_4)$$

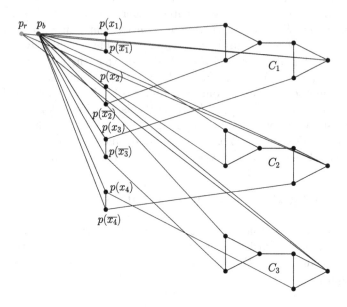

Fig. 2. The full embedding for a given 3-SAT formula.

red, green and blue. Consider Fig. 1(a). It shows the variable-gadgets. The points representing a variable x_i and its negation (say, $p(x_i)$ and $p(\overline{x_i})$, respectively) are adjacent to each other, making n pairs altogether. No two points in different pairs are adjacent to each other. A separate point p_b is wlog assumed to be blue and made adjacent to all the other points in the variable gadgets. So, each variable point must have exactly one red and one green point. For variable points, let green and red represent an assignment of true and false, respectively. The point p_b is also adjacent to a separate point p_r assumed to be red.

Now consider Fig. 1(b). It shows a clause gadget. Suppose that the points p_1, p_2 and p_6 can be coloured only with green and red. Then p_9 can be coloured with green if and only if at least one of p_1, p_2 and p_6 are coloured with green. To prevent p_9 from being coloured red or blue, p_9 is made adjacent to p_r and p_b.

The whole embedding corresponding to the 3-SAT formula is shown in Fig. 2. The points p_r and p_b are the same for all variables and clauses. For each clause gadget, the points corresponding to p_1, p_2 and p_6 are in the respective variable gadgets. Thus, for n variables and m clauses, ξ has $2n + 6m + 2$ points in total.

2.2   Transformation to a Point Visibility Graph

Consider the following transformation of ξ into a new embedding ζ (Fig. 3(a)) for a given 3-SAT formula θ. We use some extra points called *dummy points* to act as blockers during the transformation.

$$\theta = (x_1 \vee \overline{x_2} \vee x_3) \wedge (\overline{x_1} \vee x_2 \vee \overline{x_4}) \wedge (x_2 \vee x_3 \vee x_4)$$

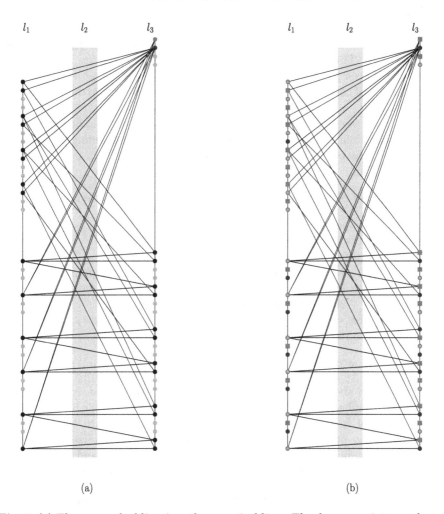

(a)   (b)

Fig. 3. (a) The new embedding ζ on three vertical lines. The dummy points are shown in gray. (b) A 3-colouring of ζ representing $x_1, x_2, \overline{x_3}$ and $\overline{x_4}$ assigned 1 in θ. The red points, blue points and green points are represented by squares, dark circles and light circles respectively. (Color figure online)

(a) All points of ξ are embedded on two vertical lines l_1 and l_3.

(b) Two points p_r and p_b are placed on l_3 above all other points of ξ, followed by two dummy points.

(c) Each pair of variable gadget points are embedded as consecutive points on l_1.

(d) Separating a variable gadget from the next variable gadget are two dummy points.

(e) For every clause gadget, the points corresponding to p_5 and p_9 are on l_1, separated by two dummy points.
(f) For every clause gadget, the points corresponding to p_3, p_4, p_7 and p_8 are on l_3, in the vertical order from top to bottom. The points p_3 and p_4 are consecutive. The points p_7 and p_8 are consecutive. There are two dummy points between p_4 and p_7.
(g) The points of consecutive clause gadgets are separated by two dummy points each.
(h) Let l_2 be a vertical line lying between l_1 and l_3. On l_2, embed points to block all visibility relationships other than those corresponding to edges in ξ. Perturb the points of l_1 and l_3 so that each point in l_2 blocks exactly one pair of points.

The total number of points needed in the new embedding is as follows:

- p_r and p_b are 2 points.
- Variable gadgets are $2n$ points.
- Clause gadgets are $2m$ points on l_1 and $4m$ points on l_3.
- There are $2n + 4m - 2$ dummy points on l_1 and $4m$ dummy points on l_3.
- Thus, there are $4n + 6m - 2$ points on l_1 and $8m + 2$ points on l_3.
- There are $9m + 2n$ edges from ξ between l_1 and l_3.
- Thus, there are $(4n + 6m - 2)(8m + 2) - (9m + 2n)$ points on l_2 to block the visibility of the rest of the pairs.

Lemma 1. *The above construction can be achieved in polynomial time.*

Proof. As shown above, the number of points used is polynomial. All the points of l_1 and l_3 are embedded on lattice points. The intersections of the line segments joining points of l_1 and l_3 are computed, and l_2 is chosen such that none of these intersection point lies on l_2. To block visibilities, the intersection of the line segments joining points of l_1 and l_3 with l_2 are computed and blockers are placed on the intersection points. All of this is achievable in polynomial time. □

Lemma 2. *The PVG of ζ can be 5-coloured if and only if θ has a satisfying assignment.*

Proof. Suppose that θ has a satisfying assignment. Then the points of ζ obtained from ξ can be coloured with red, blue and green. The pairs of consecutive dummy points that are adjacent to a green and red point, can be coloured with red and green. Otherwise they can be coloured with blue and green or red. The points on l_2 are coloured alternately with two different colours (Fig. 3(b)).

Now suppose that ζ has a 5-colouring. All points of l_2 are visible from all points of l_1 and l_3. The points of l_2 must be coloured at least with two colours. This means, the points of the graph induced by ξ are coloured with at most 3 colours, which is possible only when θ is satisfiable. □

We have the following theorem.

Theorem 1. *The problem of deciding whether the visibility graph of a given point set is 5-colourable, is NP-complete.*

Proof. A 5-colouring of a point visibility graph can be verified in polynomial time. Thus, the problem is in NP. On the other hand, 3-SAT can be reduced to the problem. Given a 3-SAT formula θ, by Lemmas 1 and 2, a point set ζ can be constructed in time polynomial in the size of θ such that ζ can be 5-coloured if and only if θ has a satisfying assignment. Thus, the problem is NP-complete. □

3   Colouring a Point Set with Small Clique Number

In Sect. 2 we proved that the problem of deciding whether a PVG is 5-colourable, is NP-complete. The remaining problem is that of characterizing 4-colourable PVGs. A possible approach is to identity a minimal set of non 4-colourable graphs such that every PVG that is not 4-colourable contains one of these as an induced subgraph. For non 2-colourable PVGs, the triangle is the only such graph, and for non 3-colourable PVGs, K_4 is the only one. In this section, we identify some examples of such graphs.

3.1   Graphs with $\omega(G) = 4$ and $\chi(G) = 5$

In general, graphs with small clique numbers can have arbitrarily large chromatic numbers. In fact, there exist triangle free graphs with arbitrarily high chromatic numbers due to the construction of Mycielski [13]. Pfender showed that for a PVG with $\omega(G) = 6$, $\chi(G)$ can be arbitrarily large [15]. But it is not known whether the chromatic number of a PVG is bounded, if its clique number is only 4 or 5. Kára et al. [11] showed that a PVG with clique number 4 can have chromatic number 5 (Fig. 4(a)). They then generalized the example to prove that there is an exponential function f such that for a family of PVGs, the identity $\chi(G) = f(\omega(G))$ holds for all graphs in the family.

We first give a simpler example of a PVG (say, G_1) that has clique number 4 but chromatic number 5 (Fig. 4(b)). To construct G_1, we consider a P_2 and C_5, and form its join, i.e. we join all the vertices of P_2 with all the vertices of C_5. The graph G_1 (Fig. 4(b)), has $\omega(G_1) = 4$ and $\chi(G_1) = 5$ and can be generalized to a family of graphs with the same property. Let G_k be the graph formed by joining all the vertices of P_{2k} with all the vertices of C_{2k+3}. We have the following lemma.

Lemma 3. *G_k is a PVG that requires five colours.*

Proof. Embed $2k + 2$ consecutive vertices of C_{2k+3} on a line, and embed the other vertex of C_{2k+3} (say, u) outside of the line. Now embed the $2k$ vertices of P_{2k} on a line to block u from all but first and last vertices on the line. Hence we get a visibility embedding of G_k.

Now we prove that G_k requires five colours. In G_k, all the vertices of P_{2k} are adjacent to all the vertices of C_{2k+3}, so that the total number of colours required is $\chi(P_{2k}) + \chi(C_{2k+3})$. Since $\chi(P_{2k}) = 2$ and $\chi(C_{2k+3}) = 3$, we have the required result.   □

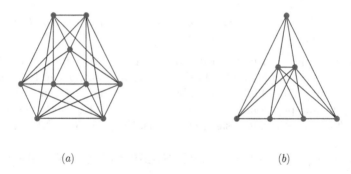

Fig. 4. (a)Kára et al's example for $\omega(G) = 4$ and $\chi(G) = 5$. (b) Our example for $\omega(G) = 4$ and $\chi(G) = 5$.

If for a graph H, $\omega(H) = 4$ and H has a graph from the above family as a subgraph, then $\chi(H) = 5$. However, if H has the join of C_{2k+3} and P_2 as an induced subgraph, which can be checked in polynomial time, then too $\chi(H) = 5$. Fig. 5 shows three particular minimal graphs on nine vertices with $\omega(G) = 4$ and $\chi(G) = 5$, apart from the example of Kára et al., which do not have the join of C_{2k+3} and P_2 as an induced subgraph. Unlike G_1, it is not clear to us whether or not these graphs can be generalized to families of graphs with the same property. If there are only finitely many such PVGs, then the problem of 4-colouring PVGs can be solved in polynomial time.

3.2   A Graph with $\omega(G) = 4$ and $\chi(G) = 6$

The main question remaining is whether PVGs with maximum clique size 4 have bounded chromatic number. Here we construct a graph G' with $\omega(G') = 4$ and $\chi(G') = 6$ (Fig. 6). We construct G' directly as a visibility embedding, as follows.

1. Consider three horizontal lines l_1, l_2 and l_3 parallel to each other.
2. On the first line, embed ten points $\{p_1, p_2, \ldots, p_{10}\}$ from left to right.
3. From left to right embed the points $q_1, \ldots, q_4$ on l_3.
4. Join with line segments the pairs (q_1, p_1), (q_1, p_4), (q_2, p_2), (q_2, p_5).
5. Join with line segments the pairs (q_3, p_6), (q_3, p_9), (q_4, p_7), (q_4, p_{10}).
6. Starting from the right of q_4, embed the points $r_1, b_1, r_2, b_2, \ldots, r_{10}$ on l_3.
7. Join each r_i with $1 \leq i \leq 5$ with p_1, p_3 and p_{i+5}.
8. Join each r_i with $6 \leq i \leq 10$ with p_1, p_4 and p_i.
9. Embed points on l_2 such that only the adjacencies described from steps 4 to 8 hold.

We have the following lemmas.

Lemma 4. *The clique number of G' is 4.*

Proof. By construction, the points on l_1 and l_3 together induce a triangle free graph. The points of l_2 can contribute at most two more points to cliques induced by the points on l_1 and l_3. So, G' has cliques of size at most four.   □

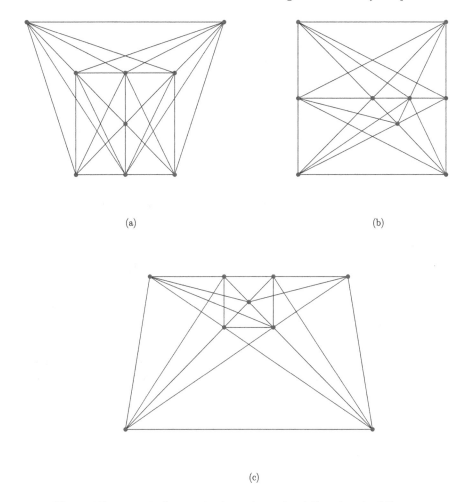

(a)   (b)

(c)

Fig. 5. Three particular minimal graphs with $\omega(G) = 4$ and $\chi(G) = 5$.

Lemma 5. *The chromatic number of G' is 6.*

Proof. Each point on l_2 is adjacent to every point on l_1 and l_3, so the points on l_2 require two colours which are absent from the points of l_1 and l_3. So, it suffices to show that the graph induced by points on l_1 and l_3 is not three colourable. Suppose that the points $p_1, \ldots p_5$ have only two colours (say, C_1 and C_2). This means that they are coloured with C_1 and C_2 alternately, and q_1 and q_2 must be coloured with two extra colours, C_3 and C_4 respectively.

Now suppose that all three of C_1, C_2 and C_3 occur among $p_1, \ldots p_5$. Similarly, all three of C_1, C_2 and C_3 occur among $p_6, \ldots p_{10}$, for otherwise the previous argument is applicable to q_3, q_4 and $p_6, \ldots p_{10}$. Also, p_1 and p_3, or p_1 and p_4 must have two distinct colours. On the other hand, three distinct colours must also occur among $p_6, \ldots p_{10}$. But for every p_i, $6 \leq i \leq 10$, there are two points

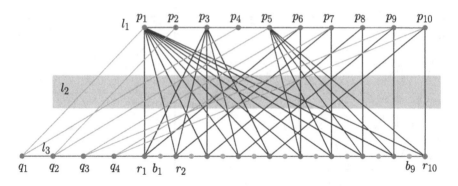

Fig. 6. A point visibility graph with clique number four but chromatic number six.

among $r_1, r_2, \ldots, r_{10}$ that are adjacent to p_i and one pair among $\{p_1, p_3\}$ and $\{p_1, p_4\}$. So, at least one of the points among $r_1, r_2, \ldots, r_{10}$ must have the fourth colour. $\qquad\square$

4   Concluding Remarks

We have shown that there is a PVG with clique number four but chromatic number six. However, we do not know whether a PVG with clique number four can have a greater chromatic number or not. We have also shown that the 5-colour problem is hard on PVGs, but the 4-colour problem on PVGs still remains unsolved.

Acknowledgements. The authors are grateful to the referees for their valuable comments and specially pointing out reference [15].

References

1. Cardinal, J., Hoffmann, U.: Recognition and complexity of point visibility graphs. In: Symposium of Computational Geometry, pp. 171–185 (2015)
2. Chazelle, B., Guibas, L.J., Lee, D.T.: The power of geometric duality. BIT **25**, 76–90 (1985)
3. Cibulka, J., Kyncl, J., Valtr, P.: On planar point sets with the pentagon property. In: Proceedings of the Twenty-Ninth Annual Symposium on Computational Geometry, pp. 81–90 (2013)
4. De Berg, M., Cheong, O., Kreveld, M., Overmars, M.: Computational Geometry, Algorithms and Applications, 3rd edn. Springer, Heidelberg (2008)
5. Diwan, A.A., Roy, B.: Partitions of planar point sets into polygons. In: Proceedings of the 28th Canadian Conference on Computational Geometry, pp. 147–154 (2016)
6. Edelsbrunner, H., O'Rourke, J., Seidel, R.: Constructing arrangements of lines and hyperplanes with applications. SIAM J. Comput. **15**, 341–363 (1986)
7. Garey, M.R., Johnson, D.S.: Computers and Intractability: A Guide to the Theory of NP-Completeness. W.H Freeman and Company, New York (1979)

8. Ghosh, S.K.: Visibility Algorithms in the Plane. Cambridge University Press, New York (2007)
9. Ghosh, S.K., Goswami, P.P.: Unsolved problems in visibility graphs of points, segments and polygons. ACM Comput. Surv. **46**(2), 22:1–22:29 (2013)
10. Ghosh, S.K., Roy, B.: Some results on point visibility graphs. Theor. Comput. Sci. **575**, 17–32 (2015)
11. Kára, J., Pór, A., Wood, D.R.: On the chromatic number of the visibility graph of a set of points in the plane. Discret. Comput. Geom. **34**(3), 497–506 (2005)
12. Lozano-Perez, T., Wesley, M.A.: An algorithm for planning collision-free paths among polyhedral obstacles. Commun. ACM **22**, 560–570 (1979)
13. Mycielski, J.: Sur le coloriage des graphes. Colloq. Math. **3**, 161–162 (1955)
14. Payne, M.S., Pór, A., Valtr, P., Wood, D.R.: On the connectivity of visibility graphs. Discret. Comput. Geom. **48**(3), 669–681 (2012)
15. Pfender, F.: Visibility graphs of point sets in the plane. Discret. Comput. Geom. **39**(1), 455–459 (2008)
16. Roy, B.: Point visibility graph recognition is NP-hard. Int. J. Comput. Geom. Appl. **26**(1), 1–32 (2016)
17. Shapiro, L.G., Haralick, R.M.: Decomposition of two-dimensional shape by graph-theoretic clustering. IEEE Trans. Pattern Anal. Mach. Intell. **PAMI-1**, 10–19 (1979)

Decomposing Semi-complete Multigraphs and Directed Graphs into Paths of Length Two

Ajit A. Diwan$^{(\boxtimes)}$ and Sai Sandeep

Department of Computer Science and Engineering, Indian Institute of Technology
Bombay, Powai, Mumbai 400076, India
{aad,saisandeep}@cse.iitb.ac.in

Abstract. A P_3-decomposition of a graph is a partition of the edges of the graph into paths of length two. We give a simple necessary and sufficient condition for a semi-complete multigraph, that is a multigraph with at least one edge between each pair of vertices, to have a P_3-decomposition. We show that this condition can be tested in strongly polynomial-time, and that the same condition applies to a larger class of multigraphs. We give a similar condition for a P_3-decomposition of a semi-complete directed graph. In particular, we show that a tournament admits a P_3-decomposition iff its outdegree sequence is the degree sequence of a simple undirected graph.

Keywords: Path decomposition · Semi-complete · Multigraph

1   Introduction

A decomposition of a graph (multigraph, directed graph) G is a collection of edge-disjoint subgraphs $G_1, G_2, \ldots, G_k$ of G, whose union is G. A decomposition in which each subgraph G_i is isomorphic to a fixed graph H, is called an H-decomposition of G.

Decompositions of complete graphs have been studied for a long time, in the context of combinatorial designs. A (v, k, λ)-design is a collection of k element subsets of a v-element set, such that every pair of distinct elements is contained in exactly λ subsets. A (v, k, λ)-design can be easily seen to be equivalent to a K_k-decomposition of the λ-complete multigraph of order v, in which there are λ edges between each pair of vertices. Perhaps the most significant result on designs is Wilson's theorem [13], that if v, k, λ satisfy some obvious necessary divisibility conditions, and if v is large enough for fixed k and λ, then a (v, k, λ)-design always exists.

There are many other problems on decomposition of complete graphs. The Oberwolfach problem [1] was to characterize 2-regular graphs of order $2n + 1$ that decompose K_{2n+1}. An old conjecture of Ringel [11] states that K_{2n+1} has a T-decomposition for any tree T with n edges. Another conjecture of Gyarfas [6]

© Springer International Publishing AG 2017
D. Gaur and N.S. Narayanaswamy (Eds.): CALDAM 2017, LNCS 10156, pp. 166–176, 2017.
DOI: 10.1007/978-3-319-53007-9_15

states that if $T_1, T_2, \ldots, T_n$ are trees with T_i having i edges, then K_{n+1} has a decomposition into subgraphs $G_1, G_2, \ldots, G_n$ such that G_i is isomorphic to T_i, for $1 \leq i \leq n$.

On the other hand, for general graphs, Dor and Tarsi [4] showed that the H-decomposition problem is NP-Complete for any fixed graph H having a connected component with at least 3 edges. Brys and Lonc [2], showed that H-decomposition is polynomial-time solvable for graphs H in which each component has at most 2 edges.

In this paper, we consider decomposition problems for semi-complete multigraphs and directed graphs. A multigraph or directed graph is said to be semi-complete if there is at least one edge between every pair of vertices. These are more general than complete graphs, but with sufficient structure that may enable decomposition problems to be solved efficiently. In particular, we consider the simplest non-trivial decomposition problem, that of decomposing a graph into paths of length 2.

A P_3-decomposition of an arbitrary graph or directed graph can be found in polynomial-time by a reduction to the perfect matching problem in the line graph. In the case of multigraphs also, the problem can be solved in strongly polynomial-time by a reduction to the b-matching problem. However, the line graphs may contain $\Omega(n^2)$ vertices and $\Omega(n^3)$ edges for a graph with n vertices. This gives an inefficient algorithm, although it runs in polynomial-time. In the case of simple undirected graphs, a much simpler characterization is given by the classical theorem of Kotzig [9], that a P_3-decomposition exists iff each connected component has an even number of edges. No such simple condition is known for multigraphs or directed graphs, although some partial results can be found in [3, 10].

We give a simpler condition, not involving the line graph, and a faster algorithm to decide whether a given semi-complete multigraph or directed graph has a P_3-decomposition. In the case of multigraphs, we show that the same condition applies to a larger class of multigraphs, characterized by forbidden induced subgraphs. Our hope is that studying the simplest case of P_3-decomposition would help in generalizing some of the classical results on decomposition of complete graphs to semi-complete multigraphs and directed graphs.

2   Multigraphs

We consider a multigraph of order n to be the complete graph on n vertices with a non-negative integer weight w_{ij}, called the multiplicity, assigned to each edge ij. A multigraph is said to be semi-complete if $w_{ij} > 0$ for all edges ij. We call it semi-complete to emphasize the fact that the multiplicities of edges could be different, although all are positive. The size of a multigraph is the sum of the weights of all edges in the complete graph. The degree of a vertex u in a multigraph G is the sum of weights of edges incident with u, and is denoted $d_G(u)$. The underlying simple graph of a multigraph is obtained by deleting all edges of weight 0 and ignoring weights of other edges.

Let $\mathcal{P}_3$ denote the set of all paths of order 3 in the complete graph. For an edge ij, let $\mathcal{P}_{ij}$ denote the set of paths of order 3 that contain the edge ij. A P_3-decomposition of a multigraph is a function $f : \mathcal{P}_3 \to \mathbb{N}$, such that for every edge ij,

$$\sum_{P \in \mathcal{P}_{ij}} f(P) = w_{ij}.$$

A matching M in a graph is a collection of disjoint edges. For any matching M, let $N(M)$ denote the set of edges in the graph that are not in M but have an endpoint in common with some edge in M. For any subset S of edges, let $V(S)$ denote the subset of vertices that are incident with some edge in S. Let $w(S) = \sum_{e \in S} w_e$ denote the sum of weights of edges in S.

Lemma 1. *A semi-complete multigraph G of even size has a P_3-decomposition iff for every matching S in G, $w(N(S)) \geq w(S)$.*

Proof. Suppose G has a P_3-decomposition f. Any path of order 3 that contains an edge in S must contain an edge in $N(S)$. Also, it cannot contain any other edge in S. Therefore,

$$w(S) = \sum_{e \in S} w_e = \sum_{e \in S}\left(\sum_{P \in \mathcal{P}_e} f(P)\right) \leq \sum_{e \in N(S)}\left(\sum_{P \in \mathcal{P}_e} f(P)\right) = w(N(S).$$

To prove the converse, we use Tutte's 1-factor theorem [12]. Consider a new graph $L(G)$ which contains w_{ij} distinct vertices V_{ij} corresponding to each edge ij in G. For every path i, j, k in G, all vertices in V_{ij} are adjacent to all vertices in V_{jk}. It is then easy to see that G has a P_3-decomposition iff $L(G)$ has a perfect matching. Given a perfect matching in $L(G)$, let $f(i, j, k)$ be the number of vertices in V_{ij} that are matched to vertices in V_{jk}, and vice versa.

A subset S of vertices in $L(G)$ is said to be compatible if for all edges ij in G, either $S \cap V_{ij} = \emptyset$ or $S \cap V_{ij} = V_{ij}$. We identify a compatible subset S of vertices in $L(G)$ by the subset $E(S)$ of edges ij in G for which $S \cap V_{ij} = V_{ij}$.

If G does not have a P_3-decomposition, then $L(G)$ does not have a perfect matching. By Tutte's theorem, there exists a subset S of vertices in $L(G)$ such that $O(L(G) - S) > |S|$, where $O(L(G) - S)$ is the number of connected components of odd order in $L(G) - S$. Let S be a minimal such set. Note that since all vertices in V_{ij} have the same neighbors in $L(G)$, S must be a compatible set. If for some vertices $x, y \in V_{ij}$, $x \in S$ but $y \notin S$, then $O(L(G) - (S \setminus \{x\})) \geq O(L(G) - S) - 1$, contradicting the minimality of S. Similarly, we can assume that the vertex set of any non-trivial component of $L(G) - S$, and the set of all isolated vertices in $L(G) - S$, are compatible sets. Let A be the set of isolated vertices in $L(G) - S$. Then $E(A)$ is a matching in G.

Let C be the vertex set of any non-trivial component of $L(G) - S$. No edge in $E(A)$ can have an endpoint in common with an edge in $E(C)$. If an edge in $E(S)$ has both endpoints in $V(E(C))$, we can remove all vertices corresponding to this edge from S, and get a smaller set that violates Tutte's condition, contradicting the minimality of S. Thus we can assume that $E(C)$ contains all edges in the subgraph of G induced by $V(E(C))$.

Now if there are two non-trivial components in $L(G) - S$, say with vertex sets C_1, C_2, then $V(E(C_1))$ and $V(E(C_2))$ must be disjoint sets. Since G is semi-complete, there exists an edge in G joining a vertex in $V(E(C_1))$ to a vertex in $V(E(C_2))$. This edge must be in $E(S)$. Again, removing vertices corresponding to this edge from S gives a smaller set violating Tutte's condition.

Therefore $L(G) - S$ contains at most one non-trivial component containing vertices corresponding to edges in some induced subgraph of G. The number of odd components in $L(G) - S$ is therefore at most $w(E(A)) + 1$ while every edge in $E(S)$ is incident with some vertex in $V(E(A))$. Thus $|S| = w(N(E(A)))$. Since G has even size and hence $L(G)$ has even order, we must have $O(L(G) - S) \geq |S| + 2$, which implies $w(E(A)) + 1 \geq w(N(E(A))) + 2$, or $w(E(A)) > w(N(E(A)))$, giving the required matching. $\square$

While Lemma 1 gives a characterization of semi-complete multigraphs that have a P_3-decomposition, it cannot be directly used to get an efficient algorithm to test whether a given semi-complete multigraph has such a decomposition. We strengthen Lemma 1 to get a more efficient strongly polynomial-time algorithm. We show that instead of checking the condition in Lemma 1 for all matchings S, it is sufficient to check it for all matchings that are contained in any fixed maximum weight matching.

Theorem 1. *Let G be a semi-complete multigraph of even size and let M be a matching with maximum weight in G. G has a P_3-decomposition iff for every subset of edges $S \subseteq M$, $w(N(S)) \geq w(S)$.*

Proof. The necessity of the condition follows from Lemma 1. To prove sufficiency, suppose for contradiction, that the condition is satisfied but G does not have a P_3-decomposition. By Lemma 1, there exists a matching S in G such that $w(N(S)) < w(S)$. Let S be a minimal such matching. We claim that S must be contained in M. If not, let $A = S \setminus M$ and $B = S \cap M$. By minimality of S, we must have $w(N(B)) \geq w(B)$, and hence $w(N(A) \setminus N(B)) < w(A)$. Let $M_1 = M \cap (N(A) \setminus N(B))$. Then $(M \setminus M_1) \cup A$ is a matching in G of weight greater than M, a contradiction. $\square$

Theorem 1 can be used to get an $O(n^4 \log n)$ time algorithm to test whether a given semi-complete multigraph has a P_3-decomposition. First we find an arbitrary maximum matching M in the weighted complete graph, using Edmond's algorithm [5]. This takes $O(n^3)$ time. Now construct a flow network as follows. For every edge $a \in M$ add a node v_a, and for every edge $b \notin M$, add a node v_b. If edges a and b have a common endpoint, add a directed edge from v_a to v_b with infinite capacity. Add a source node s and directed edges from s to v_a of capacity w_a, for all edges $a \in M$. Add a sink node t, and edges from v_b to t of capacity w_b, for all edges $b \notin M$. Now, it is easy to verify that G satisfies the condition in Theorem 1 iff this network has a maximum flow value of $w(M)$.

Since the network has $O(n^2)$ nodes and edges, the maximum flow can be found in $O(n^4 \log n)$ time [8].

We give another simple application of Lemma 1.

Theorem 2. *Any semi-complete multigraph of order n, even size and maximum multiplicity at most $n - 2$, is P_3-decomposable. The bound on the multiplicity is the best possible.*

Proof. If G is such a multigraph, then for any matching S of size k in G, $w(S) \leq k(n - 2)$. However, $w(N(S)) \geq |N(S)| = 2k(k - 1) + 2k(n - 2k) = 2k(n - k - 1)$. Since $k \leq n/2$, we have $w(S) \leq w(N(S))$, and Lemma 1 implies G has a P_3-decomposition. If n is even, and G has a perfect matching with edges of weight $n - 1$, and all other edge weights are 1, then G does not have a P_3-decomposition. $\qquad\qquad\square$

It is possible that similar results could hold for other decompositions of semi-complete multigraphs, with different bounds on the multiplicity.

We next show that Theorem 1 applies to a much larger class of multigraphs. Note that if Lemma 1 holds for a class of multigraphs, then so does Theorem 1, since the proof of the theorem does not use any properties of the multigraph G.

Let $\mathcal{F}$ be the set of graphs with exactly 2 connected components, each of which is either an odd cycle or the claw $K_{1,3}$. Let $\mathcal{G}$ be the set of graphs that do not contain any induced subgraph isomorphic to any graph in $\mathcal{F}$.

Theorem 3. *Let G be any multigraph of even size whose underlying simple graph belongs to the class $\mathcal{G}$. Let M be any maximum weight matching in G. Then G has a P_3-decomposition iff for any subset $S \subseteq M$, $w(N(S)) \geq w(S)$.*

Proof. We only need to show that Lemma 1 holds for this class of multigraphs. We follow the same argument, and choose a set S of vertices in $L(G)$ such that $O(L(G) - S) > |S|$, the number of non-trivial components in $L(G) - S$ is minimum, and subject to this, S is minimal. Following the same argument, each non-trivial component of $L(G) - S$ contains vertices corresponding to edges in some induced subgraph of G. Suppose there are 2 non-trivial components with vertex sets C_1, C_2. If there is an edge with positive weight in G joining some vertex in $V(E(C_1))$ to a vertex in $V(E(C_2))$, we can use the same argument and get a smaller set S with the same properties.

Suppose there is no such edge and let G_1 and G_2 be the underlying simple graphs of the subgraphs of G induced by $V(E(C_1))$ and $V(E(C_2))$, respectively. If both G_1 and G_2 are either not bipartite or contain a vertex of degree at least 3, then we can find an induced subgraph of $G_1 \cup G_2$ that is in $\mathcal{F}$, contradicting the assumption. Otherwise, at least one of G_1, G_2, say G_1, is bipartite and has maximum degree 2. Then the edges in $E(C_1)$ can be partitioned into two matchings. Take the matching with smaller weight, and add vertices in $L(G)$ corresponding to edges in it to the set S. The vertices corresponding to edges in the other matching will now be isolated vertices. The new set S will still satisfy $O(L(G) - S) > |S|$, but will have fewer non-trivial components in $L(G) - S$, contradicting the original choice of S. Therefore there must exist an edge in $E(S)$

joining a vertex in $V(E(C_1))$ to a vertex in $V(E(C_2))$, and we can complete the argument as before.   □

Note that the graphs in the family $\mathcal{F}$ themselves satisfy the hypothesis of Theorem 3 but do not have a P_3-decomposition. They are thus the minimal graphs for which Theorem 3 fails. The class $\mathcal{G}$ of graphs includes some well-known classes such as complete multipartite graphs, split graphs, complements of planar graphs etc. Theorem 3 is analogous to an old result of Fulkerson et al. [7], who proved an Erdos-Gallai type condition for the existence of an f-factor in certain multigraphs. In particular, the forbidden induced subgraphs in their case were graphs with two disjoint odd cycles, while we need the claw also in this case. Although their proof is quite different, it can also be proved using our approach, by reducing the f-factor problem to a perfect matching problem in the standard way. Again, for this class of multigraphs, the condition for existence of an f-factor can be checked in strongly polynomial-time, using network flows, as shown in [7].

3   Directed Graphs

A directed graph without self-loops is said to be semi-complete if for every pair of distinct vertices u, v, at least one of the ordered pairs $(u, v), (v, u)$ is an edge in the graph. A directed graph is said to be oriented if for any two vertices u, v, at most one of $(u, v), (v, u)$ is an edge in the graph. An oriented semi-complete directed graph is called a tournament.

The outdegree of a vertex u in a directed graph is the number $d^+(u)$ of vertices v such that (u, v) is an edge in the graph. The underlying multigraph $M(D)$ of a directed graph D is obtained by ignoring the directions of edges in the directed graph. If both (u, v) and (v, u) are edges in D, then the edge uv has multiplicity two in $M(D)$. We will denote the edges in $M(D)$ also by the same ordered pair as the corresponding edge in D, although these are undirected edges. A P_3-decomposition of a directed graph is a partition of the edges of the graph into directed paths of length two.

Lemma 2. *An oriented graph D has a P_3-decomposition iff there exists a spanning subgraph H of $M(D)$, such that $d_H(u) = d^+(u)$ for all vertices u in D.*

Proof. Suppose D has a P_3-decomposition. Let H be the spanning subgraph of $M(D)$ defined by the set of edges (u, v), such that u, v, w is a path in the decomposition for some vertex w. For every edge (u, v) in the directed graph, either (u, v) is an edge in H, or there exists a vertex w such that w, u, v is a path in the decomposition, in which case (w, u) is an edge in H. Thus the degree of u in H is exactly $d^+(u)$.

Conversely, suppose there exists such a subgraph H. For any vertex u, consider the set of incoming edges (w, u) that are in H. If there are k such edges, there must be exactly k outgoing edges (u, v) that are not edges in H. We pair up these edges arbitrarily to form paths of length two in the directed graph. Doing this for all vertices u, gives a P_3-decomposition of D.   □

Corollary 1. *A tournament has a P_3-decomposition iff its outdegree sequence is the degree sequence of a simple undirected graph.*

Corollary 1 gives simple necessary and sufficient conditions, in terms of the outdegree sequence, for a tournament to have a P_3-decomposition. These also give a simple $O(n^2)$ time algorithm to decide whether a given tournament has a decomposition.

Lemma 2 does not hold if the directed graph has 2-cycles, since pairing an arbitrary incoming edge with an arbitrary outgoing edge may lead to a 2-cycle rather than a path of length 2. However, we show that for semi-complete directed graphs, this can be avoided in almost all cases.

A pair of vertices u, v is said to be an isolated pair in a semi-complete directed graph if both (u, v) and (v, u) are edges in the graph, and for every vertex $w \notin \{u, v\}$, $(u, w), (w, v)$ are edges but $(w, u), (v, w)$ are not edges in the graph. Clearly, a semi-complete directed graph with an isolated pair of vertices cannot have a P_3-decomposition, since the edge (u, v) is not contained in any path of length two.

Theorem 4. *Let D be a semi-complete directed graph that is not a complete directed graph on 3 vertices and does not contain an isolated pair of vertices. Then D has a P_3-decomposition iff there exists a spanning subgraph H of $M(D)$ such that for every vertex u, $d_H(u) = d^+(u)$.*

Proof. One part of the theorem follows in the same way as Lemma 2. If there exists a P_3-decomposition of D, take the first edge of each path in the decomposition in H. This satisfies the required property.

Conversely, suppose there exists such a subgraph H. As before, consider the set of incoming edges (w, u) at a vertex u that are in H. If there are at least two such edges, we can pair them with outgoing edges at u that are not in H, such that (w, u) is not paired with (u, w), for any vertex w. The pairs of edges thus form paths of length two. The only case when this is not possible is if for some vertex w, (w, u) is the only incoming edge at u in H, and (u, w) is the only outgoing edge not in H. In this case we call u a bad vertex and w the partner of u.

Choose a subgraph H of $M(D)$ such that $d_H(u) = d^+(u)$ for all vertices u, and the number of bad vertices is minimum. If there are no bad vertices, we get a P_3-decomposition.

Let u be a bad vertex with partner w. If we replace the edge (w, u) by the edge (u, w) in H, in the resulting graph H', u is no longer a bad vertex. The choice of H implies that w must be a bad vertex in H'. Thus all outgoing edges at w must be in H, and no incoming edge at w is in H.

Case 1. Suppose there is a vertex v such that for some $x \in \{u, w\}$, both (x, v) and (v, x) are edges in D. Without loss of generality, assume $x = u$. Then the edge (u, v) must be in H, but (v, u) is not in H. Replacing (u, v) by (v, u) in H, gives a subgraph H' with the same degrees as H. Since u is not a bad vertex in H', v must be a bad vertex in H'. Let v' be the partner of v in H'. Then $v' \neq u$ and suppose $v' \neq w$. If the edge (u, v') is present in D, it must also be present in

H and H', since u is a bad vertex in H with partner w. But this contradicts the fact that v' is the partner of v in H'. A similar argument holds if the edge (v', u) is present in D. Since D is semi-complete, one of these must hold, and we get a contradiction. Therefore, the only remaining possibility is that $v' = w$. Thus $\{u, v, w\}$ induces a complete directed graph in D and a triangle in H. Note that for all $x \in \{u, v, w\}$ and $y \notin \{u, v, w\}$, any edge (x, y) in D must be in H, while (y, x) will not be in H.

Case 1.1. Suppose there is a vertex $x \in \{u, v, w\}$ such that for some vertex $y \notin \{u, v, w\}$, both (x, y) and (y, x) are edges in D. Consider the subgraph H' obtained by replacing the edge (x, y) by (y, x) and choosing edges $(p, x), (x, q), (p, q)$ for $\{p, q\} = \{u, v, w\} \setminus \{x\}$. In this subgraph, none of $\{u, v, w\}$ can be a bad vertex, since none of them has exactly one incoming edge in H'. Also, y cannot be a bad vertex. If both (y, p) and (y, q) are edges in D, then neither of them is in H', so y must have at least two incoming edges in H'. If say (y, p) is not an edge in D, then (p, y) is an incoming edge at y in H', and even if it is the only incoming edge in H' at y, it can be paired with an outgoing edge at y that is not in H'. Thus H' has fewer bad vertices than H, a contradiction.

Case 1.2. We may now assume that for every vertex $x \in \{u, v, w\}$ and $y \notin \{u, v, w\}$, exactly one of the edges $(x, y), (y, x)$ is in D. We say a vertex $y \notin \{u, v, w\}$ is of type 1 if there are at least two edges (x, y) for $x \in \{u, v, w\}$ in D, otherwise it is of type 2.

Suppose a, b are two distinct vertices of type 1, and without loss of generality (a, b) is an edge in D. Suppose (a, b) is not in H. Since a, b are of type 1, we can find two distinct vertices $p, q \in \{u, v, w\}$ such that (p, a) and (q, b) are edges in D and also in H. Consider the subgraph H' obtained by replacing the edges (p, a) and (q, b) by (a, b), and the edges in the triangle induced by $\{u, v, w\}$ by the edges $(p, q), (q, p), (p, r), (q, r)$, where $r \in \{u, v, w\} \setminus \{p, q\}$. None of $\{u, v, w\}$ can be a bad vertex in this subgraph, and neither can any of a, b, since both have an incoming edge in the subgraph such that the oppositely directed edge is not an edge in D. Again we get a subgraph with fewer bad vertices.

A similar argument holds if there are two vertices a, b of type 2, and (a, b) is an edge in D and also in H. We can find two vertices $p, q \in \{u, v, w\}$ such that $(a, p), (b, q)$ are edges in D but not in H. Replacing the edge (a, b) in H by the edges $(a, p), (b, q)$, and the edges in the triangle $\{u, v, w\}$ by $(p, r), (q, r)$, gives a subgraph with fewer bad vertices.

We may now assume that any edge in D whose endpoints are of type 1 is an edge in H, while any edge in D with endpoints of type 2 is not in H. Let A be the subset of vertices of type 1 and B the subset of vertices of type 2 and let $|A| = n_1, |B| = n_2$. Then

$$\sum_{x \in B} d_H(x) = \sum_{x \in B} d^+(x) \geq 2n_2 + n_2(n_2 - 1)/2 = n_2(n_2 + 3)/2$$

since there are at least 2 outgoing edges from each vertex in B to $\{u, v, w\}$ and B itself induces a semi-complete directed graph. Since B is an independent set in H, and there are at most n_2 edges in H joining vertices in B to $\{u, v, w\}$, there are at least $n_2(n_2 + 1)/2$ edges in H joining vertices in B to vertices in A.

Suppose there are m edges in the subgraph of D induced by A. Then

$$\sum_{x \in A} d^+(x) = \sum_{x \in A} d_H(x) \geq 2m + 2n_1 + n_2(n_2 + 1)/2 \ .$$

However, a vertex in A has at most one outgoing edge to $\{u, v, w\}$, and at most n_2 edges to B. Therefore

$$\sum_{x \in A} d^+(x) \leq m + n_1 + n_1 n_2 \ .$$

This implies

$$m \leq n_1 n_2 - n_1 - n_2(n_2 + 1)/2 \ .$$

But since D is semi-complete $m \geq n_1(n_1 - 1)/2$. This implies

$$(n_1 - n_2)^2 \leq -n_1 - n_2 \ .$$

This is possible only if $n_1 = n_2 = 0$, which implies D is a complete directed graph of order 3.

Case 2. Now suppose that for every vertex $x \in \{u, w\}$ and $y \notin \{u, w\}$, exactly one of the edges $(x, y), (y, x)$ is present in D. We say a vertex $y \notin \{u, w\}$ is of type 1 if neither (y, u) nor (y, w) are edges in D, of type 2 if both of them are edges, of type 3 if (y, w) is an edge but not (y, u), and of type 4 otherwise.

Suppose there exists a vertex a of type 3 and a vertex b of type 4. Without loss of generality, we may assume (a, b) is an edge in D. If (a, b) is in H, we can replace it by the edges (a, w) and (b, u) and delete the edge (w, u) from H, to get a subgraph H' with the same degrees. None of the vertices a, b, u, w can be a bad vertex in H', contradicting the choice of H. On the other hand, if (a, b) is not in H, we can replace the edges (u, a) and (w, b) by the edges (a, b) and (u, w), to get a graph with fewer bad vertices. We may therefore assume, without loss of generality, that there are no vertices of type 4.

Suppose a is a vertex of type 2 or 3 and b is a vertex of type 2. If either edge (a, b) or (b, a) is in H, we can replace it by edges (a, w) and (b, u), and delete the edge (w, u) to get a subgraph with fewer bad vertices. Similarly, suppose a is a vertex of type 1 and b is a vertex of type 1 or 3. If either edge (a, b) or (b, a) is not an edge in H, we can replace the edges $(u, b), (w, a)$ by this edge, and add the edge (u, w) to get a subgraph with fewer bad vertices.

Let A be the set of vertices of type 1, B the vertices of type 2, and C the vertices of type 3. Let $|A| = n_1$, $|B| = n_2$, and $|C| = n_3$. As argued in the previous case,

$$\sum_{x \in B} d_H(x) = \sum_{x \in B} d^+(x) \geq n_2(n_2 + 3)/2 \ .$$

All edges in H incident with a vertex in B must also be incident with a vertex in A. Let m be the number of edges in the subgraph of D induced by A, and m_1 the number of edges in D with one endpoint in A and one in C. Then

$$\sum_{x \in A} d^+(x) = \sum_{x \in A} d_H(x) \geq 2m + 2n_1 + m_1 + n_2(n_2 + 3)/2 \ .$$

But
$$\sum_{x \in A} d^+(x) \leq m + m_1 + n_1 n_2$$

which implies
$$m \leq n_1 n_2 - 2n_1 - n_2(n_2 + 3)/2 .$$

Again, since $m \geq n_1(n_1 - 1)/2$, we get
$$(n_1 - n_2)^2 \leq -3n_1 - 3n_2.$$

This is possible only if $n_1 = n_2 = 0$, which implies all vertices are of type 3, and thus $\{u, w\}$ is an isolated pair in D.

This completes the proof of the theorem.   □

Theorem 4 also gives an efficient algorithm to test whether a given semi-complete directed graph has a P_3-decomposition. We can first test for isolated pairs of vertices in $O(n^2)$ time. The theorem of Fulkerson, Hoffman and McAndrew [7], gives a simple necessary and sufficient condition for a semi-complete multigraph to contain a spanning subgraph with specified degrees. Further such a subgraph can be found using network flows. The network constructed has $O(n)$ nodes and $O(n^2)$ edges. Thus the maximum flow can be found in $O(n^3)$ time [8]. After finding the subgraph H, if it exists, we can modify it to remove bad vertices, if any. Following the proof of Theorem 4, each removal takes $O(n^2)$ time. Thus, overall, finding the P_3-decomposition can be done in $O(n^3)$ time.

4   Conclusion

In this paper we have given simple necessary and sufficient conditions for semi-complete multigraphs and directed graphs to have a P_3-decomposition. We have also shown that the conditions can be checked in strongly polynomial-time using network flows. However, we believe there should be faster algorithms to check these conditions, without resorting to general network flows.

We would also like to consider more general decomposition problems for semi-complete multigraphs and directed graphs. We conclude with a specific problem: When can a d-regular semi-complete multigraph of even order be decomposed into d perfect matchings? A necessary condition is that a subgraph induced by any subset of k vertices must contain at most $d\lfloor k/2 \rfloor$ edges. We leave open the question of whether this condition is sufficient.

References

1. Alspach, B.: The Oberwolfach problem. In: The CRC Handbook of Combinatorial Designs, pp. 394–395. CRC Press, Boca Raton (1996)
2. Brys, K., Lonc, Z.: Polynomial cases of graph decomposition: a complete solution of Holyers problem. Discrete Math. **309**, 1294–1326 (2009)

3. Diwan, A.A.: P_3-decomposition of directed graphs. Discrete Appl. Math. (to appear). http://dx.doi.org/10.1016/j.dam.2016.01.039
4. Dor, D., Tarsi, M.: Graph decomposition is NP-complete: a complete proof of Holyers conjecture. SIAM J. Comput. **26**, 1166–1187 (1997)
5. Edmonds, J.: Maximum matching and a polyhedron with 0,1-vertices. J. Res. Natl. Bureau Stan. Sect. B **69**, 125–130 (1965)
6. Gyarfas, A., Lehel, J.: Packing trees of different order into K_n. In: Combinatorics (Proceedings of Fifth Hungarian Colloquium, Keszthely, 1976), vols. I, 18, Colloquium Mathematical Society, Janos Bolyai, pp. 463–469. North-Holland, Amsterdam (1978)
7. Fulkerson, D.J., Hoffman, A.J., McAndrew, M.H.: Some properties of graphs with multiple edges. Can. J. Math. **17**, 166–177 (1965)
8. King, V., Rao, S., Tarjan, R.: A faster deterministic maximum flow algorithm. J. Algorithms **17**(3), 447–474 (1994)
9. Kotzig, A.: From the theory of finite regular graphs of degree three and four. Casopis. Pest. Mat. **82**, 76–92 (1957)
10. Kouider, M., Maheo, M., Brys, K., Lonc, Z.: Decomposition of multigraphs. Discuss. Math. Graph Theor. **18**, 225–232 (1998)
11. Ringel, G.: Problem 25. In: Theory of Graphs and its Applications, Proceedings of International Symposium Smolenice 1963 Nakl, CSAV, Praha, p. 162 (1964)
12. Tutte, W.T.: The factorisation of linear graphs. J. Lond. Math. Soc. **22**, 107–111 (1947)
13. Wilson, R.M.: Decompositions of complete graphs into subgraphs isomorphic to a given graph. In: Proceedings of British Combinatorial Conference, pp. 647–659 (1975)

On Rank and MDR Cyclic Codes of Length 2^k Over Z_8

Arpana Garg and Sucheta Dutt$^{(\boxtimes)}$

Department of Applied Sciences, PEC University of Technology, Chandigarh, India
arpanapujara@gmail.com, suchetapec@yahoo.co.in

Abstract. In this paper, a set of generators (in a unique from) called the distinguished set of generators, of a cyclic code C of length $n = 2^k$ (where k is a natural number) over Z_8 is obtained. This set of generators is used to find the rank of the cyclic code C. It is proved that the rank of a cyclic code C of length $n = 2^k$ over Z_8 is equal to $n - v$, where v is the degree of a minimal degree polynomial in C. Then a description of all MHDR (maximum hamming distance with respect to rank) cyclic codes of length $n = 2^k$ over Z_8 is given. An example of the best codes over Z_8 of length 4 having largest minimum Hamming, Lee and Euclidean distances among all codes of the same rank is also given.

Keywords: Cyclic codes · Minimal degree polynomial · Rank · MDR codes

1 Introduction

The class of cyclic codes over modular rings Z_m has recently received a lot of attention from researchers and experts in the field of coding theory. This attention can be attributed to the fact that many non-linear binary codes such as Kerdock, Preparata and Goethals codes can be viewed as linear codes over Z_4 by defining gray maps from Z_4^n to F_2^{2n} [10].

The structure of cyclic codes over Z_4 of length 2^e has been given by T. Abualrub and Oehmke in [1] and a distinguished set of generators of cyclic codes over Z_4 of length 2^e has been obtained by them in [2]. A mass formula and rank of cyclic codes of length 2^e over Z_4 have been derived by Abualrub et al. in [3]. Minjia et al. have generalized the results of reference [2] of Abualrub and Oehmke to cyclic codes over the ring Z_{p^2} of length p^e for an odd prime p [14], which are further extended by Kiah et al. to cyclic codes over $GR(p^2, m)$ of length p^k [11]. Dinh and Lopez Permouth [5] have given the structure of cyclic and negacyclic codes over finite chain rings such that length of the code is relatively prime to the characteristic of the ring. They have also found the structure of negacyclic codes of length 2^t over Z_{2^m} in this paper and further extended this result to negacyclic codes of length 2^s over Galois ring $GR(2^a, m)$ in [6]. Lopez and Szabo [13] have given the structure of negacyclic codes of length p^s over $GR(p^a, m)$ for an odd prime p and $a > 1$. They have extended this result to cyclic codes of the same

© Springer International Publishing AG 2017
D. Gaur and N.S. Narayanaswamy (Eds.): CALDAM 2017, LNCS 10156, pp. 177–186, 2017.
DOI: 10.1007/978-3-319-53007-9_16

type, via an isomorphism between cyclic and negacyclic codes over $GR(p^a, m)$ of length p^s for an odd prime p. By reducing the generators, they have also proved that any ideal of $GR(2^a, m)/<x^{2^s} - 1>$ is at most $'a'$ generated. Dougherty and Park have given the structure of cyclic codes over Z_{p^e} of length $p^k n$ such that $(n, p) = 1$ using an isomorphism between $Z_{p^e}[x]/<x^n - 1>$ and a direct sum $\oplus_{i \in I} S_i$ of certain local rings which are ambient spaces for codes of length p^k over certain Galois rings [7]. The approach of Abualrub and Oehmke in [2] for finding the generators of cyclic codes of length 2^k over Z_4 is simpler and based on certain minimal degree polynomials. We take forward the approach of Abualrub et al. and give a constructive proof to show that cyclic codes of length 2^k over Z_8 are generated by at most three elements [8], and cyclic codes of length 2^k over Z_{2^m} are generated by at most m elements [9]. It is noted that this result extends to the case when 2 is replaced by an arbitrary prime p. Kaur et al. have further extended this result to cyclic codes over Galois Rings of arbitrary length [12].

In this paper, a distinguished set of generators(in a unique form) of cyclic codes of length $n = 2^k$ (where k is a natural number) over Z_8 as ideals of the ring $Z_8[x]/<x^n - 1>$ is obtained and used to find the rank of such cyclic codes. It is followed by a description of all MHDR codes over Z_8 of length 2^k. Further, the best codes over Z_8 of length 4 which have the largest minimum Hamming, Lee and Euclidean distances among all codes of the same rank are listed.

2   Preliminaries

A linear code C of length n over a finite commutative ring R is defined as an R - submodule of R^n. A cyclic code C over R of length n is a linear code such that any cyclic shift of a codeword in C is also a codeword in C, that is, whenever $(c_0, c_1, c_2, \ldots, c_{n-1})$ is in C, $(c_{n-1}, c_0, c_1, \ldots, c_{n-2})$ also belongs to C. The one-one correspondence between ideals of $R[x]/<x^n - 1>$ and cyclic codes of length n over R which associates the polynomial $\sum_{i=0}^{n-1} c_i x^i$ with the codeword $(c_0, c_1, c_2, \ldots, c_{n-1})$ is well known. The cyclic shift of a code-word $(c_0, c_1, c_2, \ldots, c_{n-1})$ corresponds to multiplication by x with the polynomial $\sum_{i=0}^{n-1} c_i x^i$ in $R[x]/<x^n - 1>$.

It is easy to see that for a positive integer m, the set $S = \{1, (x + 1), (x + 1)^2, \ldots, (x + 1)^{n-1}\}$ is a basis of $Z_{2^m}[x]/<x^n - 1>$ as a Z_{2^m} - module. Thus an element $f(x)$ of $Z_{2^m}[x]/<x^n - 1>$ can be uniquely expressed as

$$f(x) = \sum_{i=0}^{n-1} c_i (x + 1)^i \tag{1}$$

where $c_i \in Z_{2^m}$. It has been proved in [2] that $f(x) = \sum_{i=0}^{n-1} c_i (x + 1)^i$ with $c_i \in Z_2$ is a unit in $Z_2[x]$ if and only if $c_0 \neq 0$. It is also shown that $f(x) = \sum_{i=0}^{n-1} c_i (x + 1)^i$ with $c_i \in Z_4$ is a unit in $Z_4[x]$ if and only if $c_0 = 1$ or -1.

Define $\phi_1 : Z_8[x] \to Z_4[x]$ by mapping the coefficients of a polynomial in $Z_8[x]$ to Z_4 using the map $\phi_1(a) = a(mod 4), a \in Z_8$ and $\phi_2 : Z_8[x] \to Z_2[x]$, by mapping the coefficients of a polynomial in $Z_8[x]$ to Z_2 using the map $\phi_2(a) = a(mod 2), a \in Z_8$. It can be easily seen that ϕ_1, ϕ_2 are ring homomorphisms.

The rank of the code C (denoted by $rank(C)$) over Z_8 is defined as the minimum number of generators of C as a Z_8 - module [2]. Hamming weight of an element x in Z_8 is defined as

$$w_H(x) = \begin{cases} 0, & \text{if } x = 0 \\ 1, & \text{if } x \neq 0 \end{cases}$$

Lee weight of an element x in Z_8 is defined as $w_L(x) = min(x, 8 - x)$. Euclidean weight of an element x in Z_8 is defined as $w_E(x) = min(x^2, (8 - x)^2)$. The Hamming weight, Lee weight and Euclidean weight of a codeword are defined as the rational sums of the Hamming weights, Lee weights and Euclidean weights of its coordinates respectively. The Hamming distance, Lee Distance and Euclidean distance of a linear code C are defined as the minimum Hamming, Lee and Euclidean weight of non zero codewords of C and are denoted by $d_H(C)$, $d_L(C)$ and $d_E(C)$ respectively.

3 A Distinguished Set of Generators of a Cyclic Code of Length 2^k Over Z_8

We recall the following results of Lopez and Szabo [13] below.

Lemma 1 *(Lemma 4.6, [13]).* In $GR(2^a, m)[x]/<x^{2^s} - 1>$, for $t \geq 0$, $(x + 1)^{2^s + t2^{s-1}} = 2^{t+1} b_t(x)(x + 1)^{2^{s-1}} + a_t(x)$ where $b_t(x)$ is invertible and $2^{t+2}(x + 1)|a_t(x)$.

Corollary 1 *(Corollary 4.7, [13]).* In $GR(2^a, m)[x]/<x^{2^s} - 1>$, the nilpotency of $x + 1$ is $(a + 1)2^{s-1}$.

For $t = 0, m = 1, a = 3$ and $n = 2^k$, Lemma 1 reduces to the following lemma.

Lemma 2. In $Z_8[x] <x^n - 1>$, $(x+1)^n = 2b_0(x)(x+1)^{n/2} + a_0(x)$, where $b_0(x)$ is invertible and $2^2(x+1)|a_0(x)$.

We obtain the explicit values of $a_0(x)$ and $b_0(x)$ in Lemma 3 below.

Lemma 3. In $Z_8[x]/<x^n - 1>$, $(x + 1)^n = 2b_0(x)(x + 1)^{n/2} + a_0(x)$, where

1. $b_0(x) = 1$ and $a_0(x) = 0$ for $k = 1$ and
2. $b_0(x) = 1$ and $a_0(x) = 4(x + 1)^{3n/4} + 4(x + 1)^{n/4}$ for $k \geq 2$.

Proof. In $Z_8[x]/<x^n - 1>$, for $k = 1$, $(x + 1)^2 = 2(x + 1)$ implies that $b_0(x) = 1$ and $a_0(x) = 0$. This proves 1.

For $k \geq 2$ and $n = 2^k$, it can be easily proved by mathematical induction that $x^n - 1 = (x + 1)^n + 6(x + 1)^{n/2} + 4(x + 1)^{3n/4} + 4(x + 1)^{n/4}$ in $Z_8[x]/<x^n - 1>$. This implies that $(x + 1)^n = -6(x + 1)^{n/2} - 4(x + 1)^{3n/4} - 4(x + 1)^{n/4} = 2(x + 1)^{n/2} + 4(x + 1)^{3n/4} + 4(x + 1)^{n/4} = 2b_0(x)(x + 1)^{n/2} + a_0(x)$, where $b_0(x) = 1$ and $a_0(x) = 4(x + 1)^{3n/4} + 4(x + 1)^{n/4}$. This proves 2.

For $m = 1, a = 3$ and $n = 2^k$, Corollary 1 reduces to the following corollary.

Corollary 2. *The element* $(x+1)$ *in* $Z_8[x]<x^n-1>$ *is nilpotent with nilindex less than or equal to* $2n$.

Corollary 3 below is an immediate consequence of Corollary 2.

Corollary 3. *Let* $f(x) = \sum_{i=0}^{n-1} c_i(x+1)^i$ *be an element of* $Z_8[x]<x^n-1>$. *Then* $f(x)$ *is a unit in* $Z_8[x]<x^n-1>$ *if and only if* c_0 *is a unit in* Z_8.

Let C be cyclic code of length $n = 2^k$ over Z_8. Let $f(x) = q_0(x)$ be a minimal degree polynomial among all monic polynomials in C, $g(x) = 2q_1(x)$ be a minimal degree polynomial among all polynomials in C with leading coefficient 2 or 6 and $h(x) = 4q_2(x)$ be a minimal degree polynomial among all polynomials in C with leading coefficient 4. Let $\deg(f(x)) = r$, $\deg(g(x)) = s$ and $\deg(h(x)) = t$. It is obvious that $r \geq s \geq t$. A unique form for the polynomials $f(x), g(x)$ and $h(x)$ is determined in Theorem 1 below. Note that the cyclic code C may not contain any monic polynomial or any polynomial with leading coefficient 2 or 6.

Theorem 1. *Let* C *be a cyclic code of length* $n = 2^k$ *over* Z_8. *Let* $f(x), g(x)$ *(if they exists) and* $h(x)$ *be polynomials in* C *as defined above. Then*

1. $h(x) = 4q_2(x) = 4(x+1)^t$.
2. *If* C *contains polynomials with leading coefficient 2 or 6, then there exists a unique polynomial in* C *of the type* $2(x+1)^s + 4\sum_{i=0}^{t-1} \alpha_i(x+1)^i$ *of degree* s, *where* $\alpha_i \in Z_2$. *Therefore,* $g(x) = 2q_1(x)$ *can be chosen as* $2(x+1)^s + 4\sum_{i=0}^{t-1} \alpha_i(x+1)^i$, *where* $\alpha_i \in Z_2$.
3. *If* C *contains monic polynomials, then there exists a unique polynomial in* C *of the type* $(x+1)^r + 2\sum_{i=0}^{s-1} \beta_i(x+1)^i + 4\sum_{i=0}^{t-1} \gamma_i(x+1)^i$ *of degree* r, *where* $\beta_i, \gamma_i \in Z_2$. *Therefore,* $f(x) = q_0(x)$ *can be chosen as* $(x+1)^r + 2\sum_{i=0}^{s-1} \beta_i(x+1)^i + 4\sum_{i=0}^{t-1} \gamma_i(x+1)^i$, *where* $\beta_i, \gamma_i \in Z_2$.

Proof. Using Eq. (1), we write $4q_2(x) = 4\sum_{i=0}^{t} c_i(x+1)^i$ where $c_i \in Z_2$. We claim that $c_i = 0$ for $0 \leq i < t$. Suppose this is not so. Let l be the least positive integer less than t such that $c_l \neq 0$. Then $4q_2(x) = 4(x+1)^l[\sum_{i=l}^{t} c_i(x+1)^{i-l}]$. As $c_l = 1$, $\sum_{i=l}^{t} c_i(x+1)^{i-l}$ is a unit in $Z_2[x]$. Therefore $4q_2(x)[\sum_{i=l}^{t} c_i(x+1)^{i-l}]^{-1} = 4(x+1)^l \in C$. As $l < t$, and the degree t of $4q_2(x)$ is minimal among all polynomials in C with leading coefficient 4, we arrive at a contradiction. Therefore $c_i = 0$ for $0 \leq i < t$. Hence $4q_2(x) = 4c_t(x+1)^t$, where $c_t = 1$. This proves the first part of the theorem.

Now, let $2q_1(x) = 2f_0(x) + 4f_1(x)$, where $f_0(x), f_1(x) \in Z_2[x]<x^n-1>$ so that $\phi_1(2q_1(x)) = 2f_0(x)$. Note that $2f_0(x)$ is a minimal degree polynomial among all polynomials with leading coefficient 2 in $\phi_1(C)$ which is a cyclic code over Z_4. Therefore by Theorem 8 of [2], $2f_0(x)$ can be chosen as $2(x+1)^s$, which implies that $2q_1(x) = 2(x+1)^s + 4f_1(x) = 2(x+1)^s + 4\sum_{i=0}^{s} \alpha_i(x+1)^i$, where $\alpha_i \in Z_2$ (Using 1). Let $2q'(x) = 2(x+1)^s + 4\sum_{i=0}^{t-1} \alpha_i(x+1)^i$. It is easy to see that $2q'(x) \in C$. Also $\deg(2q'(x)) = \deg(2q_1(x))$. Therefore $2q'(x)$ can be chosen as a minimal degree polynomial in C among all polynomials in C with leading coefficient 2 or 6.

For uniqueness, suppose $2(x+1)^s + 4\sum_{i=0}^{t-1} l_i(x+1)^i, l_i \in Z_2$ is another polynomial of the same form belonging to C. Then $[2(x+1)^s + 4\sum_{i=0}^{t-1}\alpha_i(x+1)^i] - [2(x+1)^s + 4\sum_{i=0}^{t-1} l_i(x+1)^i] \in C$. That is $4\sum_{i=0}^{t-1}(\alpha_i - l_i)(x+1)^i \in C$. If $\alpha_i - l_i \neq 0$ for some i, then it is a non zero polynomial with leading coefficient 4 of degree at most $t-1$, which contradicts the minimality of t. Hence $\alpha_i - l_i = 0$ for all i. Therefore $2(x+1)^s + 4\sum_{i=0}^{t-1}\alpha_i(x+1)^i$ is unique in C. This completes the proof of the second part of the theorem.

Now, let $q_0(x) = f_0(x) + 2f_1(x) + 4f_2(x)$, where $f_0(x), f_1(x)$ and $f_2(x) \in Z_2[x] < x^n - 1 >$ so that $\phi_2(q_0(x)) = f_0(x)$. As $f_0(x)$ is a minimal degree polynomial among all monic polynomials in $\phi_2(C)$, which is a cyclic code over Z_2, we must have $f_0(x) = (x+1)^r$. Hence $q_0(x) = (x+1)^r + 2f_1(x) + 4f_2(x)$. Now $\phi_1(q_0(x)) = (x+1)^r + 2f_1(x)$ is a minimal degree monic polynomial in $\phi_1(C)$ which is a cyclic code over Z_4. Therefore as proved in Theorems 9 and 10 of [2], we get that $\phi_1(q_0(x)) = (x+1)^r + 2\sum_{i=0}^{s-1}\beta_i(x+1)^i$ with $\beta_i \in Z_2$. Hence $q_0(x) = (x+1)^r + 2\sum_{i=0}^{s-1}\beta_i(x+1)^i + 4\sum_{i=0}^r \gamma_i(x+1)^i = (x+1)^r + 2\sum_{i=0}^{s-1}\beta_i(x+1)^i + 4\sum_{i=0}^{t-1}\gamma_i(x+1)^i + 4\sum_{i=t}^r \gamma_i(x+1)^i$, where $\gamma_i \in Z_2$.

Let $q''(x) = (x+1)^r + 2\sum_{i=0}^{s-1}\beta_i(x+1)^i + 4\sum_{i=0}^{t-1}\gamma_i(x+1)^i$. Then $q''(x) = q_0(x) - 4\sum_{i=t}^r \gamma_i(x+1)^i = q_0(x) - 4(x+1)^t\sum_{i=t}^r \gamma_i(x+1)^{i-t}$ belongs to C. Also $\deg(q''(x)) = \deg(q_0(x))$. Therefore, $q''(x)$ can be taken as the minimal degree polynomial in C among all monic polynomials in C. The uniqueness of $q''(x)$ can be proved in a similar fashion as that of $2q'(x)$ in the proof of the second part of the theorem. This completes the proof of the theorem.

It has been proved in [8] that C is generated by one or more polynomials from the set $\{f(x), g(x), h(x)\}$. This, together with Theorem 1 above, gives us the following result.

Theorem 2. *A cyclic code C of length 2^k over Z_8 is generated by one or more polynomials from the set $\{f(x), g(x), h(x)\}$ where*

1. $f(x) = q_0(x) = (x+1)^r + 2\sum_{i=0}^{s-1}\beta_i(x+1)^i + 4\sum_{i=0}^{t-1}\gamma_i(x+1)^i$ *with* $\beta_i, \gamma_i \in Z_2$
2. $g(x) = 2q_1(x) = 2(x+1)^s + 4\sum_{i=0}^{t-1}\alpha_i(x+1)^i$ *with* $\alpha_i \in Z_2$
3. $h(x) = 4q_2(x) = 4(x+1)^t$.

As the generators $f(x), g(x)$ and $h(x)$ of the cyclic code C of length $n = 2^k$ over Z_8 are unique in the form described in Theorem 2 above, these are referred to as the distinguished generators of the cyclic code C.

4   Rank of a Cyclic Code of Length 2^k Over Z_8

Using the distinguished set of generators of the cyclic code C of length $n = 2^k$ over Z_8 obtained in Sect. 3, we find in this section the rank of C by determining its minimal generating set as a Z_8 – module.

Theorem 3. *Let C be a cyclic code of length $n = 2^k$ over Z_8 such that $C = < h(x) >$. Then $S = \{h(x), xh(x), \ldots, x^{n-t-1}h(x)\}$ is a minimal generating set in C as a Z_8 – module.*

Proof. Let $c(x)$ be an element of C. Let $c(x) = h(x)a(x) = 4(x+1)^t a(x) = 4(x+1)^t[\sum_{i=0}^{u} a_i(x+1)^i + 2\sum_{i=0}^{u} b_i(x+1)^i + 4\sum_{i=0}^{u} c_i(x+1)^i]$, where u is the degree of $a(x)$ and $a_i, b_i, c_i \in Z_2$. This implies $c(x) = 4\sum_{i=0}^{u} a_i(x+1)^{t+i}$. Using Lemma 3, we get $4(x+1)^{t+i} = 0$ for $t+i \geq n$. Therefore $a(x)$ is a polynomial of degree at most $n-t-1$. This shows that S generates C as a Z_8 – module. Suppose, if possible, that $x^l h(x) = a_0 h(x) + a_1 x h(x) + \ldots + a_{l-1} x^{l-1} h(x) + a_{l+1} x^{l+1} h(x) + \ldots + a_{n-t-1} x^{n-t-1} h(x)$, $a_i \in Z_8$, for some l, $0 \leq l \leq n-t-1$. It can be easily seen that the degree of the polynomial on the right hand side can never be equal to $l+t$, which is a contradiction as the degree of $x^l h(x)$ on the left hand side is equal to $l+t$. This proves that, S is a minimal generating set of C as a Z_8 – module.

Theorem 4. *Let C be a cyclic code of length $n = 2^k$ over Z_8 such that $C = \langle g(x), h(x) \rangle$ or $C = \langle g(x) \rangle$. Then*

$$S = \{g(x), xg(x), x^2 g(x), \ldots, x^{n-s-1} g(x)\} \cup \{h(x), xh(x), x^2 h(x) \ldots, x^{s-t-1} h(x)\}$$

is a minimal generating set in C as a Z_8 – module.

Proof. It can be easily proved in a similar manner as in Theorem 3, that the set $S' = \{g(x), xg(x), \ldots, x^{n-s-1} g(x)\} \cup \{h(x), xh(x), \ldots, x^{n-t-1} h(x)\}$ generates C as a Z_8 – module. Note that, for $l \geq s-t$, $x^l h(x) = 4\sum_{i=0}^{l+t} d_i(x+1)^i = 4\sum_{i=0}^{t-1} d_i(x+1)^i + 4\sum_{i=t}^{l+t} d_i(x+1)^i$ with $d_i \in Z_2$ and $d_{l+t} = 1$. As $4(x+1)^t \in C$, we have $4\sum_{i=0}^{t-1} d_i(x+1)^i \in C$, which contradicts the minimality of t. Therefore $4\sum_{i=0}^{t-1} d_i(x+1)^i = 0$ and we obtain $x^l h(x) = 4\sum_{i=t}^{l+t} d_i(x+1)^i$ which can easily be written as a linear combination of elements of S. Hence S generates C as a Z_8 – module. It can be easily proved, as in Theorem 3, that S is a minimal generating set of C as a Z_8 – module.

Theorem 5. *Let C be a cyclic code of length $n = 2^k$ over Z_8 such that $C = \langle f(x), g(x), h(x) \rangle$ or $C = \langle f(x), g(x) \rangle$ or $C = \langle f(x), h(x) \rangle$ or $C = \langle f(x) \rangle$. Then*

$$S = \{f(x), xf(x), \ldots, x^{n-r-1} f(x)\} \cup \{g(x), xg(x), \ldots, x^{r-s-1} g(x)\}$$
$$\cup \{h(x), xh(x) \ldots, x^{s-t-1} h(x)\}$$

is a minimal generating set for C as a Z_8–module.

Proof. Let

$$S'' = \{f(x), xf(x), \ldots, x^{n-r-1} f(x)\} \cup \{g(x), xg(x), \ldots, x^{n-s-1} g(x)\}$$
$$\cup \{h(x), xh(x), \ldots, x^{s-t-1} h(x)\}$$

Proceeding as in the proof of Theorem 4, it is easy to see that S'' generates C as a Z_8 – module. Now, for $l \geq r-s$, $x^l g(x) = 2\sum_{i=0}^{l+s} c_i(x+1)^i$, where $c_i = a_i + 2b_i$ and $a_i, b_i \in Z_2$. Therefore, $x^l g(x) = 2\sum_{i=0}^{s-1} c_i(x+1)^i + 2\sum_{i=s}^{l+s} c_i(x+1)^i = 2\sum_{i=0}^{s-1} c_i(x+1)^i + [2(x+1)^s + 4\sum_{i=0}^{t-1} \alpha_i(x+1)^i][\sum_{i=s}^{l+s} c_i(x+1)^{i-s}] + 4[\sum_{i=0}^{t-1} \alpha_i(x+1)^i][\sum_{i=s}^{l+s} c_i(x+1)^{i-s}]$. As $x^l g(x) \in C$

and $2(x + 1)^s + 4\sum_{i=0}^{t-1}\alpha_i(x + 1)^i \in C$, the above equation implies that $2\sum_{i=0}^{s-1} c_i(x + 1)^i + 4\left[\sum_{i=0}^{t-1}\alpha_i(x + 1)^i\right]\left[\sum_{i=s}^{l+s} c_i(x + 1)^{i-s}\right] \in C$. This contradicts the minimality of s if $c_i = a_i + 2b_i$, $a_i \neq 0$ and we have, in this case, $x^l g(x) = [2(x + 1)^s + 4\sum_{i=0}^{t-1}\alpha_i(x + 1)^i][\sum_{i=s}^{l+s} c_i(x + 1)^{i-s}]$.

In case $c_i = 2b_i$ we have $x^l g(x) = [2(x + 1)^s + 4\sum_{i=0}^{t-1}\alpha_i(x + 1)^i][\sum_{i=s}^{l+s} c_i(x + 1)^{i-s}] + 4\sum_{i=0}^{t-1} b_i(x + 1)^i + 4(x + 1)^t\sum_{i=t}^{s-1} b_i(x + 1)^{i-t}$. This gives $4\sum_{i=0}^{t-1} b_i(x + 1)^i \in C$ which contradicts the minimality of t. Hence $4\sum_{i=0}^{t-1} b_i(x + 1)^i = 0$ and $x^l g(x) = [2(x + 1)^s + 4\sum_{i=0}^{t-1}\alpha_i(x + 1)^i][\sum_{i=s}^{l+s} c_i(x + 1)^{i-s}] + 4(x + 1)^t\sum_{i=t}^{s-1} b_i(x + 1)^{i-t}$. Note that in $x^l g(x)$ (in both the above cases) terms with degree less than s can be written as linear combination of elements of $\{h(x), xh(x), \ldots, x^{s-t-1}h(x)\}$, terms with degree greater than s and less than r can be written as a linear combination of elements of $\{g(x), xg(x), \ldots, x^{r-s-1}g(x)\}$ and terms with degree greater than r and less than n can be written as a linear combination of elements of $\{f(x), xf(x), \ldots, x^{n-r-1}f(x)\}$. This proves that S is generated by C. That S is a minimal generating set of C as a Z_8 – module can be easily verified.

Theorem 6 below is an immediate consequence of Theorems 3, 4 and 5.

Theorem 6. *The rank of a cyclic code C over Z_8 of length $n = 2^k$ is equal to $n - v$, where v is the degree of a minimal degree polynomial in C.*

5   MDR Cyclic Codes of Length 2^k Over Z_8

Shiromoto [15] has given singleton type bounds for Hamming, Lee and Euclidean distances with respect to rank of linear codes over Z_l. His results are summed up in the following theorem.

Theorem 7 [15]. *Let C be a linear code of length n over Z_l with minimum hamming weight $d_H(C)$, minimum Lee weight $d_L(C)$ and minimum Euclidean weight $d_E(C)$. Then*

$$d_H(C) \leq n - rank(C) + 1 \tag{2}$$

$$\left\lceil\frac{d_L(C) - 1}{[l/2]}\right\rceil \leq n - rank(C) \tag{3}$$

and

$$\left\lceil\frac{d_E(C) - 1}{[l/2]^2}\right\rceil \leq n - rank(C) \tag{4}$$

A code which attains the bound in (2), is called Maximum Hamming Distance Code with respect to rank (MHDR). A code which attains the bound in (3), is called Maximum Lee Distance Code with respect to rank (MLDR). A code which attains the bound in (4), is called Maximum Euclidean Distance Code with respect to rank (MEDR).

Theorem 8 below follows as an immediate consequence of Theorems 6 and 7.

Theorem 8. *Let C be a cyclic code of length $n = 2^k$ over Z_8. Then*

$$d_H(C) \leq v + 1 \tag{5}$$

$$\left\lceil \frac{d_L(C) - 1}{4} \right\rceil \leq v \tag{6}$$

and

$$\left\lceil \frac{d_E(C) - 1}{16} \right\rceil \leq v \tag{7}$$

where v is the degree of a minimum degree polynomial in C.

Salagean [16], has shown that the Hamming distance of a cyclic code C is equal to the Hamming distance of the code generated by the torsion of the smallest degree polynomial belonging to its set of generators. Therefore, if $f(x), g(x)$ and $h(x)$ are the polynomials as defined in Theorem 2, it is easy to see that Hamming distance of C is equal to the Hamming distance of the code generated by $(x + 1)^r, (x + 1)^s$ or $(x + 1)^t$ according as $f(x), g(x)$ or $h(x)$ is the minimal degree polynomial in C. This implies $d_H(C) = d_H(< (x+1)^v >)$, where v is the degree of a minimal degree polynomial in C.

We record this observation as Lemma 4 below.

Lemma 4. *Let C be a cyclic code of length 2^k over Z_8. Then $d_H(C) = d_H$ $(<(x + 1)^v>)$, where v is the degree of a minimal degree polynomial in C.*

Dinh [4] has given the Hamming distance of a cyclic code of length 2^s over a field of characteristic 2. We recall below his result as quoted by Lopez and Szabo [13].

Lemma 5 *(Corollary 4.12, [13]).* *In $\frac{GR(2,m)[x]}{<x^{2^s}-1>}$, for $0 \leq i \leq 2^s$*

$$d_H(< (x + 1)^i >) = \begin{cases} 1 & \text{if } i = 0 \\ 2 & \text{if } 1 \leq i \leq 2^{s-1} \\ 2^{k+1} & \text{if } 2^s - 2^{s-k} + 1 \leq i \leq 2^s - 2^{s-k} + 2^{s-k-1} \\ & \text{where } 1 \leq k \leq s - 1 \\ 0 & \text{if } i = 2^s \end{cases}$$

Theorem 9 below characterizes all MHDR codes of length 2^k over Z_8.

Theorem 9. *Let C be a cyclic code of length 2^k over Z_8. Let v be the degree of a minimal degree polynomial in C. Then C is a MHDR code if and only if $v \in \{0, 1, 2^k - 1\}$.*

Proof. 1. For $v = 0$, Lemmas 4 and 5 imply that $d_H(C) = 1$. Also, by Theorem 6, $rank(C) = n - 0 = n$, and by Theorem 7, singleton bound is given by $d_H(C) \leq n - n + 1 = 1$. Thus C attains its singleton bound for $v = 0$.
2. If $1 \leq v \leq 2^{k-1}$, using Lemmas 4 and 5, we see that $d_H(C) = 2$. Using Theorems 6 and 7, we see in this case that C attains its singleton bound if and only if $v + 1 = 2$, that is $v = 1$.

3. By using Lemmas 4 and 5, we get that $d_H(C) = 2^{l+1}$ for $2^k - 2^{k-l} + 1 \leq v \leq 2^k - 2^{k-l} + 2^{k-l-1}$, where $1 \leq l \leq k - 1$. Further, using Theorems 6 and 7, it can be easily seen that C attains its singleton bound if and only if $v + 1 = 2^{l+1}$. Clearly, no value of l other than $k - 1$ satisfies $v + 1 = 2^{l+1}$. Thus C is a MHDR code in this case if and only if $l = k-1$ that is $v = 2^k - 1$. Combining 1, 2 and 3 above, we get that C is a MHDR code if and only if $v \in \{0, 1, 2^k - 1\}$.

The following table gives a list of the best cyclic codes over Z_8 of length 4 that attain largest minimum Hamming, Lee and Euclidean distances among all codes of the same rank. The codes labeled with an asterisk achieve the singleton bounds (Table 1).

Table 1. Best codes of length 4 over Z_8

Sr. no.	Rank	Cyclic codes	d_H	d_L	d_E
1	3	$<4(x+1)>$	2^*	8^*	32^*
2	3	$<2(x+1)^3 + 4>$	2^*	8^*	16
3	3	$<2(x+1)^3, 4(x+1)>$	2^*	8^*	16
4	3	$<(x+1)^3 + 2(x+1)>$	2^*	8^*	16
5	3	$<(x+1)^3 + 2(x+1) + 2(x+1)^2, 4(x+1)>$	2^*	8^*	16
6	3	$<(x+1)^3 + 2(x+1) + 2(x+1)^2 + 4>$	2^*	8^*	16
7	2	$<4(x+1)^2>$	2	8	32
8	2	$<2(x+1)^3>$	2	8	16
9	2	$<2(x+1)^2 + 4 + 4(x+1)>$	2	8	16
10	2	$<2(x+1)^2 + 4(x+1)>$	2	8	16
11	2	$<2(x+1)^3 + 4(x+1), 4(x+1)^2>$	2	8	16
12	2	$<(x+1)^3 + 2(x+1) + 2(x+1)^2, 4(x+1)^2>$	2	8	16
13	2	$<(x+1)^3 + 4(x+1)>$	2	8	16
14	1	$<2(x+1)^3 + 4(x+1)>$	$4*$	8	16
15	1	$<2(x+1)^3 + 4(x+1) + 4(x+1)^2>$	4^*	8	16
16	1	$<(x+1)^3 + 2(x+1) + 2(x+1)^2>$	4^*	8	16
17	1	$<4(x+1)^3>$	4^*	16^*	64^*

6 Conclusion

The rank of a cyclic code C of length $n = 2^k$ over Z_8 is equal to $n - v$, where v is the degree of a minimal degree polynomial in C. Moreover, C is a MHDR cyclic code if and only if v belongs to the set $\{0, 1, 2^k - 1\}$.

References

1. Abualrub, T., Oehmke, R.: Cyclic codes of length 2^e over Z_4. Discret. Appl. Math. **128**(1), 3–9 (2003)
2. Abualrub, T., Oehmke, R.: On generators of Z_4 cyclic codes of length 2^e. IEEE Trans. Inf. Theory **49**(9), 2126–2133 (2003)
3. Abualrub, T., Ghrayeb, A., Oehmke, R.: A mass formula and rank of Z_4 cyclic codes of length 2^e. IEEE Trans. Inf. Theory **50**(12), 3306–3312 (2004)
4. Dinh, H.Q.: On the linear ordering of some classes of negacyclic and cyclic codes and their distance distributions. Finite Fields Appl. **14**(1), 22–40 (2008)
5. Dinh, H.Q., Lopez Permouth, S.R.: Cyclic and negacyclic codes over finite chain rings. IEEE Trans. Inf. Theory **50**(8), 1728–1744 (2004)
6. Dinh, H.Q.: Negacyclic codes of length 2^s over Galois rings. IEEE Trans. Inf. Theory **51**(12), 4252–4262 (2005)
7. Dougherty, S.T., Park, Y.H.: On modular cyclic codes. Finite Fields Appl. **13**, 31–57 (2007)
8. Garg, A., Dutt, S.: Cyclic codes of length 2^k over Z_8. Sci. Res. Open J. Appl. Sci. **2**, 104–107 (2012). Oct - 2012 World Congress on Engineering and Technology
9. Garg, A., Dutt, S.: Cyclic codes of length 2^k over Z_{2^m}. Int. J. Eng. Res. Dev. **1**(9), 34–37 (2012)
10. Hammons, A.R., Kumar, P.V., Calderbank, A.R., Sloane, N.J.A., Sole, P.: The Z_4–linearity of Kerdock, Preparata, Goethals, and related codes. IEEE Trans. Inf. Theory **40**, 301–319 (1994)
11. Kiah, H.M., Leung, K.H., Ling, S.: Cyclic codes over $GR(p^2, m)$ of length p^k. Finite Fields Appl. **14**, 834–846 (2008)
12. Kaur, J., Dutt, S., Sehmi, R.: Cyclic codes over galois rings. In: Govindarajan, S., Maheshwari, A. (eds.) CALDAM 2016. LNCS, vol. 9602, pp. 233–239. Springer, Heidelberg (2016)
13. Lopez, S.R., Szabo, S.: On the hamming weight of repeated root codes (2009). arXiv:0903.2791v1 [cs.IT]
14. Minjia, S., Shixin, Z.: Cyclic codes over the ring Z_{p^2} of length p^e. J. Electr. (China) **25**(5), 636–640 (2008)
15. Shiromoto, K.: A basic exact sequence for the Lee and Euclidean weights of linear codes over Z_l. Linear Algebra Appl. **295**, 191–200 (1995)
16. Salagean, A.: Repeated-root cyclic and negacyclic codes over a finite chain ring. Discret. Appl. Math. **154**(2), 413–419 (2006)

Group Distance Magic Labeling of C_n^r

Aloysius Godinho$^{(\boxtimes)}$ and T. Singh

Department of Mathematics, Birla Institute of Technology and Science Pilani,
K K Birla Goa Campus, NH-17B, Zuarinagar, Sancoale, Goa, India
{p2014001,tksingh}@goa.bits-pilani.ac.in
http://www.bits-goa.ac.in

Abstract. Let $G = (V, E)$ be a graph and Γ be an abelian group both of order n. For $D \subset \{0, 1, \ldots, diam(G)\}$, the D-distance neighbourhood of a vertex v in G is defined to be the set $N_D(v) = \{x \in V \mid d(x, v) \in D\}$. A bijection $f : V \to \Gamma$ is called a (Γ, D)-distance magic labeling of G if there exists an $\alpha \in \Gamma$ such that $\sum_{x \in N_D(v)} f(x) = \alpha$ for every $v \in V$. In this paper we study (Γ, D)-distance magic labeling of the graph C_n^r for $D = \{d\}$. We obtain $(\Gamma, \{d\})$-distance magic labelings of C_n^r with respect to certain classes of abelian groups. We also obtain necessary conditions for existence of such labelings.

Keywords: Group distance magic labeling · Circulant graphs

1 Introduction

All graphs considered in this paper are simple, finite graphs. For graph theoretic terminology and notation we refer to Chartrand and Lesniak [4].

The concept of *distance magic labeling* was motivated by the construction of magic squares. Suppose, we have a magic square consisting of n rows and n columns and each row sum is k. Consider a complete multipartite graph with each row of the magic square representing a partite set and we label each vertex with the corresponding integers in the magic square. We find that the sum of the labels of all vertices at distance 1 (i.e. an open neighborhood set of vertices) for each vertex is the same and is equal to $(n - 1)k$.

Let $G = (V, E)$ be a graph with n vertices and m edges. Let $f : V \to \{1, 2, \ldots, n\}$ be a bijection such that for every vertex u, $\sum_{x \in N(u)} f(x) = k$ which is a constant and independent of u. This constant value k is called the *magic constant* of the graph G and f is called a *distance magic labeling* of G. A graph which admits such a labeling is called a *distance magic graph*.

This concept was introduced by Vilfred [14] as *sigma labeling*. The same concept was independently studied by Miller et al. [11] using the terminology *1-vertex magic vertex labeling*. The same concept was introduced independently in more general way as *neighborhood magic labeling* by Acharya et al. [1]. For a survey on the topic of distance magic labeling we refer the reader to Arumugam et al. [3] and Rupnow [13].

© Springer International Publishing AG 2017
D. Gaur and N.S. Narayanaswamy (Eds.): CALDAM 2017, LNCS 10156, pp. 187–192, 2017.
DOI: 10.1007/978-3-319-53007-9_17

A generalised form of the notion of distance magic labeling introduced by O'Neal and Slater [12] as follows: Let G be a graph of order n. Let $f : V \rightarrow \{1, 2, \ldots, n\}$ be a bijective function. For $D \subset \{0, 1, \ldots, diam(G)\}$ the *D-distance neighbourhood* of a vertex v in G is defined to be the set $N_D(v) = \{x \in V \mid d(x, v) \in D\}$. For $v \in V(G)$, compute the *D-distance weight* $w_D(v)$ of v as $w_D(v) = \sum_{x \in N_D(v)} f(x)$. If $w_D(v) = k$ (a constant) for every $v \in V(G)$ then the graph G said to be *D-distance magic*, the map f is said to be a *D-distance magic labeling* and k is called the *D-distance magic constant*.

The notion of *group distance magic labeling* of graphs was introduced by Froncek in [7]. For an abelian group Γ and a graph $G = (V, E)$ of the same order, a group distance magic labeling or a Γ-distance magic labeling of G is a bijection $f : V \rightarrow \Gamma$ such that $\sum_{x \in N(v)} f(x) = \alpha \in \Gamma$ for every vertex $v \in V$. A graph G of order n is said to be *group distance magic* if it is Γ-distance magic with respect to every abelian group Γ of order n. Now we list some known results as follows:

Theorem 1 [2]. *The Cartesian product $C_m \square C_k$, $m, k \geq 3$, is a $\mathbb{Z}_{mk}$-distance magic graph if and only if km is even.*

Theorem 2 [5]. *The complete bipartite graph $K_{m,n}$ is a group distance magic graph if and only if $m + n \equiv 0, 1$ or $3 \pmod 4$.*

For the graph $G = (V, E)$ and a positive integer r the graph $G^r = (V', E')$ is defined to be the graph with $V' = V$, such that two vertices $u, v \in V'$ are adjacent if the distance between these vertices in G is at most r. For the cycle C_n the graph C_n^r is a graph of size rn and diameter $\lceil \frac{n-1}{2r} \rceil$. The following result is due to Cichacz [6]:

Theorem 3. *Let $\gcd(n, p + 1) = d$. If p is even, $n > 2p + 1$ and $n = 2kd$, then C_n^p has a $\mathbb{Z}_\alpha \times \mathcal{A}$-distance magic labeling for any $\alpha \equiv 0 \pmod{2k}$ and any abelian group $\mathcal{A}$ of order $\frac{n}{\alpha}$.*

Motivated by this, we generalise group distance magic labeling to arbitrary *D-distance neighbourhoods* as follows:

Definition 1. *Let $G = (V, E)$ be a graph and Γ be an abelian group both of order n. A bijection $f : V \rightarrow \Gamma$ is called a (Γ, D)-distance magic labeling of G if there exists an $\alpha \in \Gamma$ such that the weight $w(v) = \sum_{x \in N_D(v)} f(x) = \alpha$ for every $v \in V$, α is called the magic constant.*

Remark 1. We shall denote the vertex set of C_n^r by $\{v_0, v_1, \ldots, v_{n-1}\}$, such that the vertex v_i is adjacent to the set of vertices $\{v_{i+j}, v_{i-j} : 1 \leq j \leq r\}$. The index i in v_i is assumed to be taken modulo n. For a bijective function $f : V(C_n^r) \rightarrow \{1, 2, \ldots, n\}$, we shall use the notation u_i to denote $f(v_i)$ (i.e. $f(v_i) = u_i$).

2   Necessary Conditions for Existence of Group Distance Magic Labeling

We begin this section by investigating the existence of $(\Gamma, \{d\})$-distance magic labeling for the graph C_n^r when $d = diam(C_n^r)$.

Lemma 1. *For* $d = \lceil \frac{n-1}{2r} \rceil$, *the graph* C_n^r *is not* $(\Gamma, \{d\})$-*distance magic for any abelian group* Γ *of order* n.

Proof. Let $d = \lceil \frac{n-1}{2r} \rceil$ and Γ be an abelian group of order n such that C_n^r is $(\Gamma, \{d\})$-distance magic. Let $f : V \rightarrow \Gamma$ be a $(\Gamma, \{d\})$-distance magic labeling of C_n^r. For $v_i \in V(C_n^r)$ and $N_{\{d\}}(v_i) = \{v_{i+(d-1)r+1}, v_{i+(d-1)r+2}, \ldots, v_{i+(d-1)r+l}\}$ where $l = n - 2r(d-1) - 1$. Since f is a magic labeling, for any $i \in \{0, 1, \ldots, n-1\}$, the vertex weight $w(v_i) = w(v_{i+1})$, canceling out the common terms we get, $u_{i+(d-1)r+1} = u_{i+(d-1)r+l+1}$. Since f is bijective, $v_{i+(d-1)r+1}$ and $v_{i+(d-1)r+l+1}$ must represent the same vertex which implies that $i + (d-1)r + 1 \equiv i + (d-1)r + l + 1 (\mathrm{mod}\ n)$. Hence $l \equiv 0 (\mathrm{mod}\ n)$ which is not true. Therefore $v_{i+(d-1)r+1}$ and $v_{i+(d-1)r+l+1}$ are distinct vertices which is a contradiction. Hence the graph C_n^r is not $(\Gamma, \{d\})$-distance magic for any abelian group Γ. $\qquad \square$

Lemma 2. *Let* C_n^r *be a* $(\Gamma, \{d\})$-*distance magic graph with a magic labeling* f *for* $d < \lceil \frac{n-1}{2r} \rceil$. *Then for any* $0 \leq i \leq n - 1$ *and* $\lambda \in \mathbb{N}$,

$$u_i + u_{i+\rho} = u_{i+\lambda r} + u_{i+\lambda r+\rho} \tag{1}$$

where $\rho = r(2d - 1) + 1$.

Proof. Let $0 \leq j \leq n - 1$. Comparing $w(v_j)$ and $w(v_{j+1})$ we obtain $u_{j-dr} + u_{j+r(d-1)+1} = u_{j-(d-1)r} + u_{j+dr+1}$. Setting $i = j - dr$ and $\rho = r(2d - 1) + 1$ we obtain:

$$u_i + u_{i+\rho} = u_{i+r} + u_{i+r+\rho} \tag{2}$$

The above equation holds for every $i \in \{0, 1, \ldots, n-1\}$. Substituting $i = i + r$ in (2) we get, $u_{i+r} + u_{i+r+\rho} = u_{i+2r} + u_{i+2r+\rho}$. Hence $u_i + u_{i+\rho} = u_{i+r} + u_{i+r+\rho} = u_{i+2r} + u_{i+2r+\rho}$. Substituting $i = i + 2r$ in (2) and comparing with the previous expression we obtain $u_i + u_{i+\rho} = u_{i+3r} + u_{i+3r+\rho}$. By induction $u_i + u_{i+\rho} = u_{i+\lambda r} + u_{i+\lambda r+\rho}$ for every $\lambda \in \mathbb{N}$. This completes the proof. $\qquad \square$

Let $gcd(n, r) = a$ and $\frac{n}{a} = b$. From (1) we have the following set of equations for the graph C_n^r.

$$
\left.
\begin{aligned}
u_0 + u_{0+\rho} &= u_r + u_{r+\rho} = u_{2r} + u_{2r+\rho} = u_{3r} + u_{3r+\rho} + \ldots = u_{(b-1)r} + u_{(b-1)r+\rho} \\
u_1 + u_{1+\rho} &= u_{1+r} + u_{1+r+\rho} = u_{1+2r} + u_{1+2r+\rho} = \ldots = u_{1+(b-1)r} + u_{1+(b-1)r+\rho} \\
&\quad\ \vdots \qquad\qquad\qquad \vdots \qquad\qquad \vdots \qquad\qquad \vdots \\
u_{a-1} + u_{a-1+\rho} &= u_{a-1+r} + u_{a-1+r+\rho} = \ldots = u_{a-1+(b-1)r} + u_{a-1+(b-1)r+\rho}
\end{aligned}
\right\}
$$

Observe that if $i \equiv j \pmod{a}$, then

$$u_i + u_{i+\rho} = u_j + u_{j+\rho}. \tag{3}$$

We shall denote $u_i + u_{i+\rho} = c_i$. Therefore if f is a $(\Gamma, \{d\})$-distance magic labeling of C_n^r, there exists $c_0, c_1, \ldots, c_{a-1} \in \Gamma$ such that $u_i + u_{i+\rho} = c_{i(\bmod\, a)}$. Henceforth we shall assume the index i in c_i to be taken modulo a.

3   A Group Distance Magic Labeling of C_n^r

Theorem 4. *The bijection* $f : V(C_n^r) \to \Gamma$ *is* $(\Gamma, \{d\})$-*distance magic if and only if* $u_i + u_{i+\rho} = u_j + u_{j+\rho}$ *whenever* $i \equiv j \pmod{a}$, *where* $a = \gcd(n, r)$.

Proof. The forward implication follows from Eq. (3).

For the converse suppose $c_0, c_1, \ldots, c_{a-1} \in \Gamma$ such that $u_i + u_{i+\rho} = c_{i(\bmod\, a)}$.

$$w(v_i) = \sum_{j=1}^{r} (u_{i+(d-1)r+j} + u_{i-(d-1)r-j})$$

$$= \frac{r}{a}(c_0 + c_1 + \ldots + c_{a-1})$$

Hence $w(v_i) = \frac{r}{a}(c_0 + c_1 + \ldots + c_{a-1})$ for every $v_i \in V(C_n^r)$. Therefore the labeling f is $(\Gamma, \{d\})$-distance magic with magic constant $\frac{r}{a}(c_0 + c_1 + \ldots + c_{a-1})$.

Lemma 3. *Suppose* C_n^r *is* $(\Gamma, \{d\})$-*distance magic with a magic labeling* f *and* $a = \gcd(n, r)$. *Suppose that* $c_0, c_1, \ldots, c_{a-1} \in \Gamma$ *such that* $u_i + u_{i+\rho} = c_i \,(\bmod\, a)$. *Then for any vertex* $v_i \in V(C_n^r)$, *the following equations hold,*

$$u_{i+a\rho} = \sum_{j=1}^{a} (-1)^{j-1} c_{i+a-j} + (-1)^a u_i \tag{4}$$

$$u_{i-a\rho} = \sum_{j=0}^{a-1} (-1)^{j-1} c_{i-(a-j)} + (-1)^a u_i \tag{5}$$

Proof. Since $u_i + u_{i+\rho} = c_i$, we obtain $u_{i+\rho} = c_i - u_i$. Similarly $u_{i+\rho} + u_{i+2\rho} = c_{i+\rho}$. Since $\rho = r(2d-1) + 1 \equiv 1 \pmod{a}$, it follows that $i + \rho \equiv i + 1 \pmod{a}$. Therefore $u_{i+2\rho} = c_{i+1} - u_{i+\rho} = c_{i+1} - c_i + u_i$. Similarly $u_{i+3\rho} = c_{i+2} - c_{i+1} + c_i - u_i$ proceeding in this manner we obtain the expression $u_{i+a\rho} = c_{i+a-1} - c_{i+a-2} + \ldots + c_i - u_i$ if a is odd and the expression $u_{i+a\rho} = c_{i+a-1} - c_{i+a-2} + \ldots - c_i + u_i$ if a is even. This proves Eq. (4).

To prove (5), observe that $u_{i-\rho} + u_i = c_{i-\rho}$. Since $i - \rho \equiv i - 1 \pmod{a}$, we have $u_{i-\rho} + u_i = c_{i-1}$. Hence $u_{i-\rho} = c_{i-1} - u_i$. Similarly $u_{i-2\rho} + u_{i-\rho} = c_{i-2\rho} = c_{i-2}$. From this we get $u_{u_{i-2\rho}} = c_{i-2} - c_{i-1} + u_i$. Proceeding in this manner we obtain the expression $u_{i-a\rho} = c_{i-a} - c_{i-(a-1)} + \ldots + c_{i-1} - u_i$ if a is odd and the expression $u_{i-a\rho} = c_{i-a} - c_{i-(a-1)} + \ldots - c_{i-1} + u_i$ if a is even. This proves Eq. (5).

Theorem 5. *For the graph* C_n^r *and* $d < \lceil \frac{n-1}{2r} \rceil$, *let* $a = \gcd(n, r) \equiv 1 \pmod{2}$ *and* $\rho = r(2d-1) + 1$. *Then for an abelian group* Γ *of order* n *the graph* C_n^r *is* $(\Gamma, \{d\})$-*distance magic only if* $2a\rho \equiv 0 \pmod{n}$

Proof. Suppose C_n^r is $(\Gamma, \{d\})$-distance magic. Substituting $i = 0$ in (4) and (5) we get:

$$u_{a\rho} = u_{-a\rho} = c_{a-1} - c_{a-2} + \ldots + c_0 - u_0 \tag{6}$$

Since the labeling is one-one, $v_{a\rho}$ and $v_{-a\rho}$ must represent the same vertex. Hence $a\rho \equiv -a\rho (\text{mod } n)$ which implies that $2a\rho \equiv 0 (\text{mod } n)$. This completes the proof.

Theorem 6. *For the graph C_n^r, let d be a positive integer $< \lceil \frac{n-1}{2r} \rceil$. Let $\rho = r(2d - 1) + 1$ and $\gcd(n, \rho) = m$. If r is even and $n = 2km$, then C_n^r has a $(\mathbb{Z}_\alpha \times A, d)$-distance magic labeling for any $\alpha \equiv 0(\text{mod } 2k)$ and any abelian group A of order $\frac{n}{\alpha}$.*

Proof. Let $\frac{n}{\alpha} = q$. Let $A = \{a_0, a_1, \ldots, a_{q-1}\}$ such that $a_0 = 0$ (i.e. the identity element of A). Let $\Gamma = \mathbb{Z}_\alpha \times A$, every element $g \in \Gamma$ can be written in the form $g = (j, a_i)$ with $j \in \mathbb{Z}_\alpha$ and $a_i \in A$.

Let $V = \{v_0, v_1, \ldots, v_{n-1}\}$ be the vertex set of C_n^r. Let $X = \langle \rho \rangle$ be the subgroup of Z_n of order $2k$. Let $\{X, X+1, \ldots, X+(m-1)\}$ be the set of cosets of X in $\mathbb{Z}_n$. Let X_j denote the subset V whose subscripts belong to the coset $X + j$. Notice that $\alpha = 2kh$ for some positive integer h. Let $H = \langle 2h \rangle$ be the subgroup of Z_α of order k.

We shall define a $(\Gamma, \{d\})$ distance magic labeling $l : \{v_0, v_1, \ldots, v_{n-1}\} \rightarrow \mathbb{Z}_\alpha \times A$ such that $l(v_i) = (l_1(v_i), l_2(v_i))$ where l_1 and l_2 are maps from V into $\mathbb{Z}_\alpha$ and A respectively. First Label the vertices of X_0 as follows:

$$l(v_{2i\rho}) = (2ih, a_0), \quad l(v_{(2i+1)\rho}) = (-2ih - 1, -a_0), \quad i = 0, 1, \ldots, k-1$$

If the subscript m in v_m belongs to the coset $X + j$ then denote it by m_j. Notice that a vertex v_{m_j} belongs to X_j then the vertex v_{m_j-p-2} belongs to X_{j-1}. We label $X_1, X_2, \ldots, X_{m-1}$ recursively in the following manner:

$$l_1(v_{m_j}) = \begin{cases} l_1(v_{m_j-p-2}) + 1 & l_1(v_{m_j-j(p+2)}) \equiv 0(\text{mod } 2h) \\ l_1(v_{m_j-p-2}) - 1 & l_1(v_{m_j-j(p+2)}) \not\equiv 0(\text{mod } 2h) \end{cases}$$

$$l_2(v_{m_j}) = \begin{cases} a_{\lfloor j/k \rfloor} & l_1(v_{m_j}) \equiv 0(\text{mod } 2) \\ -a_{\lfloor j/k \rfloor} & l_1(v_{m_j}) \equiv 1(\text{mod } 2) \end{cases}$$

Clearly l is a bijection and satisfies the relation $l(v_{2i}) + l(v_{2i+\rho}) = (-1, 0)$ and $l(v_{2i+1}) + l(v_{2i+\rho+1}) = (2h - 1, 0)$ for any i.

Recall that $N(v_i) = \{v_{i-dr}, v_{i-dr+1}, \ldots, v_{i-(d-1)r-1}, v_{i+(d-1)r+1}, v_{i+(d-1)r+2}, \ldots, v_{i+dr}\}$. Since r is even it implies that

$$w(v_i) = \sum_{j=1}^{r} (l(v_{i-(d-1)r-j}) + l(v_{i-(d-1)r-j+\rho})) = r(2h - 1, 0) + r(-1, 0) = r(2h - 2, 0)$$

for any i. Hence l is $(\Gamma, \{d\})$-distance magic with magic constant $r(2h - 2, 0)$. This completes the proof.

4   Conclusion and Scope

In this paper we have studied the existence $(\Gamma, \{d\})$ of distance magic labeling for the graph C_n^r. The main objective in studying this problem is to characterise all abelian groups Γ of order n which induce a $(\Gamma, \{d\})$-distance magic labeling on C_n^r for a given $\{d\}$. In this paper we have given such labelings for some classes of abelian groups, a complete characterization is not yet known. Furthermore a graph G of order n to be $\{d\}$-*group distance magic* if there exists a $(\Gamma, \{d\})$-distance magic labeling of G for every abelian group Γ of order n. It would be interesting to investigate which of the graphs C_n^r are $\{d\}$-group distance magic for a given distance d.

Acknowledgement. Authors are thankful to the Department of Science and Technology, New Delhi for financial support through the project No. SR/S4/MS-734/11. Authors also thank the referees for their valuable comments/suggestions.

References

1. Acharya, B.D., Parameswaran, V., Rao, S.B., Singh, T.: Neighborhood Magic Graphs (2006, preprint)
2. Anholcer, M., Cichacz, S., Gorlich, A.: Note on distance magic graphs Km, n[C4], preprint
3. Arumugam, S., Froncek, D., Kamatchi, N.: Distance magic graphs-a survey. J. Indones. Math. Soc., Spec. Ed. 11–26 (2011)
4. Chartrand, G., Lesniak, L.: Graphs & Digraphs, 4th edn. Chapman and Hall, CRC, Boca Raton (2005)
5. Cichacz, S.: Note on group distance magic complete bipartite graphs. Cent. Eur. J. Math. **12**, 529–533 (2014)
6. Cichacz, S.: Group distance magic labeling of some cycle-related graphs. Australas. J. Combin. **57**, 235–243 (2013)
7. Froncek, D.: Group distance magic labeling of Cartesian product of cycles. Australas. J. Combin. **55**, 167–174 (2013)
8. Gallian, J.A.: A dynamic survey of graph labeling. Electron. J. Combin. Dyn. Surv. DS **6**, 1–145 (2005)
9. Kamatchi, N.: Distance magic and antimagic labelings of graphs. Ph.D. thesis, Kalaslingam University, Tamil Nadu, India (2013)
10. Kovar, P., Froncek, D., Kovarova, T.: A note on 4-regular distance magic graphs. Australas. J. Combin. **54**, 127–132 (2012)
11. Miller, M., Rodger, C., Simanjuntak, R.: Distance magic labelings of graphs. Australas. J. Combin. **28**, 305–315 (2003)
12. O'Neal, A., Slater, P.J.: Uniqueness of vertex magic constants. SIAM J. Discret. Math. **27**, 708–716 (2013)
13. Rupnow, R.: A survey on distance magic graphs. Master's report, Michigan Technological University (2014). http://digitalcommons.mtu.edu/etds/829
14. Vilfred, V.: Σ-labelled graph and circulant graphs. Ph.D. thesis, University of Kerala, Trivandrum, India (1994)

Broadcast Graphs Using New Dimensional Broadcast Schemes for Knödel Graphs

Hovhannes A. Harutyunyan and Zhiyuan Li[✉]

Concordia University, Montreal, QC H3G 1M8, Canada
l_zhiyua@encs.concordia.ca

Abstract. Broadcasting is an information disseminating process of distributing a message from an originator vertex v of graph $G = (V, E)$ to all of its vertices. The broadcast time of vertex v is the minimum possible time required to broadcast from v in graph G. A graph G on n vertices is called broadcast graph if broadcasting from any originator in G can be accomplished in $\lceil \log n \rceil$ time. A broadcast graph with the minimum number of edges is called minimum broadcast graph. The number of edges in a minimum broadcast graph on n vertices is denoted by $B(n)$. Finding the values of $B(n)$ is very difficult. A large number of papers present techniques to construct broadcast graphs and to obtain upper bounds on $B(n)$. In this paper, we first find new dimensional broadcast schemes for Knödel graphs, and then use them to give a general upper bound on $B(n)$ for almost all odd n.

1 Introduction

One-to-all information spreading is one of the major tasks on a modern interconnection network. This process, named *broadcasting*, originates from one node in the network, called *originator*, and finishes when every node in the network has the information. The broadcast time is one of the main measures the overall performance of a network.

Over the past half century, a long sequence of research papers study this topic under different models. These models differ at the number of originators, the number of receivers at each time unit, the distance of each call, the number of destinations in the network, and other characteristics of the network. In this paper, we focus on the classical model with the following assumptions:

- the network has only one originator;
- each call has only one informed node, the sender and one of its uninformed neighbors - the receiver;
- every call requires one time unit.

A network is modeled as a simple connected graph $G = (V, E)$, where vertex set V represents the nodes in the network, and edge set E represents the communication links.

© Springer International Publishing AG 2017
D. Gaur and N.S. Narayanaswamy (Eds.): CALDAM 2017, LNCS 10156, pp. 193–204, 2017.
DOI: 10.1007/978-3-319-53007-9_18

Definition 1. *The broadcast scheme is a sequence of parallel calls in a graph G originating from a vertex v. Each call, represented by a directed edge, defines the sender and the receiver. The broadcast scheme generates a broadcast tree, which is a spanning tree of the graph rooted at the originator.*

Definition 2. *Let G be a graph on n vertices and v be the broadcast originator in graph G, $b(G,v)$ defines the minimum time unit required to broadcast from originator v in graph G. The broadcast time of graph G, $b(G) = max\{b(G,v)|v \in V(G)\}$ defines the maximum number of time units required from any originator to broadcast in graph G.*

Note that one informed vertex can at most call one uninformed vertex in each time unit; therefore, the number of informed vertices in each time unit is at most doubled, and $b(G) \geq \lceil \log_2 n \rceil = \lceil \log n \rceil$. For convenience, we will omit the base of all logarithms throughout this paper when the base is 2.

Definition 3. *A graph G on n vertices is called broadcast graph if $b(G) = \lceil \log n \rceil$. A broadcast graph with the minimum number of edges is called minimum broadcast graph (mbg). This minimum number of edges is called broadcast function and denoted by $B(n)$.*

From the application perspective mbgs represent the cheapest graphs (with minimum number of edges) where broadcasting can be accomplished in minimum possible time.

The study of minimum broadcast graphs and broadcast function $B(n)$ has a long history. Farley et al. introduce minimum broadcast graphs in [10]. In the same paper they define the broadcast function, determine the values of $B(n)$, for $n \leq 15$ and $n = 2^k$ and prove that hypercubes are minimum broadcast graphs. Khachatrian and Haroutunian [22] and independently Dinneen et al. [7] show that Knödel graphs, defined in [23], are minimum broadcast graphs on $n = 2^k - 2$ vertices. Park and Chwa prove that the recursive circulant graphs on 2^k vertices are minimum broadcast graphs [27]. The comparison of the three classes of minimum broadcast graphs can be found in [11]. Besides these three classes, there is no other known infinite construction of minimum broadcast graphs. The values of $B(n)$ are also known for $n = 17$ [26], $n = 18, 19$ [5,32], $n = 20, 21, 22$ [25], $n = 26$ [28,33], $n = 27, 28, 29, 58, 61$ [28], $n = 30, 31$ [5], $n = 63$ [24], $n = 127$ [15] and $n = 1023, 4095$ [29].

Since minimum broadcast graphs are difficult to construct, a long sequence of papers present different techniques to construct broadcast graphs in order to obtain upper bounds on $B(n)$. Furthermore, proving that a lower bound matches the upper bound is also extremely difficult, because most of the lower bound proofs are based on vertex degree. However, minimum broadcast graphs except hypercubes and Knödel graphs on $2^k - 2$ vertices are not regular. Thus, in general the upper bounds cannot match the lower bounds based on vertex degree.

Upper bounds on $B(n)$ are given by good constructions of broadcast graphs. Farley constructs broadcast graphs recursively by combining two or three smaller broadcast graphs and shows $B(n) \leq \frac{n}{2}\lceil \log n \rceil$ [9]. This construction is generalized in [6] using up to seven small broadcast graphs. A tight asymptotic bound on $B(n) = \Theta(L(n) \cdot n)$ is given in [13] by proving that $\frac{L(n)-1}{2}n \leq B(n) \leq (L(n)+2)n$, where $L(n)$ is the number of consecutive leading 1's in the binary representation of $n - 1$. In [22], the compounding method is introduced which uses vertex cover of graphs. This method constructs new broadcast graphs by forming the compound of several known broadcast graphs. In [3] the compounding method was generalized to any n by using solid vertex cover. A compounding method using center vertices is introduced in [31] and shown to be equivalent to the method of using solid vertex cover in [8]. The authors in [17] continue on the line of compounding and introduce a method of also merging vertices. And more recently [1,16], compounding binomial trees with hypercubes improves the upper bound on $B(n)$ for many values of n.

Vertex addition is another approach to construct good broadcast graphs by adding several vertices to existing broadcast graphs [5]. [15] continues on this line and adds one vertex to Knödel graphs on $2^k - 2$ vertices. The added vertex is connected to every vertex in the dominating set of the Knödel graph. In [19], the same method is applied to generalized Knödel graphs, in order to construct broadcast graphs on any odd number of vertices.

Adhoc constructions sometimes also provide good upper bounds. This method usually constructs broadcast graphs by adding edges to a binomial tree [13,17].

Vertex deletion is studied in [5]. Several other constructions are presented in [5,12,13,17,30–32]. Lower bounds on $B(n)$ are also studied in the literature. The authors in [12] show $B(n) \geq \frac{n}{2}(\lfloor \log n \rfloor - \log(1 + 2^{\lceil \log n \rceil} - n))$, for any n. $B(n) \geq \frac{n}{2}(m - p - 1)$ is proved in [18], where m is the length of the binary representation $a_{m-1}a_{m-2}...a_1a_0$ of n and p is the index of the leftmost 0 bit. This bound is improved to be $B(n) \geq \frac{n}{2}(m - p - 1 + b)$, where $b = 0$ if $p = 0$ or $a_0 = a_1 = \cdots = a_{p-1} = 0$ and $b = 1$ otherwise.

Besides the general lower bounds, $B(n) \geq \frac{k^2(2^k-1)}{2(k+1)}$ for $n = 2^k - 1$ is shown in [24]. The lower bounds on $B(2^k - 3)$, $B(2^k - 4)$, $B(2^k - 5)$ and $B(2^k - 6)$ are given in [28]. The lower bounds on $B(2^k - 2^p)$ and $B(2^k - 2^p + 1)$, where $3 \leq p < k$ are presented in [14]. Better lower bounds on $n = 24, 25$ are given in [2]. Note that $23 \leq n \leq 25$ are the only values of $n \leq 32$ for which $B(n)$ is not known.

For more on broadcasting and gossiping see [20,21].

This paper is organized as follows. Section 2 introduces the new dimensional broadcast schemes for Knödel graph. Section 3 constructs new broadcast graphs by adding one vertex to Knödel graph. To show the minimum time broadcast in this graph newly obtained dimensional broadcast schemes are applied.

2   New Dimensional Broadcast Schemes for Knödel Graph

2.1   Knödel Graph

Knödel graph, the most important broadcast graph, gives good upper bound on $B(n)$ for even values of n, ever since it is defined by Knödel in [23].

Definition 4. *A Knödel graph on even number of vertices is* $KG_n = (V, E)$, *where* $V = \{v_i | 0 \le i \le n - 1\}$ *and* $E = \{(v_i, v_j) | i + j \equiv 2^s - 1 \mod n, 1 \le s \le \lfloor \log n \rfloor \text{ and } v_i, v_j \in V\}$.

We say that v_i and v_j are adjacent on dimension s if $i + j \equiv 2^s - 1 \mod n$. Knödel graph on n vertices with $n = 2^m - 2^k - 2a$, where $0 \le k \le m - 2$ and $0 \le 2a \le 2^{k-1}$, has the following well known properties.

- KG_n is $(m-1)$-regular. Each vertex has $m - 1 = \lfloor \log n \rfloor$ dimensional neighbors. Each edge has dimension i for some $1 \le i \le m - 1$
- KG_n is bipartite, v_i and v_j are adjacent only if i and j have different parities.
- KG_n has $\frac{n \lfloor \log n \rfloor}{2}$ edges.

[4] shows the dimensionality property of Knödel graph and defines a special types of broadcast schemes, called *dimensional broadcast scheme*. Figure 2a shows one example of Knödel graph on 14 vertices, and Fig. 1b shows the dimensional broadcast scheme. In particular, $1, 2, \cdots, m - 1$ denotes the dimensional broadcast scheme, where at time unit i, all of the informed vertices call their i-th dimensional neighbors for all $i = 1, 2, \cdots, m - 1$ and call their first dimensional neighbors at time unit m. Authors of [4] also show that any *cyclic shift* of dimensions, $s, s + 1, \cdots, m - 1, 1, 2, \cdots, s - 1, s$ is also a valid broadcast scheme for Knödel graph KG_n, and $1 \le s \le m - 1$. There are no other known dimensional broadcast schemes in Knödel graphs.

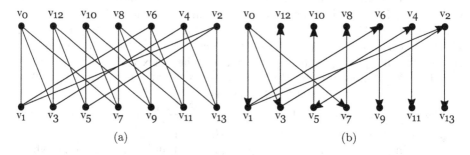

Fig. 1. Knödel graph KG_{14} and its broadcast scheme. We draw every Knödel graph bipartitely in this paper to see the dimensionality. An even vertex v_i locates its s'th dimensional neighbor at the 2^{s-1}'s position to the right of the odd vertex right under v_i (also v_i's first dimensional neighbor).

The problem of finding all dimensional broadcast schemes is a very difficult problem [4]. In this section, we describe new dimensional broadcast schemes for Knödel graph and use them to construct new broadcast graphs in Sect. 3. Our first result generalizes the basic result of [4].

Theorem 1. *Let* KG_n *be a Knödel graph on* n *vertices, where* $n = 2^m - 2^k - 2a$, $0 \le k \le m-2$, $m \ge 0$, *and* $2a < 2^k$. *Dimensional broadcast scheme* $1, 2, 3, ..., m - 1, t$, *where* $1 \le t \le k$ *is a valid broadcast scheme.*

Proof. This dimensional broadcast scheme is the same as the dimensional broadcast scheme given in [4], except the last dimension could be any $t \leq k$. To prove that this broadcast scheme is valid, we only need to show that the uninformed vertices before the last time unit can be called by their neighbors on dimension t, for any $1 \leq t \leq k$ during the last time unit m.

Since Knödel graph is regular and vertex transitive, then without loss of generality, we can assume that the originator is v_0. Since every vertex broadcasts on dimension s at time unit s for $1 \leq s \leq m - 1$, then the informed vertices after time s are $v_0, v_1, \cdots, v_{2^s-1}$. So, after $m - 1$ time units, the informed vertices are $V_i = \{v_0, v_1, \cdots, v_{2^{m-1}-1}\}$ and thus, the uninformed vertices are $V_u = \{v_{2^{m-1}}, \cdots, v_{n-1}\}$.

To prove the validity of our broadcast scheme, we must show that every vertex $v_x \in V_u$ is adjacent to a vertex in V_i on dimension t, for any $1 \leq t \leq k$. Let $x = 2^{m-1} + c$, where $0 \leq c \leq 2^{m-1} - 2^k - 2a - 1$.

Assume v_x is adjacent to v_y on dimension t. Thus, we have

$$x + y \equiv 2^t - 1 \mod n \qquad \text{by the definition of Knödel graph}$$
$$x + y = n + 2^t - 1 \qquad \text{since } x \geq 2^{m-1}$$
$$2^{m-1} + c + y = n + 2^t - 1$$
$$y = 2^{m-1} - 2^k + 2^t - 2a - c - 1$$

From the above bounds on c, we get that $2^t \leq 2^{m-1} - 2^k + 2^t - 2a - c - 1 \leq 2^{m-1} - 1$. Thus, $0 < y \leq 2^{m-1} - 1$ and $v_y \in V_i$. Therefore, each vertex $v_x \in V_u$ has a neighbor $v_y \in V_i$ on dimension t, for any $1 \leq t \leq k$. Thus, every vertex broadcasts on dimension t at the last time unit completing broadcast of KG_n in m time units. $\qquad\square$

From the proof of Theorem 1, it actually follows that during time unit m, vertices $v_{2^t}, v_{2^t+1}, \cdots, v_{2^{m-1}+2^t-2a-2^k-1}$ call vertices $v_{n-1}, v_{n-2}, \cdots, v_{2^{m-1}}$ respectively. Our next result shows that the validity of a dimensional broadcast scheme with one left shift of dimensions $1, 2, \cdots, m-1$ and then repeating dimension 1.

Theorem 2. *Let KG_n be a Knödel graph on n vertices, where $n = 2^m - 2d$, $m \geq 3$, and $2 \leq d \leq 2^{m-1}$. Dimensional broadcast scheme $m-1, 1, \cdots, m-2, 1$ is a valid broadcast scheme.*

Proof. Without loss of generality, we assume that the originator is v_0. At the first time unit, v_0 calls $v_{2^{m-1}-1}$ on dimension $m - 1$. Then, during time units $2, 3, \cdots, m-1$, v_0 informs the odd vertices of $I_1 = \{v_i | 1 \leq i \leq 2^{m-2}-1, i \text{ is odd}\}$ and the even vertices of $I'_1 = \{v_i | 0 \leq i \leq 2^{m-2} - 2, i \text{ is even}\}$, while $v_{2^{m-1}-1}$ informs the odd vertices of $I_2 = \{v_i | 2^{m-1} - 1 \leq i \leq 2^{m-1} + 2^{m-2} - 3, i \text{ is odd}\}$ and the even vertices in $I'_2 = \{v_i | 2^{m-1}-2d+2 \leq i \leq 2^{m-1}+2^{m-2}-2d, i \text{ is even}\}$. Thus, after time unit $m - 1$, the odd numbered uninformed vertices are

$$U_1 = \{v_i | 2^{m-2} + 1 \leq i \leq 2^{m-1} - 3, i \text{ is odd}\}$$
$$U_2 = \{v_i | 2^{m-1} + 2^{m-2} - 1 \leq i \leq 2^m - 2d - 1, i \text{ is odd}\}$$

and the even numbered vertices are

$$U_1' = \{v_i | 2^{m-2} \leq i \leq 2^{m-1} - 2d, \ i \text{ is even}\}$$
$$U_2' = \{v_i | 2^{m-1} + 2^{m-2} - 2d + 2 \leq i \leq 2^m - 2d - 2, \ i \text{ is even}\}.$$

We can verify that the first dimensional neighbors of the vertices in U_1, U_2, U_1', and U_2' are all in I_2', I_1', I_2, and I_1 respectively. For example, the first dimensional neighbor of vertex $v_i \in U_1$ (where $i = 2^{m-2} + x$, x is odd, and $1 \leq x \leq 2^{m-2} - 3$) is vertex v_j, where $j = n + 1 - 2^{m-2} - x = 2^m - 2d + 1 - 2^{m-2} - x = 2^{m-1} - 2d + (2^{m-2} - x + 1)$. $v_j \in I_2'$, since $4 \leq 2^{m-2} - x + 1 \leq 2^{m-2}$. Thus, at time unit m vertex $v_{2^{m-2}+x} \in U_1$, where x is odd, $1 \leq x \leq 2^{m-2} - 3$, receives the message from vertex $v_{2^{m-1}-2d+(2^{m-2}-x+1)} \in I_2'$, for all $1 \leq x \leq 2^{m-2} - 3$.

We omit the proof of the other three pairs of subsets (U_2, I_1'), (U_1', I_2'), and (U_2', I_1) since all proofs are similar to the proof above.

Thus, broadcast on dimension 1 at the last time unit completes the broadcast of KG_n in m time units.   □

Fig. 2a shows KG_{14} and its dimensional broadcast scheme $3, 1, 2, 1$.

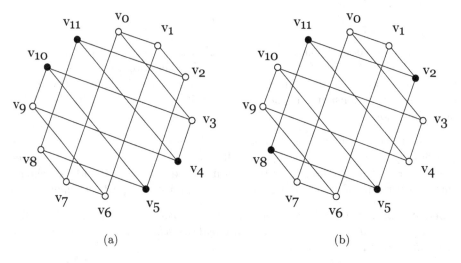

(a)   (b)

Fig. 2. Dimensional broadcast schemes for KG_{12} originating from v_0. The empty circles are the informed vertices before the last time unit. In (a), $I_1 = \{v_1, v_3\}$, $I_2 = \{v_7, v_9\}$, $I_1' = \{v_0, v_2\}$, $I_2' = \{v_6, v_8\}$, $U_1 = \{v_5\}$, $U_2 = \{v_{11}\}$, $U_1' = \{v_4\}$, and $U_2' = \{v_{10}\}$. So, the cyclic shift on dimension 3, 1, 2, and 1 gives a valid broadcast scheme. However, in (b), the cyclic shift on dimension 2, 3, 1, and 3 is not a valid broadcast scheme.

It turns out that the generalization of Theorem 1 with cyclic shifts similar to the result of [4] is difficult. Theorem 2 proves that one cyclic shift of dimensions $1, 2, \cdots, m - 1$ and then repeating the second dimension also generates a valid dimensional broadcast scheme. However, our example in Fig. 2b shows that two cyclic shifts of dimensions $1, 2, \cdots, m - 1$ with repeating the second dimension does not always generate a valid dimensional broadcast scheme. In particular, the dimensional broadcast scheme $m - 2, m - 1, 1, \cdots, m - 3, m - 1$ is not valid for KG_{12} (when $m = 4$ and $d = 2$). Again, without loss of generality the originator is v_0. In this particular case, the dimensional broadcast scheme $m - 2, m - 1, 1, \cdots, m - 3, m - 1$ is $2, 3, 1, 3$. Then, the informed vertices are $v_0, v_1, v_3, v_4, v_6, v_7, v_9$, and v_{10} before the last time unit. The uninformed vertices are v_2, v_5, v_8, and v_{11} be for the last time unit. We can clearly see that broadcast on dimension 3 cannot complete broadcasting, neither does dimension 1. So, this cyclic shift is an invalid broadcast scheme, see Fig. 2b for example.

3   A New Vertex Addition to Construct Broadcast Graphs

Since Knödel graph gives good general upper bound on $B(n)$, but only for even values of n, vertex addition method from [15] adds one more vertex and some edges to Knödel graph to obtain new broadcast graphs for odd values of n. The construction uses dimensional broadcast schemes and their cyclic shifts.

Let vertex v_s be the neighbor of a particular originator v_i on dimension s, where $1 \leq s \leq m - 1$ under valid broadcast scheme $s, s + 1, \cdots m - 1, 1, \cdots, s$ vertices v_s and v_i are idle at the last time unit. Then, the additional vertex added to the Knödel graph can be informed by vertex v_i if they are adjacent. Thus, v_i dominates all its s neighbors on different dimensions. Then, constructing a broadcast graph by adding one vertex to a Knödel graph leads to the problem of dominating set.

The same paper also introduces a dominating set of size 2^{m-2} and obtains an upper bound on $B(n)$ based on the dominating set.

$$B(n) \leq \frac{1}{2} n \lceil \log n \rceil + 2^{m-2}$$

This is the best known general upper bounds for odd n. However, if $m > 2$ is prime, m divides n, and for any integer $d < m - 1$ which is a divisor of $m - 1$, $2^d \not\equiv 1 \mod m$, KG_n has a dominating set of size $\frac{n}{m}$. Then [19] gives the better bound for these specific values of n. In particular,

$$B(n) \leq \frac{1}{2}(n - 1) \lfloor \log n \rfloor + \frac{n-1}{\lceil \log n \rceil} + \lfloor \log n \rfloor - 2$$

In this section, we follow the track given by [15, 19] and construct a broadcast graph by adding one vertex to Knödel graph. The construction improves the general bound for almost all odd values of n to $B(n) \leq \frac{1}{2}n\lfloor \log n \rfloor + \lceil \frac{1}{7}n \rceil + \lfloor \log n \rfloor$.

Our method is similar to the one in [15, 19] but using 3-distance dominating sets. This also requires more careful consideration of the connections in Knödel graph.

Definition 5. *Let* $KG_n = (V, E)$ *be a Knödel graph on* n *vertices, where* $n = 2^m - 2^k - 2d$, $3 \leq k \leq m - 2$, *and* $0 \leq d < 2^{k-1}$, *and let* $e \equiv n \mod 14$. *Define* $U = \{v_0\} \cup \{v_{n-14a} | 1 \leq a \leq \frac{n}{14}\} \cup \{v_{14a+13} | 1 \leq a \leq \frac{n}{14}\} \cup X$, *where* $X = \{v_{e-7}\}$ *if* $e \geq 8$, *and* $X = \emptyset$ *otherwise.*

Theorem 3. U *contains at least one idle vertex* u *at the last time unit under dimensional broadcast scheme* $1, 2, 3, \cdots, m - 1, 3$ *from any originator in graph* KG_n.

Proof. First, from Theorem 1, the dimensional broadcast scheme $1, 2, \cdots, m - 1, 3$ is a valid dimensional broadcast scheme for KG_n. Here $t = 3 \leq k$ as stated in Theorem 1. So, we will show that under the dimensional broadcast scheme from any originator v_i, there is a vertex $u \in U$, such that v_i informs u and its third dimensional neighbor during the first 3 time units. Then, at the last time unit, when every vertex broadcasts on the third dimension, u and its third dimensional neighbor are both idle. We partition the vertices into 14 subsets and show the connections on dimensions 1, 2, and 3 between the sets.

For all $v_i \in V$, when i is even, $i = n - 14a + 2b$, where $0 \leq a \leq \frac{n}{14}$, $0 \leq b \leq 6$, and $14a - 2b \leq n - 2$. When i is odd, $i = 14a + 2b + 1$, where $0 \leq a \leq \frac{n}{14}$, $0 \leq b \leq 6$ and $14a + 2b + 1 \leq n - 1$. Then, we have 14 cases depending on the different parities of i and values of b.

1. If i is even and $b = 0$, $i = n - 14a$ and $v_i \in U$ by definition. The broadcast originating from v_i idles v_i at the last time unit for sure. If i is odd and $b = 6$, the situation is the same.

2. If i is odd and $b = 0$, 1, or 3, $i = 14a + 1$, $14a + 3$, or $14a + 7$. These three different vertices have a common neighbor $v_{n-14a} \in U$ on dimension 1, 2, and 3 respectively. Thus, if the originator is one of these vertices, v_{n-14a} and v_{14a+7} are informed in time unit 3. And $v_j \in U$ is idle at the last time unit. If i is even and $b = 3$, 5, and 6, we have the same situation.

3. If i is even and $b = 1$, $i = n - 14a - 2$. Vertex v_i is adjacent to vertex v_{14a+3}, which is the case we discussed above. So, v_i informs v_{14a+3} at the first time unit. v_{14a+3} informs $v_{n-14a} \in U$ at the second time unit. And v_{n-14a} informs v_{14a+7} at the third time unit. Thus, $v_{n-14a} \in U$ and its third dimensional neighbor v_{14a+7} are both informed after time unit 3, and v_{n-14a} is idle at the last time unit.
If i is odd and $b = 5$, we have the same case.

4. If i is even and $b = 2$, $i = n - 14a - 4$. v_i is adjacent to vertex v_{14a+7} on dimension 2. So, we broadcast from v_i and inform v_{14a+7} at time unit 2. Then, v_{14a+7} informs v_{n-14a} at time unit 3. At the last time unit, $v_{n-14a} \in U$ is idle.
If i is odd and $b = 4$, the situation is the same.

5. If i is odd and $b = 2$, $i = 14a + 5$. v_i is adjacent to $v_{n-14a-4}$ on dimension 1. So, v_i informs $v_{n-14a-4}$ at the first time unit. Then, if we just follow Case 4, vertex $v_{n-14a} \in U$ is idle at the last time unit. Again, we have the same case for i is even and $b = 4$.

Therefore, for any originator v_i, there is always a vertex in U, which is idle at the last time unit. Figure 3 shows one example of set U.   □

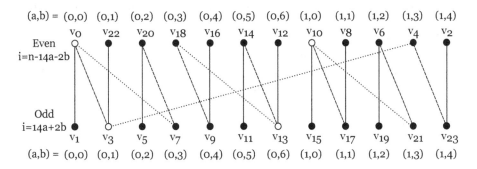

Fig. 3. Knödel graph on 24 vertices with only a part of the edges critically involved in the dimensional broadcast scheme 1, 2, and 3. Solid lines are the edges on dimension 1. Dashed lines are the edges on dimension 2. And doted lines are the edges on dimension 3. The small circles are the vertices in vertex set U. We say that, for example, $v_1 0$ covers v_6, v_8, v_{15}, v_{17}, v_{19}, and v_{21} in distance 3. So, vertex set U is a 3-distance dominating set of V. Note that v_3 is in U because $n \equiv e \mod 14$, $e = 10 \geq 8$, and $X = \{v_{10-7}\}$.

Next, we construct a new broadcast graph using the property of Knödel graph from Theorem 3 and a dimensional broadcast scheme from Theorem 1.

Let $KG_n = (V, E)$ be the Knödel graph on n vertices, where $n = 2^m - 2^k - 2a$, $3 \leq k \leq m-2$, and $a < 2^{k-1}$. Let $U = \{v_0\} \cup \{v_{n-14a} | 1 \leq a < \frac{n}{14}\} \cup \{v_{14a+13} | 1 \leq a < \frac{n}{14}\} \cup X$, where $X = \{v_{e-7}\}$ if $e \geq 8$, and $X = \emptyset$ otherwise. We add one vertex v and two types of edges E_1 and E_2 to KG_n, where $E_1 = \{(v, u) | u \in U\}$ and $E_2 = \{(v, v_i) | (v_1, v_i) \in E\}$. Then, the new graph is defined as $G = (V \cup \{v\}, E \cup E_1 \cup E_2)$.

By the definition of Knödel graph, $|E| = \frac{(m-1)n}{2}$. By the definition of vertex set U, $|E_1| = \lceil \frac{n}{7} \rceil$, if $n < 8 \mod 14$; or $\lceil \frac{n}{7} \rceil + 1$, otherwise. Since Knödel graph is $(m-1)$-regular, v_1 has $m-1$ neighbors. And one of the neighbors, v_0 is already adjacent to v, so $|E_2| = m-2$. Thus, graph G has $\frac{1}{2}(m-1)n + \lceil \frac{n}{7} \rceil + m - 2 + x \leq \frac{1}{2}(m-1)n + \lceil \frac{n}{7} \rceil + m - 1$ edges, since $x \leq 1$.

Theorem 4. *Graph G, constructed above, is a broadcast graph, and*

$$B(n) \leq \frac{1}{2}(m-1)n + \lceil \frac{n-1}{7} \rceil + \frac{1}{2}(m-1),$$

where $n = 2^m - 2^k - 2d + 1$, $3 \leq k \leq m-2$, and $0 \leq d < 2^k$.

The equation in Theorem 4 is slightly different from the equation above given for the number of edges in graph G. In Theorem 4, the number of vertices in graph G is equal to the number of vertices in Knödel graph KG_n plus one additional vertex. So, in the rest of the paper, we have KG_{n-1} instead of KG_n as a subgraph of G.

Proof. To prove the theorem, we show a broadcast scheme for any originator of graph G. Graph G has two types of vertices, the vertices in Knödel graph KG_{n-1} and the additional vertex. So, we have the following two cases.

1. If the originator $v_i \in KG_{n-1}$, by Theorem 3, we know that there is a vertex $v_k \in U$ idle at the last time unit. Since every vertex in U is adjacent to the added vertex v, vertex v_k calls v at the last time unit.
2. If the originator is the additional vertex v, v plays exactly the same roll as vertex v_1, because v is adjacent to all v_1's neighbor. And at the last time unit, vertex v_0 informs v_1 and completes the broadcasting.

Thus, the broadcast time of graph G is the same as broadcast time of KG_{n-1}. Therefore G is a broadcast graph, and since $x \leq 1$ we obtain

$$B(n) \leq \frac{1}{2}(m-1)n + \lceil \frac{n-1}{7} \rceil + \frac{1}{2}(m-1)$$

$\square$

We denote the new upper bound by NB and compare it with the previous upper bounds in [10]

$$UB_1 = \frac{n\lceil \log n \rceil}{2},$$

in [16]

$$UB_2 = (m-k+1)n - (m+k)2^{m-k} + m + k,$$

and in [15,19]

$$UB_3 = \frac{1}{2}(m-1)n + \lceil \frac{n-1}{4} \rceil.$$

Since $m = \lceil \log n \rceil$, NB is smaller than UB_1. And when $m - k + 1 > \frac{1}{2}(m-1)$, $k < \frac{1}{2}(m+3)$ or $n > 2^m - 2^{\frac{1}{2}(m+3)}$, NB is smaller than UB_2.

UB_3 is also based on vertex addition. Clearly, the new bound NB is smaller than UB_3. Therefore, the best general upper bounds on $B(n)$ for every $n \in [2^{m-1} + 1, 2^m]$ is given below:

$$B(n) \leq \begin{cases} (m-k+1)n - (m+k)2^{m-k} + m + k, \\ \qquad \text{if } 2^{m-1} + 1 \leq n \leq 2^m - 2^{\frac{1}{2}(m+3)} \text{ [16]}; \\ \frac{1}{2}(m-1)n, \\ \qquad \text{if } 2^m - 2^{\frac{1}{2}(m+3)} < n \leq 2^m \text{ for even } n \text{ [23]}; \\ \frac{1}{2}(m-1)n + \lceil \frac{n-1}{7} \rceil + \frac{1}{2}(m-1), \\ \qquad \text{if } 2^m - 2^{\frac{1}{2}(m+3)} < n \leq 2^m - 8 \text{ for odd } n \text{ (the new bound)}; \\ \frac{1}{2}(m-1)n + 2^{m-2} - \frac{1}{2}(m-1), \\ \qquad \text{if } 2^m - 8 \leq n \leq 2^m \text{ for odd } n \text{ [19]}. \end{cases}$$

4 Conclusion

This paper contains improvements in two different areas connected to broadcasting in graphs.

First, it starts a direction which considers a new dimensional broadcast schemes, where at the last time unit any of the first k dimensions can be used instead of repeating dimension 1, as in [4]. This paper also gives a small result towards the possible cyclic shifts of these dimensions.

Another direction in the paper is to use 3-distance dominating sets (instead of 1-distance dominating sets in [15,19]) to construct broadcast graphs on odd number of vertices using the above dimensional broadcast schemes in Knödel graphs. Another important future research area will be dedicated to improve the known lower bounds on $B(n)$.

References

1. Averbuch, A., Shabtai, R.H., Roditty, Y.: Efficient construction of broadcast graphs. Discret. Appl. Math. **171**, 9–14 (2014)
2. Barsky, G., Grigoryan, H., Harutyunyan, H.A.: Tight lower bounds on broadcast function for n = 24 and 25. Discret. Appl. Math. **175**, 109–114 (2014)
3. Bermond, J.-C., Fraigniaud, P., Peters, J.G.: Antepenultimate broadcasting. Networks **26**, 125–137 (1995)
4. Bermond, J.-C., Harutyunyan, H.A., Liestman, A.L., Perennes, S.: A note on the dimensionality of modified Knödel graphs. Int. J. Found. Comput. Sci. **8**(02), 109–116 (1997)
5. Bermond, J.-C., Hell, P., Liestman, A.L., Peters, J.G.: Sparse broadcast graphs. Discret. Appl. Math. **36**, 97–130 (1992)
6. Chau, S.C., Liestman, A.L.: Constructing minimal broadcast networks. J. Comb. Inf. Syst. Sci. **10**, 110–122 (1985)
7. Dinneen, M.J., Fellows, M.R., Faber, V.: Algebraic constructions of efficient broadcast networks. In: Mattson, H.F., Mora, T., Rao, T.R.N. (eds.) AAECC 1991. LNCS, vol. 539, pp. 152–158. Springer, Heidelberg (1991). doi:10.1007/3-540-54522-0_104
8. Dinneen, M.J., Ventura, J.A., Wilson, M.C., Zakeri, G.: Compound constructions of broadcast networks. Discret. Appl. Math. **93**, 205–232 (1999)
9. Farley, A.M.: Minimal broadcast networks. Networks **9**, 313–332 (1979)
10. Farley, A.M., Hedetniemi, S., Mitchell, S., Proskurowski, A.: Minimum broadcast graphs. Discret. Math. **25**, 189–193 (1979)
11. Fertin, G., Raspaud, A.: A survey on Knödel graphs. Discret. Appl. Math. **137**, 173–195 (2004)
12. Gargano, L., Vaccaro, U.: On the construction of minimal broadcast networks. Networks **19**, 673–689 (1989)
13. Grigni, M., Peleg, D.: Tight bounds on mimimum broadcast networks. SIAM J. Discret. Math. **4**, 207–222 (1991)
14. Grigoryan, H., Harutyunyan, H.A.: New lower bounds on broadcast function. In: Gu, Q., Hell, P., Yang, B. (eds.) AAIM 2014. LNCS, vol. 8546, pp. 174–184. Springer, Heidelberg (2014). doi:10.1007/978-3-319-07956-1_16

15. Harutyunyan, H.A.: An efficient vertex addition method for broadcast networks. Internet Math. **5**(3), 197–211 (2009)
16. Harutyunyan, H.A., Li, Z.: A new construction of broadcast graphs. In: Govindarajan, S., Maheshwari, A. (eds.) CALDAM 2016. LNCS, vol. 9602, pp. 201–211. Springer, Heidelberg (2016). doi:10.1007/978-3-319-29221-2_17
17. Harutyunyan, H.A., Liestman, A.L.: More broadcast graphs. Discret. Appl. Math. **98**, 81–102 (1999)
18. Harutyunyan, H.A., Liestman, A.L.: Improved upper and lower bounds for k-broadcasting. Networks **37**, 94–101 (2001)
19. Harutyunyan, H.A., Liestman, A.L.: Upper bounds on the broadcast function using minimum dominating sets. Discret. Math. **312**, 2992–2996 (2012)
20. Harutyunyan, H.A., Liestman, A.L., Peters, J.G., Richards, D.: Broadcasting and gossiping. In: Handbook of Graph Theorey, pp. 1477–1494. Chapman and Hall (2013)
21. Hedetniemi, S.M., Hedetniemi, S.T., Liestman, A.L.: A survey of gossiping and broadcasting in communication networks. Networks **18**, 319–349 (1988)
22. Khachatrian, L.H., Harutounian, H.S.: Construction of new classes of minimal broadcast networks. In: Conference on Coding Theory, Dilijan, Armenia, pp. 69–77 (1990)
23. Knödel, W.: New gossips and telephones. Discret. Math. **13**, 95 (1975)
24. Labahn, R.: A minimum broadcast graph on 63 vertices. Discret. Appl. Math. **53**, 247–250 (1994)
25. Maheo, M., Saclé, J.-F.: Some minimum broadcast graphs. Discret. Appl. Math. **53**, 275–285 (1994)
26. Mitchell, S., Hedetniemi, S.: A census of minimum broadcast graphs. J. Comb. Inf. Syst. Sci. **5**, 141–151 (1980)
27. Park, J.-H., Chwa, K.-Y.: Recursive circulant: a new topology for multicomputer networks. In: International Symposium on Parallel Architectures, Algorithms and Networks (ISPAN), pp. 73–80. IEEE (1994)
28. Saclé, J.-F.: Lower bounds for the size in four families of minimum broadcast graphs. Discret. Math. **150**, 359–369 (1996)
29. Shao, B.: On K-broadcasting in graphs. Ph.D. thesis, Concordia University (2006)
30. Ventura, J.A., Weng, X.: A new method for constructing minimal broadcast networks. Networks **23**, 481–497 (1993)
31. Weng, M.X., Ventura, J.A.: A doubling procedure for constructing minimal broadcast networks. Telecommun. Syst. **3**, 259–293 (1994)
32. Xiao, J., Wang, X.: A research on minimum broadcast graphs. Chin. J. Comput. **11**, 99–105 (1988)
33. Zhou, J., Zhang, K.: A minimum broadcast graph on 26 vertices. Appl. Math. Lett. **14**, 1023–1026 (2001)

Incremental Algorithms to Update Visibility Polygons

R. Inkulu$^{(\boxtimes)}$ and Nitish P. Thakur

Department of Computer Science and Engineering, IIT Guwahati, Guwahati, India
{rinkulu,tnitish}@iitg.ac.in

Abstract. We consider the problem of updating the visibility polygon of a point located within the given simple polygon as that polygon is modified with the incremental addition of new vertices to it. In particular, we propose the following two semi-dynamic algorithms:

- Given a simple polygon P defined with n vertices and a point $p \in P$, our preprocessing algorithm computes the visibility polygon of p in P and constructs relevant data structures in $O(n)$ time; for every vertex v added to the current simple polygon, our visibility polygon updation algorithm takes $O((k + 1) \lg n)$ time in the worst-case to update the visibility polygon of p in the new simple polygon resulted from adding v. Here, k is the change in combinatorial complexity of visibility polygon of p due to the addition of new vertex v.
- Given a simple polygon P defined with n vertices and an edge pq of P, our preprocessing algorithm computes the weak visibility polygon of pq in P and constructs relevant data structures in $O(n)$ time; for every vertex v added to the current simple polygon, our weak visibility updation algorithm takes $O((k+1) \lg n)$ time in the worst-case to update the weak visibility polygon of pq in the new simple polygon resulted from adding v. Here, k is the change in combinatorial complexity of shortest path tree rooted at p added to the change in combinatorial complexity of shortest path tree rooted at q, wherein both these changes are due to the addition of new vertex v.

1   Introduction

Let P be a simple polygon with n vertices. Two points $p, q \in P$ are said to be mutually *visible* to each other whenever the interior of line segment pq does not intersect any edge of P. For a point $p \in P$, the *visibility polygon* $VP_P(p)$ of p is the maximal set of points $x \in P$ such that x is visible to p. The problem of computing visibility polygon of a point in a simple polygon was first attempted in [7], who presented an $O(n^2)$ time algorithm. Then, ElGindy and Avis [8] and Lee [18] presented $O(n)$ time algorithms for this problem. Joe and Simpson [16] corrected a flaw in [8,18] and devised an $O(n)$ time algorithm that correctly handles winding in the simple polygon. For a polygon with holes, Suri et al. devised an $O(n \lg n)$ time algorithm in [21]. An optimal $O(n + h \lg h)$ time algorithm was

---

R. Inkulu's research is supported in part by NBHM grant 248(17)2014-R&D-II/1049.

D. Gaur and N.S. Narayanaswamy (Eds.): CALDAM 2017, LNCS 10156, pp. 205–218, 2017.
DOI: 10.1007/978-3-319-53007-9_19

given in Heffernan and Mitchell [14]. Algorithms for visibility computation amid convex sets are devised in Ghosh [9]. The algorithms under preprocess-query paradigm were studied in [1–3,5,6,13,15,23,24]. Algorithms for computing visibility graphs are given in [11].

For a line segment $pq \in P$, the *weak visibility polygon* $WVP_P(pq)$ is the maximal set of points $x \in P$ such that x is visible from at least one point belonging to line segment pq. Chazelle and Guibas [4] and Lee and Lin [19] gave $O(n \lg n)$ time algorithms for computing $WVP(pq)$. Later, Guibas et al. [12] gave an $O(n)$ time algorithm. Suri and O'Rourke [21] gave an $O(n^4)$ time algorithm for computing weak visibility polygon amid polygonal obstacles.

Ghosh [10] gives a detailed account of visibility related algorithms and the book by O' Rourke [20] gives various algorithms on art gallery problems.

Our Contribution. We propose two semi-dynamic algorithms: one to update the visibility polygon of a given point interior to a simple polygon and the other to update weak visibility polygon of an edge of a simple polygon. In particular, we devise the following two algorithms.

- Given a simple polygon P defined with n vertices and a point $p \in P$, our preprocessing algorithm computes the visibility polygon of p in P and constructs relevant data structures in $O(n)$ time; for every vertex v added to the current simple polygon, our visibility polygon updation algorithm takes $O((k+1) \lg n)$ time in the worst-case to update the visibility polygon of p in the new simple polygon resulted from adding v. Here, k is the change in combinatorial complexity of visibility polygon of p due to the addition of new vertex v.

- Given a simple polygon P defined with n vertices and an edge pq of P, our preprocessing algorithm computes the weak visibility polygon of pq in P and constructs relevant data structures in $O(n)$ time; for every vertex v added to the current simple polygon, our weak visibility updation algorithm takes $O((k+1) \lg n)$ time in the worst-case to update the weak visibility polygon of pq in the new simple polygon resulted from adding v. Here, k is the change in combinatorial complexity of shortest path tree rooted at p added to the change in combinatorial complexity of shortest path tree rooted at q, wherein both these changes are due to the addition of new vertex v.

(Both of these algorithms take $O(\lg n)$ updation time even when k is zero, hence the stated updation time complexity.)

To our knowledge, these are the first incremental algorithms to update visibility polygon as new vertices are added to simple polygon. We assume that after adding every vertex, the polygon remains as a simple polygon; the point p whose visibility polygon is updated remains interior to new simple polygon i.e., to simple polygon obtained after adding a new vertex. Further, it is assumed that every new vertex is added between two successive vertices along the boundary of the current simple polygon.

In the rest of the document, we use the following notation. The boundary of a simple polygon P is denoted with $bd(P)$. The Euclidean shortest path between two points $s, t \in P$ is denoted with $SP(s,t)$. The simple polygon at the end of previous iteration is denoted with P and the new simple polygon obtained after a new vertex to P is denoted with P'. Unless specified otherwise, simple polygon is traversed in the counterclockwise direction.

Section 2 describes an algorithm to preprocess and update the visibility polygon of a point interior to simple polygon. Further, Sect. 3 devises an algorithm to preprocess and update the weak visibility polygon of an edge of the simple polygon. The conclusions are mentioned in Sect. 4.

2   Updating $VP_P(p)$ with Vertex Insertion to P

In this Section, we consider the problem of updating $VP_P(p)$ when a new vertex is inserted to P resulting in a new simple polygon P'. With slight abuse of notation, we use VP instead of $VP_P(p)$ when p and P are clear from context. We let $v_1, \ldots, v_n$ be the vertices of P in counterclockwise direction. Similarly, we let $u_1, \ldots, u_m$ be the vertices of VP in counterclockwise direction. For any edge $u_i u_{i+1}$ of VP, if only one of $\{u_i, u_{i+1}\}$ is a vertex of P then that edge is called a *constructed edge* and the non-vertex among those two points is called a *constructed vertex*. The constructed edges of VP partition P into regions $\mathcal{R} = \{VP, R_1, R_2, \ldots, R_s\}$ such that no point q interior to R_i is visible from p for any $i \in \{1, 2, \ldots, s\}$. Also, for each R_i, there exists a constructed vertex associated to R_i. Since each vertex of P may cause one constructed edge, there can be $O(n)$ constructed edges.

The vertices of P and VP are stored in doubly linked lists L_P and L_{VP} in the order in which they appear on $bd(P)$ and $bd(VP)$. Additionally, nodes in L_{VP} and L_P which correspond to the same vertex $u_i \in VP$ contain pointers to each other. For each region R_i, vertices are stored in a balanced binary tree T_{R_i} such that a node in T_{R_i} and a node in L_P corresponding to the same vertex $v_j \in R_i$ contain pointers to each other. These balanced binary trees are useful in efficiently updating the regions that they respectively correspond with the addition of new vertices. We call the pointer from a node in L_P to a node in T_{R_i} as a *tree pointer*. For vertices $v_i \notin VP$, the tree pointer is set to null. These data structures together take $O(n)$ space.

Let v_a be the newly inserted vertex such that an existing edge $v_b v_c$ of P is replaced by edges $v_b v_a$ and $v_a v_c$. We assume that with the coordinates of v_a, a pointer to v_b in list L_P is also supplied as input so that v_a occurs after v_b in the counterclockwise order of vertices that occur along the boundary of new simple polygon P'. As mentioned, we assume that after adding v_a, simple polygon P continues to be a simple polygon and that p is not exterior to new simple polygon P'. We first list the algorithms to preprocess the given simple polygon before describing the algorithm to update the visibility polygon VP.

Preprocessing P

Using the algorithm of Lee [18], the visibility polygon VP of $p \in P$ is computed in $O(n)$ time. Further, balanced binary tree T_{R_i} corresponding to each R_i is constructed. Since the vertices of R_i are along $bd(P)$, computing each such T_{R_i} takes time linear in the number of vertices in R_i. Therefore, preprocessing phase takes $O(n)$ time.

Updating $VP(p)$

The following cases are individually handled and we later prove that these cases are exhaustive. The cases are listed based on the relative location of v_b and v_c with respect to VP together with considering any two disjoint regions R_i, R_j in $\mathcal{R}$.

(a) $v_b, v_c \in VP$ (Fig. 1(a))
(b) $v_b \notin VP$ and $v_c \in VP$, or vice versa (Fig. 1(b))
(c) $v_b, v_c \notin VP$ and $v_b \in R_i, v_c \in R_j$ for $R_i \neq R_j$ (Fig. 1(c))
(d) $v_b, v_c \notin VP$ and $v_b, v_c \in R_i$ (Fig. 1(d))

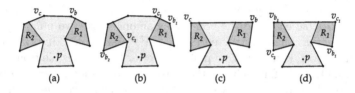

(a)   (b)   (c)   (d)

Fig. 1. Illustrating cases based on the locations of v_b and v_c with respect to VP

Further, these cases are subdivided to consider whether the newly inserted vertex v_a is located interior or exterior to P (simple polygon just before adding v_a). We first consider the cases (a)–(d) when v_a is located interior to P. Finally, we give an algorithm to identify these four cases. We make use of join, split and insert operations on the balanced binary trees corresponding to regions, each of which take $O(\lg n)$ time [22].

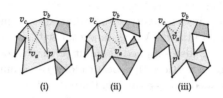

(i)   (ii)   (iii)

Fig. 2. Illustrating subcases of case (a) in which $v_a, v_b, v_c \in VP$

Case (a) $(v_b, v_c \in VP)$. There are three subcases (Fig. 2): (i) $v_b v_a$ intersects $p v_c$, (ii) $v_a v_c$ intersects $p v_b$, or (iii) $v_a \in \triangle p v_b v_c$.

In subcase (i), we add v_c to a new region R_{new}. Then, starting from v_c, we traverse the vertices $u_i \in VP$ in counterclockwise direction until we reach a u_i such that pu_i does not intersect $v_b v_a$ (Fig. 3). This implies that v_a lies in $\triangle p u_i u_{i-1}$ and we add a new constructed vertex x_{new} by extending $p v_a$ on to $u_i u_{i-1}$. The constructed vertex v_a and x_{new} together define the region R_{new}: every vertex that occurs while traversing P' in counterclockwise direction from v_c to x_{new}; not that none of these vertices are visible from p.

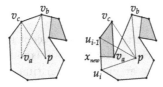

Fig. 3. Illustrating case (a)(i): v_a lies in $\triangle p u_i u_{i-1}$

Now consider the case when pu_i intersects $v_b v_a$ (Fig. 4). This implies that u_i is not visible from p. The point u_i can either be a vertex of P or a constructed vertex of VP. If it is a constructed vertex and part of constructed edge $u_{i-1} u_i$ with corresponding region R_j, then $\triangle v_a v_b v_c$ cannot intersect R_j and we union R_j with R_{new}.

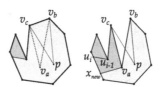

Fig. 4. Illustrating case (a)(i): $\triangle v_a v_b v_c$ cannot intersect $u_i u_{i-1}$

If u_i is a constructed vertex and part of constructed edge $u_i u_{i+1}$ with corresponding region R_j, we check if $v_b v_a$ intersects $u_i u_{i+1}$. If $v_b v_a$ intersects $u_i u_{i+1}$ (Fig. 5), this implies that $v_a \in R_j$ and we join R_{new} with R_j. We then delete u_i from VP and add a constructed vertex x_{new} found by extending $p u_{i+1}$ on to $v_b v_a$. If $v_b v_a$ does not intersect $u_i u_{i+1}$ (Fig. 6), then we delete u_i and u_{i+1} from VP and add u_{i+1} to R_{new}. The region R_j is unioned (joined) with R_{new}. Finally, if u_i is not a constructed vertex, we just delete u_i from VP and add it to R_{new}.

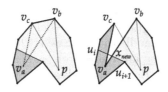

Fig. 5. Illustrating case (a)(i): $v_a v_b$ intersects $u_i u_{i+1}$

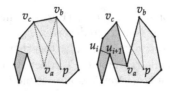

Fig. 6. Illustrating case (a)(i): $v_a v_b$ does not intersects $u_i u_{i+1}$

It can be seen that every vertex traversed except the last one is deleted from VP. So, the number of vertices deleted from VP equals one less than the number of vertices traversed. We take at most $O(\lg n)$ time for processing each vertex u_i. Hence, when k vertices are deleted, k changes are made to the VP and it takes $O(k \lg n)$ time.

The operations in subcase (ii) are similar to subcase (i) except that we traverse vertices $u_i \in VP$ in clockwise direction. This again takes $O(k \lg n)$ time.

In subcase (iii), v_a is visible from p hence it is added to VP and no other changes are required. This takes $O(1)$ time. This completes all the subcases of case (a).

Case (b) ($v_b \notin VP$ **and** $v_c \in VP$). We assume that $v_b \in R_i$ and $v_c \in VP$ (Fig. 7). If v_c is a reflex vertex and v_c is the only point on $v_b v_c$ visible from p, then the analysis is similar to case (d) (see below). Otherwise, let x_i be a constructed vertex on $v_b v_c$ (Fig. 7). Similar to case (a), there are three subcases: (i) $v_b v_a$ intersects $p v_c$, (ii) $v_a v_c$ intersects $p x_i$, or (iii) $v_a \in \triangle p v_c x_i$.

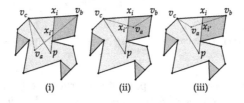

Fig. 7. Illustrating case (b): $v_b \in R_i$ and $v_c \in VP$

Subcase (i) under this case is similar to subcase (i) of case (a). The only difference being that we add a constructed vertex $x_{i'}$ to VP by finding the intersection of $p x_i$ with $v_b v_a$. Further, we delete x_i from VP. The worst-case time involved is $O(k \lg n)$ time.

In subcase (ii), we find a constructed vertex $x_{i'}$ and add it to VP by finding the intersection of $p x_i$ with $v_a v_c$. Further, we delete x_i from VP and add v_a to R_i. The worst-case time to handle this subcase is $O(\lg n)$ time.

In subcase (iii), v_a and $x_{i'}$ are visible from p, hence they are added to VP. Also, we delete x_i from VP. Hence, this takes $O(1)$ time in the worst-case.

Case (c) ($v_b, v_c \notin VP$ **and** $v_b \in R_i, v_c \in R_j$ **s.t.** $i \neq j$). Let $v_b \in R_i, v_c \in R_j$ and let x_i and x_j be the respective constructed vertices on $v_b v_c$. We again have three subcases (Fig. 8): (i) $v_b v_a$ intersects $p x_j$, (ii) $v_a v_c$ intersects $p x_i$ or (iii) $v_a \in \triangle p x_i x_j$. These subcases are similar to case (b). Subcase (i) and (ii) require $O(\lg n)$ time while subcase (iii) requires $O(1)$ time to update VP along with regions R_i and R_j.

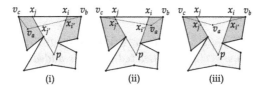

(i)   (ii)   (iii)

Fig. 8. Illustrating case (c): $v_b \in R_i$ and $v_c \in R_j$ s.t. $i \neq j$

Case (d) $(v_b, v_c \notin VP$ **and** $v_b, v_c \in R_i)$. Let $v_b, v_c \in R_i$ and let x_i be the corresponding constructed vertex. We have two subcases (Fig. 9): (i) $v_a \in R_i$ or (ii) $v_a \notin R_i$.

In subcase (i), no change is made to VP but we need to add v_a to T_{R_i} (Fig. 9(i)). This is done by following the tree pointers of v_b and v_c into T_{R_i}; hence, take $O(\lg n)$ time.

In subcase (ii) (Figs. 9(ii) and (iii)), we assume a portion of $v_a v_c$ is visible from p (when a portion of $v_b v_a$ is visible from p, it can be handled similarly). Let x_i be the constructed vertex corresponding to R_i and let $v_i x_i$ be the constructed edge wherein $v_i \in P$ is a reflex vertex. This leads to splitting R_i into two regions. Accordingly, we split T_{R_i} at v_b. This results in two trees: $T_{R_{i1}}$ with vertices on $bd(P)$ lying between x_i and v_b and $T_{R_{i2}}$ with vertices on $bd(P)$ lying between v_c and v_i. The line segments $v_a v_c$ and $v_i x_i$ intersect at a new constructed vertex $x_{i'}$ which is associated with $T_{R_{i2}}$ and $x_{i'}$ is added to VP after removing x_i. Note that after taking $T_{R_{i1}}$ as an initial tree, the procedure is similar to case (a) subcase (i). This completes the case (d). Note that this case takes $O(k \lg n)$ time in the worst-case.

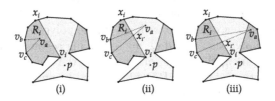

(i)   (ii)   (iii)

Fig. 9. Illustrating subcases of case (d): (i) $v_a \in R_i$ and (ii), (iii) $v_a \notin R_i$.

When $v_a \notin P$, each of the four cases (a) to (d) can be handled similar to case (c). Each of these cases take $O(\lg n)$ time. To identify whether $v_a \in P$ or $v_a \notin P$ we can check if v_b, v_c, v_a turns left or right respectively.

Distinguishing Cases (a)–(d). To determine the applicable cases (a)–(d), we follow the tree pointers associated with v_b and v_c in L_P. For v_b, if the tree pointer is not null, then it points to a node in the tree, say T_{R_b}. We follow the parent pointers from this node to reach the root of T_{R_b} say r_1, which takes $O(\lg n)$ time. Similarly, r_2 is found if the tree pointer for v_c is not null. The appropriate case is trivially identified using these tree pointers.

Lemma 1. *The cases (a)–(d) are exhaustive. Further, the corresponding sub-cases are exhaustive as well.*

Lemma 2. *In cases (a)–(d), a vertex $u_i \in VP$ is deleted if and only if it is not visible from p and a new vertex/constructed vertex is added only to VP whenever it is visible from p.*

Lemma 3. *The worst-case time taken by algorithm to update visibility polygon is $O((k+1)\lg n)$.*

Proof. Identifying the appropriate case requires $O(\lg n)$ time. Each case requires $O(k \lg n)$ time. Since k equals to zero in case (d) of subcase (i), updating the visibility polygon takes $O((k+1)\lg n)$ time.   □

Theorem 1. *Let $VP_P(p)$ be the visibility polygon of a point p in a simple polygon P. Let v_a be the new vertex inserted to P such that an existing edge $v_b v_c$ is replaced by edges $v_a v_b$ and $v_b v_c$, resulting in a new simple polygon P'. After preprocessing P and $VP_P(p)$, assuming that p is not exterior to P', the visibility polygon $VP_P(p)$ can be updated in $O((k+1)\lg n)$ time, wherein k is the number of vertices that require to be inserted to and/or deleted from $VP_P(p)$ to obtain $VP_{P'}(p)$.*

3   Updating $WVP_P(p)$ with Vertex Insertion to P

In this Section, we consider the problem of updating weak visibility polygon $WVP_P(pq)$ of an edge pq when a new vertex is inserted into P. With slight abuse of notation, we denote $WVP_P(pq)$ with WVP when pq and P are clear from context. We let $v_1, \ldots, v_n$ be the vertices of P in counterclockwise direction. Similarly, let $u_1, \ldots, u_m$ be the vertices of WVP in counterclockwise direction. When pq is a line segment contained within a simple polygon Q, we can partition Q into two simple polygons Q_1 and Q_2 by extending segment pq on both sides until it hits $bd(Q)$. Then, the WVP of pq in Q is the union of WVP of pq in Q_1 and Q_2. Hence, to update WVP in Q, we need to update the WVP in Q_1 and Q_2. Therefore, it suffice to consider the case when pq is an edge of P. For any edge $u_i u_{i+1}$ of WVP, if only one of $\{u_i, u_j\}$ is a vertex of P then that edge is called a *constructed edge* and the non-vertex among these two points is called a *constructed vertex*. The constructed edges of WVP partition P into regions $\mathcal{R} = \{WVP, R_1, R_2, \ldots, R_s\}$ such that no point q interior to R_i is visible from any point $r \in pq$ for any $i \in \{1, 2, \ldots, s\}$. Also, for each R_i, there exists a constructed vertex associated to R_i. Similar to visibility polygon of a point, since each vertex of P can cause at most one constructed edge, there can be $O(n)$ constructed edges.

We review the algorithm by Guibas et al. [12] which is used in computing the initial $WVP(pq)$ as well as to update the same. First, the *shortest path tree* $SPT(p)$ is computed: $SPT(p)$ is the union of $SP(p, v_i)$ for every $v_i \in P$. Then, the depth first traversal is performed over $SPT(p)$. If the shortest path to a child v_j of v_i makes a right turn at v_i, then a segment is constructed by extending $v_k v_i$ to intersect $bd(P)$ (v_k is parent of v_i in $SPT(p)$). The portion of P lying on the right side of such constructed segments does not belong to $WVP(pq)$ and hence it is removed from P. Let P'' be the resulting polygon. A similar procedure is performed with respect to q on P'', resulting in $WVP(pq)$. From the algorithm of Guibas et al. [12], as new vertex is added to P, it is apparent that to update the $WVP(pq)$, we need to update $SPT(p)$ and $SPT(q)$ and remove the vertices which are not part of $WVP(pq)$.

For updating $SPT(p)$ and $SPT(q)$, we use the algorithm by Kapoor and Singh [17]. Their algorithm is divided into three phases: first phase computes segment e_i of $SPT(p)$ intersecting $\triangle v_a v_b v_c$; in the second phase, $SPT(p)$ is updated to include the endpoints z_i of e_i that are not visible; and in the final phase, the algorithm updates $SPT(p)$ to include every children of all such z_i. As a result, [17] updates $SPT(p)$ in $O(k \lg(n/k))$ time in the worst-case; here, k is the number of changes made to $SPT(p)$.

A *funnel* consists of a vertex r, called the *root*, and a segment xy, called the *base* of the funnel. The *sides* of the funnel are $SP(r, x)$ and $SP(r, y)$. In [17], $SPT(p)$ is stored as a set of funnels. For a funnel F with root r and base xy, the segments from r to x and from r to y are stored in a balanced binary search trees T_1 and T_2 respectively. The root nodes of T_1 and T_2 contain pointers to each other. It can be seen that every edge e that is not a boundary edge lies in exactly two funnels say F_i and F_j. Then the nodes in F_i and F_j corresponding to e contains pointers to each other. The whole of this data structure require $O(n)$ space.

Like in previous Section, the vertices of P and WVP are stored in doubly linked lists L_P and L_{WVP} in the order in which they appear on the $bd(P)$ and $bd(WVP)$. Additionally, nodes in L_{WVP} and L_P which correspond to the same vertex $u_i \in WVP$ contain pointers to each other. For each region R_i, vertices are stored in a balanced binary tree T_{R_i} such that a node in T_{R_i} and a node in L_P corresponding to the same vertex $v_j \in R_i$ contain pointers to each other. We call the pointer from a node in L_P to a node in T_{R_i} as a *tree pointer*. For vertices $v_i \notin WVP$, the tree pointer is set to null. All of these data structures, together with the ones to store funnels, take $O(n)$ space.

In our algorithm, we first update $SPT(p)$ and $SPT(q)$. Then, with the depth first traversal starting from v_a in the updated $SPT(p)$ and $SPT(q)$, we remove the regions that are not entirely visible from pq. While removing vertices in these regions, we also remove the corresponding nodes in $SPT(p)$ and $SPT(q)$ so that only the vertices visible from pq are left. Like in previous Section, the tree pointers are used to identify which of the four cases are applicable in a given situation while the algorithm splits/joins regions as it proceeds.

All the assumptions mentioned in previous Section with respect to vertex insertion are assumed to be applicable here as well.

Preprocessing

As mentioned above, we compute $SPT(p)$ and $SPT(q)$ using [12] and store them as funnels using the algorithm from [17]. We also compute WVP and the balanced binary trees for each region. The preprocessing phase takes $O(n)$ time in the worst-case.

Updating $WVP(pq)$

We have the following four cases depending on the locations of v_b and v_c and the disjoint regions $R_i, R_j \in \mathcal{R}$.

(a) $v_b, v_c \in WVP$
(b) $v_b \notin WVP$ and $v_c \in WVP$ or vice versa
(c) $v_b, v_c \notin WVP$ and $v_b \in R_i, v_c \in R_j$ such that $i \neq j$
(d) $v_b, v_c \notin WVP$ and $v_b, v_c \in R_i$

Further, depending on the location of the newly inserted vertex v_a, there are two subcases in each of these cases: (i) $v_a \notin P$, (ii) $v_a \in P$. We first consider the cases (a)–(d) when $v_a \in P$. Identifying these cases is done in a manner similar to the earlier Section. We use k_1 and k_2 to denote the number of changes required to update $SPT(p)$ and $SPT(q)$ respectively.

Case (a). When $SP(p, v_b)$ and $SP(p, v_c)$ do not intersect $\triangle v_a v_b v_c$, add v_a to $SPT(p)$ by finding a tangent from v_a to funnel with base $v_b v_c$ in $SPT(p)$. Analogously, if $SP(q, v_b)$ and $SP(q, v_c)$ do not intersect $\triangle v_a v_b v_c$, then v_a is added to $SPT(q)$ as well. Further, v_a is added to $WVP(pq)$ in place of v_b and remove v_b and v_c. No further changes are required.

We now consider the case when $SP(p, v_c)$ intersects $\triangle v_a v_b v_c$. It is possible that v_a lies in a region R_i with constructed edge $v_i x_i$. While updating $SPT(p)$, we check if one of the edges in $SPT(p)$ intersecting $\triangle v_a v_b v_c$ is $v_i x_i$. If it is, all vertices on $bd(P)$ from v_c to x_i need to be joined into a single region and the vertices on $bd(WVP)$ from v_c to x_i need to be deleted from WVP. This is because $\triangle v_a v_b v_c$ blocks the visibility of these vertices from pq. And, this is accomplished by unioning (joining) all the regions R_j that are encountered while traversing along $bd(WVP)$ from v_c. For $O(k_1)$ constructed vertices, joining takes $O(k_1 \lg n)$ time. The resultant region is in turn unioned with R_i. The point of intersection $x_{i'}$ of $v_a v_b$ and $v_i x_i$ is added as a constructed vertex to both WVP and $SPT(p)$, and R_i is associated with $x_{i'}$. To update $SPT(q)$, we find all edges e_i in $SPT(q)$ that intersect $\triangle v_a v_b v_c$ and delete all the endpoints z_i that are not visible from q along with their children from $SPT(q)$. This completes the update in this subcase.

If $v_a \in WVP$, then no constructed edge intersects $\triangle v_a v_b v_c$ and we update $SPT(p)$. Let F_i be the funnel containing v_a with root r_i and base $v_x v_y$. We then perform the depth first traversal in $SPT(p)$, starting from vertex v_a. If for any vertex v_j with parent v_i in $SPT(p)$, $SP(r_i, v_j)$ takes a right turn at vertex v_i we need to add a constructed edge. This is done by finding the leftmost child $v_{j'}$

of v_i for which $SP(v_a, v_{j'})$ still takes a right turn at v_i. Let v_k be the parent of v_i in $SPT(p)$, then the constructed vertex x_i is found by extending $v_k v_i$ on to $v_{j'} v_{j'+1}$. This results in a new region R_i.

To add vertices to this region and delete them from WVP, we traverse $bd(WVP)$ in counterclockwise direction and add every vertex from $v_{j'}$ to v_i. Every region R_j that is encountered in this traversal are unioned with R_i. As there can be $O(k_1)$ constructed vertices in this traversal along $bd(WVP)$, it requires $O(k_1 \lg n)$ time. Finally, the constructed vertex x_i is added to WVP and $SPT(p)$. To update $SPT(q)$, we find all edges e_i in $SPT(q)$ that intersect $\triangle v_a v_b v_c$. If the hidden endpoint z_i of an edge e_i has a non-null tree pointer (it lies in some region R_j), then z_i is deleted along with its children. Then, similar to the depth first traversal in $SPT(p)$ above, we perform the depth first traversal in the leftover $SPT(q)$ to update WVP. This completes the update. Note that whole of this algorithm takes $O(k_2 \lg n)$ time in the worst-case.

When $SP(p, v_b)$ intersects $\triangle v_a v_b v_c$, we first update $SPT(q)$ and steps analogous to above are performed. Further, we update $SPT(p)$, and finally we have the updated WVP. Since both $SPT(p)$ and $SPT(q)$ are updated, our algorithm takes $O(k_1 \lg n + k_2 \lg n)$ time in the worst-case.

Case (b). We assume that $v_b \notin WVP$ and $v_c \in WVP$. Let $v_b \in R_i$ with constructed vertex x_i and constructed edge $v_i x_i$. If $v_a v_c$ intersects $v_i x_i$ at $x_{i'}$, then $v_a \in R_i$. We update WVP by deleting x_i and adding $x_{i'}$. Similarly, we update $SPT(p)$ and $SPT(q)$ by deleting x_i and adding $x_{i'}$. Finally, we add v_a to R_i. Overall, handling this subcase takes $O(\lg n)$ time.

When $v_a \notin R_i$, we find the intersection of $v_a v_b$ with $v_i x_i$ at $x_{i'}$. We update WVP by deleting x_i and adding $x_{i'}$. Similarly, we update $SPT(p)$ and $SPT(q)$ by deleting x_i and adding $x_{i'}$. This case is then similar to case (a) where v_b is replaced by $x_{i'}$; hence, takes $O(k_1 \lg n + k_2 \lg n)$ time.

Case (c). Let $v_b \in R_i$ and let $v_c \in R_j$. Also, let x_i and x_j be the respective constructed vertices on $v_b v_c$. Then we have three subcases which are similar to case (b) when $v_a \in R_i$. Here, we additionally update $SPT(p)$ and $SPT(q)$ by adding the new constructed vertices (intersection points $x_{i'}$ and $x_{j'}$ of $\triangle v_a v_b v_c$ with the constructed edges). This additional step takes $O(\lg n)$ time.

Case (d). Let $v_b, v_c \in R_i$ and x_i be the corresponding constructed vertex and $v_i x_i$ be the constructed edge. If $v_a \in R_i$ then no change is made to WVP but we need to add v_a to T_{R_i} which takes $O(\lg n)$ time. If $v_a \notin R_i$, we assume a portion of $v_a v_b$ is visible from pq. (When a portion of $v_a v_c$ is visible from pq, it is handled analogously.) Then the polygonal region R_i is split into two regions. Hence, we split the tree T_{R_i} as well. This results in two trees: $T_{R_{i1}}$ with vertices from v_i to v_b along the counterclockwise traversal of $bd(P)$; and $T_{R_{i2}}$ with vertices from v_c to x_i along the counterclockwise traversal of $bd(P)$. Let $v_a v_b$ and $v_a v_c$ respectively intersect $v_i x_i$ at points x_1 and x_2 respectively. We add x_1 and x_2 to $SPT(p)$ and $SPT(q)$. Also, we associate x_1 as a constructed vertex with $T_{R_{i1}}$. This case is then similar to case (a) with $\triangle v_a x_1 x_2$ except that in the last step, we join tree $T_{R_{i2}}$ with a newly formed tree. This completes the case (d). And, this case require $O(k_1 \lg n + k_2 \lg n)$ time.

When $v_a \notin P$, each of the four cases (a) to (d) can be handled similar to case (c). Each of these cases again take $O(\lg n)$ time in the worst-case.

Distinguishing Cases (a)–(d). To determine the appropriate case among cases (a)–(d), we follow the tree pointers associated with v_b and v_c in L_P. For v_b, if the tree pointer is not null, then it points to a node in a tree say T_{R_b}. We follow the parent pointers from this node to reach the root of T_{R_b}, say r_1, which takes $O(\lg n)$ time. Similarly, r_2 is found if the tree pointer for v_c is not null. The appropriate case can thus be found using these tree pointers. To identify whether $v_a \in P$ or $v_a \notin P$, we can check if v_b, v_c, v_a turns left or right respectively.

Lemma 4. *The cases (a)–(d) are exhaustive. Further, the corresponding subcases are exhaustive as well.*

Lemma 5. *In updating, a vertex $u_i \in WVP$ is deleted if and only if it is not weakly visible from pq and a (constructed) vertex is added whenever it is weakly visible from pq.*

Lemma 6. *The worst-case time in updating the weak visibility polygon is $O((k+1)\lg n)$.*

Proof. Identifying the appropriate case requires $O(\lg n)$ time. Each case takes $O(k_1 \lg n + k_2 \lg n) = O(k \lg n)$ time. Note that in case (d) subcase (i), $k = 0$. Since k equals to zero in subcase (i) of case (d), overall time complexity for updating the weak visibility polygon is $O((k+1)\lg n)$.   □

Theorem 2. *Let $WVP_P(pq)$ be the weak visibility polygon of an edge pq of a simple polygon P. Let v_a be the new vertex inserted to P such that an existing edge $v_b v_c$ is replaced by edges $v_a v_b$ and $v_b v_c$, resulting in a new simple polygon P'. After preprocessing P and $WVP_P(pq)$, the weak visibility polygon $WVP_P(pq)$ is updated in $O((k+1)\lg n)$ time, wherein k is the total number of changes required to update $SPT(p)$ and $SPT(q)$ in adding new vertex v_a to P.*

4   Conclusions

We gave algorithms for dynamically maintain the visibilty polygon from a point $p \in P$ and weak visibility polygon from an edge $pq \in P$ when a new vertex is added to simple polygon P. To our knowledge, this is first result in this direction. Our semi-dynamic algorithm to update the visibility polygon is output-sensitive: the update time complexity is just a log multiplication factor away from the changes required to update the visibility polygon. The other semi-dynamic algorithm to update the weak visibility polygon of an edge pq relies on the changes required in updating the shortest path trees rooted at p and q. We see lots of scope for future work in devising dynamic algorithms in the context of visibility, art gallery problems, minimum link paths, and geometric shortest paths.

Acknowledgements. The authors wish to acknowledge the anonymous reviewer for the valuable comments which has improved the quality of the paper.

References

1. Aronov, B., Guibas, L.J., Teichmann, M., Zhang, L.: Visibility queries and maintenance in simple polygons. Discret. Comput. Geom. **27**(4), 461–483 (2002)
2. Asano, T., Asano, T., Guibas, L.J., Hershberger, J., Imai, H.: Visibility of disjoint polygons. Algorithmica **1**(1), 49–63 (1986)
3. Bose, P., Lubiw, A., Munro, J.I.: Efficient visibility queries in simple polygons. Comput. Geom. **23**(3), 313–335 (2002)
4. Chazelle, B., Guibas, L.J.: Visibility and intersection problems in plane geometry. Discret. Comput. Geom. **4**, 551–581 (1989)
5. Chen, D.Z., Wang, H.: Visibility and ray shooting queries in polygonal domains. Comput. Geom. **48**(2), 31–41 (2015)
6. Chen, D.Z., Wang, H.: Weak visibility queries of line segments in simple polygons. Comput. Geom. **48**(6), 443–452 (2015)
7. Davis, L.S., Benedikt, M.L.: Computational models of space: isovists and isovist fields. Comput. Graph. Image Proces. **11**(1), 49–72 (1979)
8. ElGindy, H.A., Avis, D.: A linear algorithm for computing the visibility polygon from a point. J. Algorithms **2**(2), 186–197 (1981)
9. Ghosh, S.K.: Computing the visibility polygon from a convex set and related problems. J. Algorithms **12**(1), 75–95 (1991)
10. Ghosh, S.K.: Visibility Algorithms in the Plane. Cambridge University Press, New York (2007)
11. Ghosh, S.K., Mount, D.M.: An output-sensitive algorithm for computing visibility graphs. SIAM J. Comput. **20**(5), 888–910 (1991)
12. Guibas, L.J., Hershberger, J., Leven, D., Sharir, M., Tarjan, R.E.: Linear-time algorithms for visibility and shortest path problems inside triangulated simple polygons. Algorithmica **2**, 209–233 (1987)
13. Guibas, L.J., Motwani, R., Raghavan, P.: The robot localization problem. SIAM J. Comput. **26**(4), 1120–1138 (1997)
14. Heffernan, P.J., Mitchell, J.S.B.: An optimal algorithm for computing visibility in the plane. SIAM J. Comput. **24**(1), 184–201 (1995)
15. Inkulu, R., Kapoor, S.: Visibility queries in a polygonal region. Comput. Geom. **42**(9), 852–864 (2009)
16. Joe, B., Simpson, R.: Corrections to Lee's visibility polygon algorithm. BIT Numer. Math. **27**(4), 458–473 (1987)
17. Kapoor, S., Singh, T.: Dynamic maintenance of shortest path trees in simple polygons. In: Chandru, V., Vinay, V. (eds.) FSTTCS 1996. LNCS, vol. 1180, pp. 123–134. Springer, Heidelberg (1996). doi:10.1007/3-540-62034-6_43
18. Lee, D.T.: Visibility of a simple polygon. Comput. Vis. Graph. Image Proces. **22**(2), 207–221 (1983)
19. Lee, D.T., Lin, A.K.: Computing the visibility polygon from an edge. Comput. Vis. Graph. Image Proces. **34**(1), 1–19 (1986)
20. O'Rourke, J.: Art Gallery Theorems and Algorithms. Oxford University Press Inc., New York (1987)
21. Suri, S., O'Rourke, J.: Worst-case optimal algorithms for constructing visibility polygons with holes. In: Proceedings of the Symposium on Computational Geometry, pp. 14–23 (1986)
22. Tarjan, R.E.: Data Structures and Network Algorithms. Society for Industrial and Applied Mathematics, Philadelphia (1983)

23. Vegter, G.: The visibility diagram: a data structure for visibility problems and motion planning. In: Proceedings of Scandinavian Workshop on Algorithm Theory, pp. 97–110 (1990)
24. Zarei, A., Ghodsi, M.: Efficient computation of query point visibility in polygons with holes. In: Proceedings of the Symposium on Computational Geometry, Pisa, Italy, 6–8 June 2005, pp. 314–320 (2005)

Liar's Domination in 2D

Ramesh K. Jallu and Guatam K. Das[(✉)]

Department of Mathematics, Indian Institute of Technology Guwahati,
Guwahati, India
{j.ramesh,gkd}@iitg.ernet.in

Abstract. In this paper we consider Euclidean liar's domination problem, a variant of dominating set problem. In the Euclidean liar's domination problem, a set $\mathcal{P} = \{p_1, p_2, \ldots, p_n\}$ of n points are given in the Euclidean plane. For $p \in \mathcal{P}$, $N[p]$ is a subset of $\mathcal{P}$ such that for any $q \in N[p]$, the Euclidean distance between p and q is less than or equal to 1. The objective of the Euclidean liar's domination problem is to find a subset $D(\subseteq \mathcal{P})$ of minimum size having the following properties: (i) $|N[p_i] \cap D| \geq 2$ for $1 \leq i \leq n$, and (ii) $|(N[p_i] \cup N[p_j]) \cap D| \geq 3$ for $i \neq j, 1 \leq i, j \leq n$. We first propose a simple $O(n \log n)$ time $\frac{63}{2}$-factor approximation algorithm for the liar's domination problem. Next we propose approximation algorithms to improve the approximation factor to $\frac{732}{k}$ for $3 \leq k \leq 183$, and $\frac{846}{k}$ for $3 \leq k \leq 282$. The running time of the algorithms is $O(n^{k+1}\Delta)$, where $\Delta = \max\{|N[p]| : p \in \mathcal{P}\}$.

Keywords: Unit disk graph · Approximation algorithm · Dominating set · Liar's dominating set

1 Introduction

Let $G = (V, E)$ be a graph. For a vertex $v \in V$, we define $N_G(v) = \{u \in V \mid (v, u) \in E\}$ and $N_G[v] = N_G(v) \cup \{v\}$, open and closed neighborhoods of v in G, respectively. A subset D of V is a **dominating set** *if for each vertex* $v \in V$, $|N_G[v] \cap D| \geq 1$, *that is, every vertex in V is either in D or adjacent to at lest one vertex in D.* We say that a vertex v is dominated by u in G, if $(v, u) \in E$ and u is in a dominating set of G (here, we call v as dominatee and u as dominator). *A subset D of V is a k-**tuple dominating set** if each vertex $v \in V$ is dominated by at least k vertices in D i.e., $|N_G[v] \cap D| \geq k$ for each $v \in V$.* The minimum cardinality of a k-tuple dominating set of a graph G is known as k-tuple domination number of G. A subset D of V is a **liar's dominating set** *if (i) for every $v \in V$, $|N_G[v] \cap D| \geq 2$, and (ii) for every distinct pair of vertices u and v, $|(N_G[u] \cup N_G[v]) \cap D| \geq 3$.* Liar's domination problem in a graph $G = (V, E)$ asks to find a liar's dominating set of G with minimum size. The minimum cardinality of a liar's dominating set of a graph G is known as liar's domination number of G. Every 3-tuple dominating set is a liar's dominating set as it satisfies both the conditions, so the liar's domination number lies between 2-tuple and 3-tuple domination numbers.

ⓒ Springer International Publishing AG 2017
D. Gaur and N.S. Narayanaswamy (Eds.): CALDAM 2017, LNCS 10156, pp. 219–229, 2017.
DOI: 10.1007/978-3-319-53007-9_20

The liar's dominating problem is a variant of dominating set problem. Dominating set and its variants have been extensively studied for last two decades due to its wide range of applications, e.g., networks, operation research, etc. [4]. Consider the following real life scenario: we have an art gallery and our objective is to protect the paintings from an intruder such as a thief, or a saboteur. Assume that the gallery has multiple entrances and each entrance is a possible location for an intruder. A protection device such as a sensor, or a camera placed at an entrance can not only detect (and report precise location) the presence of the intruder entering through it, but also at all the entrances that are visible from it. We can model the gallery as a graph. A vertex in the graph represents an entrance and an edge corresponds to their respective entrances visibility one from the other. Now, the goal is to place the protection devices as minimum as possible subject to the intrusion of an intruder at any entrance is detected and reported. The goal can be achieved by finding a minimum dominating set (MDS) of the graph and placing the devices at all the vertices in the MDS. The protection devices are prone to failure. If any one of the devices placed is failed to detect the presence of the intruder, then we must place protection devices at all the vertices of a minimum 2-tuple dominating set of the graph. Hence, the set of protection devices in the security system is a single fault tolerant set. Due to transmission error it may so happen that all the protection devices detect the location correctly but some of these devices can lie while reporting. Assume that at most one protection device in the closed neighborhood of the intruder can lie (misreport). That is, one of the protection devices can misreport any entrance in its closed neighborhood as the intruder location. In this scenario, in order to protect the gallery we must place the protection devices at a minimum liar's dominating set of the graph.

2   Related Work

The liar's domination problem is introduced by Slater in 2009 and showed that the problem is NP-hard for general graphs [10]. Later, Roden and Slater showed that the problem is NP-hard even for bipartite graphs [9]. Panda and Paul [7] proved that the problem is NP-hard for split graphs and chordal graphs. They also proposed a linear time algorithm for computing a minimum cardinality liar's dominating set in a tree. Bishnu et al. [2] proved that the liar's domination problem is W[2]-hard for general graphs. Recently, Alimadadi et al. [1] provided characterization of graphs and trees for which liar's domination number is $|V|$ and $|V|-1$, respectively. For connected graphs with girth (the length of a shortest cycle) at least five, they obtained an upper bound for the ratio between the liar's domination number and the 2-tuple domination number. Panda et al. [6] studied approximability of the problem and presented $3 + 2\ln(\Delta(G) + 1)$-factor approximation algorithm, where $\Delta(G)$ is the maximum degree of the graph. Panda and Paul [8] considered the problem for proper interval graphs and proposed a linear time algorithm for computing a minimum cardinality liar's dominating set. The problem is also studied for bounded degree graphs, and p-claw free graphs [6].

2.1 Our Contribution

Due to lack of good approximation algorithms for the liar's domination problem for general graphs in the literature, several researchers consider the problem for a set of special kind of graphs [6,8]. In this paper we consider the geometric version of the liar's domination problem and we call it as *Euclidean liar's domination problem.*

In the Euclidean liar's domination problem we are given a set $\mathcal{P}$ of n points in the plane. For $p \in \mathcal{P}$, $N[p]$ is a subset of $\mathcal{P}$ such that for any $q \in N[p]$, the Euclidean distance between p and q is less than or equal to 1. Some times we say $N[p]$ as the closed neighborhood of p in the rest of the paper. We define $\Delta = \max\{|N[p]| : p \in \mathcal{P}\}$. *The objective of the Euclidean liar's domination problem is to find a minimum size subset D of $\mathcal{P}$ such that (i) D is a 2-tuple dominating set i.e., for every point in $\mathcal{P}$ there exists at least two points in D which are at most distance one, and (ii) for every distinct pair of points p_i and p_j in $\mathcal{P}$, $|(N[p_i] \cup N[p_j]) \cap D| \geq 3$, in other words, the number of points in D that are within unit distance with points in the closed neighborhood union of p_i and p_j is at least three.* Recently Jallu and Das proved that the Euclidean liar's domination problem is NP-hard [5]. Unlike in general graphs, Euclidean liar's domination problem admits constant factor approximation algorithms.

We present a simple $\frac{63}{2}$-factor approximation algorithm in $O(n \log n)$ time followed by a $\frac{732}{k}$-factor approximation algorithm in $O(n^{k+1}\Delta)$ time for $3 \leq k \leq 183$. We then extend the idea of $\frac{732}{k}$-factor approximation algorithm to get a $\frac{846}{k}$-factor approximation algorithm in $O(n^{k+1}\Delta)$ time for $3 \leq k \leq 282$.

2.2 Organization of the Paper

We propose a $\frac{63}{2}$-factor approximation algorithm for the Euclidean liar's domination problem in Sect. 3. In Sect. 4, we propose a $\frac{732}{k}$-factor approximation algorithm for $3 \leq k \leq 183$. We further improve the approximation factor to $\frac{846}{k}$ for $3 \leq k \leq 282$. Finally we conclude the paper in Sect. 5.

3 $\frac{63}{2}$-Factor Approximation Algorithm

Let $\mathcal{R}$ be the rectangular region containing the point set $\mathcal{P}$. The objective of this section is to find a Euclidean liar's dominating set $D(\subseteq \mathcal{P})$ such that $|D| \leq \frac{63}{2}|OPT|$, where OPT is an optimum solution. We partition the region $\mathcal{R}$ into regular hexagonal cells with side length $\frac{1}{2}$ (see Fig. 1(a)). We have chosen side length of a cell as $\frac{1}{2}$ to make the maximum distance between any two points lying in the cell is at most one. The following observations are true as the largest diagonal of each cell is of unit length.

Observation 1. *All the points inside a cell can be covered by a unit radius disk centered at any point in that cell (see Fig. 1(b)).*

Observation 2. [3] *A unit disk centered at a point in a cell can not cover the points lying in more than 12 cells simultaneously.*

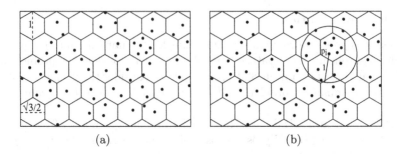

(a)   (b)

Fig. 1. (a) Hexagonal partition of the rectangular region containing the points and, (b) A unit disk centered at a point p_i circumscribes the cell in which p_i lies.

The pseudo code to find a liar's dominating set for the points in $\mathcal{P}$ is given in Algorithm 1. For any two points p, q in $\mathcal{P}$, if $q \in N[p]$, then we say that q is a neighbor of p (some times we say that q is covered by p) and vice versa. Since, for any liar's dominating set D, $|(N[p_i] \cup N[p_j]) \cap D| \geq 3$ for all $p_i, p_j \in \mathcal{P}$, we assume that $|\mathcal{P}| \geq 3$ and $|N[p]| \geq 2$.

Algorithm 1. Liar's_Dominating_Set $(\mathcal{P})$

Input: The set $\mathcal{P} = \{p_1, p_2, \ldots, p_n\}$ of n points.
Output: A liar's dominating set D of $\mathcal{P}$
1: Let $\mathcal{R}$ be the minimum size axis parallel rectangle containing points in $\mathcal{P}$
2: $D = \emptyset$
3: Partition the region $\mathcal{R}$ into regular hexagonal cells of side length $\frac{1}{2}$.
4: **for** (each non-empty cell H_i in the partition) **do**
5:     $n_i \leftarrow$ the number of points in H_i
6:     **if** $(n_i \geq 3)$ **then**
7:         Choose any three arbitrary points $p_i, p_j,$ and p_k from H_i.
8:         $D = D \cup \{p_i, p_j, p_k\}$.
9:     **else**
10:        **if** $(n_i = 2)$ **then**
11:            Let p_i and p_j are the points in H_i.
12:            $D = D \cup \{p_i, p_j, p_k\}$, where $p_k \in N[p_i] \cup N[p_j]$ and is different from p_i and p_j.
13:        **else**$(n_i = 1)$
14:            Let p_i be the point in H_i.
15:            $D = D \cup \{p_i, p_j, p_k\}$, where p_j and p_k are two neighbors of p_i, if exists, other wise, p_j is a neighbor of p_i and p_k is a neighbor of $p_j (i \neq k)$.
16:        **end if**
17:    **end if**
18: **end for**
19: Return D.

Lemma 1. *D returned by Algorithm 1 is a liar's dominating set of $\mathcal{P}$.*

Proof. Let H be a non-empty cell in the hexagonal partion of $\mathcal{R}$ and $p_i \in H \cap \mathcal{P}$. If $|H \cap \mathcal{P}| \geq 2$, then we choose at least two points from $H \cap \mathcal{P}$ as part of D in our algorithm (see line numbers 8 and 12 in Algorithm 1). Therefore, $|N[p_i] \cap D| \geq 2$ by Observation 1. Now, if $|H \cap \mathcal{P}| = 1$, then D contains p_i and one more point of $\mathcal{P}$, which is a neighbor of p_i (see line number 15 in Algorithm 1). So, in this case also $|N[p_i] \cap D| \geq 2$. Therefore, $|N[p_i] \cap D| \geq 2$ for any $p_i \in \mathcal{P}$.

Now we prove that every pair of points p_i and $p_j (p_i \neq p_j)$ in $\mathcal{P}$ have at least three points of D in their closed neighborhood union.

Case 1. p_i and p_j belong to the same cell, say H_i.
  (a) H_i contains more than two points of $\mathcal{P}$: D contains three points from H_i (see line number 8 in Algorithm 1). p_i and p_j may or may not be part of these three points. In either case $|(N[p_i] \cup N[p_j]) \cap D| \geq 3$ holds by Observation 1.
  (b) H_i contains only two points of $\mathcal{P}$: in this case both p_i and p_j must be in D and one of its neighbors also must be in D (see line number 12 in Algorithm 1). Hence $|(N[p_i] \cup N[p_j]) \cap D| \geq 3$.
Case 2. p_i and p_j belong to distinct cells, say H_i and H_j, respectively.
  (a) $|(H_i \cup H_j) \cap \mathcal{P}| \geq 3$ i.e., either $|H_i \cap \mathcal{P}| \geq 2$ and $|H_j \cap \mathcal{P}| \geq 1$, or $|H_i \cap \mathcal{P}| \geq 1$ and $|H_j \cap \mathcal{P}| \geq 2$: D contains at least 3 points from $(H_i \cup H_j) \cap \mathcal{P}$ (see line numbers 8, 12, and 15 in Algorithm 1). Therefore, $|(N[p_i] \cup N[p_j]) \cap D| \geq 3$ (by Observation 1).
  (b) $|(H_i \cup H_j) \cap \mathcal{P}| = 2$ i.e., $|H_i \cap \mathcal{P}| = 1$ and $|H_j \cap \mathcal{P}| = 1$: in this case, the algorithm chooses p_i and p_j as members of D along with at least one neighbor of p_i (resp. p_j) (see line number 15 in Algorithm 1). Therefore, in this case also $|(N[p_i] \cup N[p_j]) \cap D| \geq 3$.

Thus the lemma.   □

Theorem 1. *The approximation factor and the running time of the proposed algorithm (Algorithm 1) are $\frac{63}{2}$ and $O(n \log n)$ respectively.*

Proof. Let OPT be an optimum solution for the point set $\mathcal{P}$. Consider a non-empty cell H_i in the hexagonal partition. The points in H_i can be covered by the points in H_i itself or/and also by the points in at most 18 surrounding cells (see Fig. 2). Let $p_i \in H_i$. Observe that the number of points in OPT dominating (covering) the point p_i is at least 2 (due to first condition of liar's domination). Hence, the points in OPT dominating the point p_i, say p_j and p_k, should be from H_i or/and its 18 surrounding cells (shown as solid cells in Fig. 2), and these 18 cells are the cells that are at most unit distance away from H_i.

The point p_j (resp. p_k) can cover points in at most 12 cells including H_i (by Observation 2). Hence, the points coverd by p_j (resp. p_k) must lie within the unit radius disk centered at p_j (resp. p_k). However they have some common cell covered. The points p_k and p_j cover more cells when they have less number of common cells covered. One such possible scenario happens when p_j and p_k

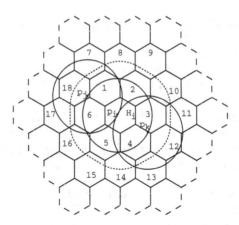

Fig. 2. Cells within unit distance (solid cells) from H_i.

are the end points of the diameter of the unit radius disk centered at p_i. In this case they can have at most three common cells, and cover points in 21 distinct cells. Our algorithm chooses at most 3 points for each cell. Hence approximation factor is $\frac{21 \times 3}{2} = 31.5$.

For a point $p_i \in \mathcal{P}$, computing its cell number can be done in constant time as we know the co-ordinate positions of the points. We can store the non-empty grid cells in a data structure such as balanced binary tree. For a point $p_i \in \mathcal{P}$ we compute its cell number and check for the presence of the cell in the data structure. If the cell is not present, we insert the cell in the data structure. Hence, we have in hand that how many and which points of $\mathcal{P}$ are in a cell. After processing the points we just need go through the data structure to report the required points. For one point we spend $O(\log n)$ time, hence for n points it takes $O(n \log n)$ time. Thus the running time for Algorithm 1 follows.   □

4   Improving the Approximation Factor

A 37-hexagon is a combination of 37 adjacent regular hexagonal cells such that one cell is surrounded by 36 other cells (see Fig. 3(a)). Let us consider the partition of the rectangular region $\mathcal{R}$ into 37-hexagons such that no point of $\mathcal{P}$ lies on the boundary of any 37-hexagon in the partition, and a 4 coloring scheme of it (see Fig. 3(b)). Consider a 37-hexagon colored with A. It's adjacent six 37-hexagons are colored B, C, and D (in Fig. 3(b) different patterns have been shown), such that opposite pair of 37-hexagons receive the same color.

Lemma 2. *In the 4-coloring scheme, the minimum distance between any two same colored 37-hexagons is greater than or equal to 5.*

Proof. Let H' and H'' be two same colored 37-hexagons in the partition and 37-hexagon H lies between H' and H''. Draw a maximum radius circle that can

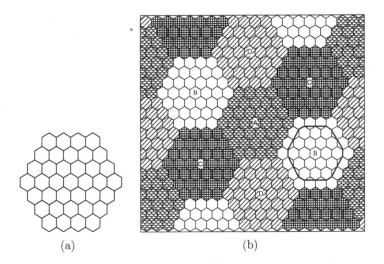

Fig. 3. (a) A single 37-hexagon, (b) A four coloring scheme of the hexagonal grid

fit entirely in H. This circle must touch the common boundary of H and H', H and H''. Draw a line segment between two points where the circle touches the boundaries. Observe that the segment is the diameter of the circle and its length is 5. Thus the lemma follows. □

Consider a 37-hexagon H. H can be viewed as a 19-hexagon, say H', surrounded by 18 cells. Let us consider the convex hull overlay, say CH, of the set of corners of H' (shown as loop in Fig. 3(b)). Observe that the maximum distance between any two points in the convex hull overlay CH is at most $\frac{5\sqrt{3}}{2}(> 4)$. Let $S = CH \cap \mathcal{P}$, $S' = \{p \in \mathcal{P} \mid \delta(p, q) \leq 1 \; for \; q \in S\}$, and $S'' = \{p \in \mathcal{P} \mid \delta(p, q) \leq 1 \; for \; q \in S'\}$, where $\delta(p, q)$ denotes the Euclidean distance between p and q. In this subsection, we first propose an approximation algorithm to find a liar's dominating set $S_H \subseteq S''$ for S'. Next, we use the above approximation result to design an approximation algorithm to find out a liar's dominating set for $\mathcal{P}$. Let $OPT_{S'_H}$ denotes an optimal liar's dominating set of S'.

Lemma 3. $|OPT_{S'_H}| \leq 183$.

Proof. The points in $S' \setminus S$ lie in at most 24 cells around H by definition of S' (i.e., one layer around H). Hence, the points in S' can span over 61 cells. If we choose at most three points for each cell, we get a liar's dominating set. Thus, the cardinality of $OPT_{S'_H}$ can not be more than 183. □

Lemma 4. If H_1 and H_2 are two same colored 37-hexagons, then $OPT_{S'_{H_1}} \cap OPT_{S'_{H_2}} = \emptyset$.

Proof. The proof follows from Lemma 2. □

The detailed pseudo code to find a liar's dominating set for the points lying in a given 37-hexagon H is given in Algorithm 2. For a given k $(3 \leq k \leq 183)$, we choose all possible $t = 1, 2, \ldots, k - 1$ combinations of points in S'' to find a 2-tuple dominating set of S'. For each combination, we check the combination of points is a 2-tuple dominating set or not (Line number 5). While considering the subsets, if there exists a subset S_H that is a 2-tuple dominating set, then for every distinct pair of points p_i and p_j in S', we check whether $|(N[p_i] \cup N[p_j]) \cap S_H| \geq 3$ or not (Line number 7). A subset S_H satisfying the above condition is reported and is a liar's dominating set for S'. In the algorithm Boolean variable $flag$ is used to ensure that the set returned by the algorithm is a feasible solution.

Algorithm 2. Liar's_Dominating_Set (H, k)

Input: Point set $\mathcal{P}$, a 37-hexagon H and an integer $3 \leq k \leq 183$.
Output: A liar's dominating set of size $\leq k - 1$ for the set S' from S'' (if exists),
    otherwise a set $S_H(\subseteq S'')$ of size at most 183.
1: Obtain sets S, S', and S''.
2: $flag = 0$.
3: **for** $(t = 1, 2, \ldots, k - 1)$ **do**
4:    **for** (each combination S_H of t points in S'') **do**
5:        **if** $(|S_H \cap N[p_i]| \geq 2 \; \forall p_i \in S')$ **then**
6:            **for** (every distinct pair of points p_i and p_j in S') **do**
7:                **if** $(|(N[p_i] \cup N[p_j]) \cap S_H| \geq 3)$ **then**
8:                    $flag = 1$.
9:                **else**
10:                    $flag = 0$.
11:                    **break**/*break the for loop in line number 6 */
12:                **end if**
13:            **end for**
14:        **end if**
15:        **if** $(flag = 1)$ **then**
16:            Return S_H.
17:            **break**/*break the for loop in line number 3 */
18:        **end if**
19:    **end for**
20: **end for**
21: $S_H \leftarrow \emptyset$
22: **if** $(flag = 0)$ **then**
23:    Consider any three points from each non-empty cell corresponding to S' and
        add it to S_H (if a non-empty cell contains less than three points we choose
        the remaining points from its neighbors as in Algorithm 1).
24: **end if**
25: Return S_H.

Lemma 5. *For a given 37-hexagon H, Algorithm 2 produces a solution S_H for the set S' from S'' with size is at most $\frac{183}{k} \times |OPT_H|$, where $3 \leq k \leq 183$ and OPT_H is an optimum solution for the points lying in H.*

Proof. If the algorithm can not produce a solution of size $k - 1$ for given k ($3 \leq k \leq 183$), then $|OPT_H| \geq k$. Observe that, our algorithm may produce a solution S_H whose size is 183, in the worst. Hence $\frac{|S_H|}{|OPT_H|} \leq \frac{183}{k}$. $\square$

Lemma 6. *Algorithm 2 runs in $O(n^{k+1}\Delta)$ time, where $\Delta = \max\{|N[p]| : p \in \mathcal{P}\}$ and $3 \leq k \leq 183$.*

Proof. Algorithm 2 chooses all possible $k-1$ combinations for a given k. If any of these combinations satisfies liar's domination conditions, Algorithm 2 reports the combination of points. If it is not possible to find a solution of size $k-1$, the algorithm chooses at most three points for each non-empty cell (like in Algorithm 1). Steps 4–19 in Algorithm 2 can be done in $O(n^{t-1})(O(\Delta)+O(n^2\Delta)) = O(n^{t+1}\Delta)$. Hence, steps 3–20 take $\sum_{t=1}^{k-1} O(n^{t+1})\Delta$ time. Steps 1 and 22 can be done in $O(n \log n)$ time. Therefore the total running time of Algorithm 2 is $O(n^{k+1}\Delta)$. Thus the running time result follows. $\square$

We consider each 37-hexagon and compute a feasible solution (using Algorithm 2) for the points lying in it. Two same colored 37-hexagons can be solved independently as the minimum distance between them is greater than 4. Let S_j be the union of solutions generated by the algorithm the 37-hexagons colored j, for $j \in \{A, B, C, D\}$. The set $\mathcal{S} = \bigcup_{j \in \{A,B,C,D\}} S_j$ is reported.

Theorem 2. *$\mathcal{S}$ is a liar's dominating set of $\mathcal{P}$.*

Proof. In Algorithm 2, for a 37-hexagon H, we find a liar's dominating set S_H for S' from S'' (see the definition of S' and S'' defined previously in this section). Now, $S = \bigcup_{H \in \{\text{all 37-hexagons in } \mathcal{R}\}} S_H$. Thus, the theorem follows. $\square$

Theorem 3. *The 4-coloring scheme gives a $\frac{732}{k}$-factor approximation algorithm for the Euclidean liar's domination problem in the plane and runs in $O(n^{k+1}\Delta)$ time, where $3 \leq k \leq 183$.*

Proof. By Lemma 2, any two same colored 37-hexagons are greater than five units apart. Therefore, we can solve them independently. Let OPT_j be the union of solutions in optimal solution for the 37-hexagons colored j, for $j \in \{A, B, C, D\}$. Also let OPT be an optimum solution for the point set $\mathcal{P}$, hence, $|OPT_j| \leq |OPT|$ for each $j \in \{A, B, C, D\}$ and $OPT = OPT_A \cup OPT_B \cup OPT_C \cup OPT_D$. Observe that, OPT_j is the optimum for color class j, where we dominate the sets S' with respect to the group of 37-hexagons of color j using the points from S'' and not just S'. The 4-coloring scheme reports the set $\mathcal{S}$, which is the union of the solutions for all 37-hexagons. Therefore, $|\mathcal{S}| \leq \frac{183}{k}(|OPT_A| + |OPT_B| + |OPT_C| + |OPT_D|)$. Implies, $|\mathcal{S}| \leq \frac{732}{k} \times |OPT|$ as $|OPT_i| \leq |OPT|$ for each $i \in \{A, B, C, D\}$.

The running time result follows from Lemma 6 and the fact that a point in $\mathcal{P}$ participates a finite number of times in the algorithm. Hence the theorem. $\square$

4.1   Further Improvement of the Approximation Factor

The best approximation factor that can be achieved using the algorithm proposed in Sect. 4 is 4 for $k = 183$. We can extend the idea used for 37-hexagonal partition further to get best approximation factor 3. A 66-hexagon is a combination of 66 cells arranged in six rows and each row contains 11 cells (see Fig. 4(a)). We partition the rectangular region $\mathcal{R}$ containing the points in $\mathcal{P}$ into 66-hexagons such that no point of $\mathcal{P}$ lies on the boundary and consider a 3-coloring scheme of it (see Fig. 4(b)).

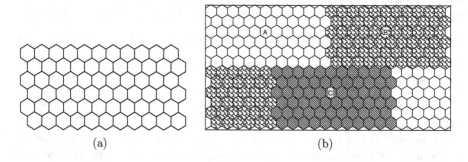

(a)   (b)

Fig. 4. (a) A single 66-hexagon, (b) A three coloring scheme of the hexagonal grid

By using the technique in Sect. 4, we can get $\frac{282}{k}$-factor approximation algorithm for the points lying in a 66-hexagon, where $3 \leq k \leq 282$, and we have the following theorem.

Theorem 4. *The 3-coloring scheme gives a $\frac{846}{k}$-factor approximation algorithm for the Euclidean liar's domination problem in the plane and the algorithm runs in $O(n^{k+1}\Delta)$ time, where $3 \leq k \leq 282$.*

5   Conclusion

In this paper we first proposed a simple $O(n \log n)$ time $\frac{63}{2}$-factor approximation algorithm for the Euclidean liar's domination problem. Next, we proposed approximation algorithms to improve the approximation factor to $\frac{732}{k}$ for $3 \leq k \leq 183$, and $\frac{846}{k}$ for $3 \leq k \leq 282$. The running time of the algorithms are $O(n^{k+1}\Delta)$.

Acknowledgement. The authors would like to thank Dr. Subhabrata Paul for introducing the liar's domination problem in graphs.

References

1. Alimadadi, A., Chellali, M., Mojdeh, D.A.: Liar's dominating sets in graphs. Discret. Appl. Math. **211**, 204–210 (2016)
2. Bishnu, A., Ghosh, A., Paul, S.: Linear kernels for k-tuple and liar's domination in bounded genus graphs. arXiv preprint arXiv:1309.5461 (2013)
3. De, M., Das, G.K., Carmi, P., Nandy, S.C.: Approximation algorithms for a variant of discrete piercing set problem for unit disks. Int. J. Comput. Geom. Appl. **23**(06), 461–477 (2013)
4. Haynes, T.W., Hedetniemi, S., Slater, P.: Fundamentals of Domination in Graphs. CRC Press, Boca Raton (1998)
5. Jallu, R.K., Das, G.K.: Hardness of Liar's Domination on Unit Disk Graphs. https://arxiv.org/abs/1611.07808 (2016)
6. Panda, B.S., Paul, S., Pradhan, D.: Hardness results, approximation and exact algorithms for liar's domination problem in graphs. Theor. Comput. Sci. **573**, 26–42 (2015)
7. Panda, B., Paul, S.: Liar's domination in graphs: complexity and algorithm. Discret. Appl. Math. **161**(7), 1085–1092 (2013)
8. Panda, B., Paul, S.: A linear time algorithm for liar's domination problem in proper interval graphs. Inf. Process. Lett. **113**(19), 815–822 (2013)
9. Roden, M.L., Slater, P.J.: Liar's domination in graphs. Discret. Math. **309**(19), 5884–5890 (2009)
10. Slater, P.J.: Liar's domination. Networks **54**(2), 70–74 (2009)

Structured Instances of Restricted Assignment with Two Processing Times

Klaus Jansen and Lars Rohwedder[✉]

University of Kiel, 24118 Kiel, Germany
{kj,lro}@informatik.uni-kiel.de

Abstract. For the RESTRICTED ASSIGNMENT Problem the best known polynomial algorithm is a 2-approximation by Lenstra, Shmoys and Tardos [7]. Even for the case with only two different processing times the ratio above has merely been improved by a tiny margin [2].

In some cases where the restrictions on one or both job sizes are somewhat structured, simple combinatorial algorithms are known that provide better approximation ratios than the algorithm above.

In this paper we study two classes of structured restrictions, that we refer to as the Inclusion Chain Class and the Two Partition Class. We present a 1.5-approximation for each of them.

Keywords: Approximation · Scheduling · Restricted assignment

1 Introduction

In the RESTRICTED ASSIGNMENT Problem we are given a set of m machines M and a set of n jobs J. Each job $j \in J$ has a processing time (also referred to as size) $p(j) \in \mathbb{R}_{>0}$ and assignment restrictions $M(j) \subseteq M$. We want to assign each job j to exactly one machine in $M(j)$ in a way that the makespan is minimized. The makespan refers to the maximum load among all machines. The load on a machine is the sum of the processing times of the jobs assigned to it.

For RESTRICTED ASSIGNMENT the best known polynomial algorithm is a 2-approximation. It was published by Lenstra, Shmoys and Tardos for the more general problem of SCHEDULING WITH UNRELATED MACHINES [7]. On the negative side they show, that no approximation guarantee better than 3/2 can be achieved unless P = NP. This holds even for the special case of RESTRICTED ASSIGNMENT. Since then neither side has been improved. For the case of two different processing times recently a polynomial $(2 - \delta)$-approximation was published for some very small δ [2].

On a different note Svensson has found a $(33/17 + \epsilon)$-estimation algorithm for an arbitrary small ϵ [8]. Unlike an approximation, an estimation does not produce a solution, but merely estimates the value of the optimal solution. The

Research was in part supported by German Research Foundation (DFG) project JA 612/15-1.

D. Gaur and N.S. Narayanaswamy (Eds.): CALDAM 2017, LNCS 10156, pp. 230–241, 2017.
DOI: 10.1007/978-3-319-53007-9_21

result was later improved to $11/6 + \epsilon$ [6]. In the case of two processing times the best known estimation ratio is $5/3$ [5].

We consider it an interesting area of research to investigate the case where we have two job sizes and one of them has somewhat structured restrictions. A notable example for this is the constraint $|M(j)| \leq 2$ for all big jobs j (small job restrictions are arbitrary). This class has been shown to admit a 1.5-approximation and this result is tight [4]. If any number of job sizes is allowed and this cardinality constraint exists on all jobs, then this problem is known as Graph Balancing [3].

Another example is given by Chakrabarty and Khanna [1]. They study a class, in which all big jobs are allowed on all machines. Their approximation algorithm has the additional requirement that the optimal makespan equals the size of the big job. For this they guarantee a solution of value at most $3/2 \cdot OPT + \epsilon$ where ϵ is the size of the small jobs.

Our Contribution. We improve the result above, in which big jobs are allowed on all machines and small jobs can have restrictions of arbitrary form. For this fundamental class we give an approximation that does not rely on the optimal makespan being equal to the size of the big job, as is the case in the result mentioned above. We also improve the approximation guarantee to $3/2$.

In fact we achieve this for two classes of instances that are slightly more general. Our techniques involve an abstraction of the problem as the maximization of a particular monotone, submodular set function, for which we prove a couple of general properties. We note that there cannot be a polynomial algorithm with approximation ratio better than $4/3$ for these classes, unless P = NP (first shown in [1]; see Theorem 3).

Theorem 1 (Inclusion Chain). *Given an instance of the 2-valued* RESTRICTED ASSIGNMENT *problem and for all big jobs j_1, j_2 it holds that $M(j_1) \subseteq M(j_2)$ or $M(j_2) \subseteq M(j_1)$, a 1.5-approximation can be computed in polynomial time.*

Theorem 2 (Two Partition). *Given an instance of the 2-valued* RESTRICTED ASSIGNMENT *problem, $A, B \subseteq M$ with $A \dot\cup B = M$, and for all big jobs j it holds that $M(j) = A$ or $M(j) = B$, a 1.5-approximation can be computed in polynomial time.*

2   Preliminaries

2.1   Notation

Let A be a set and $a \in A, b \notin A$. We will use the shortcut $A + b$ as the set containing b and the elements in A, i.e. $A \cup \{b\}$. For $A \backslash \{a\}$ we will use $A - a$ respectively. For two sets A, B we may express their union as $A \dot\cup B$ to additionally indicate that A and B are disjoint, i.e. $A \cap B = \emptyset$.

Job sizes are scaled such that the big jobs have a size of 1. Note that this is not an additional requirement, but merely simplifies notation. The size of the small jobs is denoted by ϵ. We will denote by J_B the jobs of size 1 and by J_S those of size ϵ.

2.2 Guessing the Makespan

A common idea that we will embrace is to guess the optimal makespan. Although NP-hard to determine, we can often assume that our algorithm knows it. The guessed makespan, which we denote by t is treated as an additional input. The algorithm should produce a schedule with makespan at most $1.5 \cdot t$ for any $t \geq OPT$. If $t < OPT$ we do not care about the output, but expect the algorithm to terminate in polynomial time.

Note that $OPT = K + K' \cdot \epsilon$ for some $K, K' \in \{0, \dots, n\}$. We can try all candidates for OPT and choose the lowest one for which we get a valid solution. This is then guaranteed to be of value at most $1.5 \cdot OPT$.

Simple Cases. There are methods to obtain a solution of value $OPT + 1$ [7]. For $OPT \geq 2$ this yields a 1.5-approximation. Hence, for the algorithms we design it is sufficient to consider guesses with $t < 2$. In particular, we can assume that no two big jobs fit on the same machine in the optimal solution.

2.3 Inapproximability

Theorem 3 (Inapproximability [1]). *It is NP-hard to approximate the 2-valued* RESTRICTED ASSIGNMENT *Problem with a ratio better than 4/3, even if the big jobs are allowed on all machines.*

This reduction was first shown in [1]. In the variation that we give, we use $\epsilon = 1/3$, whereas in the original reduction an arbitrary small ϵ is used. Then however the bound would only be 7/6.

Proof. This reduction uses the Vertex Cover Problem in cubic graphs: It is NP-hard to determine for a cubic graph G and a $k \in \mathbb{N}$, whether a vertex cover of size k exists.

Let $\epsilon := 1/3$. Create one machine for each vertex in G. For every edge (u, v) in the graph add a small job j with $M(j) = \{u, v\}$. Finally add $|M| - k$ large jobs j with $M(j) = M$. This instance of Restricted Assignment has a solution with makespan 1 if and only if there is a vertex cover of size at most k:

First assume there is a solution with makespan 1. We will use as our vertex cover all those machines to which no big job is assigned. The size of this cover is $|M| - (|M| - k) = k$. We know that every small job is assigned to one of these machines; Hence every edge is incident to one of the corresponding vertices.

Now assume that there is a vertex cover of size k. We assign the big jobs to the other machines. By definition we know, that every edge is incident to a vertex in the cover; Therefore every small job is permitted on at least one machine, which has no big job assigned. We schedule the small jobs on any of these machines. Since G is cubic, no machine is assigned more than 3 small jobs and the load is at most 1.

Note that if no schedule exists with makespan 1, the makespan has to be at least 4/3. We conclude that determining whether an instance has a schedule with makespan less than 4/3 is at least as hard as solving the original cover problem. □

3   Overview of Techniques

In our setting the optimal solutions can fit at most one big job on each machine. For our approximation we will also assign at most one big job to each machine. This means the problem can be reduced to finding a particular subset X of machines, on which one big job is processed each, and $Y = M \backslash X$, on which none are. These sets need to satisfy that there actually is such an assignment of big jobs and there needs to be a schedule for the small jobs with a load of at most $1.5 \cdot t - 1$ on machines in X and $1.5 \cdot t$ on others, where t is the guessed makespan. This adds up to a total load of at most $1.5 \cdot t$ on each machine, which is exactly what we desire.

Indeed for fixed X, Y it is easy to find a schedule for the small jobs, if it exists, by solving the maximum flow problem in the network in Fig. 1. More generally this can be used to assign a maximum number of small jobs complying with the given capacities.

The capacities have to be rounded to integers in order to find an integral flow: On machines in X we allow at most $\lfloor (t - 1/2)/\epsilon \rfloor$ small jobs each. On machines in Y we allow $\lfloor t/\epsilon \rfloor$. This is actually stronger than what we need for a 1.5-approximation (which is $\lfloor (3/2 \cdot t - 1)/\epsilon \rfloor$ and $\lfloor (3/2 \cdot t)/\epsilon \rfloor$ respectively).

However, in none of our arguments it seems beneficial to raise the capacity above $\lfloor t/\epsilon \rfloor$. We will denote the maximum flow problem with the capacities above by $N(Y)$. By $\omega(Y)$ we denote the optimum of $N(Y)$, i.e., the maximal number of small jobs that can be assigned. A solution for $N(Y)$ will be interpreted as

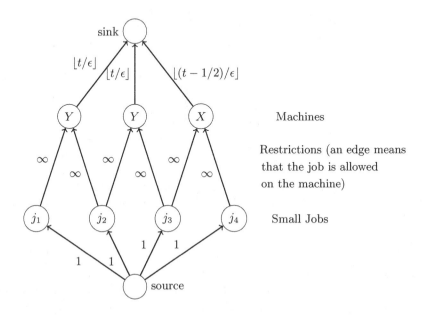

Fig. 1. Flow network for assigning small jobs

a function $f : M \rightarrow \mathcal{P}(J_S)$, that specifies which jobs are assigned to which machines. For all $S \subseteq M$ we define $f(S) := \bigcup_{b \in S} f(b)$.

Next, we will show some fundamental properties of the set function ω, which will be used in the analysis of our algorithms.

Lemma 1. *Let* $S, T \subseteq M$. *Let* f_S *be an optimal solution for* $N(S)$. *Then there exists an optimal solution* f_T *for* $N(T)$ *such that for every* $a \in M$ *it holds that* $f_S(a) \subseteq f_T(a)$ *or* a *is saturated in* f_T *(that is* $|f_T(a)| = \lfloor t/\epsilon \rfloor$ *if* $a \in T$ *and* $|f_T(a)| = \lfloor (t - 1/2)/\epsilon \rfloor$ *otherwise).*

The intuition behind this is that f_T is chosen in a way, that it is similar to f_S. Figure 2 illustrates it with an example.

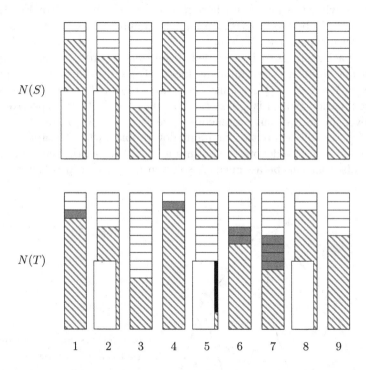

Fig. 2. Visualization of Lemma 1. In this example not all small jobs, that are placed on machine 5 in $N(S)$, can be placed there in $N(T)$. However the machine 5 is then guaranteed to be saturated.

Proof. Let $f_T^{(0)}$ be an optimal solution for $N(T)$. For $i \in \mathbb{N}$ construct $f_T^{(i)}$ incrementally from $f_T^{(i-1)}$ by applying the following operation. Choose a machine $a \in M$ and a job $j \in f_S(a) \backslash f^{(i-1)}(a)$, where a is not saturated; then set $f_T^{(i)}(a) := f_T^{(i-1)}(a) + j$ and $f_T^{(i)}(b) := f_T^{(i-1)}(b) - j$ for all $b \in M - a$.

After each of these steps the number of differently assigned jobs strictly decreases. Thus, after a bounded number of steps we can no longer perform the operation. This means we have for every $a \in M$ that $f_T^{(i)}(a) \supseteq f_S(a)$ or a is saturated in $f_T^{(i)}$. Clearly $f_T^{(i)}$ remains optimal, since we don't remove any jobs from the solution. □

A similar result to Theorem 4 was used in [1]. Their proof, however, relies on $OPT = 1$, which ours does not need. Moreover our proof uses integral assignments, whereas theirs uses fractional assignments of small jobs. This results in a better approximation ratio, since no rounding is necessary at a later time.

Theorem 4. *Let Y^* be the machines on which no big job is scheduled in the optimal solution. Let $Y \subseteq M$ with $|Y| = |Y^*|$ and $\omega(Y) < \omega(Y^*)$. Then there exists a machine $b \in Y \backslash Y^*$ such that $\omega(Y - b) = \omega(Y)$.*

Proof. Define $B := Y \backslash Y^*, C := Y^* \backslash Y$ and $A := Y^* \cap Y \cup (M \backslash (Y^* \cup Y))$. Note that $|B| = |C|$ and $A \dot\cup B \dot\cup C = M$.

Let f^*, f be optimal solutions for $N(Y^*)$ and $N(Y)$ respectively. Assume w.l.o.g. that every machine a is saturated in f or $f^*(a) \subseteq f(a)$ (Lemma 1). We may also assume that $|f^*(a)| \le \lfloor t/\epsilon \rfloor$, if $a \in Y^*$, and $|f^*(a)| \le \lfloor (t-1)/\epsilon \rfloor$ otherwise, since the optimal makespan is t.

Let us first show that $|f^*(c)| - |f(c)| \le \lfloor t/\epsilon \rfloor - \lfloor (t-1/2)/\epsilon \rfloor$ for every $c \in C$. Recall that the capacity on c is $\lfloor (t-1/2)/\epsilon \rfloor$ in $N(Y)$.

If $|f(c)| = \lfloor (t-1/2)/\epsilon \rfloor$, then the claim is obviously true. If on the other hand $|f(c)| < \lfloor (t-1/2)/\epsilon \rfloor$, then by definition of f we get $f^*(c) \subseteq f(c)$. By linearity we have

$$|f^*(C)| - |f(C)| \le (\lfloor t/\epsilon \rfloor - \lfloor (t-1/2)/\epsilon \rfloor) \cdot |C|. \tag{1}$$

Since on machines in A the capacity is the same in $N(Y)$ and $N(Y^*)$, with the same argument as above we get $|f(A)| \ge |f^*(A)|$. Also note that $|f(M)| = \omega(Y) < \omega(Y^*) = |f^*(M)|$. Therefore,

$$\begin{aligned}
|f(B)| - |f^*(B)| &= (|f(M)| - |f(A)| - |f(C)|) \\
&\quad - (|f^*(M)| - |f^*(A)| - |f^*(C)|) \\
&= (|f(M)| - |f^*(M)|) + (|f^*(A)| - |f(A)|) \\
&\quad + (|f^*(C)| - |f(C)|) \\
&< (|f^*(A)| - |f(A)|) + (|f^*(C)| - |f(C)|) \\
&\le (\lfloor t/\epsilon \rfloor - \lfloor (t-1/2)/\epsilon \rfloor) \cdot |C| \\
&= (\lfloor t/\epsilon \rfloor - \lfloor (t-1/2)/\epsilon \rfloor) \cdot |B|.
\end{aligned}$$

Recall f^* assigns at most $\lfloor (t-1)/\epsilon \rfloor$ small jobs to a machine in B. This implies

$$\begin{aligned}
|f(B)| &< |f^*(B)| + (\lfloor t/\epsilon \rfloor - \lfloor (t-1/2)/\epsilon \rfloor) \cdot |B| \\
&\le (\lfloor (t-1)/\epsilon \rfloor + \lfloor t/\epsilon \rfloor - \lfloor (t-1/2)/\epsilon \rfloor) \cdot |B|.
\end{aligned}$$

We conclude that a machine $b \in B$ exists with

$$|f(b)| < \lfloor (t-1)/\epsilon \rfloor + \lfloor t/\epsilon \rfloor - \lfloor (t-1/2)/\epsilon \rfloor.$$

Since $|f(b)|$ is integral, it holds that

$$|f(b)| \leq \lfloor (t-1)/\epsilon \rfloor + \lfloor t/\epsilon \rfloor - (\lfloor (t-1/2)/\epsilon \rfloor + 1)$$
$$\leq (t-1)/\epsilon + t/\epsilon - (t-1/2)/\epsilon$$
$$= (t-1/2)/\epsilon.$$

Again, since $|f(b)|$ is integral, $|f(b)| \leq \lfloor (t-1/2)/\epsilon \rfloor$ holds. This implies that f is also a feasible solution for $N(Y - b)$ and thus $\omega(Y - b) = \omega(Y)$.   □

Definition 1. A set function $\mu : \mathcal{P}(E) \to \mathbb{R}$ is called monotone, if for all $A \subseteq B \subseteq E$

$$\mu(A) \leq \mu(B).$$

μ is *submodular*, if we have for all $A, B \subseteq E$

$$\mu(A) + \mu(B) \geq \mu(A \cup B) + \mu(A \cap B).$$

An equivalent definition of submodularity is that for all $A \subseteq B \subseteq E$ and $b \in E \backslash B$

$$\mu(A + b) - \mu(A) \geq \mu(B + b) + \mu(A).$$

Theorem 5. ω *is a monotone, submodular function.*

The proof for Theorem 5 is omitted to conserve space.

4   Inclusion Chain Class

We now apply the previous results to the greedy algorithm (see Algorithm 1). Clearly the solution Y satisfies $\omega(Y) = \omega(M)$ at any state of the algorithm and therefore all small jobs can be assigned. Left to be shown is that all big jobs can be assigned as well. We will first prove that in the Inclusion Chain Case a particular condition is sufficient (Lemma 2) and then show that Y, as obtained by the algorithm, satisfies it (implied by Lemma 3).

Lemma 2. *Let Y^* be the machines on which no big job is scheduled in the optimal solution, let $\{b_1, \ldots, b_m\} := M$ such that $b_i \in M(j) \Rightarrow b_{i'} \in M(j)$ for every $j \in J_B$ and $i' < i$.*

Let $Y \subseteq M$ such that $|Y \backslash \{b_i, \ldots, b_m\}| \leq |Y^* \backslash \{b_i, \ldots, b_m\}|$ for all $i \in \{1, \ldots, m\}$. Then there is an assignment for the big jobs to $M \backslash Y$ where each machine is assigned at most one big job.

Proof. Consider an assignment of a maximum number of big jobs to $M \backslash Y$ with at most one job on each machine. Suppose towards contradiction that there is a $j \in J_B$ that could not be assigned. Assume w.l.o.g. that $M(j)$ is maximal in a sense that if a $j' \in J_B$ with $M(j') \supsetneq M(j)$ is assigned to a machine in $M(j)$ then we exchange j and j'.

Algorithm 1. Greedy Algorithm

```
1    Order  M = {b₁,...,bₘ}  such  that  bᵢ ∈ M(j) ⇒ bᵢ' ∈ M(j)
2         for all  j ∈ J_B  and  i' < i;
3    Y ← M;
4    for  i ∈ {1,...,m}  do
5       if  ω(Y - bᵢ) = ω(Y)  then
6          Y ← Y - bᵢ;
7       end
8    end
9    return  Y;
```

Let $i \in \{1, \dots, m+1\}$ such that $M(j) = \{b_1, \dots, b_{i-1}\}$. Then

$$|Y \cap M(j)| = |Y \backslash \{b_i, \dots, b_m\}| \leq |Y^* \backslash \{b_i, \dots, b_m\}| = |Y^* \cap M(j)|.$$

Thus $|(M \backslash Y) \cap M(j)| \geq |(M \backslash Y^*) \cap M(j)|$. We will now bound the number of big jobs that are assigned to $(M \backslash Y) \cap M(j)$. Since none of the jobs $j' \in J_B$ with $M(j') \supsetneq M(j)$ are assigned to $M(j)$, this number is at most

$$|\{j' \in J_B : M(j') \subseteq M(j), j' \neq j\}| \leq |(M \backslash Y^*) \cap M(j)| - 1$$
$$\leq |(M \backslash Y) \cap M(j)| - 1.$$

Hence there is a machine in $(M \backslash Y) \cap M(j)$ to which j could have been assigned.

$\square$

Lemma 3. *Let Y^* be the machines on which no big job is scheduled in the optimal solution, let $S \subseteq Y \subseteq M$ with $|Y \backslash S| > |Y^* \backslash S|$. Then there exists a machine $b \in Y \backslash (Y^* \cup S)$ with $\omega(Y - b) = \omega(Y)$.*

Proof. Let $Y' := (Y \backslash S) \cup (Y^* \cap S)$. Then

$$|Y'| = |Y \backslash S| + |Y^* \cap S| > |Y^* \backslash S| + |Y^* \cap S| = |Y^*|.$$

Let $B := Y' \backslash Y^* = Y \backslash (Y^* \cup S)$. We will show that in B there is an element b as in the lemma. Choose a Y'' with $Y' \backslash B = Y' \cap Y^* \subseteq Y'' \subsetneq Y'$ and $|Y''| = |Y^*|$.

Case 1: $\omega(Y'') = \omega(Y')$. Let $b \in B \backslash Y''$. Such a machine b exists, since otherwise $B \subseteq Y''$ and therefore $Y' = Y' \backslash B \cup B \subseteq Y''$ which contradicts its definition. Then

$$\omega(Y) - \omega(Y - b) \leq \omega(Y') - \omega(Y' - b) \tag{2}$$
$$\leq \omega(Y') - \omega(Y'') \tag{3}$$
$$= 0.$$

In (2) we use submodularity and in (3) monotonicity of ω (Fig. 3).

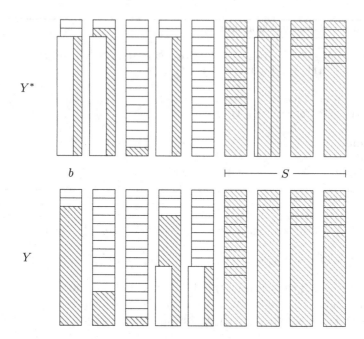

Fig. 3. Visualization of Lemma 3. Lemma 3 says that since $|Y\backslash S| > |Y^*\backslash S|$ there has to be a machine b that can be removed from Y without affecting the small jobs.

Case 2: $\omega(Y'') < \omega(Y')$. Then $\omega(Y'') < \omega(Y^*)$ and by Theorem 4 there exists a $b \in Y''\backslash Y^* \subseteq B$ with $\omega(Y'' - b) = \omega(Y'')$. With submodularity we get $\omega(Y - b) = \omega(Y)$. □

To apply the above lemmata to our algorithm, observe that after each iteration i of the algorithm we have $|Y\backslash\{b_{i+1},\ldots,b_m\}| \le |Y^*\backslash\{b_{i+1},\ldots,b_m\}|$ (∗). Suppose otherwise, Then with Lemma 3 and by setting $S := \{b_{i+1},\ldots,b_m\}$ we get: At iteration i the set S is fully contained in Y and therefore there would be a $b \in Y\backslash(Y^*\cup S) \supseteq \{b_1,\ldots,b_i\}\cap Y$ with $\omega(Y-b) = \omega(Y)$. Therefore the algorithm would have removed b from Y. With (∗) and Lemma 3 we conclude that we can obtain a 1.5-approximation from Y.

5   Two Partition Class

Consider the following case. There are $A, B \subseteq M$ with $A \dot\cup B = M$. For every $j \in J_B$ we have $M(j) = A$ or $M(j) = B$. Under these circumstances it is sufficient to find a $Y \subseteq M$ with $\omega(Y) = \omega(M)$, $|A\backslash Y| \ge |\{j \in J_B : M(j) = A\}|$ and $|B\backslash Y| \ge |\{j \in J_B : M(j) = B\}|$. Algorithm 2 does exactly this.

In the analysis we will make use of the following property of ω:

Lemma 4. *Let $T_1,\ldots,T_k \subseteq M$ and $T_i = T_{i-1} + c$ for some $c \in M\backslash T_{i-1}$ or $T_i = T_{i-1} - b$ for some $b \in T_{i-1}$ with $\omega(T_{i-1} - b) = \omega(T_{i-1})$ for all $i \in \{2,\ldots,k\}$.*
Let $d \in M\backslash T_1$. Then $\omega(T_i + d) \ge \omega(T_1 + d)$ for all $i \in \{1,\ldots,k\}$.

Algorithm 2. Greedy Algorithm with Reparation

```
1    Y ← M ;
2    for b ∈ A do
3        if ω(Y − b) = ω(Y) then
4            Y ← Y − b;
5        end
6    end
7    while |B\Y| < |{j ∈ J_B : M(j) = B}|
8        let b ∈ Y ∩ B with ω(Y ∪ A − b) = ω(M);
9        Y ← Y − b;
10       while ω(Y) < ω(M)
11           if ∃c ∈ Y ∩ B with ω(Y − c) = ω(Y) then
12               Y ← Y + b − c;
13               break;
14           else if ∃c ∈ Y ∩ A with ω(Y − c) = ω(Y) then
15               Y ← Y − c;
16           else
17               let c ∈ A\Y with ω(Y + c) > ω(Y);
18               Y ← Y + c;
19           end
20       end
21   end
22   return Y ;
```

Proof. This lemma follows directly from monotonicity and submodularity: For $i = 1$ this is trivial. Let $i \in \{2, \ldots, k\}$ with $\omega(T_{i-1} + d) \geq \omega(T_1 + d)$. If $T_i = T_{i-1} + c$ for some $c \in M \backslash T$, then by monotonicity

$$\omega(T_i + d) \geq \omega(T_{i-1} + d) \geq \omega(T_1 + d).$$

If $T_i = T_{i-1} - b$ for some $b \in T_{i-1}$ and $\omega(T_{i-1} - b) = \omega(T_{i-1})$, then by submodularity $\omega(T_{i-1} - b + d) - \omega(T_{i-1} - b) \geq \omega(T_{i-1} + d) - \omega(T_{i-1})$; Hence

$$\omega(T_{i-1} - b + d) \geq \omega(T_{i-1} + d) + \omega(T_{i-1} - b) - \omega(T_{i-1})$$
$$\geq \omega(T_1 + d) + \omega(T_{i-1}) - \omega(T_{i-1})$$
$$= \omega(T_1 + d).$$

□

Theorem 6. *The set Y as returned by Algorithm 2 satisfies*

1. $|A\backslash Y| \geq |\{j \in J_B : M(j) = A\}|$,
2. $|B\backslash Y| \geq |\{j \in J_B : M(j) = B\}|$, and
3. $\omega(Y) = \omega(M)$.

Proof. We notice that the first time line 7 is reached we have $\omega(Y) = \omega(M)$ and $|A\backslash Y| \geq |\{j \in J_B : M(j) = A\}|$. This is due to properties of the greedy algorithm which is used. Hence, the goal of lines 7–21 is to establish $|B\backslash Y| \geq |\{j \in J_B : M(j) = B\}|$ without compromising the two properties above. It is easy to see that in each iteration of the outer loop (7–21) $|B\backslash Y|$ increases by 1: The only modifications to Y that affect $B \cap Y$ are those in lines 9 and 12. Line 9 is executed exactly once in each outer loop iteration and line 12 is executed at most once. This means if the inner loop (10–20) terminates, then $B \cap Y' = B \cap Y - b$ or $B \cap Y' = B \cap Y - b + b - c = B \cap Y - c$, where Y is the state before an outer loop iteration and Y' afterwards. b and c are the machines chosen at line 8 and 11 respectively. Therefore, if the inner loop always terminates and $|A\backslash Y| \geq |\{j \in J_B : M(j) = A\}|$ and $\omega(Y) = \omega(M)$ are never compromised, then the algorithm is correct.

Fact 61. *At the end of each iteration of the outer loop (7–21), $\omega(Y) = \omega(M)$.*

Let Y_0 be the state of Y before an iteration and Y_2 after an iteration of the outer loop. We will show that if $\omega(Y_0) = \omega(M)$, then $\omega(Y_2) = \omega(M)$. Since the first time the outer loop is entered we do have $\omega(Y_0) = \omega(M)$, the claim follows directly.

If the inner loop (and with it the outer loop iteration) ends because of a violation of $\omega(Y) < \omega(M)$, we are done. In the other case the outer loop ends at line 13. Let c be as chosen in line 11 and let $Y_1 = Y_2 + c - b$. That means Y_1 is the state of Y right before line 12. Y_1 is derived from $Y_0 - b$ only by removing elements without loss (line 15) or adding elements (line 18). By Lemma 4 we know that $\omega(Y_1 + b) \geq (Y_0 - b + b) = \omega(M)$. From submodularity we get

$$0 \leq \omega(M) - \omega(Y_2) = \omega(Y_1 + b) - \omega(Y_1 + b - c) \leq \omega(Y_1) - \omega(Y_1 - c) = 0.$$

Fact 62. *At any point we have $|A\backslash Y| \geq |\{j \in J_B : M(j) = A\}|$.*

Suppose for contradiction that $|A\backslash Y| \leq |\{j \in J_B : M(j) = A\}|$ right before line 18. This is the only situation, which could break the above property. Then

$$|M\backslash Y| = |A\backslash Y| + |B\backslash Y|$$
$$\leq |\{j \in J_B : M(j) = A\}| + |\{j \in J_B : M(j) = B\}| = |J_B|.$$

This means $|Y| \geq |M| - |J_B| \geq |Y^*|$ and by Theorem 4 there is a $b \in Y\backslash Y^*$ with $\omega(Y - b) = \omega(Y)$; Hence one of the other cases (line 11 or 14) applies and line 18 was never reached.

Fact 63. *In line 8 such a machine b always exists.*

This follows directly from Lemma 3.

Fact 64. *In line 17 such a machine c always exists.*

Note that $\omega(Y) < \omega(M)$, since otherwise the inner loop would have terminated. Also note that by choice of Y (line 8–9) we have $\omega(Y \cup A) = \omega(M)$. Now suppose that for all $c \in A \backslash Y$ we have $\omega(Y + c) \le \omega(Y)$. Then by submodularity we also have $\omega(Y \cup A) \le \omega(Y) < \omega(M)$.

Fact 65. *The inner loop (10–20) terminates after a polynomial number of iterations.*

Throughout 10–20 $\omega(Y)$ never decreases and in line 18 it increases. Since $\omega(Y)$ is bounded by $|J_S|$, this means the else case in lines 16–18 can only be entered $|J_S|$ times before the loop must terminate.

Moreover, the case in lines 14–19 can only be entered $|J_B|$ times before either case 11–13 has to be entered (which terminates the loop immediately) or case 16–18 is entered. □

References

1. Chakrabarty, D., Khanna, S.: A special case of restricted assignment makespan minimization. In: 11th Workshop on Models and Algorithms for Planning and Scheduling Problems (MAPSP) (2013)
2. Chakrabarty, D., Khanna, S., Li, S.: On $(1, \epsilon)$-restricted assignment makespan minimization. In: Proceedings of the Twenty-Sixth Annual ACM-SIAM Symposium on Discrete Algorithms, SODA 2015, pp. 1087–1101. SIAM (2015). http://dl.acm.org/citation.cfm?id=2722129.2722202
3. Ebenlendr, T., Krčál, M., Sgall, J.: Graph balancing: a special case of scheduling unrelated parallel machines. In: Proceedings of the Nineteenth Annual ACM-SIAM Symposium on Discrete Algorithms, SODA 2008, Society for Industrial and Applied Mathematics, Philadelphia, PA, USA, pp. 483–490 (2008). http://dl.acm.org/citation.cfm?id=1347082.1347135
4. Huang, C., Ott, S.: A combinatorial approximation algorithm for graph balancing with light hyper edges. In: 24th Annual European Symposium on Algorithms, ESA 2016, 22–24 August 2016, pp. 49:1–49:15, Aarhus, Denmark (2016). http://dx.doi.org/10.4230/LIPIcs.ESA.2016.49
5. Jansen, K., Land, K., Maack, M.: Estimating the makespan of the two-valued restricted assignment problem. In: 15th Scandinavian Symposium and Workshops on Algorithm Theory, SWAT 2016, 22–24 June 2016, Reykjavik, Iceland, pp. 24:1–24:13 (2016). http://dx.doi.org/10.4230/LIPIcs.SWAT.2016.24
6. Jansen, K., Rohwedder, L.: On the configuration-LP of the restricted assignment problem. In: Proceedings of the Twenty-Eighth Annual ACM-SIAM Symposium on Discrete Algorithms, SODA 2017, Barcelona, Spain, 16–19 January 2017 (2017, to appear)
7. Lenstra, J.K., Shmoys, D.B., Tardos, E.: Approximation algorithms for scheduling unrelated parallel machines. Math. Program. **46**(3), 259–271 (1990). http://dx.doi.org/10.1007/BF01585745
8. Svensson, O.: Santa claus schedules jobs on unrelated machines. In: Proceedings of the Forty-Third Annual ACM Symposium on Theory of Computing, STOC 2011, pp. 617–626. ACM, New York (2011). http://doi.acm.org/10.1145/1993636.1993718

Elusiveness of Finding Degrees

Dishant Goyal[1], Varunkumar Jayapaul[2(✉)], and Venkatesh Raman[3]

[1] Indian Institute of Technology, Jodhpur, Rajasthan, India
dishant.in@gmail.com
[2] Chennai Mathematical Institute, Chennai, India
varunkumarj@cmi.ac.in
[3] The Institute of Mathematical Sciences, HBNI, Chennai, India
vraman@imsc.res.in

Abstract. We address the complexity of finding a vertex with specific (or maximum) (out)degree in undirected graphs, directed graphs and in tournaments in a model where we count only the probes to the adjacency matrix of the graph. Improving upon some earlier bounds, using adversary arguments, we show that the following problems require $\binom{n}{2}$ probes to the adjacency matrix of an n node graph:

- determining whether a given directed graph has a vertex of outdegree k (for a non-negative integer $k \leq (n+1)/2$);
- determining whether an undirected graph has a degree 0 or 1 vertex;
- finding the maximum (out)degree in a directed or an undirected graph, and
- finding all vertices with the maximum outdegree in a tournament.

A property of a simple graph is elusive, if any algorithm to determine the property requires all the relevant probes to the adjacency matrix of the graph in the worst case. So the above results imply that determining whether a directed graph has a vertex of (out)degree k (for a non-negative integer $k \leq (n+1)/2$) or an undirected graph has a vertex of degree 0 or 1 vertex are elusive properties.

In contrast, we show that one can find a maximum outdegree in a tournament using at most $\binom{n}{2} - 1$ probes. By substantially improving a known lower bound, we show that, for this problem $\binom{n}{2} - 2$ probes are necessary if n is odd, and $\binom{n}{2} - n/2 - 2$ probes are necessary if n is even. For determining the existence of a vertex with degree $k > 1$ in an undirected graph, we give a lower bound of $.42n^2$ improving on the earlier lower bound of $.25n^2$.

Keywords: Elusive · k-degree · Maximum degree · Adversary arguments

1 Introduction

The topic of the paper is on a query model which was quite popular based on a conjecture from 70's which is yet unproven. We consider simple (directed or undirected) graphs, where there are no loops and there is at most one (directed

© Springer International Publishing AG 2017
D. Gaur and N.S. Narayanaswamy (Eds.): CALDAM 2017, LNCS 10156, pp. 242–253, 2017.
DOI: 10.1007/978-3-319-53007-9_22

or undirected appropriately) edge between any pair of vertices, which can be represented using $\binom{n}{2}$ entries of a matrix. A graph property is said to be elusive (or evasive) [8] if any algorithm to determine the property requires probing all $\binom{n}{2}$ entries of the adjacency matrix in the worst case. In this model, only the number of queries is only counted, and the algorithm is allowed to do any other operations for free.

Several graph properties are known to be elusive, e.g. having a clique of size k or a coloring with k colors [2] or having atleast one edge in the graph. A property is monotone if it remains true when edges are added. For example a non-empty graph stays nonempty even after adding an edge. Rosenberg [8] conjectured that any deterministic algorithm must query at least a constant fraction of entries in the adjacency matrix in the worst case, to determine if the graph has a given non-trivial monotone graph property. This was proven by Rivest and Vuillemin [7]. A stronger version of the conjecture called Aanderaa-Karp-Rosenberg conjecture was also formulated [8], which stated that exactly $\binom{n}{2}$ probes are needed to determine whether a monotone property is elusive. The stronger Aanderaa-Karp-Rosenberg conjecture remains unproven. See [5] for more information on recent developments.

One can also define elusiveness for other problems on graphs (not necessarily properties, for example, finding a vertex with maximum degree) and for properties of directed graphs [4]. There are also non-monotone properties that are elusive. Hougardy and Wagler [3] showed that the property of being perfect, though not a monotone property, is still an elusive property. In this paper, we show that another non-monotone property, the graph having a vertex of outdegree $k \leq (n+1)/2$ in directed graphs, is elusive. This improves an earlier lower bound of $n(n-1-k)/2$ [1] for $k > 1$. For $k = (n-1)/2$, for example, the earlier bound was $n(n-1)/4$ whereas we improve the bound to $\binom{n}{2}$ for all values of $k \leq (n+1)/2$. Existence of a vertex with degree k is not a monotone property, since adding an edge in the graph may make the graph to lose all its vertices with degree exactly k.

We also address the complexity of finding a vertex of degree k in an undirected graph. We show that determining whether an undirected graph has a vertex of degree 0 or 1 is elusive, while for larger k, we could show a lower bound of $.42n^2$ improving the previous lower bound of $.25n^2$. We also show that finding the maximum degree in an undirected graph or a directed graph requires $\binom{n}{2}$ queries to the adjacency matrix.

It is known [1] that finding a maximum outdegree vertex in a tournament requires $\binom{n}{2} - 2n + 3$ queries, but it was not known whether one needs $\binom{n}{2}$ probes. We show that one can find a maximum outdegree vertex in $\binom{n}{2} - 1$ queries, and improve the lower bound to $\binom{n}{2} - 2$ when n is odd and $\binom{n}{2} - n/2 - 2$ when n is even. All our lower bounds are shown using simple, but subtle adversary arguments.

1.1 Organization of the Paper

In Sect. 2, we give some definitions and conventions and some results we use. In Sect. 3, we show results for finding degree k vertices in directed or undirected graphs. In Sect. 4, we show results for finding the maximum degree as well as a maximum degree vertex in directed or undirected graphs. In Sect. 5, we show improved lower bounds and upper bounds for finding a maximum degree vertex in a tournament. Section 6 lists some interesting open problems.

2 Definitions and Conventions

All our graphs (directed or undirected) are simple graphs that have no loops and have at most one edge between any pair of vertices. The outdegree of a vertex u in a directed graph is the number of vertices v such that there is a directed edge from u to v, and the indegree of vertex is defined analogously. Sometimes we say degree to mean outdegree when dealing with directed graphs. For a subset S of vertices of a graph, the induced subgraph on S ($G[S]$) is the graph with vertex set S, and the edge set that contains pairs of vertices of S that are edges in the graph G. In a directed graph/tournament, when we say that a vertex u wins vertex v, we mean that there is an edge directed from u to v. By the same token, we say that v looses to u. A tournament is an orientation of a complete graph, i.e. a tournament is a directed graph in which there is exactly one directed edge between every pair of vertices. A round-robin over the tournament (or a subtournament) involves finding the direction of all the edges in the (sub)tournament. A regular tournament is a tournament in which every vertex has the same indegree and the same outdegree. The following result is well known.

Lemma 1. *If a tournament on n vertices is regular, then n is odd. Furthermore, for every odd integer n, there exists a regular tournament on n vertices [6].*

The adjacency matrix of a graph on n vertices is an n by n matrix that has its (i, j)-th entry 1 whenever there is an edge (i, j) and 0 otherwise. In the case of directed graphs if (i, j) is a (directed) edge, then the (i, j)-th entry has 1 and the (j, i)-th entry has a -1. Clearly, in undirected graphs, the matrix is symmetric, and in directed graphs, the matrix is anti-symmetric.

All our proofs use an adversary argument. Here an adversary maintains the adjacency matrix, and the algorithm probes the adversary for entries of the matrix. The adversary has a fixed strategy and creates entries in the matrix, when they are queried by the algorithm. Our argument to prove a lower bound of $f(n)$ typically has the form, 'if the algorithm does not probe $f(n)$ entries, then the adversary has enough options on the rest of the graph to prove the algorithm wrong, whatever the algorithm answers'.

3   Finding Vertices of Degree k

3.1   Directed Graphs

Earlier known lower bound [1] for finding a outdegree k vertex in a directed graph is $n(n-1-k)/2$ for any $0 < k \leq n-1$. For $0 < k \leq (n+1)/2$, we improve the bound to show the following.

Theorem 1. *Any algorithm to determine whether a given directed graph has a vertex with outdegree k $(0 < k \leq (n+1)/2)$ on n vertices, requires $\binom{n}{2}$ probes to the adjacency matrix.*

Proof. The adversary constructs the digraph in following manner depending on the value of k:

1. k *is even and at least 2.* The adversary partitions the graph into two sets of vertices A and B. A contains $2(k-1)+1$ vertices. B contains $n-2k+1$ vertices. Adversary maintains a regular tournament (that exists by Lemma 1) on A (each vertex has outdegree and indegree exactly $(k-1)$) and the induced subgraph on B is empty (edgeless). There is no edge directed from A towards B. For every vertex $x \in B$, the adversary would add directed edges from x to the first $k-1$ vertices of A it is probed with. This way, it is clear that there is no vertex in the graph with outdegree k.
   In order to determine that no vertex in A is a outdegree-k vertex, the algorithm is required to probe at least $n-k$ non-outdegree edges of each vertex in A. This would make any algorithm to probe the entire subtournament on A and every edge between A and B. For otherwise, the adversary can flip/add an edge out of that vertex to make that vertex of degree k consistent with its answers for all vertices. To prove that no vertex in B is an outdegree k vertex, the algorithm has to probe all pairs within B, as otherwise, the adversary can add an edge to make a vertex of outdegree k.
2. k *is odd and $k \geq 3$.* The adversary partitions the graph into two sets A and B by taking $2k-2$ vertices in A and the rest in B. Adversary maintains a regular subtournament A' inside A on $2k-3$ vertices (each vertex within A' has indegree and outdegree $(k-2)$) and the remaining 1 vertex z of A will have all edges directed to the vertices of the regular subtournament. The rest of the construction for edges across A and B and within B are the same as in the previous case.
   In order to claim that no vertex in A' is a degree-k vertex, the algorithm is required to probe at least $n-k$ non-outdegree edges of each vertex in A'. This would make the algorithm to probe the entire subtournament on A' and every edge between A' and $B \cup \{z\}$. Although z already may have degree more than k, queries with z are required to verify whether any vertex in B or A' can get degree k or not.
   As in the previous case, the algorithm has to probe all pairs within B to prove that no vertex in B is an outdegree k vertex.

3. $k = 1$. The adversary maintains an empty (edgeless) graph and answers the queries according to it. If the algorithm leaves some edge unprobed between a pair of vertices, say u and v, then following two cases happen:

   (a) If the algorithm declares any vertex other than u and v as 1-outdegree vertex, then adversary would trivially show that the algorithm is wrong, as it has never given any edge, till this point.

   (b) If the algorithm declares u (without loss of generality) as a 1-outdegree vertex, then adversary would add an edge $v \to u$, making vertex v the unique vertex of outdegree 1 proving the algorithm wrong. □

For $k = 0$, we improve the earlier known lower bound [1] of $n^2/4$ to show the following.

Theorem 2. *Determining whether a given directed graph on n vertices has a vertex with outdegree 0 requires $\binom{n}{2}$ probes to the adjacency matrix.*

Proof. For $n = 2$, it is straightforward to see that the only edge in the graph has to be probed, to determine whether any of the two vertices has outdegree 0. For $n \geq 3$, the adversary strategy is as follows: For the first query between two vertices, say x and y, the adversary answers that the graph has an edge $y \to x$ and adds y to a set W which is a set maintained by the adversary, that contains all vertices with outdegree at least 1. The vertex x is now marked as a sink vertex s. The label of the sink vertex is not fixed and may change over the course of the algorithm's execution.

If the algorithm asks a query (u, w) for a vertex $w \in W$, and $u \notin W$, and $u \neq s$, the adversary answers that there is no edge unless w is the only vertex of W with which u has not been probed, in which case it directs the edge from u to w and adds u to W. For a query involving a pair of vertices (u, v), $u, v \notin W \cup \{s\}$, the adversary answers that there is no edge.

For a query between sink s and y, the adversary answers that there is no edge unless y is the last vertex (not necessarily in W) with which sink s has not been probed, in which case the adversary gives an edge from s to y and adds s to W. If in the process y gets indegree 1 (and outdegree 0) then y plays the role of new sink and marked as s, otherwise there is no sink. This strategy allows the adversary to maintain the following invariants:

- The set W contains vertices whose outdegree so far is exactly one.

  Proof. Each vertex gets its first outdegree while entering W, after which point it only get indegrees.

- Every probe between vertices in W has been already made by the algorithm.

  Proof. If there is an unprobed edge between two vertices u and v, where $v \in W$, then u will not get an outdegree edge, and thus it would not be present in W.

- There is at most one vertex outside W, referred to as sink s (if it exists) that has indegree exactly one.

- Vertices other than those in $W \cup \{s\}$ have outdegree and indegree 0.

It is clear that all the above invariants are maintained by the answers of the adversary. The adversary never gives an edge, if the edge between two vertices both of which are outside W is made. Thus, to confirm whether a vertex has 0 outdegree or not, it has to compare with vertices in W. The only time when the algorithm can discard a vertex for a candidate of 0-degree vertex, is when it is added to W, after it gets its first outdegree.

Suppose an algorithm has not queried an edge between vertices u and v, and it answers that $w(\neq u, v)$ is a 0-outdegree vertex. If $w \notin W$, then there is at least one vertex (say w') with which it has not made a query, and its outdegree thus far is 0. So the adversary can give an edge $w \rightarrow w'$ and prove the algorithm wrong. If $w \in W$, then the algorithm is wrong, since all vertices in W have at least one outdegree.

If the algorithm outputs u as a 0-outdegree vertex (without loss of generality), then the adversary can give the edge $u \rightarrow v$ and contradict the algorithm. Thus all pairs of queries have to be made, to determine whether there is any vertex with outdegree 0.   □

For $k > (n+1)/2$, we improve the known lower bound of $n(n-k-1)/2$ to show the following.

Theorem 3. *For $k > (n+1)/2$, determining whether there is a vertex with outdegree k in a directed graph on n vertices requires at least $(n-k-1)(n+k)/2$ probes to the adjacency matrix.*

Proof. The adversary maintains two sets A and B. A contains $k+1$ vertices, with no edge between any pair of vertices within A. B contains $n-k-1$ vertices, with no edge between any pair of vertices within B. For any vertex in B, the adversary will direct the first $k-1$ vertices of A with which it is probed towards A, and will answer that there is no edge with the remaining two vertices of A (if probed). The algorithm is required to probe all edges out of each vertex of B to prove that none of them is of degree k (as otherwise, the adversary can create edges to prove the algorithm wrong). The number of queries probed is $(k+1)(n-k-1) + (n-k-1)(n-k-2)/2 = (n-k-1)(n+k)/2$.   □

3.2 Undirected Graphs

Similar to the adversary for determining outdegree 1 vertex in a directed, graph, we can show the following by maintaining an adversary that doesn't give any edge between pairs of vertices. If some edge is unprobed, then the adversary can keep the option of making both of them a degree 1 vertex.

Theorem 4. *Determining whether a given undirected graph has a degree 1 vertex is an elusive property.*

For degree 0, the adversary is similar to the adversary for outdegree 0 in the case of directed graphs. We show the following theorem.

Theorem 5. *Determining whether an undirected graph on n vertices has a degree 0 vertex requires all $n(n-1)/2$ probes to the adjacency matrix of the graph.*

Proof. For the first query between two vertices, adversary would add an undirected edge and place both of them in a set W. After that, if the algorithm asks a query (u, w) for a vertex $w \in W$, and $u \notin W$, the adversary answers that there is no edge unless w is the only vertex of W with which u has not been probed, in which case it gives an edge between u and w and adds u to W. For a query involving a pair of vertices (u, v), $u, v \notin W$, the adversary answers that there is no edge.

Thus the adversary maintains that

- vertices in W have degree at least 1,
- all pairs of vertices in W have been probed, and
- all vertices outside W have degree 0.

So the algorithm can not declare any vertex of W as one of degree 0 as it will be wrong otherwise. If $V(G) \setminus W \neq \emptyset$, and if the algorithm declares a vertex of $V(G) \setminus W$ as a vertex of degree 0, then the algorithm can make the degree of that vertex at least 1, as that vertex has still some edge unprobed (as otherwise it will be in W). If the algorithm declares that there is no vertex of degree 0, then $W = V$, and in that case, all edges between pairs of vertices in $W = V$ have been probed due to the invariant maintained by adversary.   □

For $k > 1$, we are unable to show whether the property is elusive or not, but we improve the known lower bound from $n^2/4$ [1] to show the following.

Theorem 6. *Finding a degree k vertex in an undirected graph on n vertices requires at least $(0.42)n^2$ probes to the adjacency matrix of the graph.*

Proof. Without loss of generality we can assume that $k \leq \lfloor (n-1)/2 \rfloor$. For $k \geq (n-1)/2$, one can simply find a vertex with degree $n - 1 - k$ in the complement graph which can be obtained online from the adjacency matrix by interpreting 1s as 0s and vice versa. Adversary has a choice to answer in one of following two ways, which it decides based on the value of k:

1. $k \leq (3 - \sqrt{5})n/2$: Adversary constructs the graph as follows:
   The vertices are divided into two sets A and B. A consists of $k + 2$ vertices. B consists of $n - k - 2$ vertices. Each vertex of B has an edge with every vertex of A. The subgraph of the set B contains no edge and the subgraph of the set A is complete and hence every vertex of A has degree at least $k + 1$. For each vertex in B, adversary would give edges to the first $k - 1$ vertices of A with which it is probed, and answer the last three edges with set A as empty, to make each vertex of B of degree $k - 1$. To prove that no k degree vertex exists in B, the algorithm would probe every edge of subgraph B.
   Hence the algorithm has to probe all edges incident on vertices of B which is $(n - k - 2)(k + 2) + (n - k - 2)(n - k - 3)/2 = (n - k - 2)(n + k + 1)/2$ probes.
2. $k > (3 - \sqrt{5})n/2$: Adversary constructs the graph as follows:
   The graph is divided into two sets of vertices A and B. A contains $n - k + 1$ vertices and B contains $k - 1$ vertices. The subgraph on each of the sets A

and B is complete. Each vertex of B has degree at least $k - 2$ and each vertex of A has degree of at least $k + 1$ (because $k \leq \lfloor (n - 1)/2 \rfloor$).

In order to claim that some vertex $x \in B$ is of degree $< k$, the algorithm is required to probe between x and at least $n - k$ vertices of A. Adversary would add an edge between x and the last 3 vertices of A it is probed with. This would make x a degree $k + 1$ vertex. Adversary would do the same for each such vertex of B. Now, the algorithm has to probe the whole subgraph on B, to prove that none of the vertex in B is a degree k vertex.

In order to prove that none of vertex in A is a degree k vertex, the algorithm would probe at least $k + 1$ edges of each vertex of A. As there can be at most $3k - 3$ edges between A and B, the algorithm is required to probe the subgraph on A for the remaining edges. Algorithm would stop, when each vertex of A has degree of $k + 1$. Suppose e is the number of edges that algorithm is forced to probe inside the set A, to make each vertex of A degree $k + 1$. Then, $2e + 3(k - 1) \geq (n - k + 1) * (k + 1)$ and so $e \geq (n - k + 1)(k + 1)/2 - 3(k - 1)/2$. Thus the algorithm would probe at least $(k - 1)(n - k + 1) + (k - 1)(k - 2)/2 + (n - k + 1)(k + 1)/2 - (3k - 3)/2 = (k(3n - 2k - 2) - n + 4)/2$ probes.

The minimum value of complexity of both strategies is when $k = (3 - \sqrt{5})n/2$, which gives a lower bound of $.42n^2$ probes. □

4   Finding Maximum Degrees

4.1   Directed Graphs

The adversary for $k = 1$ in Theorem 1 also gives the following

Theorem 7. *For a directed graph on n vertices, any algorithm would require to probe at least $n(n - 1)/2$ queries to the adjacency matrix to find a maximum outdegree vertex.*

The same adversary implies that finding the maximum outdegree (not just a vertex with the maximum outdegree) requires $\binom{n}{2}$ probes.

Corollary 1. *Finding the maximum outdegree in a directed graph on n vertices, requires all $n(n - 1)/2$ queries to the adjacency matrix.*

4.2   Undirected Graphs

The adversary strategy used to show that finding a degree 1 vertex in an undirected graph is an elusive property (Theorem 4) shows the following.

Theorem 8. *Finding the maximum degree in an undirected graph requires all $n(n - 1)/2$ edge queries in the worst case.*

However, if we don't care about the maximum degree, and is sufficient to find a vertex with maximum degree, then, we can show the following.

Theorem 9. *Finding a maximum degree vertex in an undirected graph on n vertices, has a lower and upper bound of $n(n-1)/2 - 1$.*

Proof. Lower bound: Whenever a query comes for any pair of vertices x and y, the adversary answers that there is no edge between x and y, except when all of the remaining unprobed edges share the same vertex, in which case the adversary would work as follows: If the query between vertices u and v is the last query, after which all queries will have a common vertex x, then adversary would add an edge (u, v) to the graph, and not give any more edges to x. Now, the algorithm has to probe all the edges of x except the last one, to prove that vertex u or v is indeed a maximum degree vertex. For, if the algorithm omits two edges out of x, the adversary can make x the unique maximum degree vertex.

If such a case does not happen i.e., if there exist two unprobed edges that do not share any vertex, say u_1v_1 and u_2v_2. Then the following two cases arise:

> **Case 1:** If the algorithm declares any vertex x other than u_1,v_1,u_2 and v_2 as a maximum outdegree vertex, then adversary would add an edge u_1v_1 to make vertex u_1 as the maximum degree vertex, contradicting the algorithm's claim.
>
> **Case 2:** Without loss of generality, if the algorithm declares vertex u_1 as maximum degree vertex, then the adversary would add edge u_2v_2 to make vertex u_2 of degree 1, contradicting the algorithm's claim.

Upper bound: The algorithm would probe all edges except the last edge, say between u and v. If there exists some vertex x with degree larger than that of any other vertex including u and v, then the algorithm declares x as a maximum degree vertex, as both u and v cannot get degree more than that of x, even if edge uv exists.

Without loss of generality, if u has degree greater or equal to that of any other vertex, then the algorithm would declare u as a maximum degree vertex (since presence or absence of an edge between u and v won't affect the fact that u has maximum degree). If both u and v are of the largest degree, then either of u and v can be output by algorithm as a maximum degree vertex. ☐

5   Tournaments

A tournament is a directed graph in which there is exactly one directed edge between every pair of vertices. Balasubramanian et al. [1] gave a $2kn$ lower bound and $4kn$ upper bound for finding a vertex of outdegree k ($0 < k \le (n-1)/2$). Bridging this gap is still open for general value of k. In this section we deal with finding the maximum outdegree vertex in a tournament. The previous known lower bound for finding a maximum outdegree vertex in tournament was $\binom{n}{2} - 2n + 3$ [1]. We give an improved lower bound and also show that such a vertex can be found without probing all the edges in the tournament.

5.1   Lower Bound for Finding Maximum Outdegrees

Theorem 10. *Finding a vertex with the maximum outdegree in a tournament on n vertices requires $\binom{n}{2} - 2$ edge-probes if n is odd and $\binom{n}{2} - n/2 - 2$ edge-probes if n is even.*

Proof. When n is odd, the adversary starts by maintaining a regular tournament on n vertices where each vertex has outdegree and indegree $(n-1)/2$. Our first claim is the following.

Claim. Suppose the algorithm declares a vertex x as one with maximum outdegree. Then the algorithm must have ensured that the remaining $(n-1)$ vertices have indegree at least $(n-1)/2$.

Otherwise, the adversary can flip an unprobed edge coming into one of these vertices, say y, and make y the unique vertex with maximum outdegree proving the algorithm wrong. 0 This claim already shows a lower bound of $(n-1)$ $(n-1)/2$. We improve the bound further by modifying the adversary as follows. The adversary keeps track of vertices with indegree $(n-1)/2$, and when the $(n-1)^{th}$ vertex z is about to get indegree $(n-1)/2$, the adversary flips the edge coming into z. Let y be the nth vertex, now every vertex other than y and z are known to have indegree $(n-1)/2$ and z has indegree $(n-1)/2 - 1$. Now the adversary will continue to answer according to its initial choice of orientations (of the regular tournament) making z the vertex with the unique maximum outdegree. But to rule out y as a candidate, the algorithm has to ensure that y has indegree at least $(n-1)/2 - 1$ (as otherwise it can flip an edge coming into y to make it a unique vertex with maximum outdegree). All these indegree edges are disjoint, but it is possible that the edge that was going to come into z that got flipped was actually from y counting for the indegree of y. Hence, the algorithm has to perform at least $(n-1)(n-1)/2 + (n-1)/2 - 2 = (n-1)n/2 - 2$ comparisons.

Suppose n is even. The adversary constructs a tournament that has two sets of vertices: X is a set of $n/2$ vertices with outdegree $n/2$ each and Y is a set of $n/2$ vertices with outdegree $n/2 - 1$ each. Before declaring v which is a vertex of X as a vertex with maximum outdegree, the algorithm has to ensure that every other vertex has indegree at least $n/2 - 1$ as otherwise the adversary can make one of those vertices to have outdegree $n/2 + 1$ by flipping an unprobed incoming edge to one of them, thus contradicting the algorithm. This proves a lower bound of $(n-1)(n/2 - 1)$. As before, we further improve the lower bound as follows. The algorithm flips the edge coming into the $(n-1)^{th}$ vertex (say u) that is about to get indegree $n/2 - 1$ so that it and the nth vertex (say v) are candidates for maximum outdegree. Now the adversary answers in such a way that both u and v dont get more than $n/2 + 1$ outdegree, thus the algorithm has to ensure that both get indegree at least $n/2 - 2$. This, as before, results in a lower bound of $(n-2)(n/2 - 1) + 2(n/2 - 2) = n(n-1)/2 - n/2 - 2$.   □

Theorem 11. *Finding the maximum outdegree (along with a vertex with the maximum outdegree) in a tournament on n vertices requires at least $\binom{n}{2}$ comparisons if n is odd, and $\binom{n}{2} - n/2$ comparisons if n is even.*

Proof. It is evident that one cannot find out the maximum outdegree of a graph, without actually knowing which vertex has the maximum outdegree, although the converse is not true. If n is odd, the adversary works with a regular tournament, and if some edge is not probed, it can flip the edge and change the maximum outdegree.

If n is even, the adversary works with a tournament where $n/2$ vertices have outdegree $n/2$, and other $n/2$ vertices have outdegree $n/2 - 1$. Any algorithm which declares the maximum outdegree(which is $n/2$ in this case) along with the vertex which has that degree (say u) has to ensure that all the vertices including u have indegree at least $n/2 - 1$, because if a vertex, say v, with indegree less than $n/2 - 1$ exists, then the adversary will make v have outdegree $n/2 + 1$, and thus contradict the algorithm.   □

Corollary 2. *Any algorithm to determine all vertices with the maximum outdegree requires $\binom{n}{2}$ edge-probes.*

Proof. For odd sized graph, it follows from Theorem 11. For even sized tournaments, the adversary is similar to the one shown in Theorem 11. There are $n/2$ maximum degree vertices and each of the $n/2$ maximum outdegree vertices must have their all edges probed, otherwise the adversary could alter any edge to prove the algorithm wrong. Also, the algorithm needs to probe all indegrees of the remaining $n/2$ vertices to prove that none of them is a max-outdegree vertex. Thus algorithm has to probe all indegree edges of each vertex of the tournament.   □

5.2   Upper Bound for Finding a Maximum Outdegree

Theorem 12. *Finding a vertex with the maximum outdegree in a tournament on n vertices, requires at most $\binom{n}{2} - 1$ edge-probes for $n > 2$.*

Proof. Consider the subtournament T on some $n-1$ vertices and perfom a round robin tournament by making all possible comparisons between them taking $(n-1)(n-2)/2$ edge-probes. Let x be the remaining vertex. Two possible cases arise for T.

1. T is a regular subtournament: In this case, every vertex in T has outdegree/indegree $(n-2)/2$.
   Choose an arbitrary vertex y of T. Probe x with all vertices of T except y. We will argue that we have sufficient information to determine a vertex with the maximum outdegree. Let z be a vertex in $T - \{y\}$, which is a vertex with the maximum degree among vertices in T. At this point,
   (a) if z lost to x, then x is a maximum outdegree vertex.
   (b) if z won against x and outdegree of x is less than the outdegree of z, then z is a maximum outdegree vertex.
   (c) if z won against x and outdegree of x is at least the outdegree of z, then output x as the maximum outdegree vertex.

2. T is not a regular subtournament:- Let y be a vertex of T with minimum outdegree in T, which is at most $n/2 - 2$. Probe x with all vertices of T except y. Let z be a vertex in $T - \{y\}$, which is the vertex with the maximum outdegree among vertices of $T - \{y\}$. At this point,
   (a) if outdegree of x is less than outdegree of z, then output z as the maximum outdegree vertex.
   (b) if outdegree of x is the same or more than outdegree of z, then output x as the maximum outdegree vertex.

The correctness of the algorithm is clear, and in both cases, we have probed at most $n(n - 1)/2 - 1$ edges.   □

Note that the gap between the current upper bound (Theorem 12) and the lower bound (Theorem 10) for finding a maximum degree vertex is 1 when n is odd and is $n/2 + 2$ when n is even.

6   Conclusions and Open Problems

We have addressed the complexity of finding vertices of degree k or maximum degree in a directed or an undirected graph or a tournament in a model where only probes to the adjacency matrix are counted. While our adversaries are simple, they are still non-trivial and they narrow the gap between upper and lower bounds substantially. We end with the following open problems.

1. Is $n(n - 1)/2 - 1$ the optimal bound for finding a maximum degree vertex in an odd sized tournament and is $n(n - 1)/2 - n/2 - 1$ the optimal bound for finding a maximum degree vertex in an even sized tournament? We believe that the answer is yes when n is odd and no for n is even. For example, when $n = 4$, we can find a vertex with the maximum outdegree in three probes (simply by declaring the winner of a knock-out tournament). Extending this argument/bound for larger even sized tournaments would be interesting.
2. Is determining whether an undirected graph has a degree k vertex an elusive property for $k > 1$?

References

1. Balasubramanian, R., Raman, V., Srinivasaragavan, G.: Finding scores in tournaments. J. Algorithms **24**(2), 380–394 (1997)
2. Bollabas, B.: Complete subgraphs are elusive. J. Comb. Theor. B **21**, 1–7 (1976)
3. Hougardy, S., Wagler, A.: Perfectness is an elusive graph property. SIAM J. Comput. **34**(1), 109–117 (2004)
4. King, V.: A lower bound for the recognition of digraph properties. Combinatorica **10**(1), 53–59 (1990)
5. Miller, C.A.: Evasiveness of graph properties and topological fixed-point theorems. Found. Trends Theoret. Comput. Sci. **7**(4), 337–415 (2011)
6. Moon, J.W.: Topics on Tournaments. Reinhart and Winston, Holt (1968)
7. Rivest, R.L., Vuillemin, J.: On recognizing graph properties from adjacency matrices. Theoret. Comput. Sci. **3**(3), 371–384 (1976)
8. Rosenberg, A.L.: On the time required to recognize properties of graphs: a problem. SIGACT News **5**(4), 15–16 (1973)

Maximum Weighted Independent Sets
with a Budget

Tushar Kalra, Rogers Mathew$^{(\boxtimes)}$, Sudebkumar Prasant Pal, and Vijay Pandey

Department of Computer Science and Engineering, Indian Institute of Technology,
Kharagpur 721302, West Bengal, India
tushar11nitjkalra@gmail.com , rogersmathew@gmail.com, sudebkumar@gmail.com,
vijayiitkgp13@gmail.com

Abstract. Given a graph G, a non-negative integer k, and a weight function that maps each vertex in G to a positive real number, the *Maximum Weighted Budgeted Independent Set (MWBIS) problem* is about finding a maximum weighted independent set in G of cardinality at most k. A special case of MWBIS, when the weight assigned to each vertex is equal to its degree in G, is called the *Maximum Independent Vertex Coverage (MIVC)* problem. In other words, the MIVC problem is about finding an independent set of cardinality at most k with maximum coverage.

Håstad in [Clique is hard to approximate within $n^{1-\epsilon}$. *Foundations of Computer Science*, 1996.] showed that there is no $\frac{1}{n^{1-\epsilon}}$-factor approximation algorithm for the well-known Maximum Weighted Independent Set (MWIS) problem, where $\epsilon > 0$, assuming NP-hard problems have no randomized polynomial time algorithms. Being a generalization of the MWIS problem, the above-mentioned inapproximability result applies to MWBIS too. Due to the existence of such inapproximability results for MWBIS in general graphs, in this paper, we restrict our study of MWBIS to the class of bipartite graphs. We show that, unlike MWIS, the MIVC (and thereby the MWBIS) problem in bipartite graphs is NP-hard. Then, we show that the MWBIS problem admits an easy, greedy $\frac{1}{2}$-factor approximation algorithm in the class of bipartite graphs, which matches the integrality gap of a natural LP relaxation. This rules out the possibility of any LP-based algorithm that uses the natural LP relaxation to yield a better factor of approximation.

Keywords: Independent set · Partial vertex cover · Coverage · Approximation algorithm · NP-hard · Inapproximability

1 Introduction

1.1 Problem Definition

Let G be a graph and let $w : V(G) \to \mathbb{R}^+$ be a function that assigns positive real numbers as weights to the vertices of G. Under this assignment of weights, for any set $S \subseteq V(G)$, we define the weight of S, denoted by $w(S)$, as the sum

© Springer International Publishing AG 2017
D. Gaur and N.S. Narayanaswamy (Eds.): CALDAM 2017, LNCS 10156, pp. 254–266, 2017.
DOI: 10.1007/978-3-319-53007-9_23

of the weights of the vertices in S. The famous *Maximum Weighted Independent Set (MWIS) problem* is about finding an independent set of vertices in G that has the highest weight amongst all the independent sets in G. In this paper, we study a budgeted version of the well-studied MWIS problem, namely *Maximum Weighted Budgeted Independent Set* (MWBIS) problem.

Definition 1. *Given a graph G, a weight function $w : V(G) \rightarrow \mathbb{R}^+$, and a positive integer k, the* MWBIS *problem is about finding an independent set of size at most k in G that has the highest weight amongst all independent sets of size at most k in G.*

1.2 Related Work

A more general problem (also known by the same name, MWBIS,) was introduced and studied in the context of special graphs like trees, forests, cycle graphs, interval graphs, and planar graphs in [5] where each vertex in the given graph G has a cost associated with it and the problem is about finding an independent set of total cost at most C, where C is a part of the input, in G that has the highest weight amongst all such independent sets. Apart from this work, to the best of our knowledge, not much is known about MWBIS.

Given a graph G, we know that the *Vertex Cover (VC) problem* is about finding the minimum number of vertices that cover all the edges of G. Several variants of the VC problem has been studied in the literature. We discuss about a couple of them here. For a positive integer t, the *Partial Vertex Cover (PVC) problem* is about finding the minimum number of vertices that cover at least t distinct edges of G. In the year 1998, Burroughs and Bshouty introduced and studied the problem of partial vertex cover [9]. In this paper, the authors gave a 2-factor approximation algorithm by rounding fractional optimal solutions given by an LP relaxation of the problem. Bar-Yehuda in [6] came up with another 2-approximation algorithm that relied on the beautiful 'local ratio' method. A primal-dual algorithm achieving the same approximation factor was given in [11]. In [10], it was shown that the PVC problem on bipartite graphs is NP-hard.

Another popular variant of the VC problem is the *Maximum Vertex Coverage* (MVC) problem. Given a graph G and a positive integer k, the *MVC problem* is about finding k vertices that maximize the number of distinct edges covered by them in G. Ageev and Sviridenko in [1] gave a 3/4-approximation algorithm for the MVC problem. An approximation algorithm, that uses a semidefinite programming technique, based on a parameter whose factor of approximation is better than 3/4 when the parameter is sufficiently large was shown in [14]. Apollonio and Simeone in [3] proved that the MVC problem on bipartite graphs is NP-hard. The same authors in [2] gave a 4/5 factor approximation algorithm for MVC on bipartite graphs that exploited the structure of the fractional optimal solutions of a linear programming formulation for the problem. The authors of [10] improved this result to obtain an 8/9 factor approximation algorithm for MVC on bipartite graphs.

1.3   MIVC Problem - A Special Case of MWBIS

Let us come back to our problem - the MWBIS problem. In this problem, what happens if the weight function given maps each vertex to its degree in G? That is, let $w(v) = deg(v)$, $\forall v \in V(G)$. We call this the *Maximum Independent Vertex Coverage (MIVC) problem.*

Definition 2. *Given a graph G, a weight function $w : V(G) \to \mathbb{R}^+$ defined as, for each vertex $v \in V(G)$, $w(v) = deg(v)$, and a positive integer k, the MIVC problem is about finding an independent set of size at most k in G that has the highest weight amongst all independent sets of size at most k.*

Observe that the MIVC problem, as its name suggests, can also be seen as a variant of the MVC problem where the k vertices that we choose need to be an independent set. This observation gives us the following alternate definition.

Definition 3. *Given a graph G and a positive integer k, the MIVC problem is about finding at most k independent vertices that maximize the number of edges covered by them in G.*

1.4   IP Formulation of MWBIS

Let G be a graph on n vertices, where $V(G) = \{v_1, \ldots, v_n\}$, and let $w : V(G) \to \mathbb{R}^+$ be a weight function given. Let k be a positive integer given. Let $\mathcal{C}$ denote the set of all maximal cliques in G. In order to formulate MWBIS on G as an integer program, let us assign a variable x_i for each $v_i \in V(G)$, which is allowed $0/1$ values. The variable x_i will be set to 1 if and only if the vertex v_i is picked in the independent set.

$$Maximize \sum_{i \in [n]} w(v_i) \cdot x_i \tag{1}$$

$$s.t. \sum_{i=1}^{n} x_i \le k$$

$$\sum_{i : v_i \in C} x_i \le 1, \quad \forall C \in \mathcal{C}$$

$$x_i \in \{0, 1\}, \quad \forall i \in [n].$$

The constraint $\sum_{i=1}^{n} x_i \le k$ ensures that not more than k vertices are picked. The constraint $\sum_{i : v_i \in C} x_i \le 1$, for each maximal clique $C \in \mathcal{C}$, ensures that the set of selected vertices is an independent set. Replacing $w(v_i)$ with $deg(v_i)$ in the objective function of (1) yields the IP formulation of MIVC for G.

Later in Sect. 2.3, we consider the LP-relaxation of this integer program. We show an instance of the MIVC problem (a bipartite graph) to illustrate the fact that the integrality gap of this LP relaxation is not 'good'. This rules out the possibilities of using this relaxation to obtain good LP-based approximation algorithms for the MIVC problem (and thereby for the MWBIS problem) in bipartite graphs.

1.5   Hardness Results

It is well-known that the MWIS problem cannot be approximated to a constant factor in polynomial time, unless $P = NP$. Håstad [15] showed that there is no $\frac{1}{n^{1-\epsilon}}$-factor approximation algorithm for MWIS, where $\epsilon > 0$, assuming NP-hard problems have no randomized polynomial time algorithms. For every sufficiently large Δ, there is no $\Omega(\frac{\log^2 \Delta}{\Delta})$-factor polynomial time approximation algorithm for MWIS in a degree-Δ bounded graph, assuming the unique games conjecture and $P \neq NP$ [4]. As MWBIS is a generalization of MWIS, these results hold true for the MWBIS problem too. But, what about the special case of MWBIS - the MIVC problem? Below, we show that it is hard to approximate MIVC within a certain factor. We prove this by giving an approximation factor preserving reduction from another problem, namely 3-Maximum Independent Set (3-MIS) problem. The 3-*MIS problem* is about finding an independent set of maximum cardinality in a given 3-regular graph.

Theorem 1. *There is no $\left(\frac{139}{140} + \epsilon\right)$-factor approximation algorithm for the MIVC problem, assuming $P \neq NP$, where $\epsilon > 0$.*

Proof. It was shown in (Statement (v) of Theorem 1 in) [7] that there is no $\left(\frac{139}{140} + \epsilon\right)$-factor approximation algorithm for the 3-MIS problem, assuming $P \neq NP$, where $\epsilon > 0$. If there existed a polynomial time f-factor approximation algorithm $A(G, k)$ for the MIVC problem, where $f > 139/140$, then we could use it as an f-factor approximation algorithm for the 3-MIS problem as follows. Given a positive integer k and a 3-regular graph G on n vertices as input, we run $A(G, n)$. This algorithm will return an independent set of cardinality at least f times the size of a maximum independent set in G.

We define decision versions of the MWBIS and the MIVC problems as:-

D-MWBIS = {$<G, w, k, t>$ | G is an undirected graph, $w : V(G) \to \mathbb{R}^+$ is a
weight function and G contains an independent set of cardinality
at most k whose weight is at least t}.

D-MIVC = {$<G, k, t>$ | G is an undirected graph and G contains a set of
at most k independent vertices that cover at least t distinct
edges}.

Since the MIS problem and the VC problem are known to be NP-hard, it is not surprising to see that the D-MIVC (and thereby the D-MWBIS) problem is NP-hard for general graphs. The following is an easy corollary to Theorem 1

Corollary 1. *D-MIVC problem is NP-hard.*

1.6   Motivation

The main motivation behind studying the MWBIS problem is the fact that it is a natural generalization of a very well-studied problem in the context of graph algorithms and approximation algorithms, the MWIS problem. MWIS problem finds application in wireless networks, scheduling, molecular biology, pattern recognition, coding theory, etc. MWBIS finds application in most scenarios where MWIS is used. Refer [5] to know about applications of MWBIS in 'job scheduling in a computer' and in 'selecting non-interfering set of transmitters'.

1.7   Our Contribution

We know that the MWIS problem on bipartite graphs is polynomial time solvable (since a minimum weighted vertex cover in a bipartite graph can be found in polynomial time using 'maximum flow' techniques). Intuitively, the MWBIS problem, being a budgeted variant of the MWIS, is also expected to follow the same behaviour. But, contrary to our intuition, we prove the following theorem in Sect. 2.1.

Theorem 2. *D-MIVC problem on bipartite graphs is NP-hard.*

This motivated us into looking at approximation algorithms. In Sect. 2.2, we give an $O(nk)$ time, $\frac{1}{2}$-factor greedy approximation algorithm for the MWBIS problem on bipartite graphs. We give a tight example to the algorithm. In Sect. 2.3, we consider the LP relaxation of the integer program for MWBIS given in Sect. 1.4. We show that the integrality gap for this LP is upper bounded by $\frac{1}{2} + \epsilon$ for bipartite graphs, where ϵ is any number greater than 0. In other words, no LP-based technique, that uses this natural LP relaxation of MWBIS, is going to give us a better factor approximation algorithm for the MWBIS problem on bipartite graphs.

1.8   Notational Note

Throughout the paper, we consider only finite, undirected, and simple graphs. For a graphs G, we shall use $V(G)$ to denote its vertex set and $E(G)$ to denote its edge set. For any $S \subseteq V(G)$, we shall use $|S|$ to denote the cardinality of the set S. If a weight function $w : V(G) \to \mathbb{R}^+$ is given, then $||S||$ shall be used to denote the sum of the weights of the vertices in S. Otherwise, $||S||$ will denote the number of edges in G having at least one endpoint in S. For any vertex v in the graph under consideration, we shall use $deg(v)$ to denote its degree in the graph. For any positive integer n, we shall use $[n]$ to denote the set $\{1, \ldots, n\}$.

2   MWBIS in Bipartite Graphs

We begin by first showing that D-MIVC on bipartite graphs is NP-hard in Sect. 2.1. We give a 1/2 factor approximation algorithm for MWBIS in Sect. 2.2 and show in Sect. 2.3 that LP-based methods are unlikely to yield better approximation algorithms.

2.1 Proof of Theorem 2

We show that D-MIVC problem on bipartite graphs is NP-hard by reducing from $(n-4, k)$-CLIQUE problem. The $(n-4, k)$-CLIQUE problem is the famous clique problem on a graph, where the graph under consideration is an $(n-4)$-regular graph on n vertices.

$(n-4, k)$-CLIQUE = $\{<G, k> \mid G$ is an $(n-4)$-regular graph on n vertices that contains a clique of size at least k, where $k < \frac{n}{2}\}$.

The $(3, k)$-Independent Set $((3, k)$-IS) problem is about deciding whether there exists an independent set of size at least k or not in a given 3-regular graph. It was shown in [12] that the $(3, k)$-IS problem is NP-hard. This means that the $(n-4, k)$-CLIQUE problem is also NP-hard. Note that when $k \geq \frac{n}{2}$, the $(3, k)$-IS problem is about deciding whether the given graph is bipartite or not and is therefore polynomial-time solvable. Below we outline our reduction from the $(n-4, k)$-CLIQUE problem to D-MIVC.

Construction 3. *Let n be a positive integer greater than 11 and let $r = n - 4$. Given an r-regular graph G with $V(G) = \{v_1, \ldots, v_n\}$ and $E(G) = \{e_1, \ldots, e_m\}$, we construct a bipartite graph H with bipartition $\{A, B\}$, where $A = \{a_1, \ldots, a_m\}$, $B = \beta \cup \Pi$ with $\beta = \{b_1, \ldots, b_n\}$ and $\Pi = \{p_{i,j} \mid i \in [m], j \in [r-3]\}$. The edge set $E(H) = \{a_i p_{i,j} \mid i \in [m], j \in [r-3]\} \cup \{a_i b_j \mid$ edge e_i is incident on vertex v_j in $G\}$. Note that, $deg(a_i) = 2 + (r-3) = r - 1$, $deg(b_i) = r$, and every $p_{i,j}$ is a pendant vertex.*

Observation 4. *In Construction 3, the subgraph of H induced on the vertex set $A \cup \beta$ is isomorphic to the incidence graph of G. Thus, H is isomorphic to the incidence graph of G with $r-3$ pendant vertices, namely the $p_{i,j}$ vertices hanging down from each a_i.*

In Lemma 1, we prove that if the graph G contains a clique of size k, then there exist some $k + x$ independent vertices in H that cover at least $kr + x(r-1)$ edges, where $x = m - (kr - \binom{k}{2})$. In Lemma 5, we prove the reverse implication of this statement.

Lemma 1. *If there exists a clique of size k in the r-regular graph G given in Construction 3, then there exists $k + x$ independent vertices in the bipartite graph H constructed that cover at least $kr + x(r-1)$ edges, where $x = m - (kr - \binom{k}{2})$.*

Proof. Without loss of generality, let us assume that the set of vertices $S = \{v_1, \ldots, v_k\}$ form a k-clique in G. It is easy to verify that $\|S\| = kr - \binom{k}{2}$ and therefore the number of edges not incident on the vertices of S in G is $x = m - \|S\| = m - (kr - \binom{k}{2})$. Without loss of generality, let these x edges be $e_1, \ldots, e_x$. Let $T = \{a_1, \ldots, a_x, b_1, \ldots, b_k\}$. Clearly, T is a set of independent vertices in H and $\|T\| = kr + x(r-1)$.

Before we prove Lemma 5, we prove a few supporting lemmas, namely Lemmas 2 to 4, that give us some insight into the structure of G and H.

Lemma 2. *Suppose there exists no clique of size k in the r-regular graph G given in Construction 3. Let i be an integer such that $0 \leq i \leq r - k$. Let $p_{k+i}(G) = \max\{m - \|S\| : S \subseteq V(G), |S| = k + i\}$. Then,*

(i) $p_k(G) \leq x - 1$,
(ii) for every $i \in [r - k]$, $p_{k+i}(G) \leq p_{k+i-1}(G) - (r - (k + i - 2))$.

Proof. For every i, let S_{k+i}, which is a subset of $V(G)$, be a set such that $|S_{k+i}| = k + i$ and $m - \|S_{k+i}\| = p_{k+i}(G)$.

(i) Since G does not contain any k-clique, $\|S_k\| \geq kr - (\binom{k}{2} - 1)$ and thus $p_k(G) = m - \|S_k\| \leq m - (kr - (\binom{k}{2} - 1)) = x - 1$.

(ii) Let $i \in [r - k]$. Since G does not contain any k-clique, $\exists v \in S_{k+i}$ such that v has at most $k + i - 2$ neighbours in S_{k+i} and therefore at least $r - (k + i - 2)$ neighbours outside S_{k+i}. Let $X_{k+i-1} = S_{k+i} \setminus \{v\}$. Then, $\|S_{k+i}\| \geq \|X_{k+i-1}\| + r - (k + i - 2) \geq \|S_{k+i-1}\| + r - (k + i - 2)$ (from the definition of S_{k+i-1}). Therefore, $p_{k+i}(G) = m - \|S_{k+i}\| \leq m - (\|S_{k+i-1}\| + r - (k + i - 2)) \leq p_{k+i-1}(G) - (r - (k + i - 2))$.

Below, we introduce a couple of new definitions. Consider the bipartite graph H with bipartition $\{A, B\}$, where $B = \beta \uplus \Pi$, constructed from G in Construction 3. Let i be an integer such that $0 \leq i \leq r - k$.

Definition 4. *For any set $S \subseteq \beta$, let $q(S) := \max\{|X| : X \subseteq A$ and no vertex in X is adjacent with any vertex in $S\}$. Then, we define $q_{k+i}(H) := \max\{q(S) : S$ is a $(k + i)$-sized subset of $\beta\}$.*

Definition 5. *Let $\mathcal{I}_{k+i} = \{I \subseteq V(H) \mid I$ is a $(k + x)$-sized independent set, and $|I \cap \beta| = k + i\}$, where $x = m - (kr - \binom{k}{2})$. Then, we define $s_{k+i}(H) := \max\{\|I\| : I \in \mathcal{I}_{k+i}\}$.*

Suppose the graph G does not contain any k-clique. Then, from Observation 4, Lemma 2, and Definition 4, we have $p_{k+i}(G) = q_{k+i}(H)$. This gives us the following lemma.

Lemma 3. *Suppose there exists no clique of size k in the r-regular graph G given in Construction 3. Consider the bipartite graph H constructed from G. We have,*

(i) $q_k(H) \leq x - 1$,
(ii) for every $i \in [r - k]$, $q_{k+i}(H) \leq q_{k+i-1}(H) - (r - (k + i - 2))$.

Lemma 4. *Suppose there exists no clique of size k in the r-regular graph G given in Construction 3. Consider the bipartite graph H constructed from G. Then, $s_r(H) < s_{r-1}(H) < \cdots < s_k(H) < kr + x(r - 1)$, where $x = m - (kr - \binom{k}{2})$.*

Proof. Let i be an integer such that $0 \leq i \leq r - k$. Let S_{k+i} be a $(k + x)$-sized independent set in H having exactly $k + i$ vertices from the set β such that $\|S_{k+i}\|$ is maximum among all such independent sets. That is, $\|S_{k+i}\| = s_{k+i}(H)$ (from the definition of $s_{k+i}(H)$). Recall that $S_{k+i} \subseteq V(H) = A \uplus \beta \uplus \Pi$ and from its definition we have $|S_{k+i} \cap (A \cup \Pi)| = x - i$. Since the degree of any vertex in A (which is $r - 1$) is greater than the degree of any vertex in Π (which is 1), we have $||S_{k+i}||$ maximized when $S_{k+i} \cap A$ is maximized. Then, from the definition of S_{k+i} and by Definition 4, we get $|S_{k+i} \cap A| = q_{k+i}(H)$. Therefore,

$$
\begin{aligned}
s_{k+i}(H) &= ||S_{k+i}|| \\
&= ||S_{k+i} \cap \beta|| + ||S_{k+i} \cap A|| + ||S_{k+i} \cap \Pi|| \\
&= |S_{k+i} \cap \beta|r + |S_{k+i} \cap A|(r-1) + |S_{k+i} \cap \Pi| \\
&= (k+i)r + (q_{k+i}(H))(r-1) + (k+x) - (k+i+(q_{k+i}(H))). \quad (2)
\end{aligned}
$$

By Statement (i) of Lemma 3, we have $q_k(H) = x - a$, where $a \geq 1$. Therefore, $s_k(H) = kr + (x-a)(r-1) + a \leq kr + (x-1)(r-1) + 1 < kr + x(r-1)$.

Let i be an integer such that $0 \leq i < r - k$. Then,

$$
\begin{aligned}
s_{k+i}(H) - s_{k+i+1}(H) &= (q_{k+i}(H) - q_{k+i+1}(H))(r-2) - (r-1) \quad \text{(from Eqn. (2))} \\
&\geq (r - (k+i-1))(r-2) - (r-1) \quad \text{(by (ii) in Lemma 3)} \\
&\geq 2(r-2) - (r-1) \quad \text{(since } i \leq r - k - 1) \\
&\geq 5 \quad \text{(since } r = n - 4 \text{ and } n > 11).
\end{aligned}
$$

Lemma 5. *If the r-regular graph G given in Construction 3 does not contain any clique of size k, then no set of $k + x$ independent vertices in the bipartite graph H constructed covers $kr + x(r-1)$ edges or more, where $x = m - (kr - \binom{k}{2})$.*

Proof. Suppose the graph G given in Construction 3 does not contain any clique of size k. We shall then show that no set of $k + x$ independent vertices in the bipartite graph H constructed from G covers $kr + x(r-1)$ edges or more.

Let Y be an independent set of vertices of cardinality at most $k + x$ in H. Then, $Y = Y_A \uplus Y_\beta \uplus Y_\Pi$, where $Y_A = Y \cap A$, $Y_\beta = Y \cap \beta$, and $Y_\Pi = Y \cap \Pi$. The contribution of any vertex $v \in Y$ to the coverage $||Y||$ is its degree and this is highest when $v \in Y_\beta$ (then, $deg(v) = r$) and lowest when $v \in Y_\Pi$ (in this case, $deg(v) = 1$). When $v \in Y_A$, $deg(v) = r - 1$. Below, we shall prove that $||Y|| < kr + x(r-1)$. The proof is split into 2 cases based on the cardinality of Y_β.

Case 1 $(|Y_\beta| < k)$. Let $|Y_\beta| = k - a$, where $a \geq 1$. Then, $||Y|| = |Y_\beta|r + |Y_A|(r-1) + |Y_\Pi| \leq (k-a)r + (x+a)(r-1) \leq (k-1)r + (x+1)(r-1) = kr + x(r-1) - 1$.

Case 2 $(|Y_\beta| \geq k)$. We split this case into two subcases:-

(i) $|Y_\beta| \leq r \ (= n - 4)$.
   Let $0 \leq i \leq r - k$. Let $|Y_\beta| = k + i$. Then, by Definition 5 and Lemma 4, $||Y|| \leq s_{k+i} < kr + x(r-1)$.

(ii) $|Y_\beta| > r \ (= n - 4)$.
   Since $|\beta| = n$, by definition of Y_β, $|Y_\beta| \leq n$. Let $i \in \{0, \ldots, 3\}$. Let $|Y_\beta| = n - i$. The set Y being an independent set in H, by Observation 4, the vertices in Y_A correspond to edges that are 'outside' the set of vertices in G

that correspond to Y_β. As the cardinality of Y_β is close to n (in fact, at least $n - 3$), we get $|Y_A| \leq i$. We are now prepared to estimate the coverage of Y.

$$\begin{aligned} ||Y|| &= |Y_\beta|r + |Y_A|(r - 1) + |Y_\Pi| \\ &\leq (n - i)r + i(r - 1) + ((k + x) - n) \\ &\leq nr + k + x - n \quad \text{(since } i = 0 \text{ yields the highest value)} \end{aligned}$$

In order to prove subcase (ii), we need to show that

$$\begin{aligned} kr &+ x(r - 1) > nr + k + x - n \\ &\text{i.e. } x(r - 2) > (n - k)(r - 1) \quad \text{(rearranging terms)} \\ &\text{i.e. } x > (n - k)\left(1 + \frac{1}{r - 2}\right) \end{aligned}$$

Since $r = n - 4$ and $n > 11$, we have $r \geq 8$. Thus, in order to prove this subcase it is enough to prove the following inequality:-

$$x > \frac{7}{6}(n - k).$$

We know that $x = m - \left(kr - \binom{k}{2}\right) = \frac{nr}{2} - \left(kr - \frac{k^2 - k}{2}\right) = \frac{nr - 2kr + k^2 - k}{2} = \frac{r(n-k) - kr + k^2 - k}{2} = \frac{r(n-k) - k(n-4) + k^2 - k}{2} = \frac{(r-k)(n-k) + 3k}{2} > \frac{((n-4)-k)(n-k)}{2} \geq \frac{(n-7)(n-k)}{4}$ (since $k \leq \frac{n-1}{2}$. See the definition of $(n - 4, k)$-CLIQUE in the beginning of this section.). Since $n > 11$, we get $x \geq \frac{5}{4}(n - k) > \frac{7}{6}(n - k)$.

Lemmas 1 and 5 imply Theorem 2.

2.2   A $\frac{1}{2}$-Approximation Algorithm

Consider the following greedy approximation algorithm for the MWBIS problem in bipartite graphs.

Algorithm 1. $\frac{1}{2}$-approximation algorithm for the MWBIS problem on bipartite graphs

---

INPUT: A bipartite graph G with bipartition $\{A, B\}$, a weight function $w : V(G) \to \mathbb{R}^+$, and a positive integer k.

OUTPUT: An independent set of size at most k.

1: Let S_A denote the set of k highest weight vertices in A. In the case when $k > |A|$, $S_A = A$. Similarly, let S_B denote the set of k highest weight vertices in B. In the case when $k > |B|$, $S_B = B$. Find S_A and S_B.
2: If $|| S_A || \geq ||S_B||$, return S_A. Otherwise, return S_B.

We claim that Algorithm 1 is a $\frac{1}{2}$-factor approximation algorithm for MWBIS in bipartite graphs. Without loss of generality, let us assume that Algorithm 1 returns the set S_A. Let O be an optimal solution for the MWBIS problem on the bipartite graph G under consideration. Let $O_A = O \cap A$ and $O_B = O \cap B$. Then $\|O\| = \|O_A\| + \|O_B\| \leq \|S_A\| + \|S_B\| \leq 2\|S_A\|$. Hence, we prove the claim. It is easy to see that Algorithm 1 runs in $O(nk)$ time.

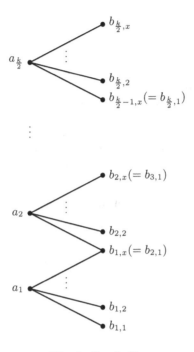

Fig. 1. Graph H_1

Tight Example. Below we describe the construction of a bipartite graph H that is a tight example to our analysis of Algorithm 1. Let H_1 be the connected, bipartite graph given in Fig. 1. Let H_2 be a graph isomorphic to H_1 (one can imagine H_2 to be a mirror copy of Fig. 1 drawn right above it). We obtain the connected, bipartite graph H from the union of H_1 and H_2 by adding an edge connecting a pendant vertex in H_1 with a pendant vertex in H_2. More formally, $V(H) = V(H_1) \cup V(H_2)$ and $E(H) = E(H_1) \cup E(H_2) \cup \{h_1, h_2\}$, where h_1 and h_2 are any two pendant vertices in H_1 and H_2, respectively. We claim that H is a tight example for Algorithm 1. Consider the MIVC problem on graph H with parameter k. The optimal solution chooses all the degree x vertices. This gives a solution of size kx. Since each part in H has exactly $\frac{k}{2}$ vertices of degree x and the remaining vertices are each of degree at most 2, our algorithm returns a solution of size at most $\frac{kx}{2} + \frac{k}{2}(2) = \frac{kx}{2} + k$. Thus, the approximation ratio is at most $\frac{\frac{kx}{2}+k}{kx} = \frac{1}{2} + \frac{1}{x}$. This ratio approaches $\frac{1}{2}$ as x tends to infinity.

Extending Algorithm 1 to General Graphs. Algorithm 1 can be generalized to an arbitrary graph G (that is not necessarily bipartite) provided we are given a proper vertex coloring of G. Suppose we are given a proper coloring of the vertices of G using p colors that partitions the vertices into p color classes namely $C_1, \ldots, C_p$. Let $S_1, \ldots, S_p$ denote sets of k highest weight vertices in color classes $C_1, \ldots, C_p$, respectively. For any S_i, if $k > |C_i|$, then $S_i = C_i$. Find a set S_i such that $\|S_i\| \geq \|S_j\|$, $\forall j \in [p]$. Return S_i.

This algorithm gives us a $\frac{1}{p}$-approximation for MWBIS. Suppose it returns a set S_j as the output. Let O be an optimal solution for the MWBIS problem on G. Let $O_i = O \cap C_i$ for $i \in [p]$. Then $\|O\| = \|O_1\| + \ldots + |O_p\| \leq p\|S_j\|$. It is easy to see that the algorithm runs in $O(nk)$ time.

We know that we can properly color any graph G with $\Delta + 1$ colors using a greedy coloring algorithm, where Δ denotes the maximum degree of G. This gives us a $\frac{1}{\Delta+1}$-approximation algorithm for MWBIS on G. Since d-degenerate graphs can be properly colored using $d + 1$ colors in linear time, this algorithm is a $\frac{1}{d+1}$-factor approximation algorithm for such graphs.

2.3 LP Relaxation and a Matching Integrality Gap

Let us take another look at Integer Program (1) given in Sect. 1.4 for the MWBIS problem on G, where G is a graph with $V(G) = \{v_1, \ldots, v_n\}$ and $w : V(G) \to \mathbb{R}^+$. Recall, we use $\mathcal{C}$ to denote the set of all maximal cliques in G. We obtain the LP relaxation of this program by changing the domain of variable x_i from $x_i \in \{0, 1\}$ to $x_i \geq 0$.

$$Maximize \sum_{i \in [n]} w(v_i) \cdot x_i \tag{3}$$

$$s.t. \sum_{i=1}^{n} x_i \leq k$$

$$\sum_{i:v_i \in C} x_i \leq 1, \quad \forall C \in \mathcal{C}$$

$$x_i \geq 0, \quad \forall i \in [n].$$

We now present an instance of the MIVC problem to illustrate that the integrality gap of the above LP is upper bounded by $\frac{1}{2} + \epsilon$ (where ϵ is any number greater than 0) for bipartite graphs.

Example 1. Let H be a bipartite graph with bipartition $\{A, B\}$, where $|B| = k(k-1)+1 = p$ and $|A| = (k-1)p+1$. Let $A = \{a_{i,j} \mid i \in [p], j \in [k-1]\} \cup \{a_0\}$ and $B = \{b_1, \ldots, b_p\}$. The edge set $E = \{a_0 b_i \mid i \in [p]\} \cup \{a_{i,j} b_i \mid i \in [p], j \in [k-1]\}$. Note that $\deg(a_0) = p$, $\deg(b_i) = k$, and $\deg(a_{i,j}) = 1$. For every $v \in V(H)$, let $w(v) = deg(v)$. Figure 2 illustrates a drawing of H with $k = 3$.

The best integral solution to the above LP for H is to select vertex a_0 and $k-1$ vertices of degree 1 from the set $\{a_{i,j} \mid i \in [p], j \in [k-1]\}$. The coverage by these k vertices is $deg(a_0) + (k-1) = p + (k-1) = k(k-1) + 1 + (k-1) = k^2$.

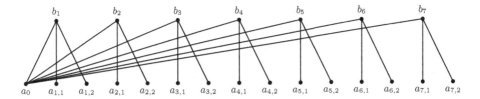

Fig. 2. A drawing of the bipartite graph H described in Example 1 with $k = 3$.

The x-variables in the LP have the following associations: x_0 is associated with vertex a_0, x_i with b_i, and $x_{i,j}$ with $a_{i,j}$. One possible fractional solution to the above LP for H is to assign $x_0 = \frac{k-1}{k}$, $x_i = \frac{1}{k}$, $\forall i \in [p]$, and $x_{i,j} = 0, \forall i \in [p], \forall j \in [k-1]$. This is a feasible solution to the above LP as $x_0 + \sum_{i \in [p], j \in [k-1]} x_{i,j} + \sum_{i \in p} x_i = \frac{k-1}{k} + 0 + p\frac{1}{k} = k$ and sum of the x-values of every maximal clique is at most 1. The number of edges covered by this fractional solution is $(\frac{k-1}{k} + \frac{1}{k})p + (\frac{1}{k}p(k-1)) = p(1 + \frac{k-1}{k}) = (k(k-1)+1)(\frac{2k-1}{k})$. Hence the upper bound for integrality gap of the above LP established by this example is $\frac{k^2}{(k(k-1)+1)(\frac{2k-1}{k})} = \frac{k^3}{(k^2-k+1)(2k-1)} = \frac{k^3}{2k^3-3k^2+3k-1} = \frac{1}{2-\frac{3}{k}+\frac{3}{k^2}-\frac{1}{k^3}} = \frac{1}{2-\frac{1}{k}(3+\frac{1}{k^2}-\frac{3}{k})}$. This value approaches $1/2$ as k tends to infinity. Thus, the integrality gap of the above LP is upper bounded by $\frac{1}{2} + \epsilon$ (where ϵ is any number greater than 0) for bipartite graphs.

3   Concluding Remarks

In Sect. 2.2, we propose an easy, greedy $\frac{1}{2}$-approximation algorithm for MWBIS on bipartite graphs, whose factor of approximation matches with the integrality gap of a natural LP relaxation of the problem. It would be interesting to see if one can improve the factor of approximation using a more clever and sophisticated approach.

Suppose we have a t-factor approximation algorithm $A(G)$ for finding MWIS in a graph G having n vertices. Then, we can obtain a $\frac{kt}{n}$-factor approximation algorithm for MWBIS in G, where k denotes the budget. Let $A(G)$ return an independent set S as output. Let S' denote the set of k highest weight vertices in S. If $k > |S|$, then let $S' = S$. Return S'. Let O, O' be optimal solutions for the MWIS and the MWBIS problem in G, respectively. We have $||S'|| \geq \min(||S||, \frac{k||S||}{|S|}) \geq \frac{k||S||}{n} \geq \frac{kt||O||}{n} \geq \frac{kt||O'||}{n}$. Boppana and Halldórsson in [8] gave an $\Omega(\log^2 n/n)$-factor approximation algorithm for MWIS. This yields an $\Omega(k \log^2 n/n^2)$-factor algorithm for MWBIS. For bounded degree graphs, Halldórsson in [13] gave an $\Omega(\sqrt{\log \Delta}/\Delta)$-factor approximation algorithm for MWIS. This gives an $\Omega(k\sqrt{\log \Delta}/n\Delta)$-factor approximation algorithm for MWBIS. Designing better approximation algorithms for MWBIS in general graphs is another natural open question.

References

1. Ageev, A.A., Sviridenko, M.I.: Approximation algorithms for maximum coverage and max cut with given sizes of parts. In: Cornuéjols, G., Burkard, R.E., Woeginger, G.J. (eds.) IPCO 1999. LNCS, vol. 1610, pp. 17–30. Springer, Heidelberg (1999). doi:10.1007/3-540-48777-8_2
2. Apollonio, N., Simeone, B.: Improved approximation of maximum vertex coverage problem on bipartite graphs. SIAM J. Discret. Math. **28**(3), 1137–1151 (2014)
3. Apollonio, N., Simeone, B.: The maximum vertex coverage problem on bipartite graphs. Discret. Appl. Math. **165**, 37–48 (2014)
4. Austrin, P., Khot, S., Safra, M.: Inapproximability of vertex cover and independent set in bounded degree graphs. In: 24th Annual IEEE Conference on Computational Complexity, CCC 2009, pp. 74–80. IEEE (2009)
5. Bandyapadhyay, S.: A variant of the maximum weight independent set problem. CoRR, abs/1409.0173 (2014)
6. Bar-Yehuda, R.: Using homogeneous weights for approximating the partial cover problem. J. Algorithms **39**(2), 137–144 (2001)
7. Berman, P., Karpinski, M.: On some tighter inapproximability results (extended abstract). In: Wiedermann, J., Emde Boas, P., Nielsen, M. (eds.) ICALP 1999. LNCS, vol. 1644, pp. 200–209. Springer, Heidelberg (1999). doi:10.1007/3-540-48523-6_17
8. Boppana, R., Halldórsson, M.M.: Approximating maximum independent sets by excluding subgraphs. BIT Numer. Math. **32**(2), 180–196 (1992)
9. Bshouty, N.H., Burroughs, L.: Massaging a linear programming solution to give a 2-approximation for a generalization of the vertex cover problem. In: Morvan, M., Meinel, C., Krob, D. (eds.) STACS 1998. LNCS, vol. 1373, pp. 298–308. Springer, Heidelberg (1998). doi:10.1007/BFb0028569
10. Caskurlu, B., Mkrtchyan, V., Parekh, O., Subramani, K.: On partial vertex cover and budgeted maximum coverage problems in bipartite graphs. In: Diaz, J., Lanese, I., Sangiorgi, D. (eds.) TCS 2014. LNCS, vol. 8705, pp. 13–26. Springer, Heidelberg (2014). doi:10.1007/978-3-662-44602-7_2
11. Gandhi, R., Khuller, S., Srinivasan, A.: Approximation algorithms for partial covering problems. J. Algorithms **53**(1), 55–84 (2004)
12. Garey, M.R., Johnson, D.S., Stockmeyer, L.: Some simplified NP-complete graph problems. Theor. Comput. Sci. **1**(3), 237–267 (1976)
13. Halldórsson, M.M.: Approximations of weighted independent set and hereditary subset problems. J. Graph Algorithms Appl. **4**(1), 1–16 (2000)
14. Han, Q., Ye, Y., Zhang, H., Zhang, J.: On approximation of max-vertex-cover. Eur. J. Oper. Res. **143**(2), 342–355 (2002)
15. Håstad, J.: Clique is hard to approximate within $n^{1-\epsilon}$. In: Proceedings of the 37th Annual Symposium on Foundations of Computer Science, pp. 627–636. IEEE (1996)

Demand Hitting and Covering of Intervals

Datta Krupa R., Aniket Basu Roy, Minati De, and Sathish Govindarajan[⊠]

Indian Institute of Science, Bangalore, India
{datta.krupa,aniket.basu,minati,gsat}@csa.iisc.ernet.in

Abstract. Hitting and Covering problems have been extensively studied in the last few decades and have applications in diverse areas. While the hitting and covering problems are NP-hard for most settings, they are polynomial solvable for intervals. Demand hitting is a generalization of the hitting problem, where there is an integer demand associated with each object, and the demand hitting set must contain at least as many points as the demand of each object. In this paper, we consider the demand hitting and covering problems for intervals that have no containment. For the unweighted setting, we give a simple greedy algorithm. In the weighted setting, we model this problem as a min-cost max flow problem using a non-trivial reduction and solve it using standard flow algorithms.

Keywords: Hitting set · Intervals · Demand hitting set · Min-cost max flow

1 Introduction

Hitting and Covering are classical problems that have been extensively studied in the last few decades and have been used in applications in diverse areas. In these problems, we are given a ground set X of m elements and a collection R of n subsets of X (called ranges). The hitting set problem is to compute a minimum sized subset $H \subseteq X$ such that each range in R contains (is "hit" by) at least one element in H. The covering problem is to compute a minimum sized subset $S \subseteq R$ of ranges such that each element in X is contained (is "covered" by) in at least one range of S. Computing the minimum hitting/covering set is NP-hard even for very restricted settings (for example, when X is a set of points and R is collection of subsets obtained using unit squares in the plane). However, the minimum hitting/covering set can be computed in polynomial time for intervals on the real line [7].

Intervals are widely studied structures with applications in many areas. Interval graphs that capture the intersection pattern of intervals are an important graph class. Since intervals are simple objects, the interval graph is highly structured. Thus, most optimization problems on interval graphs like independent set, vertex cover, dominating set, maximum clique, etc. can be solved efficiently using combinatorial algorithms in polynomial time [7,8]. In fact, these optimization problems can be also solved using Linear Programming as the constraint matrix is totally unimodular.

© Springer International Publishing AG 2017
D. Gaur and N.S. Narayanaswamy (Eds.): CALDAM 2017, LNCS 10156, pp. 267–280, 2017.
DOI: 10.1007/978-3-319-53007-9_24

Demand hitting set problem is a generalization of the hitting problem, where there is an integer demand associated with each range in R (object). $H \subseteq X$ is a demand hitting set if each range (objects) in R contains (is "hit" by) at least as many elements in H as its demand. Note that if the demand is one for all objects, we get the standard hitting set problem. A special case of the demand hitting set is the k-*hitting set* problem where the demand of all the intervals is k. Similarly, the demand covering problem is a generalization of the covering problem, where there is an integer demand associated with each element in X (point). $S \subseteq R$ is a demand covering if each element in X is contained in ("covered by) at least as many objects in S as its demand. Note that if the demand is one for all points, we get the standard covering problem. In the unweighted setting, the objective is to minimize the size of the solution. In the weighted setting, the objective is to minimize the sum of weights of entities in the solution.

Related Work. Variants and generalizations of the hitting and covering have been recently studied. A generalization of hitting set problem is studied by Even et al. [6]. In this problem, each point has a capacity, representing the maximum number of intervals it can hit and a weight to pay if that point is included in solution. The solution is an association of a point to every interval. For the weighted case, they give $\mathcal{O}(n^2m^2(m+n))$ time dynamic programming algorithm where n and m are the number of intervals and the number of points respectively.

Similarly, the covering problems have been generalized and studied in the geometric context but to a greater extent. Chekuri et al. [5] gave an $\mathcal{O}(\log OPT)$-approximation algorithm when the range space has bounded VC-dimension and unit weights. Bansal and Pruhs [1] gave a constant factor approximation for pseudo-disks in the plane. A more general version of the multi-cover problem called capacitated covering problem is studied in [2,4].

Our Contributions: In this paper, we study the demand hitting and covering problem where X is a set of points and R is a set of intervals on the real line. We present the following results:

1. For the unweighted demand hitting problem, we give a simple greedy algorithm. The greedy algorithm is a natural generalization of the one known for unweighted hitting set. We give a $\mathcal{O}(m+n)$ time implementation of this greedy algorithm assuming that the points and intervals are sorted.
2. For the weighted demand hitting problem, we model the problem as a min-cost max flow problem using a non-trivial reduction and solve it using standard flow algorithms. We first consider the k-hitting problem and model it as a min cost k-flow problem. The correctness of the reduction is argued using an elegant structural lemma of decomposing a k-hitting set into k disjoint hitting sets. Finally, we extend the reduction and arguments for the weighted demand hitting set.

Note that our algorithms for demand hitting (unweighted and weighted) make an assumption that there is *no containment* among the intervals. However for the k-hitting special case (unweighted and weighted), the *no containment* assumption is not required since if there is containment we can eliminate it by removing the larger interval. k-hitting the smaller interval would also k-hit the removed larger interval.

The algorithms presented for unweighted and weighted demand hitting can be appropriately adapted to solve the corresponding demand covering problems. We give the details of the covering algorithms in the full version.

The idea of modeling an interval optimization problem as a flow problem is motivated from Carlisle and Lloyd [3]. They model the maximum k-independent set/k-coloring problem using flows. While this problem has only intervals as input, the demand hitting has both intervals and points as inputs. Thus, the modeling and the correctness of reduction is quite involved for demand hitting.

2   Unweighted Demand Hitting Set

In this section, we consider the demand hitting set problem and give an efficient greedy algorithm. We first describe the demand hitting set problem.

Problem Definition 1 (Unweighted Demand Hitting Set). *Let* $P = \{p_1, p_2, \ldots, p_m\}$ *be a set of* m *points,* $I = \{I_1, I_2, \ldots, I_n\}$ *be a set of* n *intervals with a positive integer demand* d_k *associated with an interval* $I_k \in I$. *A subset of points* $H \subseteq P$ *is called a demand hitting set of* I *if each interval* $I_k \in I$ *contains at least* d_k *points in* H.

The unweighted demand hitting set problem is to compute a demand hitting set $H \subseteq P$ *of* I *of minimum size.*

Greedy Algorithm

Demand hitting set problem is a generalization of the hitting set problem. The natural strategy is to generalize the greedy algorithm of the unweighted hitting set to the unweighted demand hitting set. The greedy algorithm does the following: Pick the first ending interval I_k and pick the rightmost d_k points that hits I_k. For every point picked, decrement the demand by 1 for all the intervals hit by that point. Delete I_k and the picked points and continue this process until the demand of all the intervals are satisfied.

Naively implementing the above algorithm would require a demand counter for every interval and for every point picked decrement the demands of all the intervals containing that point. So the running time will be $\Theta(\text{sum of demands} + m + n)$. A point can be contained in $\Theta(n)$ intervals, i.e., we may need to do $\Theta(n)$ decrements for a point. The demand hitting set can contain $\Theta(m)$ points. Thus the running time of the naive implementation of the greedy algorithm is $\Theta(mn)$.

We now give a linear time implementation of the above greedy algorithm under the assumption that *no interval is contained in any other interval* and the points and intervals are in sorted order. Note that when interval containment is forbidden, there is a natural linear ordering of intervals from left to right and the sorted order of intervals with respect to starting or ending point is the same.

The key idea of the implementation is the following: When a point p_i is picked in the solution, mark the first interval starting after p_i, rather than decrementing the demand counter of all the intervals hit by p_i. This marking strategy implicitly keeps track of all the intervals hit by p_i.

The following data structures are used by the algorithm:

- An array M of size n that tracks the markings of the first starting interval for each selected point. All entries in M are initialized to 0.
- An array T of size n that keeps track of the satisfied demand of the current interval due to the points in the current solution. All entries in T are initialized to 0.
- A stack S to pick the rightmost required number of points to satisfy the remaining demand of current interval.

The algorithm processes the intervals $I_1, I_2, \ldots, I_n$ in their natural linear order. When interval I_k is processed, the algorithm maintains the invariant that T_k is the number of points in current solution that hits I_k. The following steps are performed when interval I_k is processed.

(1) $d = d_k - T_k$ is the number of additional points needed to satisfy I_k's demand. We pick the rightmost d points using the stack S as follows:
   - Push all the points which lie after I_{k-1}'s right end point and lie before or equal to I_k's right end point, to stack S.
   - Pop d points from stack S and add them to the solution H.
(2) Increment T_k by d as d points that hit I_k have been included in the solution.
(3) For every point p_i included in the solution, mark the first interval starting after p_i, i.e., I_{s_i}, by adding one to M_{s_i}.
(4) Update T_{k+1} to $T_k - M_{k+1}$. Here M_{k+1} is the number of points in solution that hit I_k but not I_{k+1}. So now T_{k+1} is the number of points in the updated solution hitting I_{k+1}.

Feasibility of the Algorithm

We have to prove that the solution of our algorithm is feasible, i.e., it picks at least d_k points for interval $I_k, 1 \leq k \leq n$. During the iteration for I_k, the algorithm picks $d = d_k - T_k$ points from stack S and adds them to the solution H. The algorithm maintains the invariant that at the start of the iteration for I_k, T_k is the number of points in current solution that hits I_k.

Claim. At the start of the iteration for I_k, T_k is the number of points in the current solution that hits I_k.

Proof. We give a proof by induction. For $k = 1$, the claim is trivially true as the current solution is ϕ and $T_1 = 0$ by initialization. Assume that for $k = r - 1$, the claim is true, i.e., T_{r-1} is the number of points hitting I_{r-1} at the start of iteration for I_{r-1}.

During the iteration for I_{r-1}, if additional points are needed ($d > 0$) then additional points are picked from stack and T_{r-1} is incremented (step 2). So T_{r-1} stores the number of points hitting I_{r-1}.

The points of the current solution that hits I_{r-1} can be categorized into two types: (1) points that do not hit I_r and (2) points that hit I_r.

Whenever a point p_i of type 1 is selected in the solution H during any of the iterations for interval $I_l, l \leq r - 1, I_r$ is the first starting interval after p_i and thus M_r is incremented by 1 (step 3).

Before updating T_r to $T_{r-1} - M_r$ in the iteration for I_{r-1}, T_{r-1} stores the number of points hitting I_{r-1} (by induction hypothesis) and M_r stores the number of points hitting I_{r-1} of type 1, i.e., those points not hitting I_r. After update T_r stores the number of points in the current solution that hit I_r.   □

Optimality of the Algorithm

The greedy algorithm is a natural generalization of the standard greedy algorithm for hitting set problem. The optimality of the standard greedy algorithm for hitting set is shown by comparing the greedy solution with the optimal solution and arguing that greedy algorithm "stays ahead" of the optimal (see Chap. 4.1 in [8] for details). The optimality of our algorithm can be argued in a similar manner since our algorithm picks the rightmost required number of points to satisfy the demand of I_k.

Running Time

Assuming that the intervals and points are already sorted, we can preprocess and store the first starting interval after every point in linear time. The algorithm traverses the intervals and points in the sorted order once. Each point is pushed and popped from stack S at most once. For each point picked in solution, two counters in T array are updated. Since size of solution is $\mathcal{O}(m)$, the total running time of our algorithm is $\mathcal{O}(n + m)$.

3   Weighted k-Hitting Set

In this section, we consider the weighted k-hitting set problem and give a flow based algorithm for efficiently solving it.

Definition 1 (Min Cost k-Flow). *Let G be a directed graph where each edge has a capacity and weight associated with it. Min-cost k-flow sends k units of flow from source node to destination node, satisfying capacity constraints and minimizing the sum of weights of edges having positive flow.*

Refer to [9] for more details.

Problem Definition 2 (Weighted k-Hitting Set). *Let $P = \{p_1, p_2, \ldots, p_m\}$ be a set of m points with a real weight w_i associated with each point $p_i \in P$, $I = \{I_1, I_2, \ldots, I_n\}$ be a set of n intervals and k be a positive integer. A subset of points $H \subseteq P$ is called a k-hitting set of I if every interval in I contains at least k points in H.*

The weighted k-hitting set problem is to compute a k-hitting set $H \subseteq P$ of I of minimum cost, i.e., sum of weights of points in H is minimized.

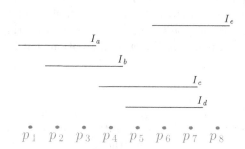

Fig. 1. Example intervals and points for k-hitting

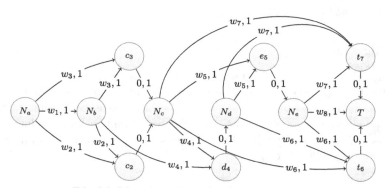

Edge label (w_i, c): w_i is weight of p_i, c is capacity.

Fig. 2. k-hitting flow graph for intervals and points in Fig. 1

Algorithm

We describe a flow based algorithm for the weighted k-hitting set assuming that there is no containment among intervals in I. Note that if an interval is contained in another interval, k-hitting the smaller interval would k-hit the larger interval as well. Thus, if there is containment, we preprocess and remove the larger interval.

The main idea of the algorithm is the following: Construct a graph whose nodes correspond to intervals in I and edges correspond to points in P. Compute a min cost k-flow in this graph (see Definintion 1). Points corresponding to edges carrying this flow gives the optimal k-hitting set.

We now describe the construction of the graph G. Sort the intervals in I based on their right end points. For every interval $I_l \in I$, create a node N_l in G. Order the nodes horizontally according to the sorted order of the corresponding intervals I_l (see Figs. 1 and 2).

Let s_i be the index of first starting interval after point p_i. For each point p_i in I_l, add a directed edge (N_l, N_{s_i}) with weight w_i and capacity 1. If there is no interval starting after p_i, add a directed edge (N_l, T) with weight w_i and capacity 1, where T is the sink node.

Whenever a node N_l has more than one incoming edge corresponding to the same point p_i, create an additional node l_i in G and direct all such edges into l_i. Add a directed edge (l_i, N_l) with weight 0 and capacity 1.

Add a source node S and add a directed edge (S, N_1) with weight 0, capacity k. (see Fig. 2 for graph G corresponding to example given in Fig. 1).

Compute a min cost k-flow in G using the algorithm given in [9]. Since the capacities are integers, there exists a maximum flow where flow in each edge is integral (see Chap. 7.2 in [8]). The points in P corresponding to the edges that carry a unit flow gives the solution to the weighted k-hitting set problem. Note that the creation of an additional node l_i ensures that there will be a unit flow in at most one edge corresponding to a point. Thus a point will be chosen at most once in-the solution.

Feasibility and Optimality

We now argue that the solution returned by this algorithm is feasible and optimal. We first prove an important structural lemma about k-hitting sets that will be used later to show feasibility and optimality.

Lemma 1. *Any k-hitting set for I can be split into k disjoint 1-hitting sets for I.*

Proof. Let H be any k-hitting set for I. We split H into sets H_1 and H', such that H_1 is a 1-hitting set for I and H' is a $(k-1)$-hitting set for I. The 1-hitting set H_1 is obtained by using the standard greedy hitting set algorithm (Algorithm 1) to hit intervals in I using points from H. Note that an 1-hitting set is the same as standard hitting set. We argue that the remaining points H' forms a $(k-1)$-hitting set for I.

Algorithm 1. Split k-hitting set

1: Sort I on right end point and sort H.
2: Pick the first ending interval I_l.
3: Pick the rightmost point $p_i \in H$ hitting I_l. Add p_i to H_1. Remove p_i from H.
4: Associate p_i to I_l and remove all the intervals hit by p_i.
5: Repeat Line 2 to Line 4 on the remaining intervals.
6: The leftover points in H are H'.

Observation 1. *An interval of I that is associated with a point by Algorithm 1 (Line 4) contains exactly one point in H_1 as this interval has not been hit by previous picked points and Algorithm 1 chooses the rightmost point to hit this interval.*

If an interval loses only one point to H_1 it will have $k - 1$ points left in H' since H is a k-hitting set. Consider an interval I_l that loses more than one point to H_1. Let p_i and p_j, $i < j$ be the rightmost two points of H_1 in interval I_l. By Observation 1, I_l is not associated with any point in H_1 as it contains more than one point in H_1.

Let point p_j be associated with I_r. By Observation 1, p_j is the only point from H_1 that is contained in I_r, implying that p_i is not contained in I_r. Since p_j is the rightmost point in I_r, there exist at least $k - 1$ points of H' that lie to the left of p_j and to the right of p_i. These $k - 1$ points of H' that hit I_r also hit I_l. Hence H' is $(k - 1)$-hitting set.

Apply Algorithm 1 recursively on H' to get k disjoint 1-hitting sets. □

Lemma 2. *An S-T path in the graph G corresponds to a 1-hitting set and vice versa.*

Proof. In the proof we do not refer to any additional nodes l_i in the S-T path as additional nodes have exactly one out-going edge which would be part of any path passing through this additional node.

Path to Hitting Set: Consider an S-T path in G. Let the edges in the path be $(S, N_1), (N_{b_1}, N_{b_2}), (N_{b_2}, N_{b_3}), \ldots, (N_{b_i}, N_{b_{i+1}}), \ldots, (N_{b_t}, T)$, where $b_1, b_2, \ldots,$ $b_t \in [1, n], b_1 = 1$. Let $p_{a_i} \in P$ be the point corresponding to edge $(N_{b_i}, N_{b_{i+1}})$. By construction, p_{a_i} hits I_{b_i}. Consider the nodes between N_{b_i} and $N_{b_{i+1}}$ in the left-to-right order in which the nodes are arranged. Since $I_{b_{i+1}}$ is the first starting interval after p_{a_i} and there is no containment of intervals in I, point p_{a_i} hits all the intervals $I_{b_i}, I_{b_i+1}, \ldots, I_{b_{i+1}-1}$. Applying the above argument to all the edges in S-T path, we see that the points corresponding to the edges in S-T path hits all the intervals in I.

Hitting Set to a Path: Given a hitting set H, apply greedy Algorithm 1 to associate points in H to interval in I. The S-T path in G is constructed as follows: Start with the edge (S, N_1). Since I_1 is the first ending interval in the first iteration of Algorithm 1, there is a point $p \in H$ associated with I_1. The next edge in the S-T path is the outgoing edge from N_1 associated with point p. Continue this process and we argue next that node T would be reached.

In the path, an edge corresponding to p_i goes to the first starting interval after p_i, but Algorithm 1 chooses the first ending interval. Since there is no containment among intervals in I, the first starting interval after p_i is also the first ending interval starting after p_i. The intervals ending before p_i has been hit in the previous iterations of Algorithm 1, i.e., the first starting interval after p_i is also the first ending interval chosen by Algorithm 1 among the intervals which are not hit. Hence, in the S-T path, every node reached has an associated point. At the end, when the last point in H is picked, there is no interval starting after this point. Thus the edge corresponding to the last point goes to node T, completing the S-T path. $\square$

Feasibility: We prove that the points in P corresponding to any k-flow in G forms a feasible k-hitting set for I.

Let F be any k-flow in G. F has value k in edge (S, N_1) and is 0–1 valued in all the other edges of G. Moreover, F can be decomposed along k paths which are edge disjoint except for edge (S, N_1) (see Chap. 7.6 in [8] for details on path decomposition). By Lemma 2, each of these k-paths corresponds to a 1-hitting set. Note that the additional nodes in G ensures that there will be a unit flow in at most one edge corresponding to a point. Thus a point will appear at most once in the k 1-hitting sets. Hence, a k-flow in the graph corresponds to k disjoint 1-hitting sets which is a feasible k-hitting set.

Optimality: We prove that the solution returned by the flow based algorithm is optimal by showing that we can get a k-flow in G using points of any k-hitting set of I.

Let H be any k-hitting set of I. By Lemma 1, H can be decomposed into k disjoint 1-hitting sets $H_1, H_2, \ldots, H_k$. By Lemma 2, each H_i corresponds to an S-T path $P_i, 1 \le i \le k$. Note that the paths P_i share the edge (S, N_1) and are edge disjoint otherwise. A k-flow F is constructed by sending k units of flow from S to T, one unit of flow along each of the S-T paths P_i. The k-flow F is feasible since F has value k in edge (S, N_1) and is 0–1 valued in all the other edges of G (by edge disjointness property of paths P_i).

By the above argument we can construct a feasible k-flow corresponding to the optimal k-hitting set. Since our algorithm computes the min-cost k-flow among all feasible k-flows, the hitting set returned by our algorithm is optimal.

Lemma 3. *Given a min-cost k-flow in G, the points corresponding to the edges having unit flow form the optimal weighted k-hitting set for the intervals in I.*

Running Time

Sorting intervals and points takes $\mathcal{O}(n \log n)$ and $\mathcal{O}(m \log m)$ time respectively. There is a node for every interval. Number of additional nodes corresponding to points is $\mathcal{O}(m)$ as the first starting interval after each point is unique. Total number of nodes is $\mathcal{O}(m+n)$ and number of edges is $\mathcal{O}(mn)$. We can preprocess and store the first starting interval for every point in linear time, i.e., the graph can be constructed in $\mathcal{O}(mn)$ time.

Flow graph is acyclic as every edge goes from left to right, i.e., for every edge (N_l, N_q) $q > l$. Algorithm in [9] computes min cost k-flow in acyclic graph in time $\mathcal{O}(kS(n))$, where $S(n)$ is the time for computing shortest S-T path in directed graphs. Using Dijkstra's algorithm with Fibonacci heap implementation, the total running time of the algorithm is $\mathcal{O}(k(mn + (m+n)\log(m+n))) = \mathcal{O}(kmn)$. Since k is at most m, the total running time is $\mathcal{O}(nm^2)$.

4   Weighted Demand Hitting Set

In this section, we consider the weighted demand hitting set problem for intervals without containment and give a flow based algorithm for this problem.

Problem Definition 3 (Weighted Demand Hitting Set). *Let $P = \{p_1,$ $p_2, \ldots, p_m\}$ be a set of m points with a real weight w_i associated with each point $p_i \in P$, $I = \{I_1, I_2, \ldots, I_n\}$ be a set of n intervals with a positive integer demand d_l associated with an interval $I_l \in I$. A subset of points $H \subseteq P$ is called a demand hitting set of I if each interval $I_l \in I$ contains at least d_l points in H.*

The weighted demand hitting set problem is to compute a demand hitting set $H \subseteq P$ of I of minimum cost, i.e., sum of weights of points in H is minimized.

Note that the weighted demand hitting set is a generalization of the weighted k-hitting (demand is k for all intervals) discussed in the previous section.

Algorithm

We now describe a flow based algorithm for computing the optimal demand hitting set under the assumption that there is no containment among intervals in I.

The algorithm is a generalization of the algorithm for computing the weighted k-hitting set described in the previous section.

The main idea of the algorithm is the following: Construct a graph whose nodes corresponds to intervals in I and edges corresponds to points in P. Compute a min cost k-flow in this graph, where k is the maximum demand among all the intervals. Points corresponding to edges carrying this flow gives the optimal demand hitting set.

We now describe the construction of the graph G. Sort the intervals in I based on their right end point. For every interval $I_l \in I$, create two nodes N_l', N_l in G. Order the N_l nodes and N_l' nodes horizontally according to the sorted order of right end points of the corresponding intervals I_l (see Figs. 3 and 4).

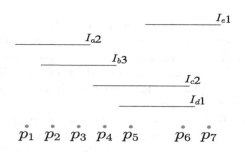

Fig. 3. Example intervals and points for demand hitting

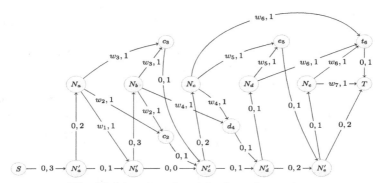

Edge label (w_i, c): w_i is weight of p_i, c is capacity.

Fig. 4. Demand-hitting flow graph for Fig. 3

For each l, $1 \leq l \leq n$, add a directed edge (N'_l, N_l) with weight 0 and capacity d_l and a directed edge (N'_l, N'_{l+1}) with weight 0 and capacity $k - d_l$. Node N'_{n+1} is the sink node which we shall denote as node T. Let s_i be the index of first starting interval after point p_i. For each point p_i in I_l, add a directed edge (N_l, N'_{s_i}) with weight w_i and capacity 1. If there is no interval starting after p_i, add a directed edge (N_l, T) with weight w_i and capacity 1.

Whenever a node N'_l has more than one incoming edge corresponding to the same point p_i, create an additional node l_i in G and direct all such edges into l_i. Add a directed edge (l_i, N'_l) with weight 0 and capacity 1.

Add a source node S and add a directed edge (S, N'_1) with weight 0, capacity k. (see Fig. 4 for graph G corresponding to example given in Fig. 3)

Let k be the maximum demand among all the intervals. Compute a min cost k-flow in G using the algorithm given in [9]. Since the capacities of all the edges are integers, the flow values in each edge is also integral.

The points in P corresponding to the edges that carry a unit flow forms the optimal weighted demand hitting set. Note that the creation of an additional node l_i ensures that there will be a unit flow in at most one edge corresponding to a point. Thus a point will be chosen at most once in the solution.

Feasibility and Optimality

We now argue that the solution returned by this algorithm is feasible and optimal. We argue that any k-flow is a feasible demand hitting set and any demand hitting set is a feasible k-flow.

Lemma 4. *If the points corresponding to an S-T path in G does not hit interval I_l, then the edge (N'_l, N'_{l+1}) has to be present in the path.*

Proof. In an S-T path there are three possibilities to "cross" a node N_l. In the first two cases we argue that definitely I_l will be hit.

- Taking edge (N'_l, N_l), visiting N_l and then taking an out-going edge from N_l. The point corresponding to out-going edge will hit I_l.
- N_l can be "crossed" by taking an edge corresponding to p_i, which goes from a previous node to a node after N_l. In this case we show that p_i hits I_l as well. Consider an edge (N_h, N'_q) corresponding to p_i which "crosses" N_l. From construction p_i hits I_h. Consider the nodes between N_h and N_q. Since I_q is the first starting interval after p_i and there is no containment of intervals in I, point p_i hits all the intervals $I_h, I_{h+1}, \ldots, I_l, \ldots, I_{q-1}$. Thus p_i hits I_l.
- N_l can be "crossed" by taking the edge (N'_l, N'_{l+1}). Since we have reached N'_l, points corresponding to edges till N'_l do not hit I_l. In this case there is a possibility that I_l will not be hit by points corresponding to the edges in the path. For example, if the S-T path takes edges $(N'_q, N'_{q+1}), \forall q \geq l$ then I_l is not hit. □

Feasibility: We prove that the points in P corresponding to any k-flow in G forms a feasible k-hitting set for I.

Lemma 5. *Any k-flow in the graph corresponds to a feasible demand hitting set.*

Proof. Let F be any k-flow in G. F has integer flow value in all the edges of G. Thus, F can be decomposed along k S-T paths $P_1, P_2 \ldots, P_k$, each having unit flow (see Chap. 7.6 in [8] for details on path decomposition). Since the capacity of edge (N'_l, N'_{l+1}) is $k - d_l$, at most $k - d_l$ paths among $P_1, P_2, \ldots, P_k$ contains this edge. Therefore, the remaining d_l S-T paths among $P_1, P_2, \ldots, P_k$ does not contain this edge. By Lemma 4, I_l is hit by a point corresponding to some edge in each of these d_l paths. □

Optimality: We prove that the solution returned by the flow based algorithm is optimal by showing that we can get a k-flow in G using points of any demand hitting set of I.

Lemma 6. *Any demand hitting set corresponds to a feasible k-flow.*

Proof. Given a demand hitting set H, we decompose H into sets $H_1, \ldots, H_k$, where each H_i corresponds to a unit flow S-T path in the graph. The decomposition of demand hitting set is similar to the decomposition of k-hitting set (Algorithm 1), i.e., pick the first ending interval and pick the rightmost point in it. But directly doing this might not give a decomposition into k sets. In Algorithm 2, we select a subset of intervals whose current demand is the maximum and apply the above decomposition procedure. Refer Algorithm 2.

We claim that Algorithm 2 runs for exactly k iterations. This is because the demand of an interval gets decremented by at most one in an iteration. The left most interval with the maximum demand needs at least k iterations, before its demand becomes zero. Also in each iteration all the intervals with the maximum demand gets its demand decremented by one. Thus, Algorithm 2 runs for exactly k iterations and produces k S-T paths.

Algorithm 2. Demand hitting to k paths

Given a demand hitting set H for I with demands D, it is decomposed into sets $H_1, \ldots, H_k$ where each H_i corresponds to a S-T path in the graph.

1: Sort I based on their right end point.
2: $t \leftarrow 1$.
3: **while** there exists a demand in D that is not 0 **do**
4:     Let $I' \subseteq I$ be the set of intervals in I with maximum demand.
       Use points in H to hit intervals I' :
5:     **while** I' is not empty **do**
6:         Let I_l be the first ending interval in I'.
7:         Pick the rightmost $p_i \in H$ which hits I_l .
8:         Add p_i to H_t and remove p_i from H.
9:         $\forall I_q \in I$ hit by p_i decrement d_q if it has not been decremented in current iteration (t).
10:        If $I_q \in I'$ then remove I_q from I'.
11:    **end while**
       S-T path:
12:    **for all** $I_l \leftarrow I_1, \ldots, I_n$ **do**
13:        **if** p_i is picked for I_l **then**
14:            Take edges (N'_l, N_l) and the out-going edge from N_l corresponding to p_i.
15:        **else**
16:            Take edge (N'_l, N'_{l+1}).
17:        **end if**
18:    **end for**
19:    $t \leftarrow t + 1$.
20: **end while**

Construct a k-flow F by sending one unit flow in each of the k S-T paths returned by Algorithm 2.

Claim. The k-flow F is feasible.

Proof. The flow along edge (S, N'_1) is k since all the k S-T paths contains this edge. There are three types of edges in G.

- Edges (N'_l, N_l): I_l can be present in I' in at most d_l iterations of Algorithm 2. Edge (N'_l, N_l) will be picked in path only when I_l is be present in I'. The capacity of this edge is not violated since its capacity is d_l.
- Edges corresponding to points: A point $p_i \in H$ is picked exactly once during the execution of Algorithm 2. Thus, the edge corresponding to p_i appears once in the k S-T paths and the capacity of this edge is not violated.
- Edges (N'_l, N'_{l+1}): This edge is present in an S-T path returned by Algorithm 2 if the points corresponding to the edges in the sub path from S to N'_l do not hit I_l. This condition happens in at most $k - d_l$ iterations of Algorithm 2 as H is a valid demand hitting set for I. Thus, the edge (N'_l, N'_{l+1}) is present in at most $k - d_l$ S-T paths and the capacity of this edge is not violated.   □

The min cost k-flow solution is a feasible demand hitting set by Lemma 5. By Lemma 6, an optimal demand hitting set corresponds to a feasible k-flow in the graph. Since our algorithm computes the min-cost k-flow among all feasible k-flows, the demand hitting set returned by our algorithm is optimal.

Lemma 7. *Given a min-cost k-flow in G, the points corresponding to the edges having unit flow form the optimal weighted demand hitting set for the intervals in I.*

Running Time

The running time analysis of the demand hitting algorithm is exactly the same as the k-hitting algorithm described in Sect. 3. The total running time of the algorithm is $\mathcal{O}(kmn)$. Since k is at most m, the total running time is $\mathcal{O}(nm^2)$.

References

1. Bansal, N., Pruhs, K.: Weighted geometric set multi-cover via quasi-uniform sampling. In: Epstein, L., Ferragina, P. (eds.) ESA 2012. LNCS, vol. 7501, pp. 145–156. Springer, Heidelberg (2012). doi:10.1007/978-3-642-33090-2_14
2. Bansal, N., Pruhs, K.: The geometry of scheduling. SIAM J. Comput. **43**(5), 1684–1698 (2014)
3. Carlisle, M.C., Lloyd, E.L.: On the k-coloring of intervals. Discrete Appl. Math. **59**(3), 225–235 (1995). http://www.sciencedirect.com/science/article/pii/0166218X9580003M
4. Chakrabarty, D., Grant, E., Könemann, J.: On column-restricted and priority covering integer programs. In: Eisenbrand, F., Shepherd, F.B. (eds.) IPCO 2010. LNCS, vol. 6080, pp. 355–368. Springer, Heidelberg (2010). doi:10.1007/978-3-642-13036-6_27
5. Chekuri, C., Clarkson, K.L., Har-Peled, S.: On the set multicover problem in geometric settings. ACM Trans. Algorithms (TALG) **9**(1), 9 (2012)
6. Even, G., Levi, R., Rawitz, D., Schieber, B., Shahar, S.M., Sviridenko, M.: Algorithms for capacitated rectangle stabbing and lot sizing with joint set-up costs. ACM Trans. Algorithms **4**(3), 34:1–34:17 (2008). http://doi.acm.org/10.1145/1367064.1367074
7. Golumbic, M.C.: Algorithmic Graph Theory and Perfect Graphs, vol. 57. Elsevier, Amsterdam (2004)
8. Kleinberg, J., Tardos, É.: Algorithm Design. Pearson Education India, Delhi (2006)
9. Tarjan, R.E.: 8. network flows. In: Data Structures and Network Algorithms, Chap. 8, pp. 97–112. SIAM (1983). http://epubs.siam.org/doi/abs/10.1137/1.9781611970265.ch8

Exact and Parameterized Algorithms
for (k, i)-Coloring

Diptapriyo Majumdar[1(✉)], Rian Neogi[2], Venkatesh Raman[1],
and Prafullkumar Tale[1]

[1] The Institute of Mathematical Sciences, HBNI, Chennai, India
{diptapriyom,vraman,pptale}@imsc.res.in
[2] NIIT University, Neemrana, Rajasthan, India
rianneogi@gmail.com

Abstract. Graph coloring problem asks to assign a color to every vertex such that adjacent vertices get different color. There have been different ways to generalize classical graph coloring problem. Among them, we study (k, i)-coloring of a graph. In (k, i)-coloring, every vertex is assigned a set of k colors so that adjacent vertices share at most i colors between them. The (k, i)-chromatic number of a graph is the minimum number of total colors used to assign a proper (k, i)-coloring. It is clear that $(1, 0)$-coloring is equivalent to the classical graph coloring problem. We extend the study of exact and parameterized algorithms for classical graph coloring problem to (k, i)-coloring of graphs. Given a graph with n vertices and m edges, we design algorithms that take
- $\mathcal{O}(2^{kn} \cdot n^{\mathcal{O}(1)})$ time to determine the $(k, 0)$-chromatic number.
- $\mathcal{O}(4^n \cdot n^{\mathcal{O}(1)})$ time to determine the $(k, k\text{-}1)$-chromatic number.
- $\mathcal{O}(2^{kn} \cdot k^{im} \cdot n^{\mathcal{O}(1)})$ time to determine the (k, i)-chromatic number.

We prove that (k, i)-coloring is fixed parameter tractable when parameterized by the size of the vertex cover or the treewidth of the graph. We also provide some observations on (k, i)-colorings on perfect graphs.

1 Introduction

We investigate efficient parameterized and exact exponential algorithms for a variant of graph coloring called (k, i)-coloring. Given a positive integer k, and a non-negative integer $i \leq k$, the (k, i)-coloring of a graph is a an assignment of a set of k colors to every vertex such that every pair of adjacent vertices shares at most i colors [10]. Motivation to pursue this problem comes from coding theory. An (n, d, w) constant-weight binary code is a set of binary vectors of length n, such that each vector contains w ones and any two vectors differ in at most d positions. One of the most basic questions in coding theory is given n, w, d, what is the largest possible size of an (n, d, w) constant-weight binary code? This question has been studied for almost five decades and remains open for any value of n, w, d [1]. It has been proved that the largest possible size of an (n, d, w) constant-weight binary code is closely related to the (k, i)-coloring of a complete

© Springer International Publishing AG 2017
D. Gaur and N.S. Narayanaswamy (Eds.): CALDAM 2017, LNCS 10156, pp. 281–293, 2017.
DOI: 10.1007/978-3-319-53007-9_25

graph on n vertices [5]. If the total number of distinct colors used by a (k, i)-coloring is q, then we say that the graph G has a proper (k, i)-coloring using q colors. Notice that setting $k = 1, i = 0$ gives the classical graph coloring problem. The minimum number of colors needed to assign (single) color to vertices so that every adjacent pair gets different colors is called *chromatic number* of a graph G and it is denoted by $\chi(G)$. We denote the minimum number of colors needed to assign a proper (k, i)-coloring of graph G as $\chi_k^i(G)$ and call it (k, i)-*chromatic number* of a graph. The precise definition of the problem is given below.

(k, i)-COLORING

Input: Graph G, integer q

Question: Does there exist a (k, i)-coloring of G using at most q colors?

This problem was first introduced by Mendez-Diaz and Zabala [10]. In the same paper authors provided a heuristic approach and a linear programming model for this problem. There are other types of tuple coloring problems generalizing the classical graph coloring problem. For arbitrary k and for $i = 0$, it is called *k-tuple coloring*. This idea of tuple coloring was also independently introduced by Hilton et al. [2], Stahl [18], Bollobas and Thomason [4]. Tuple-Coloring was generalized in a different way by Brigham and Dutton [6] where every vertex is assigned k colors and adjacent vertices share *exactly* i colors.

Computing the chromatic number of a graph has been considered as one of the notoriously difficult problems. This generalization makes the problem even harder. There exists a polynomial time algorithm to compute the chromatic number of a perfect graph. In the case of (k, i)-chromatic number, polynomial time algorithms are known only in case of simple cycles, cactus [5] and bipartite graphs [10]. No polynomial time algorithm is known for finding (k, i)-chromatic number even on well structured graphs like complete graphs for all values of n, k, i. We prove some simple connections of this parameter to the (standard) chromatic number and initiate a study of exact and parameterized complexity of the problem under different parameterizations. A brute force algorithm for testing if $\chi(G) \leq q$ takes $q^n \cdot n^{\mathcal{O}(1)}$ time. A series of improvements led to the current best runtime of $\mathcal{O}(2^n \cdot n^{\mathcal{O}(1)})$ time [13] to compute $\chi(G)$. Similarly, a brute force exact algorithm to determine whether $\chi_k^i(G) \leq q$ will run through all possible q colorings which will assign an arbitrary set of k colors to a vertex. Then, a vertex can be assigned $\binom{q}{k}$ color-sets and there are n vertices in the graph. So, this brute force algorithm will take $\binom{q}{k}^n \cdot n^{\mathcal{O}(1)}$ time. For $(k, 0)$-coloring and $(k, 1)$-coloring, we provide an improved exact exponential algorithm (using efficient algorithm for the classical set cover problem) and then generalize it for any (k, i)-coloring. We provide the following exact algorithms in Sect. 4. Given a graph with n vertices and m edges, we design algorithms that take

- $\mathcal{O}(2^{kn} \cdot n^{\mathcal{O}(1)})$ time to determine the $(k, 0)$-chromatic number.
- $\mathcal{O}(4^n \cdot n^{\mathcal{O}(1)})$ time to determine the $(k, k\text{-}1)$-chromatic number.
- $\mathcal{O}(2^{kn} \cdot k^{im} \cdot n^{\mathcal{O}(1)})$ time to determine the (k, i)-chromatic number.

Concerning parameterized complexity results, we first observe that for standard parameterization (where the parameter is the number of colors), classical graph coloring is para-NP-hard (see Sect. 2 for definitions). So, it is clear that $(1, 0)$-COLORING is also para-NP-hard when it is parameterized by the number of colors. But, it is not clear whether (k, i)-COLORING is para-NP-hard for any other values of k and i when the parameter is the number of colors. We follow the modern trend and resort to structural parameterizations where the problem is parameterized by some structure in the input. Specifically we consider the (k, i)-COLORING problem parameterized by the size of *vertex cover* of the graph. We also give efficient FPT algorithm for the problem on bounded treewidth graphs.

We organize this paper as follows. We state preliminaries and terminologies regarding (k, i)-coloring in Sect. 2. In Sect. 3, we provide some observations about (k, i)-coloring on perfect graphs and prove a conjecture stated in [10]. Sections 4 and 5 contain exact and fixed parameter tractable algorithms for (k, i)-coloring respectively.

2   Preliminaries

All graphs considered here are finite, undirected and simple. For a graph G, its vertex set is denoted by $V(G)$ and its edge set is denoted by $E(G)$. Vertex u is said to be adjacent to vertex v is $uv \in E(G)$. For a vertex $v \in V(G)$, its open neighborhood, $N_G(v)$, is the set of all vertices adjacent to it. The closed neighborhood $N_G[v] = \{v\} \cup N_G(v)$. We drop the subscript if it is clear from context. For a set $X \subseteq V(G)$, the subgraph of G induced by X is denoted by $G[X]$ and it is defined with vertex set X and edge set $\{uv \in E(G) : u, v \in X\}$. The subgraph obtained after deleting X is denoted by $G \setminus X$. Number of vertices in a maximum induced clique of a graph G is denoted by $\omega(G)$. K_n denotes a clique on n vertices. For a positive integer q, we denote the set $\{1, 2, \ldots, q\}$ by $[q]$. The family of all the k-sized subsets of $[q]$ is denoted by $[q]^k$. A function $f : V(G) \to [q]$ is called a coloring function of graph G. If for all edges uv, $f(u) \neq f(v)$, we say that f is proper coloring of graph G. The smallest integer q for which it is possible to properly color all vertices of graph G is called its *chromatic number* and it is denoted by $\chi(G)$. (k, i)-coloring is a generalization of proper coloring and defined as follows.

Definition 1. *A coloring function* $f : V(G) \to [q]^k$ *is called proper-(k, i)-coloring of a graph G if for any edge $uv \in E(G)$, $|f(u) \cap f(v)| \leq i$.*

In COLORING problem, input is a graph G, integer q and the question is whether G can be properly colored using at most q colors. Analogously, in (k, i)-COLORING problem, input is a graph G, integer q and the question is whether G can be (k, i)-colored using at most q colors. Notice that, we consider the case when k, i are fixed constants and not part of input. In SET COVER problem, input is a universe U and a family $\mathcal{F}$ of its subsets and the question is to find cardinality of minimum sized subset $\mathcal{F}'$ of $\mathcal{F}$ which *covers* U. $\mathcal{F}'$ covers U if every element of U is present in at least one set in $\mathcal{F}'$.

Parameterized Complexity: The goal of parameterized complexity is to find ways of solving NP-hard problems more efficiently than brute force by associating a small parameter to each instance. Parameterization of a problem is assigning a positive integer parameter p to each input instance. We say that a parameterized problem is Fixed-Parameter Tractable (FPT) if there is an algorithm that solves the problem in time $f(p) \cdot |I|^{\mathcal{O}(1)}$, where $|I|$ is the size of the input and f is an arbitrary computable function depending only on the parameter p. Such an algorithm is called an FPT algorithm, and the runtime of the algorithm is also sometimes called as FPT running time. A parameterized problem is said to be in the class para-NP if it has a nondeterministic algorithm with FPT running time. To show that a problem is para-NP-hard, we need to show that the problem is NP-hard even when the parameter takes a value from a finite set of positive integers. For example COLORING problem parameterized by solution size is para-NP-hard as determining 3-colorability of a graph NP-hard. We refer interesting reader to [9], [11] for further discussions on parameterized complexity.

Structural Parameterization: *Vertex cover* of a graph is set $X \subseteq V(G)$ such that for every edge uv at least one of u or v is contained in X. In other words, $G - X$ is an independent set. A *tree decomposition* of a graph G is a pair $\mathcal{T} = (T, \{X_t\}_{t \in V(T)})$, where T is a tree whose every node t is assigned a vertex subset $X_t \subseteq V(G)$, called a bag, such that the following three conditions hold: (i) $\bigcup_{t \in V(T)} X_t = V(G)$. (ii) For every $uv \in E(G)$, there exists a node t of T such that bag X_t contains both u and v. (iii) For every $u \in V(G)$, the set $T_u = \{t \in V(T) \mid u \in X_t\}$ induces a connected subtree of T. The width of tree decomposition $\mathcal{T} = (T, \{X_t\}_{t \in V(T)})$ equals $\max_{t \in V(T)}\{|X_t| - 1\}$. The treewidth of a graph G, denoted by $tw(G)$, is the minimum possible width of a tree decomposition of G.

3 Elementary Results

In this section we state some observations related to (k, i)-coloring of general graph and on perfect graphs.

3.1 (k, i)-Coloring on General Graph

We omit the simple proof of the following two observations.

Observation 1. *For a given graph G and its (k, i)-coloring function $f :$ $V(G) \rightarrow [q]^k$, let $C \subseteq [q]$ be the set of any $i + 1$ colors. If $U := \{u \in V(G)| C \subseteq f(u)\}$ then U is an independent set in the graph.*

Observation 2. *For a given graph G and its induced subgraph H, $\chi_k^i(H) \leq \chi_k^i(G)$.*

Observation 3 ($\star$). [1] *For a given graph G, the following bounds hold –*

1. $2k - i \leq \chi_k^i(G)$ *when G has an edge.*
2. $\chi_k^i(G) \leq \chi_{k-i}^0(G) + i$.
3. $\chi_k^0(G) \leq k \cdot \chi_1^0(G)$.

3.2 Perfect Graphs and (k, i)-Coloring

Perfect graphs were defined by Berge in 1960 [3] as follows:

Definition 2. *Graph G is a perfect graph if for each of its induced subgraphs H, $\chi(H) = \omega(H)$.*

A *hole* is an induced cycle of length at least four. An *antihole* is a graph whose complement is a hole. It is easy to see that if G is perfect graph then it does not contain an induced hole or an antihole of length greater than or equal to 5. Berge conjectured the following statement in 1961 which has been resolved in a celebrated result in 2002 [8].

Strong Perfect Graph Conjecture: *G is a perfect graph if and only if G does not have induced odd holes or odd anti-holes of length greater than or equal to 5.*

In [10], authors have introduced a concept of (k, i)-perfect graphs. We first define (k, i)-clique number which will be used in defining (k, i)-perfect graphs.

Definition 3. *The (k, i)-clique number of a graph G is the (k, i)-chromatic number of its largest induced clique and it is denoted by $\omega_k^i(G)$.*

In other words, $\omega_k^i(G) = \chi_k^i(K_{\omega(G)})$. (k, i)-perfect graphs are defined as follows.

Definition 4. *A graph G is a (k, i)-perfect graph if for each of its induced subgraphs H, $\chi_k^i(H) = \omega_k^i(H)$.*

For $(k, i) = (1, 0)$ this definition coincides with the Berge's definition. Authors of [10] conjectured the following statement and proved the *if* implication.

Conjecture 5.1: *G is $(k, 0)$-perfect for $k \geq 1$ if and only if G does not have induced odd holes or antiholes of length greater than or equal to 5.*

Proposition 1 (Lemma 5.1 of [10]). *The odd holes of length greater than or equal to 5 and their complements are not $(k, 0)$-perfect graphs.*

With the following Lemma and using the proof of strong perfect graph conjecture, we prove the reverse direction concluding that this conjecture is true.

Lemma 1 ($\star$). *If G is a perfect graph then $\chi_k^i(G) = \chi_k^i(K_{\omega(G)})$.*

Lemma 2 ($\star$). *If graph G does not have induced odd holes or odd antiholes of length greater than or equal to 5 then G is $(k, 0)$-perfect graph.*

---

[1] Results marked with a $\star$ have their proofs in the full version of this paper.

4   Exact Algorithms

For a given graph G, integer q and a coloring function $f : V(G) \to [q]^k$, one can check whether or not f is a proper (k, i)-coloring of G in $\mathcal{O}(|E(G)| \cdot k)$ time. For a given integer q, there are $\binom{q}{k}$ many choices of k-tuples, which a function f can assign to a vertex v in $V(G)$. Hence the number of different coloring functions is $\binom{q}{k}^n$. By Observation 3, $\chi_k^i(G) \leq k\chi(G)$ and we know that $\chi(G) \leq n$. Brute force algorithm exhaustively searches through the all possible coloring functions $f : V(G) \to [q]^k$ for values of $q \in [kn]$ and returns the minimum value of q for which it finds a valid (k, i)-coloring of graph G. This algorithm runs in time $\mathcal{O}(2^{nk \log(nk)} \cdot n^{\mathcal{O}(1)})$. We present an exact algorithm which runs in time $\mathcal{O}(2^{kn} n^{\mathcal{O}(1)})$ to find $(k, 0)$-chromatic number of graph G. We generalize the idea to present an algorithm running in time $\mathcal{O}(2^{kn} \cdot k^{im} n^{\mathcal{O}(1)})$ to find (k, i)-chromatic number. This algorithm out performs the brute force algorithm mentioned above when the number of edges in graph are linearly bounded by the number of vertices. Finally, we present an algorithm running in time $\mathcal{O}(4^n)$ to find $(k, k\text{-}1)$-chromatic number of a graph (which is an NP-complete problem [10]). Note that running time of this algorithm is independent of k.

4.1   Computing $(k, 0)$-Chromatic Number

In $(k, 0)$-coloring, adjacent vertices should be assigned disjoint color-sets. Such coloring is also known as k-tuple coloring and it is proved to be NP-complete for any value of $k \geq 3$ [14].

For a given graph G, construct an auxiliary graph G' as follows: Graph G' contains k copies of graph G indexed by integers $\{1, 2, \ldots, k\}$. Every vertex $u \in V(G)$ has its k copies $\{u_1, u_2, \ldots, u_k\}$ in graph G'. Construct clique, K_k^u, in G' on the vertices $\{u_1, u_2, \ldots, u_k\}$ for every vertex $u \in V(G)$. If there is an $uv \in E(G)$, make each vertex in K_k^u adjacent with every vertex in K_k^v. In other words, $K_k^u \cup K_k^v$ is a clique of size $2k$. This finishes the construction. Note that the graph G' has kn vertices.

Lemma 3. $\chi_k^0(G) = \chi(G')$.

Proof. Let $f' : V(G') \to [\chi(G')]$ be an optimal proper coloring of graph G'. We define a function $f : V(G) \to [\chi(G')]^k$ as $f(u) = \{f'(u_1) \mid u_1 \in K_k^u\}$. We argue that f is a valid $(k, 0)$-coloring of G. Since f' is a proper coloring of G', for $u_1, u_2 \in K_k^u$, $f'(u_1) \neq f'(u_2)$ and hence $|f(u)| = k$ for all $u \in V(G)$. Suppose f is not a valid coloring and there exists edge $uv \in E(G)$ such that $|f(u) \cap f(v)| > 0$. Hence there exists two vertices $u_1 \in K_k^u$ and $v_1 \in K_k^v$ such that $f'(u_1) = f'(v_1)$. Since $uv \in E(G)$, by construction $K_k^u \cup K_k^v$ is a clique. This is contradiction to the fact that f' is proper coloring of G'. This proves that $\chi_k^0(G) \leq \chi(G')$.

We now prove that $\chi(G') \leq \chi_k^0(G)$. Let $f : V(G) \to [q]^k$ be an optimal (k, i)-coloring for graph G where $q = \chi_k^0(G)$. We construct function $f' : V(G') \to [q]$ by constructing a bijective map between the vertices in K_k^u and $f(u)$. For any edge $u'v' \in E(G')$, either $u'v' \in K_k^u$ for some u or $u' \in K_k^u$ and $v' \in K_k^v$ and

$uv \in E(G)$. In first case, $f'(u) \neq f'(v)$ as f' is bijection from K_k^u to $f(u)$. In second case, since $|f(u) \cap f(v)| = 0$, and $f'(u) \in f(u), f'(v) \in f(v), f'(u) \neq f'(v)$. This implies that $\chi(G') \leq \chi_k^0(G)$ which concludes the proof. $\qquad \square$

Proposition 2 [16]. *For an n-vertex graph G, there exists an algorithm running in time $\mathcal{O}(2^n \cdot n^{\mathcal{O}(1)})$ which computes its chromatic number.*

Combining Lemma 3 with Proposition 2, we obtain the following result.

Theorem 1. *For an n-vertex graph G, there exists an algorithm running in time $\mathcal{O}(2^{kn} \cdot n^{\mathcal{O}(1)})$ which computes its $(k,0)$-chromatic number.*

4.2 Computing (k,i)-Chromatic Number

We generalize the idea used in the above construction to obtain the (k,i)-chromatic number of the given graph G. Now instead of one, we construct $\mathcal{O}(k^{2im})$ many auxiliary graphs each of which is on kn vertices.

For every edge $e = uv \in E(G)$, select an index set $I_e \subseteq [k]$ of cardinality i. Let $(I_1, I_2, \ldots, I_m)$ be a m-tuple of indices selected. For every such m-tuple, we first construct an auxiliary graph G' as in Sect. 4.1. If $uv \in E(G)$ then delete an edge $u_l v_l$ from graph G' for all $l \in I_e$. Let $\mathcal{G}$ be the set of different graphs created using this operation. The number of such m-tuples are bounded by $\binom{k}{i}^m$ and hence $|\mathcal{G}| \leq \mathcal{O}(k^{im})$. Notice that if $uv \in E(G)$ then there are at most i-many vertices in K_k^u which are not adjacent to some vertex in K_k^v.

Lemma 4 ($\star$). $\chi_k^i(G) = \min_{G' \in \mathcal{G}} \{\chi(G')\}$.

Combining Lemma 4 with Proposition 2 and the bound on $\mathcal{G}$, we obtain the following result.

Theorem 2. *For an n-vertex, m-edges graph G, there exists an algorithm running in time $\mathcal{O}(2^{kn} \cdot k^{im} \cdot n^{\mathcal{O}(1)})$ which computes its (k,i)-chromatic number.*

4.3 Computing $(k, k-1)$-Chromatic Number

For a given graph G, we construct an instance of SET COVER by setting $U = V(G)$ and $\mathcal{F}$ as the family of all independent sets of $V(G)$. Notice that $|\mathcal{F}| \leq 2^n$. Let r be the cardinality of a minimum solution of the SET COVER instance for $(U, \mathcal{F})$.

Claim. If $k - i = 1$ and q is the smallest integer such that $r \leq \binom{q}{i+1}$ then $\chi_k^i(G) = q$.

Proof. Suppose $\mathcal{F}' \subseteq \mathcal{F}$ is an optimum solution for SET COVER of cardinality r. Define an injective function $\psi : \mathcal{F}' \to [q]^{i+1}$ by assigning each set S in $\mathcal{F}'$ an $i+1$ sized set from $[q]$. Since $r \leq \binom{q}{i+1}$, such injective function is possible. We now define a function $f : V(G) \to [q]^k$ as $f(v) = \psi(S)$ where S is any set in $\mathcal{F}'$ such that $u \in S$. Since $\mathcal{F}'$ is a cover of $V(G)$, for every vertex u this function assigns k

colors to each vertex. We now prove that this is indeed a proper (k, i)-coloring of graph. Suppose not, then there exists an edge uv such that $|f(u) \cap f(v)| \geq i + 1$. Since $|f(u)| = |f(v)| = k = i + 1$, this implies $f(u) = f(v)$. By construction it implies that u, v are contained in same set S. This is contradiction to the fact that S is an independent set. Hence f is proper (k, i)-coloring of graph G. This implies that $\chi_k^i(G) \leq q$.

Suppose there exists a (k, i)-coloring $f' : V(G) \to [q']$ of graph G using $q' < q$ colors. By Observation 1, for any set X of $[q']$ which is of cardinality $i + 1$, set $U := \{u \mid X \subseteq f'(u)\}$ is an independent set of graph G. We say that color set X characterizes vertex set U. Construct $\mathcal{F}'' = \{U \mid \exists X \subseteq [q]$ of cardinality $i + 1$ which characterizes $U\}$. Since there are at most $\binom{q}{i+1}$ such color set X, $|\mathcal{F}''| \leq \binom{q'}{r+1} < r$ as q is the smallest integer such that $r \leq \binom{q}{i+1}$. This contradicts the fact that r is cardinality of a minimum solution of SET COVER. Hence our assumption is wrong and $q \leq \chi_k^i(G)$ which completes the proof.   □

Proposition 3 [13]. *For any given instance $(U, \mathcal{F})$ of* SET COVER *problem, there exists an algorithm which solves it in time $\mathcal{O}(2^n n |\mathcal{F}|)$ where $|U| = n$.*

Combining the above claim, Proposition 3 and using the bound that $|\mathcal{F}| \leq 2^n$, we get following result.

Theorem 3. *For an n-vertex graph G, there exists an algorithm running in time $\mathcal{O}(4^n \cdot n^{\mathcal{O}(1)})$ which computes its $(k, k\text{-}1)$-chromatic number.*

5   Fixed-Parameter Algorithms

Parameterization of a problem is assigning a positive integer, called parameter, to each of its input instance. One of the most natural choice for a parameter is the solution size which in this case is the number of colors needed for (k, i)-coloring. For a given graph G, it is NP-hard to determine whether it can be colored with at most 3 colors. This implies that COLORING parameterized by number of colors in para-NP-hard. Hence we can not expect (k, i)-COLORING to be FPT when parameterized by the number of colors. But, COLORING is FPT when parameterized by several structural properties of the input graph. Notion of treewidth was introduced by Roberson and Seymour. It is known that COLORING parameterized by treewidth of graph and number of colors is FPT. Structural Parameterizations of classical graph coloring problem was studied by Jansen and Kratsch [15] (also studied in [7,12]). They proved that when input is a graph G with its vertex cover $Y \subseteq V(G)$ and the parameter is $|Y|$, finding the chromatic number of G is FPT. In this section, We generalize these results of classical graph coloring problem to (k, i)-coloring problem.

5.1 (k, i)-Coloring Parameterized by Vertex Cover

In this sub-section, we present an FPT algorithm for finding $\chi_k^i(G)$ when parameterized by size of a vertex cover of the input graph.

In case of structural parameters, sometimes it is necessary to demand a witness of the required structure as part of the input. However, when the size of a vertex cover is the parameter, this is not a serious demand. If given only a input graph, one can find a 2-approximation of the minimum vertex cover of the input graph G(pp 11,[17]). Thus, we may assume that we are solving the following problem.

(k, i)-COLORING   **Parameter:** $|Y|$
Input: Graph $G, Y \subseteq V(G)$ such that Y is a vertex cover of G
Output: $\chi_k^i(G)$

Theorem 4. *For an n-vertex graph G and its vertex cover Y, there exists an algorithm running in time $\mathcal{O}(2^{k|Y|\log(k|Y|)} \cdot kn^2)$ which computes $\chi_k^i(G)$.*

Proof. For a given graph G on n vertices, the algorithm iterates over all possible (k, i)-colorings of $G[Y]$. Since $\chi_k^i(G[Y]) \leq k \cdot |Y|$, there are $\mathcal{O}(2^{k|Y|\log(k|Y|)})$ many such possible colorings. For every valid (k, i)-coloring f of $G[Y]$, we extend this coloring function to the rest of the graph in the greedy fashion. For every vertex $u \in V(G) \setminus Y$, f assigns k smallest colors to u such that for any $v \in N(u)$, $|f(u) \cap f(v)| \leq i$. Since u is in an independent set, all of its neighbors have been assigned colors before function assigns k colors to u. This extension of valid coloring can be computed in $\mathcal{O}(kn^2)$ time to obtain a (k, i)-coloring of graph G. The algorithm returns the minimum number of colors used over all the valid (k, i)-coloring of graph G. The running time of this algorithm is $\mathcal{O}(2^{k|Y|\log(k|Y|)} \cdot kn^2)$ which is FPT when parameterized by cardinality of vertex cover. We now argue the correctness of the algorithm.

If the algorithm returns q as the minimum number of colors used over all the valid (k, i)-colorings of graph G, by construction it is clear that $\chi_k^i(G) \leq q$. We now prove that $q \leq \chi_k^i(G)$ using contradiction. Suppose $\chi_k^i(G) < q$. This implies that for every (k, i)-coloring f of $V(G)$ which is obtained as extension of valid (k, i)-coloring of $G[Y]$, there exists a vertex u such that $q \in f(u)$. Let $f^* : V(G) \rightarrow [\chi_k^i(G)]^k$ is a optimum (k, i)-coloring of graph G. Since we are iterating over all possible coloring of $G[Y]$, one of them is $f^*|_Y$. Let f' is an greedy extension of $f^*|_Y$ to entire graph. Hence there exists a vertex v such that $q \in f'(v)$. By Observation 2, $\chi_k^i(G[N[v]]) \leq \chi_k^i(G)$. Since f' is obtained greedily as extension of $f^*|_Y$, for every k-sized set X of $\{1, 2, \ldots, \chi_k^i(G)\}$, there exists $u \in N(v)$ such that $|X \cap f^*(u)| \geq i + 1$ which forced algorithm to use a color in $\{\chi_k^i(G), \ldots, q\}$ while constructing extension of $f^*|_Y$. This contradicts the fact that $f^*|_{N[v]}$ is a valid (k, i)-coloring of $N[v]$ which uses at most $\chi_k^i(G)$ colors. $\square$

5.2 q-(k, i)-Coloring Parameterized by Treewidth

In this sub-section, we present an FPT algorithm for finding whether $\chi_k^i(G)$ is at most q when parameterized by treewidth of input graph. Notice that, unlike previous section, we assume that q is fixed and it is not part of input. Formally, we study the following problem.

---

q-(k, i)-COLORING   **Parameter:** tw
Input: Graph G with its tree decomposition $\mathcal{T}$ of width tw
Output: Is $\chi_k^i(G) \leq q$?

We know that given a tree decomposition $\mathcal{T}' = (T', \{Y_t\}_{t \in V(T')})$, it can be converted into a *nice tree decomposition* $\mathcal{T} = (T, \{X_t\}_{t \in V(T)})$ in polynomial time (for definition and other details, see Chapter 7 of [9]) where every node is one of the following types and has at most 2 children. For a *nice tree decomposition*, we distinguish one vertex r of T which will be the root of T.

Root Node: r is the root node where $X_r = \emptyset$.
Leaf Node: If $t \in V(T)$ is a leaf node, then $X_t = \emptyset$.
Introduce Node: If $t \in V(T)$ is an introduce node then t' is the only child of t in T and $X_t = X_{t'} \cup \{u\}$ where $u \notin X_{t'}$.
Forget Node: If $t \in V(T)$ is a forget node then t' is the only child of t in T and $X_t = X_{t'} \setminus \{u\}$ where $u \in X_{t'}$.
Join Node: If $t \in V(T)$ is a join node then t_1 and t_2 are the children of t in T and $X_t = X_{t_1} = X_{t_2}$.

We compute and store two values for every node $t \in V(T)$. These are $\mathcal{C}(t)$ and $\mathcal{D}(t)$ and they are defined as follows.
    $\mathcal{C}(t) = \{f : X_t \to [q]^k | f \text{ is a proper } (k, i)\text{-coloring of } G[X_t]\}$.
    $\mathcal{D}(t) = \{f \in \mathcal{C}(t) | f \text{ is extendable to a proper } (k, i)\text{-coloring of } G_t\}$.
We can compute $\mathcal{C}(t)$ for every $t \in V(T)$ independent of their children. But, $\mathcal{D}(t)$ needs to be computed by using $\mathcal{D}(t_1), \mathcal{D}(t_2)$ where t_1, t_2 are children of t.

Leaf Node: When a node $t \in V(T)$ is a *leaf node*, then $X_t = \emptyset$. So, $\mathcal{C}(t) = \{\emptyset\}$. $\mathcal{D}(t) = \mathcal{C}(t)$.

Introduce Node: When a node $t \in V(T)$ is an *introduce node* with only child t', and let $X_t = X_{t'} \cup \{u\}$ for $u \notin X_{t'}$.
$\mathcal{D}(t) = \{f \in \mathcal{C}(t) | \exists g \in \mathcal{D}(t') \text{ such that } g \equiv f|_{X_{t'}}\}$. Correctness is clear from construction as only feasible colorings are stored and all of them extend to a feasible coloring of the induced subgraph.

Forget Node: When a node $t \in V(T)$ is a *forget node* with only child t' and let $X_t = X_{t'} \setminus \{u\}$ for $u \in X_{t'}$. We say that $\mathcal{D}(t)$ is the projection of all the members of $\mathcal{D}(t')$ at t. Formally $\mathcal{D}(t) = \{f \in \mathcal{C}(t) | f \equiv g|_{X_t} \text{ where } g \in \mathcal{D}(t')\}$. Correctness of this is clear because $G_t = G_{t'}$.

Join Node: When a node $t \in V(T)$ is a *join node* with children t_1 and t_2, then $X_t = X_{t_1} = X_{t_2}$. We say $\mathcal{D}(t) = \mathcal{D}(t_1) \cap \mathcal{D}(t_2)$.
It is clear that if $f \in \mathcal{D}(t)$, then $f \in \mathcal{D}(t_1)$ and $f \in \mathcal{D}(t_2)$. So, $\mathcal{D}(t) \subseteq \mathcal{D}(t_1) \cap \mathcal{D}(t_2)$ as (k, i)-coloring is feasible for induced subgraphs. G_{t_1} and G_{t_2} are induced subgraphs of G_t. We now justify that $f \in \mathcal{D}(t_1) \cap \mathcal{D}(t_2) \subseteq \mathcal{D}(t)$. Let $f \in \mathcal{D}(t_1) \cap \mathcal{D}(t_2)$. Then f is a proper (k, i)-coloring in G_{t_1} and also G_{t_2}. By connectivity property (Property $T3$ in Chapter 7 of [9]) of tree decomposition, we know that there is no edge between two vertices one of which is in $G_{t_1} \setminus X_{t_1}$ and the other is in $G_{t_2} \setminus X_{t_2}$. So, $f \in \mathcal{D}(t)$ as well.

Now, we describe how to compute $\mathcal{D}(t)$ from $\mathcal{D}(t_1)$ and $\mathcal{D}(t_2)$ where t_1, t_2 are the children of t. $\mathcal{C}(t)$ can be computed in $\binom{q}{k}^{tw+1}$ time for every $t \in V(T)$ and this is independent of its children in the tree decomposition. We have the following lemma.

Lemma 5. *For every $t \in V(T)$, $\mathcal{D}(t)$ and $\mathcal{C}(t)$ can be computed in $\mathcal{O}^*((\binom{q}{k})^{tw})$ time.*

Proof. We prove this statement for each type of nodes.

Leaf Node: $t \in V(T)$ is a *leaf node*. Then, $|X_t| = 0$ and hence it is trivial as $\mathcal{D}(t) = \mathcal{C}(t)$.

Introduce Node: Let $t \in V(T)$ be an *introduce node* where $X_t = X_{t'} \cup \{u\}$. u is the only new vertex in where a color of k tuple has to be assigned. For every $R \in \binom{[q]}{k}$, for every $f' \in \mathcal{D}(t')$, we check if f' can be extended to $f : X_t \to [q]^k$ by assigning R to t. This takes $\binom{q}{k} \cdot |\mathcal{D}(t')| = \binom{q}{k}^{tw+1}$ time.

Forget Node: Let $t \in V(T)$ be a *forget node* where $X_t = X_{t'} \setminus \{u\}$. u is the vertex which was in t' but not in t. Then, we copy all the colorings of $\mathcal{D}(t')$ to $\mathcal{D}(t)$ where color tuple of the vertex u is not mentioned and then remove the redundant copies. Removing redundant copies can also be done in $\mathcal{O}(|\mathcal{D}(t')| \log_2 |\mathcal{D}(t')|) = \mathcal{O}((\binom{q}{k})^{tw} \cdot poly(n, tw, k))$ time by sorting all the members of $\mathcal{D}(t)$ in lexicographic order and identifying repetitions.

Join Node: Let $t \in V(T)$ be a *join node* where $X_t = X_{t_1} = X_{t_2}$. If we compute the intersection of two sets in a very naive way, then we will spend $|\mathcal{D}(t_1)| \cdot |\mathcal{D}(t_2)|$ time. That's why we again sort both $\mathcal{D}(t_1)$ and $\mathcal{D}(t_2)$ separately and then compute the intersection in $\mathcal{O}(|\mathcal{D}(t_1)| + |\mathcal{D}(t_2)|)$ time. This procedure takes $\mathcal{O}(|\mathcal{D}(t)| \cdot \log_2 |\mathcal{D}(t)|) = \mathcal{O}((\binom{q}{k})^{tw} \cdot poly(n, tw, k))$ time as t is a join node. $\quad \square$

The following theorem follows from the above lemma.

Theorem 5. *Given an n-vertex graph G with its tree decomposition of width tw, q-(k,i)-COLORING can be solved in time $\mathcal{O}(q^{k \cdot tw} \cdot n^{\mathcal{O}(1)})$.*

Proof. From Lemma 5, $\mathcal{D}(t)$ and $\mathcal{C}(t)$ can be computed in time $\mathcal{O}((\binom{q}{k})^{tw} \cdot n^{\mathcal{O}(1)})$ time. Let the root node of the tree decomposition be r. We say that the instance is a YES-INSTANCE if and only if $\mathcal{D}(r) \neq \emptyset$. Clearly when $\mathcal{D}(r) \neq \emptyset$, there exists a proper (k,i)-coloring of G using at most q colors. But when $\mathcal{D}(r) = \emptyset$, then we see that no proper (k,i)-coloring of $G[X_r]$ is extendable to a proper coloring of G. Then it is a NO-INSTANCE. Therefore, the algorithm correctly decides in $\mathcal{O}((\binom{q}{k})^{tw} \cdot n^{\mathcal{O}(1)})$ time whether there exists (k,i)-coloring of G using q colors. $\quad \square$

6   Conclusions

We considered the (k,i)-coloring problem which is a generalization of proper coloring and is a well motivated problem from coding theory. Difficulty introduced by this generalization is evident by the fact that no polynomial time algorithm is known to optimally color a given clique. In this paper, we initiate a study of exact and parameterized algorithms for (k, i)-coloring. We provide exact algorithms running in time c^n for two cases viz $i = 0$ and $i = k - 1$. NP-hardness of (k, i)-COLORING for any $0 < i < k$ is still an open question. It is also interesting to find graph classes in which this problem can be solved in polynomial time. We prove that this problem is FPT when parameterized by *treewidth* with number of colors as a combined parameter. We also provide an FPT algorithm (without treewidth machinery) when parameterized by the size of *vertex cover* of the graph. It is an interesting open question whether we can get an FPT algorithm for (k, i)-coloring parameterized by the size of *feedback vertex set* of the graph that does not use treewidth machinery.

Acknowledgements. The second and third authors thank Debajyoti Ghosh for introducing the problem.

References

1. Agrell, E., Vardy, A., Zeger, K.: Upper bounds for constant-weight codes. IEEE Trans. Inf. Theory **46**(7), 2373–2395 (2000)
2. Hilton, A., Rado, R., Scott, S.: A (<5) color theorem for planar graph. Bull. London Math. Soc. **5**, 302–306 (1973)
3. Berge, C.: Les problemes de coloration en théorie des graphes. Publ. Inst. Stat. Univ. Paris **9**, 123–160 (1960)
4. Bollobas, B., Thomason, A.: Set colourings of graphs. Discret. Math. **25**(1), 2126 (1979)
5. Bonomo, F., Duran, G., Koch, I., Valencia-Pobon, M.: On the (k, i)-coloring of cacti and complete graphs. Ars Combinatorica (2014)
6. Brigham, R., Dutton, R.: Generalized k-tuple colorings of cycles and other graphs. J. Comb. Theory Ser. B **32**, 90–94 (1982)
7. Cai, L.: Parameterized complexity of vertex colouring. Discret. Appl. Math. **127**(3), 415–429 (2003)
8. Chudnovsky, M., Robertson, N., Seymour, P., Thomas, R.: The strong perfect graph theorem. Ann. Math. **164**, 51–229 (2006)
9. Cygan, M., Fomin, F.V., Kowalik, L., Lokshtanov, D., Marx, D., Pilipczuk, M., Pilipczuk, M., Saurabh, S.: Parameterized Algorithms, vol. 4. Springer, Heidelberg (2015)
10. Díaz, I.M., Zabala, P.: A generalization of the graph coloring problem. Investigation Operativa (1999)
11. Downey, R.G., Fellows, M.R.: Parameterized Complexity. Springer Science and Business Media, New York (2012)
12. Fiala, J., Golovach, P.A., Kratochvíl, J.: Parameterized complexity of coloring problems: treewidth versus vertex cover. Theor. Comput. Sci. **412**(23), 2513–2523 (2011)
13. Fomin, F.V., Kratsch, D.: Exact Exponential Algorithms: Texts in Theoretical Computer Science. An EATCS Series. Springer, Heidelberg (2010)
14. Irving, R.W.: NP-completeness of a family of graph-colouring problems. Discret. Appl. Math. **5**(1), 111–117 (1983)

15. Jansen, B.M.P., Kratsch, S.: Data reduction for graph coloring problems. Inf. Comput. **231**, 70–88 (2013)
16. Koivisto, M.: An $\mathcal{O}^*(2^n)$ algorithm for graph coloring and other partitioning problems via inclusion-exclusion. In: FOCS, pp. 583–590. IEEE (2006)
17. Papadimitriou, C.H., Steiglitz, K.: Combinatorial Optimization: Algorithms and Complexity. Courier Corporation, Mineola (1982)
18. Stahl, S.: n-tuple colorings and associated graphs. J. Comb. Theory Ser. B **20**, 185–203 (1976)

The Graph of the Pedigree Polytope
is Asymptotically Almost Complete
(Extended Abstract)

Abdullah Makkeh, Mozhgan Pourmoradnasseri, and Dirk Oliver Theis[✉]

Institute of Computer Science, University of Tartu, Ülikooli 17, 51014 Tartu, Estonia
{mozhgan,dotheis}@ut.ee
http://ac.cs.ut.ee/

Abstract. Graphs (1-skeletons) of Traveling-Salesman-related polytopes have attracted a lot of attention. Pedigree polytopes are extensions of the classical Symmetric Traveling Salesman Problem polytopes (Arthanari 2000) whose graphs contain the TSP polytope graphs as spanning subgraphs (Arthanari 2013). Unlike TSP polytopes, Pedigree polytopes are not "symmetric", e.g., their graphs are not vertex transitive, not even regular.

We show that in the graph of the pedigree polytope, the quotient minimum degree over number of vertices tends to 1 as the number of cities tends to infinity.

Keywords: Polytope · Extension · 1-Skeleton/Graph of a polytope · Traveling Salesman Problem

1 Introduction

Steinitz's Theorem states that 3-connected planar graphs are precisely the graphs of 3-dimensional polytopes.

Properties of graphs of polytopes of higher dimension are of interest not only in the combinatorial study of polytopes, but also in Combinatorial Optimization, and Theoretical Computer Science.

The famous Hirsch conjecture in the combinatorial study of polytopes, settled by Santos [18], concerned the diameter of graphs of polytopes.

In Combinatorial Optimization, the study of the graphs of polytopes associated with combinatorial optimization problems was initially motivated by the search for algorithms for these problems.

In Theoretical Computer Science, the theorem by Papadimitriou [16] that Non-Adjacency of vertices of (Symmetric) Traveling Salesman Problem (TSP)

D.O. Theis—Supported by the Estonian Research Council, ETAG (*Eesti Teadusagentuur*), through PUT Exploratory Grant #620, and by the European Regional Development Fund through the Estonian Center of Excellence in Computer Science, EXCS.

© Springer International Publishing AG 2017
D. Gaur and N.S. Narayanaswamy (Eds.): CALDAM 2017, LNCS 10156, pp. 294–307, 2017.
DOI: 10.1007/978-3-319-53007-9_26

polytopes is NP-complete, gave rise to similar results about other families of polytopes (cf. [1,11] and the references therein, for recent examples).

There have been particularly many attempts to understand the graph of TSP polytopes, and, where this turned out to be infeasible, of TSP-related polytopes (e.g., [22]; cf. [6,13,15,23,24] and the references therein). The presence of long cycles has been studied ([21], see also [12,14]), as has the graph density/vertex degrees (e.g., [19], see also [7,9]).

The motivation for the research in this paper was a 1985 conjecture by Grötschel and Padberg [6] — well-known in polyhedral combinatorial optimization — stating that the graph of TSP polytopes has diameter 2 (also, Problem # 36 in [20]). Already in [6], Grötschel and Padberg extend the question for the diameter to a family of TSP-related polytopes which seemed easier to understand at the time.

A more recent family of TSP-related polytopes are the *Pedigree polytopes* of Arthanari [4]. For this family of polytopes, adjacency of vertices can be decided in polynomial time [2]. Moreover, the graphs of the TSP polytopes are spanning subgraphs of the graphs of the Pedigree polytopes [3]. This is so mainly because the Pedigree polytope for n cities is an *extension*, without "hidden" vertices, of the TSP polytope (cf., [5,17]).

In this paper, we prove the following about graphs of Pedigree polytopes. Recall that the number of vertices of the TSP polytope for n cities is the number of cycles with vertex set $[n] := \{1, \ldots, n\}$, which is $(n-1)!/2$.

Theorem 1. *The minimum degree of a vertex on the Pedigree polytope for n cities is $(1 - o(1)) \cdot (n-1)!/2$ (for $n \to \infty$).*

In particular, the density graph of Pedigree polytopes is asymptotically equal to 1. Note, though, that while for TSP polytopes, these two statements are equivalent, this is not the case of Pedigree polytopes. The reason is that Pedigree extensions are not as "symmetric" as TSP polytopes (cf. [8]): For every two vertices u, v of the TSP polytope for n cities, there is an affine automorphism of the polytope mapping u to v. (Similar statements are true for monotone-TSP and graphical-TSP polytopes.) This is not true for Pedigree polytopes: Arthanari's construction removes the symmetry to a large extent.

Numerical simulations show that, even for relatively large n (say, ≈ 100), the graph of the Pedigree polytope is not complete. We have made no attempt, however, to find a non-trivial upper bound for the minimum degree.

We Now Give a Non-technical Description of the Proof of Theorem 1. Arthanari's beautiful idea of a pedigree is that of a cycle "evolving" over time: Starting from the unique cycle with node set $\{1, 2, 3\}$ at time 3, at time $n \geq 4$, the node n is added to the cycle by subdividing one of its edges. We say that n is *inserted into* that edge.

Arthanari's combinatorial condition for adjacency on the Pedigree polytope can be thought of as a process, too, with a *pedigree graph* G evolving over time. Suppose we have two evolving cycles. Let us refer to A as Alice's cycle, and to

B as Bob's cycle. At time n, Alice chooses an edge of her current cycle A (with node set $[n-1]$) and inserts her new node n into that edge to form her new cycle (with node set $[n]$). At the same time, Bob chooses an edge of his current cycle B, and inserts his new node n into that edge to form his new cycle.

The pedigree graph G may also change at time n. The new pedigree graph is either equal to the current one, or arises from the current one by adding the new vertex[1] n with incident edges. The choices of Alice and Bob determine: whether the new vertex is added or not; the number of edges incident to the new vertex n; the end vertices of these edges.

Arthanari's combinatorial characterization of adjacency on the Pedigree polytope is now this.

Theorem 2 [2]. *At all times $n \geqslant 4$, the two vertices of the Pedigree polytope for n cities corresponding to the (new) cycles A and B with node set $[n]$ are adjacent in the Pedigree polytope, if, and only if, the (new) graph G is connected.*

Theorem 1 states that, if B is a cycle chosen uniformly at random from all cycles on $[n]$, then

$$\min_{A} \mathbb{P}\big(\{ \text{ the pedigree graph is connected } \}\big) = 1 - o(1),$$

where the minimum ranges over all cycles on $[n]$. Lower bounding this quantity amounts to studying the following "connectivity game": Alice's goal is to make the graph G disconnected; whereas Bob makes uniformly random choices all the time. We prove that Alice loses with probability $1 - o(1)$. To analyze the game, we study a kind of a Markov Decision Process with state space $\mathbb{Z}_+ \times \mathbb{Z}_+$. The states are pairs (s, t), where s is the number of common edges in Alice's and Bob's cycles, and t is the number of connected components of the current pedigree graph.

In the next section, we will give rigorous statements corresponding to the hand-waving explanations above. In Sect. 3, we prove some basic facts about Bob playing randomly, and discuss the intuition of the proof of the main theorem. In Sect. 4, we introduce the Markov-Decision-Problem-ish situation that Alice finds herself in. The proof of the main theorem is sketched in Sect. 5; due to space limitations we have to refer to the upcoming journal version [10] of the paper for the details. We conclude with a couple of questions for future research which we find compelling.

2   Exact Statements of the Definitions, Facts, and Results

2.1   Cycles, One Node at a Time

Our cycles are undirected (so, e.g., there is only one cycle on 3 nodes). For ease of notation, let us say that the *positive direction* on a cycle with node set $[n]$,

---

[1] We speak of *vertices* of the pedigree graph and *nodes* of the cycles, to limit confusion.

$n \geqslant 3$, is the one in which, when starting from the node 1, the node 2 comes before the node 3; the other direction the *negative direction*. When referring to the kth edge of a cycle, we count the edges in the positive direction; the 1st one being the one incident on node 1. E.g., in the unique cycle with node set $\{1, 2, 3\}$, the 1st edge is $\{1, 2\}$, the 2nd edge is $\{2, 3\}$, and the 3rd edge is $\{3, 1\}$.

As mentioned in the introduction, Arthanari's Pedigree is a combinatorial object representing the "evolution" of a cycle "over time", and the combinatorial definition of adjacency of pedigrees makes use of that step-by-step development. The set of Pedigrees is in bijecton with the set of cycles. In our context (we do not have to associate points in space with Pedigrees), defining Pedigrees and then explaining the bijection with cycles is more cumbersome than necessary. For convenience, we use the following more convenient definitions, which mirror the definition of Pedigrees, but they use cycles only. Let us say that an *infinite cycle*[2] is a sequence $A = c_\square \in \prod_{n=3}^{\infty}[n]$. An infinite cycle A gives rise to an infinite sequence $A_\square$ of finite cycles (in the usual graph theory sense), defined inductively as follows:

- A_3 is the unique cycle with node set $\{1, 2, 3\}$;
- for $n \geqslant 3$, A_n is the cycle with node set $[n]$ which arises from adding the node n to A_{n-1} by inserting it into (i.e., subdividing) the c_{n-1}th edge.

We think of $A_\square$ as a cycle developing over time: At time n, the node n is added.

We will need to access the neighbors of node n in A_n, i.e., the ends of the edge into which n is inserted (i.e., which is subdivided) when n is added to $A_\square$. We write $\nu_A^+(n)$ for the neighbor of n in A_n following n in the positive direction, and $\nu_A^-(n)$ for the neighbor of n in A_n following n in the negative direction. The unordered pair $\nu_A(n) = \{\nu_A^+(n), \nu_A^-(n)\}$ is the c_{n-1}th edge of A_{n-1}, the one into which n was inserted.

These definitions are for $n \geqslant 4$ but extend naturally for $n = 1, 2, 3$: for $n = 3$ we let $\nu_A^+(3) = 1$, and $\nu_A^-(3) = 2$; for $n = 2$, we let $\nu_A^+(2) = \nu_A^-(2) = 1$. The equation $\nu_A(n) = \{\nu_A^+(n), \nu_A^-(n)\}$ holds for $n \geqslant 2$ (so $|\nu_A(2)| = 1$); for $n = 1$ we have $\nu_A(1) := \emptyset$.

Remark 3 (Finding $\nu(k)$ for "old" nodes k). It is readily verfied directly from the definition, that, for $k \geqslant 2$, $\nu_A^\pm(k)$ can be found as follows: start from node k and walk in positive direction. The first node smaller than k which you encounter is $\nu_A^+(k)$. Similarly, if you walk in negative direction starting from k, the first node smaller than k which you hit, is $\nu_A^-(k)$.

A pair of nodes i, j split each cycle A_n, $n > i, j$ into two (open) segments (i, j do not belong to either segment). We say that the *segment between i and j* is the one which does *not* contain the node $\min(\{1, 2, 3\}\backslash\{i, j\})$ (i.e., 1, unless

---

[2] The reason why we use this notion of "infinite cycle" is pure convenience. It does not add complexity, but it makes many statements and proofs less cumbersome. Indeed, instead of an infinite cycle, it is ok to just use a cycle whose length M is longer than all the lengths occuring in the particular argument. So instead of "let A be an infinite cycle, and consider A_k, A_ℓ, A_n" you have to say "let M be a large enough integer, A_M a cycle of length M, and A_k, A_ℓ, A_n sub-cycles of A_M". All the little arguments (e.g., Fact 8 below) have to be done in the same way.

$1 \in \{i, j\}$, in that case, 2, unless $\{1, 2\} = \{i, j\}$, in that case 3). Note that this does not depend on the choice of $n > i, j$, which justifies to say *"the segment of $A_\square$ between i and j"*.

Remark 4 (Testing/finding n with $\nu(n) = \{i, j\}$). Given a pair of nodes $\{i, j\}$ and $n' > i, j$, there exists an $n \leqslant n'$ with $\nu_A(n) = \{i, j\}$ if, and only if, the segment between i and j on $A_{n'}$ is non empty and every node in it is larger than both i and j. In that case, the smallest node, n, in the segment between i and j on $A_\square$ is the one with $\nu(n) = \{i, j\}$.

2.2   The Pedigree Graph

Two infinite cycles A, B give rise to a sequence of graphs $G_\square^{AB}$ which we call the *pedigree graphs*. We omit the superscripted A, B when possible. We speak of *vertices* of the pedigree graphs (rather than nodes). We do this to avoid confusion between the nodes of the cycles $A_\square, B_\square$ and the vertices of $G_\square^{AB}$, because the vertex set of G_n is a subset of $\{4, \ldots, n\}$, and hence of the node set of A_n and B_n. So a node $k \in [n]$ may or may not be a vertex of G_n.

The pedigree graph G_{n-1} is the subgraph of G_n induced by the vertices in $[n-1]$. In other words, G_n is either equal to G_{n-1} (if n is not a vertex), or it arises from G_{n-1} by adding the vertex n together with edges between n and vertices in $[n-1]$.

Example 5. G_1, G_2, G_3 are graphs without vertices. G_4 may be a graph without vertices, or it may consist of a single isolated vertex 4. G_5 could be a graph without vertices; a graph with a single vertex 5; a graph with two isolated vertices 4, 5, or a graph with two vertices 4, 5, linked by an edge. Check Fig. 1 for possible G_4 and G_5.

According to Arthanari [2,3] the condition for the existence of vertices is the following:

(1) A node $n \in [n]$ is a vertex of G_n, iff $\nu_A(n) \neq \nu_B(n)$.

There are several conditions for the presence of edges between the vertex n and earlier vertices. To make it easier to distinguish these, we speak of edge "types" and give the edges implicit "directions:" from A to B or from B to A. Here are the conditions for edges from n to earlier vertices.

(2) There is a *type-1 edge* "from A to B" between n and $k \in [n-1]$, if $\nu_A(n) = \nu_B(k)$. (Note that the condition implies that k is a vertex.)
(3) There is a *type-1 edge* "from B to A" Ditto, with A and B exchanged.
(4) There is a *type-2 edge* "from A to B" between n and $\ell := \max \nu_A(n)$, unless $\nu_B(\ell) \cap \nu_A(n) \neq \emptyset$. In other words, suppose the node n was inserted into the edge $\{k, \ell\}$ in A, with $k < \ell$. Now look up the end-nodes of the edge $\nu_B(\ell)$ into which ℓ was inserted when it was added to B. Unless k coincides with one of these end nodes, there is an edge between n and ℓ.
(5) *Type-2 edge* "from B to A" Ditto, with A and B exchanged.

Arthanari's theorem [2] Theorem 2 states that, if $n \geqslant 4$, and A_n, B_n are two cycles with node set $[n]$, then the two vertices of the Pedigree polytope (for n cities) corresponding to A_n and B_n are adjacent, if, and only if, G_n^{AB} is connected.

We will always think of A as "Alice's cycle" and B as "Bob's cycle".

Example 6. Going through an example will help understand the definition of a pedigree graph. Figure 1 shows two cycles A and B evolving over time $n = 3, \ldots, 10$, together with the evolving pedigree graph $G_\square^{AB}$.

$n = 3$: As mentioned above, G_3^{AB} is a graph without vertices.

$n = 4$: Alice inserts her new node 4 between into the edge $\{1,2\}$ of her cycle A_3; Bob inserts his new node 4 into the edge $\{1,3\}$ of his cycle B_3. Hence, $\{1,2\} = \nu_A(4) \neq \nu_B(4) = \{1,3\}$, so vertex 4 is added to G_3^{AB}.

$n = 5$: Alice inserts her new node 5 into the edge $\{2,4\}$ of her cycle A_4; Bob inserts his new node 5 into the edge $\{1,2\}$ of his cycle B_4. Since $\{2,4\} = \nu_A(5) \neq \nu_B(5) = \{1,2\}$, vertex 5 is added to G_4^{AB}. Let us check the edges:
- In B_4, the segment between 2 and 4 contains the node 3 which is smaller than 4. By Remark 4, there is no k with $\nu_A(5) = \nu_B(k)$, and hence no type-1 edge from A to B at this time.
- As $\nu_B(5) = \nu_A(4)$, there is a type-1 edge between 4 and 5 from B to A.
- Since $\max \nu_A(5) = 4$ and $\nu_B(4) = \{1,3\} \not\ni 2$, there is also a type-2 edge between 5 and 4 from A to B.
- MOZHGAN: Type-2 edge from B to A.

$n = 6$: Alices inserts her new node 6 into the edge $\{2,3\}$ of her cycle, Bob inserts his new node 6 into the edge $\{2,3\}$ of his cycle. Since $\{2,3\} = \nu_A(6) = \nu_B(6) = \{2,3\}$, we don't have a vertex 6 in G^{AB}.

$n = 7$: Alice throws into $\{4,5\}$, Bob throws into $\{3,4\}$. Since $\{4,5\} = \nu_A(7) \neq \nu_B(7) = \{3,4\}$, the vertex 7 is added to $G_6^{A,B}$.
- MOZHGAN: Type-1 edge from A to B.
- MOZHGAN: Type-1 edge from B to A.
- As $\max \nu_A(7) = 5$ and $\nu_B(5) = \{1,2\} \not\ni 4$, we have a type-2 edge from A to B between 7 and 5.
- As $\max \nu_B(7) = 4$ and $\nu_A(4) = \{1,2\} \not\ni 3$, there is also a type-2 edge from B to A between 7 and 4.

$n = 8$: Alice plays $\{3,6\}$, Bob chooses $\{1,4\}$. Since $\{3,6\} = \nu_A(8) \neq \nu_B(8) = \{1,4\}$, the vertex 8 is added to $G_7^{A,B}$.
- In B_7, the segment between 3 and 6 is empty (just the edge). By Remark 4, there is no k with $\nu_A(8) = \nu_B(k)$, and hence no type-1 edge from A to B.
- For the same reason (segment between 1 and 4 empty), there is no type-1 edge from B to A incident to the vertex 8.
- $\max \nu_A(8) = 6$ and $\nu_B(6) = \{2,3\} \ni 3$. So there is no type-2 edge from A to B between 8 and a smaller vertex.
- $\max \nu_B(8) = 4$ and $\nu_A(4) = \{1,2\} \ni 1$. So there is no type-2 edge from B to A between 8 and a smaller vertex.

Hence, vertex 8 is isolated in G_8.

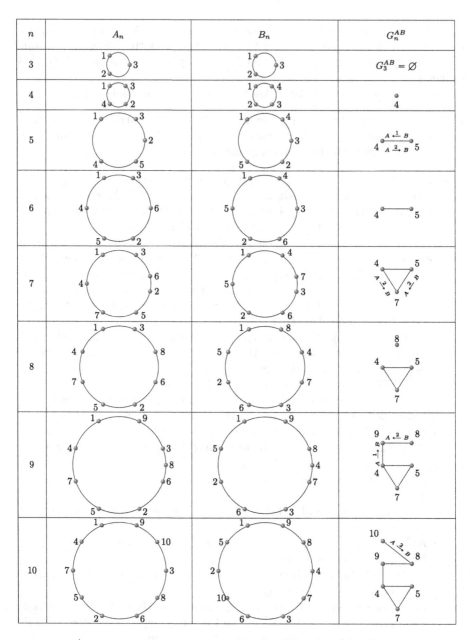

Fig. 1. Cycles $A_\square$ and $B_\square$ and corresponding $G_\square^{AB}$

$n = 9$: Alice chooses $\{1,3\}$, Bob chooses $\{1,8\}$. As $\{1,3\} = \nu_A(9) \neq \nu_B(9) = \{1,8\}$, the vertex 9 is added to $G_8^{A,B}$.

  – As $\{1,3\} = \nu_A(9) = \nu_B(4) = \{1,3\}$, there is a type-1 edge from A to B between 9 and 4.

- MOZHGAN: Type-1 edge from B to A.
- MOZHGAN: Type-2 edge from A to B.
- As $\max \nu_B(9) = 8$ and $\nu_A(8) = \{3, 6\} \not\ni 1$, there is a type-2 edge from B to A between 9 and 8.

$n = 10$: Alice chooses $\{3, 9\}$, Bob chooses $\{2, 6\}$. Since $\{3, 9\} = \nu_A(10) \neq \nu_B(10)$ $= \{2, 6\}$, the vertex 10 is added to $G_9^{A,B}$.
- MOZHGAN: Type-1 edge from A to B.
- Again, in B, the segment between 3 and 9 has a vertex (4) smaller than 9. Remark 4 gives us that there is no k $\nu_B(k) = \nu_A(10)$, so no type-1 edge from B to A is created.
- As $\max \nu_A(10) = 9$ and $\nu_B(9) = \{1, 8\} \not\ni 3$, there is a type-2 edge from A to B between 10 and 9.
- As $\max \nu_B(10) = 6$ and $\nu_A(6) = \{2, 3\} \ni 2$, no type-2 edge from B to A is created.

2.3 Rephrasing Theorem 1

We now rephrase Theorem 1, in terms of the pedigree graph. We also unravel the little-o, and move to the "Alice-and-Bob" letters for the cycles.

Theorem 7 (Theorem 1, rephrased). *For every $\varepsilon > 0$ there is an integer N such that for all $n \geqslant N$ and all cycles A_n with node set $[n]$, if B_n is drawn uniformly at random from all cycles with node set $[n]$, then*

$$\mathbb{P}\big(G_n^{AB} \text{ is connected }\big) \geqslant 1 - \varepsilon.$$

In symbols, and using infinite cycles, this reads:

$$\forall \varepsilon > 0 \; \exists N \colon \; \forall A \, \forall n \geqslant N \colon \mathbb{P}(G_n^{AB} \text{ is connected }) \geqslant 1 - \varepsilon,$$

where the probability is taken over all infinite cycles, see the next section. A close look at our proof shows that we are actually proving the following stronger statement (we don't have any use for it, though):

$$\forall \varepsilon > 0 \; \exists N \colon \; \forall A \colon \; \mathbb{P}\big(\forall n \geqslant N : G_n^{AB} \text{ is connected }\big) \geqslant 1 - \varepsilon.$$

3 Pedigree Graphs of Random Cycles

We have to reconcile uniformly random cycles with the "evolution over time" concept of pedigrees. The definition of an infinite cycle makes that very convenient, just do the same as with infinite sequences of coin tosses: Take, as probability measure on the sample space $\prod_{n=1}^{\infty}[n]$ of all infinite cycles the product of the uniform probability measures on each of the sets $[n]$, $n \geqslant 3$. We refer to the atoms in this probability space as *random infinite cycles*. The following is a basic property of product probability spaces. We will use it mostly without mentioning it.

Fact 8. *If B is a random infinite cycle, then, for each $n \geqslant 3$, the cycle B_n is uniformly random in the set of all cycles with node set $[n]$.*

Creating Isolated Vertices. The first substantial result about the connectedness of the pedigree graph, concerns the creation of isolated vertices.

As outlined in the introduction, we study the situation in which Alice chooses her edge of A_{n-1} according to a sophisticated strategy, whereas Bob always chooses a uniformly random edge of B_{n-1} to insert his node n into (which amounts to his cycle B_n being uniformly random in the set of all cycles on $[n]$, by Fact 8). In this section, we adopt a purely "random graph" perspective. For fixed A and random B, the pedigree graphs $G_\square^{AB}$ are a sequence of random graphs, with some weirdo distribution: At time n, whether the new vertex n is added or not, and if it is, how many incident edges it has, and what their end vertices are — these are all random events/variables.

For deterministic A and random B, let the random variable Y count the total number of times that an isolated vertex of the pedigree graph is created. In other words, $Y = \sum_{n=4}^\infty \mathbf{1}_{I_n}$, where I_n denotes the event that, at time n, n is added as an isolated vertex to $G_n^{A,B}$ (and $\mathbf{1}_\square$ is the indicator random variable of the event).

Lemma 9. *Whatever Alice does,* $\mathbb{E}Y = 2$.
Moreover, for every $\varepsilon > 0$, *if* $n_0 \geqslant 4/\varepsilon + 2$, *then, whatever Alice does*

$$\mathbb{P}\Big(\bigcup_{n \geqslant n_0} I_n \Big) \leqslant \varepsilon.$$

For the proof we refer to the journal version of this extended abstract [10]. To understand why the lemma is important, consider a pedigree graph at time n, just before Alice and Bob make their choices of cycle edges into which their respective new nodes n are inserted. If n is not a vertex of the new pedigree graph G_n, the number of connected components of $G_\square$ doesn't change. If n is a vertex, and and it does have incident edges, then the number of connected components can only decrease. The only way that the number of connected components of G_n can increase is if n is an isolated vertex in the new pedigree graph. Hence, Lemma 9 provides an upper bound on the expected number of connected components, uniform over n.

The Intuition. From Lemma 9, it is unlikely that the pedigree graph will have many components. Indeed, intuitively, if only 2 isolated vertices are ever created, that means that most of the time either nothing happens (no new vertex) or edges are created, ultimately reducing the number of components, so the pedigree graph is connected.

While this basic intuition is essentially correct, a closer look reveals some subtleties. First of all, Alice has a big sway in choosing the end vertices of new edges: she can pick the end vertices of type-2 edges from A to B; and she can influence the end vertices of type-1 edges (both directions).

Secondly, Bob's choices are reduced by the low degrees of the vertices. (A stronger version of (a) is proved in [10].)

Lemma 10. *The maximum degree of a vertex in a pedigree graph is at most 6:*

(a) up to 2 to vertices created in the past; and
(b) up to 6 to future vertices.

Hence, if a vertex n_0 was created as an isolated vertex or landed in a small connected component, Bob has only 4–6 shots at connecting it to another connected component. The good news is that Alice can never "shut down" a connected component completely: Bob can always extend it by one more vertex.

Lemma 11. *Let C be a connected component of the pedigree graph G_{n-1}^{AB}. There exists a $k \in C$ such that, no matter what Alice's move is at time n, Bob has a move which creates the vertex n and makes it adjacent to k.*

Proof. Take $k := \max V(C)$. Since k is a vertex, we have $|\nu_B(k) \cap \nu_A(k)| \leqslant 1$. Suppose that $\nu_B^+(k) \notin \nu_A(k)$ (the other case is symmetric). Then, the first time Bob inserts a node, say n', into the edge on the positive side of k, this will create a type-2 edge "from B to A" between n' and k. Since k is the newest vertex in its component, Bob has not yet inserted a node there, so he can insert n there, now. $\qquad\qquad\square$

However, for Bob to make a disconnected pedigree graph connected, at some time, he will have to manage to insert his new node in such a way that it has two incident edges, linking two connected components at the same time.

There is no difficulty in realizing that Alice wouldn't stand a chance against a strategically playing Bob. But we claim that the game between a clever Alice and a blindfolded Bob will turn in Bob's favour almost all of the time.

Computer simulations give another indication that some care has to be taken implementing the basic intuition: Even for n as large as 100, even if Alice's cycle is chosen uniformly *at random* instead of adversarial, the frequency (in 100000 samples) with which we saw a connected pedigree graph was only about 84%. In the remaining 16% of cases, the typical situation is that of one giant connected component containing almost every vertex, and one tiny component growing only very slowly. This indicates that even a *disinterested* Alice can do some damage.

4   The Connectivity Game

At each time, Alice moves first. As already explained, she determines the cycle A by choosing, at each time n, the edge of A_{n-1} into which her new node n will be inserted. Then Bob moves. He determines B in the same way, but (using Fact 8), he will draw the edge of B_{n-1} into which his new node n is inserted uniformly at random from all edges of B_{n-1}, and his choice is independent of his earlier choices.

We say that Bob wins, if there exists an n_0 such that for all $n \geqslant n_0$, the pedigree graph G_n^{AB} is connected. We need Bob to win "uniformly", i.e., n_0 must not depend on Alice's moves.

Let the random variable T_n denote the number of connected components in the pedigree graph G_n^{AB}. To analyze the development of the random process $T_\square$, it turns out to be useful to consider a second random process, $S_\square$. Denote by $E_n^\cap$ the set of edges that Alice's cycle and Bob's cycles have in common,

$$E_n^\cap := E(A_n) \cap E(B_n),$$

and let

$$S_n := |E_n^\cap|$$

count the number of cycle edges that Alice and Bob have in common. We will distinguish Alice's moves by whether or not she chooses a common cycle edge to place her new cycle node. The set $E_n^{\cap*}$ holds those common edges which are not incident on the edge which Alice chooses for her new node:

$$E_n^{\cap*} := \{e \in E_n^\cap \mid e \cap \nu_A(n+1) = \varnothing\};$$

we let S^* count the edges in $E^{\cap*}$:

$$S_n^* := |E_n^{\cap*}|.$$

Finally, denote by E_n^r the set of edges in Bob's cycle which are neither common nor incident on Alice's chosen edge:

$$E_n^r := \{e \in E(B_n) \backslash E_n^\cap \mid e \cap \nu_A(n+1) = \varnothing\}; \text{ and}$$
$$R_n := |E_n^r|.$$

We are now ready to state and prove the transition probabilities. They depend on whether Alice chooses, for her new node, a common edge — we refer to that as a c-move by Alice — or an edge which is in the difference $E(A_n) \backslash E_n^\cap$ — we call that a d-move.

Lemma 12. *The conditional probabilities*

$$\mathbb{P}\Big(S_{n+1} = S_n + \Delta_S \ \wedge \ T_{n+1} = T_n + \Delta_T \ \mid \ B_n\Big)$$

satisfy these bounds (entries not shown are "=0"):

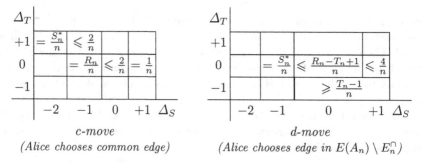

c-move
(Alice chooses common edge)

d-move
(Alice chooses edge in $E(A_n) \setminus E_n^\cap$)

The proof of this lemma requires some delicate distinguishing of cases, we refer to the journal version [10].

The proof of the main theorem now follows the following idea. From the tables in Lemma 12, you see that d-moves have chance of reducing the number of connected components — albeit a small one. Moreover, Alice cannot take a c-move only when $S_n > 0$, but c-moves have a strong tendency to reduce $S_\square$. We prove that the number of d-moves that Alice has to take are frequent enough to lead to a decrease in the number of connected components. This suffices to prove Theorem 5, along the lines sketched on page 10.

The next section gives more details of the proof.

5   Proof of Theorem 7

Using the Azuma-Hoeffding super-martingale tail bound, we can prove that, for large enough n_0, Alice has to take many d-moves between times n_0 and $2n_0$. Due to space restrictions, we have to refer to the journal version [10] for all of the proofs.

Lemma 13. *For every* $\varepsilon \in]0,1[$, *if* $n_0 \geqslant \max(900, 8\ln(1/\varepsilon))$, *and* $n_1 := 2n_0$ *then, whatever Alice does, the probability, conditioned on* B_{n_0} *and* $S_{n_0} \leqslant \ln^2 n_0$, *that among her moves at times* $n = n_0 + 1, \ldots, n_1$, *there are fewer than* $n_0/3$ *d-moves, is at most* ε.

From this, we deduce that must $T_\square$ decrease, but some sophistication is needed, because of the slow divergence of $\sum 1/n$: Indeed, between n_0 and $2n_0$, $T_\square$ decreases only with a constant probability:

Lemma 14. *Fix* $\delta := 1/42$. *If* $n_0 \geqslant \max(900, 8\ln(1/\delta))$, *and* $n_1 := 2n_0$ *then, whatever Alice does,*

$$\mathbb{P}\Big(\exists n \in \{n_0+1,\ldots,n_1\}\colon\ T_{n+1} < T_n \ \Big|\ B_{n_0}, T_{n_0} \geqslant 2, S_{n_0} \leqslant \ln^2 n_0\Big) \ \geqslant \ 1/7.$$

We can boost the probability to $1 - \varepsilon$, for arbitrary $\varepsilon > 0$, by iterating the argument $\Omega(\ln(1/\varepsilon))$ times.

Lemma 15. *Fix* $\delta := 1/42$. *For every* $\varepsilon \in]0, 1/56[$, *with* $a := 10\ln(2/\varepsilon)$, *if*

$$n_0 \geqslant \max(900, 8\ln(1/\delta), (2a)^{4/\varepsilon}, e^{6/\varepsilon}),$$

and $n_1 := 2an_0$ *then, whatever Alice does,*

$$\mathbb{P}\Big(\exists n \in \{n_0,\ldots,n_1\}\colon\ T_{n+1} < T_n \ \Big|\ B_{n_0}, T_{n_0} \geqslant 2\Big) \ \geqslant \ 1 - \varepsilon$$

Note that Lemma 15 also gets rid of the conditioning on $S_n \leqslant \ln^2 n$. We are now ready to complete the proof of the main theorem.

Proof (of Theorem 7). Let $\varepsilon' \in]0, 1/2[$ be given. Set $t := 6/\varepsilon'$. Since $T_\square$ can only increase when an isolated vertex is created, we have $T_n \leqslant Y$, for all $n \geqslant 4$, where Y is the number of isolated vertices. Hence, by Lemma 9 and Markov's inequality, we have

$$\mathbb{P}\Big(\exists n \geqslant 4: \ T_n \geqslant t + 1\Big) \leqslant \mathbb{P}(Y \geqslant t) \leqslant \mathbb{E}(Y)/t = \varepsilon'/3.$$

Now take $n_0' \geqslant 12/\varepsilon' + 2$, and large enough to apply Lemma 15 $n_0 := n_0'$ and to $\varepsilon := \varepsilon'/3t$ (note that this is less than $1/56$). Denote by a be the number defined in that lemma. Applying the lemma t times, for n_0 ranging over $n_0' + j2an_0'$, $j = 0, \ldots, t-1$, the probability that we fail at least once to obtain a decrease in the number of connected components, $T_\square$, is at most $\varepsilon'/3$. So, with probability at least $1 - 2\varepsilon'/3$, we must have $T_{n_0} = 1$ for one of these n_0's or for an n between $n_0' + (t-1)2an_0'$ and $n_0' + t2an_0'$.

Finally, since $n_0' \geqslant 12/\varepsilon' + 2$, by Lemma 9, with probability $1 - \varepsilon'/3$, $T_\square$ will not increase after n_0', and hence, with probability $1 - \varepsilon'$, will drop to 1 and stay there for all eternity. Bob wins.

6   Some Open Questions

There are two questions which we believe should be asked in the context of our result.

Firstly, are there other polytopes whose graphs are not complete, but the minimum degree is asymptotically that of a complete graph? Could that even be the case for the Traveling Salesman Problem polytope itself?

Secondly, in view of the Traveling Salesman Problem polytope, it would be interesting to find other combinatorial conditions on cycles which are implied by the adjacency of the corresponding vertices on the TSP polytope. The pedigree graph connectedness condition is derived from an extension of the TSP polytope, but maybe there are other combinatorial conditions without that geometric context. The graph resulting from such a condition might be "closer" to the actual TSP polytope graph, i.e., add fewer edges.

References

1. Aguilera, N., Katz, R., Tolomei, P.: Vertex adjacencies in the set covering polyhedron. arXiv preprint arXiv:1406.6015 (2014)
2. Arthanari, T.S.: On pedigree polytopes and hamiltonian cycles. Discret. Math. **306**, 1474–1792 (2006)
3. Arthanari, T.S.: Study of the pedigree polytope and a sufficiency condition for nonadjacency in the tour polytope. Discret. Optim. **10**(3), 224–232 (2013). http://dx.doi.org/10.1016/j.disopt.2013.07.001
4. Arthanari, T.S., Usha, M.: An alternate formulation of the symmetric traveling salesman problem and its properties. Discret. Appl. Math. **98**(3), 173–190 (2000)
5. Fiorini, S., Kaibel, V., Pashkovich, K., Theis, D.O.: Combinatorial bounds on nonnegative rank and extended formulations. Discret. Math. **313**(1), 67–83 (2013)

6. Grötschel, M., Padberg, M.W.: Polyhedral theory. In: Lawler, E.L., Lenstra, J.K., Kan, A., Shmoys, D.B. (eds.) The Traveling Salesman Problem. A Guided Tour of Combinatorial Optimization, chap. 8, pp. 251–306. Wiley (1985)

7. Kaibel, V.: Low-dimensional faces of random 0/1-polytopes. In: Bienstock, D., Nemhauser, G. (eds.) IPCO 2004. LNCS, vol. 3064, pp. 401–415. Springer, Heidelberg (2004). doi:10.1007/978-3-540-25960-2_30

8. Kaibel, V., Pashkovich, K., Theis, D.O.: Symmetry matters for sizes of extended formulations. SIAM J. Discret. Math. **26**(3), 1361–1382 (2012)

9. Kaibel, V., Remshagen, A.: On the graph-density of random 0/1-polytopes. In: Arora, S., Jansen, K., Rolim, J.D.P., Sahai, A. (eds.) APPROX/RANDOM - 2003. LNCS, vol. 2764, pp. 318–328. Springer, Heidelberg (2003). doi:10.1007/ 978-3-540-45198-3_27

10. Makkeh, A., Pourmoradnasseri, M., Theis, D.O.: On the graph of the pedigree polytope. arXiv:1611.08431 (2016)

11. Maksimenko, A.: The common face of some 0/1-polytopes with NP-complete non-adjacency relation. J. Math. Sci. **203**(6), 823–832 (2014)

12. Naddef, D.: Pancyclic properties of the graph of some 0–1 polyhedra. J. Comb. Theor. Ser. B **37**(1), 10–26 (1984)

13. Naddef, D.J., Pulleyblank, W.R.: The graphical relaxation: a new framework for the symmetric traveling salesman polytope. Math. Program. Ser. A **58**(1), 53–88 (1993). http://dx.doi.org/10.1007/BF01581259

14. Naddef, D.J., Pulleyblank, W.R.: Hamiltonicity in (0–1)-polyhedra. J. Comb. Theor. Ser. B **37**(1), 41–52 (1984)

15. Oswald, M., Reinelt, G., Theis, D.O.: On the graphical relaxation of the symmetric traveling salesman polytope. Math. Program. Ser. B **110**(1), 175–193 (2007). http://dx.doi.org/10.1007/s10107-006-0060-x

16. Papadimitriou, C.H.: The adjacency relation on the traveling salesman polytope is NP-complete. Math. Program. **14**(1), 312–324 (1978)

17. Pashkovich, K., Weltge, S.: Hidden vertices in extensions of polytopes. Oper. Res. Lett. **43**(2), 161–164 (2015)

18. Santos, F.: A counterexample to the Hirsch conjecture. Ann. Math. **176**(1), 383–412 (2012)

19. Sarangarajan, A.: A lower bound for adjacencies on the traveling salesman polytope. SIAM J. Discret. Math. **10**(3), 431–435 (1997)

20. Schrijver, A.: Combinatorial Optimization. Polyhedra and Efficiency. Algorithms and Combinatorics, vol. 24. Springer, Berlin (2003)

21. Sierksma, G.: The skeleton of the symmetric traveling salesman polytope. Discret. Appl. Math. **43**(1), 63–74 (1993)

22. Sierksma, G., Teunter, R.H.: Partial monotonizations of hamiltonian cycle polytopes: dimensions and diameters. Discret. Appl. Math. **105**(1), 173–182 (2000)

23. Theis, D.O.: A note on the relationship between the graphical traveling salesman polyhedron, the symmetric traveling salesman polytope, and the metric cone. Discret. Appl. Math. **158**(10), 1118–1120 (2010). http://dx.doi.org/10.1016/j.dam. 2010.03.003

24. Theis, D.O.: On the facial structure of symmetric and graphical traveling salesman polyhedra. Discret. Optim. **12**, 10–25 (2014). http://www.sciencedirect.com/ science/article/pii/S1572528613000625

Induced Matching in Some Subclasses of Bipartite Graphs

Arti Pandey[1(✉)], B.S. Panda[2], Piyush Dane[2], and Manav Kashyap[2]

[1] Department of Mathematics, Indian Institute of Technology Ropar,
Nangal Road, Rupnagar 140001, Punjab, India
artipandey2305@gmail.com
[2] Department of Mathematics, Indian Institute of Technology Delhi,
Hauz Khas, New Delhi 110016, India
bspanda@maths.iitd.ac.in

Abstract. For a graph $G = (V, E)$, a set $M \subseteq E$ is called a *matching* in G if no two edges in M share a common vertex. A matching M in G is called an *induced matching* in G if $G[M]$, the subgraph of G induced by M, is same as $G[S]$, the subgraph of G induced by $S = \{v \in V \mid$ v is incident on an edge of M$\}$. The MAXIMUM INDUCED MATCHING problem is to find an induced matching of maximum cardinality. Given a graph G and a positive integer k, the INDUCED MATCHING DECISION problem is to decide whether G has an induced matching of cardinality at least k. The INDUCED MATCHING DECISION problem is NP-complete on bipartite graphs, but polynomial time solvable for convex bipartite graphs. In this paper, we show that the INDUCED MATCHING DECISION problem is NP-complete for star-convex bipartite graphs and perfect elimination bipartite graphs. On the positive side, we propose polynomial time algorithms to solve the MAXIMUM INDUCED MATCHING problem in circular-convex bipartite graphs and triad-convex bipartite graphs by making polynomial reductions from the MAXIMUM INDUCED MATCHING problem in these graph classes to the MAXIMUM INDUCED MATCHING problem in convex bipartite graphs.

Keywords: Matching · Induced matching · Bipartite graphs · Graph algorithm · NP-complete

1 Introduction

Let $G = (V, E)$ be a graph. A set of edges $M \subseteq E$ is a *matching* if no two edges of M are incident on a common vertex. Vertices incident to the edges of a matching M are *saturated* by M. The MAXIMUM MATCHING problem is to find a matching of maximum cardinality in a given graph. The MAXIMUM MATCHING problem and its variations are extensively studied in literature. In this paper, we study an important variant of matchings called *induced matchings*.

The work was done when the first author (Arti Pandey) was in IIIT Guwahati.

D. Gaur and N.S. Narayanaswamy (Eds.): CALDAM 2017, LNCS 10156, pp. 308–319, 2017.
DOI: 10.1007/978-3-319-53007-9_27

A matching M in G is called an *induced matching* in G if $G[M]$, the subgraph of G induced by M, is same as $G[S]$, the subgraph of G induced by $S = \{v \in V|$ v is incident on an edge of M$\}$. A graph G with vertex set $V = \{a, b, c, d, g, h\}$ and edge set $E = \{ab, bc, cd, cg, gh, hd, ad, ac, bd, ch, gd\}$ is shown in Fig. 1. Let $M_1 = \{ab, gh\}$ and $M_2 = \{ab, cd, gh\}$. Note that M_1 is a matching as well as an induced matching in G, but M_2 is a matching but not an induced matching in G.

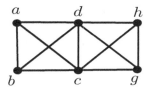

Fig. 1. Graph G.

For a graph G, the MAXIMUM INDUCED MATCHING problem is to find an induced matching of maximum cardinality in G. The maximum induced matching problem and its decision version are defined as follows:

MAXIMUM INDUCED MATCHING problem (MIMP)

Instance: A graph $G = (V, E)$.
Solution: An induced matching M in G.
Measure: Cardinality of the set M.

INDUCED MATCHING DECISION problem (IMDP)

Instance: A graph $G = (V, E)$ and a positive integer $k \leq |V|$.
Question: Does there exist an induced matching M in G such that $|M| \geq k$?

The MAXIMUM INDUCED MATCHING problem was introduced by Stock-meyer and Vazirani as "Risk-free Marriage problem" in 1982 [1]. The INDUCED MATCHING DECISION problem is NP-complete for general graphs [1], and remains so even for bipartite graphs [2] and k-regular graphs for $k \geq 4$ [3,4] (see [5] for a survey). The INDUCED MATCHING DECISION problem also remains NP-complete for bipartite graphs with maximum degree 3, and C_4-free bipartite graphs, which are two special classes of bipartite graphs [6]. On the other hand, the MAXIMUM INDUCED MATCHING problem is polynomial time solvable for many graph classes, for example chordal graphs [2], chordal bipartite graphs [7], trapezoid graphs, interval-dimension graphs and cocomparability graphs [8] etc. In this paper we study the MAXIMUM INDUCED MATCHING problem for some subclasses of bipartite graphs: perfect elimination bipartite graphs, star-convex bipartite graphs, circular-convex bipartite graphs, and triad-convex bipartite graphs. The class of circular-convex bipartite graphs was introduced by Liang and Blum [9] and has been studied recently by researchers (see [10–13]). The triad-convex bipartite graphs and star-convex bipartite graphs are studied in [12, 14–16]. The main contributions of the paper are summarized below.

1. We show that the INDUCED MATCHING DECISION problem is NP-complete for star-convex bipartite graphs and perfect elimination bipartite graphs.
2. We propose an $O(n^2m)$ time algorithm to solve the MAXIMUM INDUCED MATCHING problem in circular-convex bipartite graphs.
3. We propose an $O(n^8)$ time algorithm to solve the MAXIMUM INDUCED MATCHING problem in triad-convex bipartite graphs.

Our algorithms for the MAXIMUM INDUCED MATCHING problem in circular-convex bipartite graphs and triad-convex bipartite graphs are based on polynomial reduction for the MAXIMUM INDUCED MATCHING problem from these graph classes to convex bipartite graphs. The following result is already known for the MAXIMUM INDUCED MATCHING problem in convex bipartite graphs.

Theorem 1. [17] *The* MAXIMUM INDUCED MATCHING *problem can be solved in $O(n^2)$ time in convex bipartite graphs.*

2   Preliminaries

In a graph $G = (V, E)$, the sets $N_G(v) = \{u \in V(G) \mid uv \in E\}$ and $N_G[v] = N_G$ $(v) \cup \{v\}$ denote the *open neighborhood* and *closed neighborhood* of a vertex v, respectively. For a set $S \subseteq V$ of the graph $G = (V, E)$, the subgraph of G *induced* by S is defined as $G[S] = (S, E_S)$, where $E_S = \{xy \in E \mid x, y \in S\}$. For a set $E' \subseteq E$ of the graph $G = (V, E)$, the subgraph of G *induced* by E' is defined as $G[E'] = (V_{E'}, E')$, where $V_{E'} = \{x \in V \mid x$ is incident on an edge of $E'\}$. A graph G is said to be *chordal* if every cycle in G of length at least four has a *chord*, that is, an edge joining two non-consecutive vertices of the cycle. A graph $G = (V, E)$ is said to be *bipartite* if V can be partitioned into two disjoint sets X and Y such that every edge of G joins a vertex in X to a vertex in Y, and such a partition (X, Y) of V is called a *bipartition*. A bipartite graph with bipartition (X, Y) of V is denoted by $G = (X, Y, E)$. A bipartite graph G is said to be *chordal bipartite* if every cycle of length at least 6 has a chord.

Let $G = (X, Y, E)$ be a bipartite graph with $|X| = n_1$ and $|Y| = n_2$. G is called *convex bipartite graph* if there exists a linear ordering $<$ on X, say $x_1 < x_2 < \ldots < x_{n_1}$, such that for every vertex y in Y, $N_G(y) = \{x_i, x_{i+1}, \ldots, x_j\}$ for $1 \leq i \leq j \leq n_1$, that is, vertices in $N_G(y)$ are *consecutive* in the linear ordering $<$ on X. A set of consecutive vertices in the linear ordering $<$ on X is called an *interval*. G is called a *circular-convex bipartite graph* if there exists a circular ordering $\prec$ on X, say $x_1 \prec x_2 \prec \ldots \prec x_{n_1} \prec x_{(n_1+1)} = x_1$, such that for every vertex y in Y, either $N_G(y) = \{x_i, x_{i+1}, \ldots, x_j\}$ or $N_G(y) = \{x_j, x_{j+1}, \ldots, x_{n_1}, x_1, \ldots, x_i\}$ for $1 \leq i \leq j \leq n_1$, that is, vertices in $N_G(y)$ are *consecutive* in the circular ordering $\prec$ on X. A set of consecutive vertices in the clock-wise direction in the circular ordering $\prec$ on X is called a *circular arc* and the first vertex and the last vertex in the circular arc are called *left end point* and *right end point* of the circular arc, respectively.

A tree with exactly one non-pendant vertex is a *star*. A bipartite graph $G = (X, Y, E)$ is called a *tree-convex bipartite graph*, if a tree $T = (X, E_X)$ can be defined, such that for every vertex y in Y, the neighborhood of y induces a subtree of T. Tree-convex bipartite graphs are recognizable in linear time, and the associated tree T can also be constructed in linear-time [18]. For T a star, G is called *star-convex bipartite graphs*. For T a *triad*, that is, three paths with a common end-vertex, G is called a *triad-convex bipartite graph*. If T is a path, then G is a *convex bipartite graph*. Note that both the definitions of convex bipartite graphs are equivalent.

For a bipartite graph $G = (X, Y, E)$, an edge $uv \in E$ is a *bisimplicial edge* if $N_G(u) \cup N_G(v)$ induces a complete bipartite subgraph in G. Let $(e_1, e_2, \ldots, e_k)$ be an ordering of pairwise non-adjacent edges (no two edges have a common end vertex) of G (not necessarily all edges of E). Let S_i be the set of endpoints of edges $e_1, e_2, \ldots, e_i$ and let $S_0 = \emptyset$. Ordering $(e_1, e_2, \ldots, e_k)$ is a *perfect edge elimination ordering* for G if $G[(X \cup Y) \setminus S_k]$ has no edge and each edge e_i is bisimplicial in the remaining induced subgraph $G[(X \cup Y) \setminus S_{i-1}]$. G is a *perfect elimination bipartite graph* if G admits a perfect edge elimination ordering. The class of perfect elimination bipartite graphs was introduced by Golumbic and Goss [19]. The hierarchial relationship between subclasses of bipartite graphs is shown in Fig. 2.

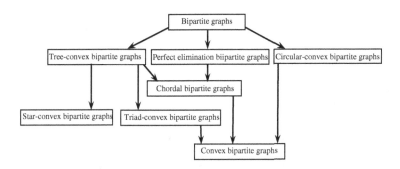

Fig. 2. The hierarchial relationship between subclasses of bipartite graphs.

3 NP-Completeness Results

In this section, we study the NP-completeness of the INDUCED MATCHING DECISION problem. The INDUCED MATCHING DECISION problem is NP-complete for bipartite graphs. We strengthen the complexity result of the INDUCED MATCHING DECISION problem, by showing that it remains NP-complete for star-convex bipartite graphs and perfect elimination bipartite graphs, two important subclasses of bipartite graphs.

3.1 Star-Convex Bipartite Graphs

In this section, we prove the hardness result for the INDUCED MATCHING DECISION problem in star-convex bipartite graphs. The following necessary and sufficient condition for a bipartite graph to be a star-convex bipartite graph will be useful in the polynomial reduction.

Lemma 1. [13] *A bipartite graph $G = (X, Y, E)$ is a star-convex bipartite graph if and only if there exists a vertex x in X such that every vertex y in Y is either a pendant vertex or is adjacent to x.*

Theorem 2. *The* INDUCED MATCHING DECISION *problem is NP-complete for star-convex bipartite graphs.*

Proof. Clearly, the INDUCED MATCHING DECISION problem is in NP for star-convex bipartite graphs. To show the NP-completeness, we give a polynomial reduction from the INDUCED MATCHING DECISION problem for bipartite graphs, which is already known to be NP-complete [2].

Given a bipartite graph $G = (X, Y, E)$, we construct a star-convex bipartite graph $H = (X_H, Y_H, E_H)$ in the following way: $X_H = X \cup \{u\}$, $Y_H = Y$, and $E_H = E \cup \{uy \mid y \in Y_H\}$. By Lemma 1, it is clear that the constructed graph is star-convex bipartite graph (as every vertex in Y_H is adjacent to the vertex $u \in X_H$). Now, the following claim is sufficient to complete the proof of the theorem.

Claim. G has an induced matching of size at least k if and only if H has an induced matching of size at least k.

Proof. The proof is omitted due to space constraint. □

Hence, the theorem is proved. □

3.2 Perfect Elimination Bipartite Graphs

In this section, we prove the hardness result for the INDUCED MATCHING DECISION problem in perfect elimination bipartite graphs. Since, the class of perfect elimination bipartite graphs is a subclass of bipartite graphs, and superclass of chordal bipartite graphs, our result reduces the complexity gap between bipartite graphs and chordal bipartite graphs.

Theorem 3. *The* INDUCED MATCHING DECISION *problem is NP-complete for perfect elimination bipartite graphs.*

Proof. Clearly, the INDUCED MATCHING DECISION problem is in NP for perfect elimination bipartite graphs. To show the NP-completeness, we give a polynomial reduction from the INDUCED MATCHING DECISION problem for bipartite graphs, which is already known to be NP-complete [2].

Given a bipartite graph $G = (X, Y, E)$ where $X = \{x_1, x_2, \ldots, x_{n_1}\}$ and $Y = \{y_1, y_2, \ldots, y_{n_2}\}$, we construct a bipartite graph $H = (X_H, Y_H, E_H)$ in the following way: For each $x_i \in X$, add a path $P_i = x_i, w_i, z_i, t_i$ of length 3. Formally $X_H = X \cup \{z_i \mid 1 \leq i \leq n_1\}$, $Y_H = Y \cup \{w_i, t_i \mid 1 \leq i \leq n_1\}$, and $E_H = E \cup \{x_i w_i, w_i z_i, z_i t_i \mid 1 \leq i \leq n_1\}$.

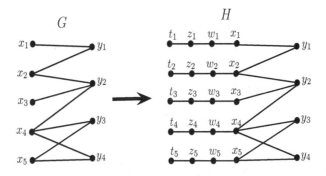

Fig. 3. An illustration to the construction of H from G.

Figure 3 illustrates the construction of H from G. Clearly H is a perfect elimination bipartite graph and $(z_1t_1, z_2t_2, \ldots, z_{n_1}t_{n_1}, x_1w_1, x_2w_2, \ldots, x_{n_1}w_{n_1})$ is perfect edge elimination ordering for H. Now, the following claim is sufficient to complete the proof of the theorem.

Claim. G has an induced matching of size at least k if and only if H has an induced matching of size at least $k + n_1$.

Proof. The proof is omitted due to space constraint.   □

Hence, the theorem is proved.   □

4   Circular-Convex Bipartite Graph

In this section we propose a polynomial time algorithm to compute a maximum induced matching in a circular-convex bipartite graph. Our algorithm is based on the reduction from circular-convex bipartite graph to convex bipartite graphs. Below we first give the construction of a convex bipartite graph from a given circular-convex bipartite graph.

Construction 1: Let $G = (X, Y, E)$ be a circular-convex bipartite graph with $|X| = n_1$ and $|Y| = n_2$. Let $e = x_iy_j \in E$. Without loss of generality, we can always assume a circular ordering $\prec$ on X, say $x_1 \prec x_2 \prec \ldots \prec x_{n_1} \prec x_{(n_1+1)} = x_1$, such that for every vertex y in Y, $N_G(y)$ is a circular arc. Now, construct the graph $G_e = (X_e, Y_e, E_e)$ as follows: $X_e = X \setminus N_G(y_j)$, $Y_e = Y \setminus N_G(x_i)$, and $E_e = \{xy \in E \mid x \in X_e, y \in Y_e\}$.

Lemma 2. G_e *is a convex bipartite graph.*

Proof. The proof is easy, and hence is omitted due to space constraint.   □

Lemma 3. *Let M be a maximum induced matching in G and $e \in M$. Let M_e be a maximum induced matching in G_e. Then $M' = M_e \cup \{e\}$ is also an induced matching in G and $|M| = |M'|$. In other words, M' is a maximum induced matching in G.*

Proof. The proof is omitted due to space constraint.   □

The detailed algorithm for finding a maximum induced matching in a circular-convex bipartite graph G is given in Fig. 4.

Algorithm 1: INDUCED-M-CIRCULAR-CONVEX()

Input: A circular-convex bipartite graph $G = (X, Y, E)$, where
$X = \{x_1, x_2, \ldots, x_{n_1}\}$ and $Y = \{y_1, y_2, \ldots, y_{n_2}\}$.
Output: A maximum induced matching M^* in G.
$M = \emptyset$, $M^* = \emptyset$;
foreach $e \in E$ **do**
  Construct G_e using Construction 1;
  Find a maximum induced matching M in G_e;
  Update $M = M \cup \{e\}$;
  if $|M^*| < |M|$ **then**
    $M^* = M$;
return M^*.

Fig. 4. Algorithm to compute a maximum induced matching in a circular-convex bipartite graph.

The following theorem directly follows from the Lemma 3 and the algorithm INDUCED-M-CIRCULAR-CONVEX.

Theorem 4. *A maximum induced matching in a circular-convex bipartite graph can be computed in $O(n^2 m)$ time.*

5   Triad-Convex Bipartite Graph

In this section, we propose a polynomial time algorithm to compute a maximum induced matching in triad-convex bipartite graphs.

Let $G = (X, Y, E)$ be a triad-convex bipartite graph with a triad $T = (X, E_X)$ defined on X, such that for every vertex $y \in Y$, $T[N_G(y)]$ is a subtree of T. Let $X = \{x_0\} \cup X_1 \cup X_2 \cup X_3$ be such that for each i, $1 \le i \le 3$, $X_i = \{x_{i,1}, x_{i,2}, \ldots, x_{i,n_i}\}$. Also suppose that $x_{i,0} = x_0$ for all i, $1 \le i \le 3$. For each i, $1 \le i \le 3$, let $P_i = x_0, x_{i,1}, x_{i,2}, \ldots, x_{i,n_i}$ is a path in $T = (X, E_X)$. Note that x_0 is a common vertex in all the three paths P_1, P_2, and P_3.

Lemma 4. *Let M be a maximum induced matching in G. Let M do not saturate x_0, but M saturates some of the neighbors of x_0. Then M saturates at most 3 neighbors of x_0.*

Proof. We prove it by contradiction. Suppose M saturates 4 neighbors of x_0, say y_{j_1}, y_{j_2}, y_{j_3}, and y_{j_4}. Also suppose that $\{x_{j_1}y_{j_1}, x_{j_2}y_{j_2}, x_{j_3}y_{j_3}, x_{j_4}y_{j_4}\} \subseteq M$. Then at least two vertices of the set $\{x_{j_1}, x_{j_2}, x_{j_3}, x_{j_4}\}$ belong to the same path P_i for some i, $1 \le i \le 3$ (see Fig. 5). Without loss of generality, we may assume that $x_{j_1}, x_{j_2} \in P_1$. Also assume that the distance between x_0 and x_{j_1} is less than the distance between x_0 and x_{j_2} in path P_1. Notice that y_{j_2} is adjacent to x_0 as well as x_{j_2}. Also, by the definition of triad-convex bipartite graph, $T[N_G(y_{j_2})]$ is a subtree of T. Hence y_{j_2} must be adjacent to x_{j_1}. But by the definition of induced matching, if $x_{j_1}y_{j_1}$ and $x_{j_2}y_{j_2}$ are edges in an induced matching, then $x_{j_1}y_{j_2} \notin E$. So, we arrive at a contradiction. □

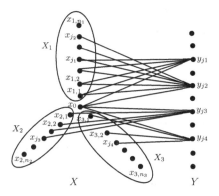

Fig. 5. A triad-convex bipartite graph.

Now let M be a maximum induced matching in G, then one of the following possibilities must occur:

(a) M does not saturate any vertex in $N_G[x_0]$.
(b) x_0 is saturated by M.
(c) M does not saturate x_0 but M saturates at most 3 neighbors of x_0. In this case, again three possibilities arise.
  - M saturates exactly one neighbor of x_0.
  - M saturates exactly two neighbors of x_0.
  - M saturates exactly three neighbors of x_0.

Now we discuss in detail that how to find maximum induced matching M in G in each of the above cases:

Case 1: M does not saturate any vertex in $N_G[x_0]$.
We construct a convex bipartite graph $G_0 = (X_0, Y_0, E_0)$ in the following way:

Construction 2: $G_0 = G[V \setminus N_G[x_0]]$.

Lemma 5. G_0 *is a convex bipartite graph.*

Proof. The proof is easy, and hence is omitted due to space constraint. □

Lemma 6. *Let M_0 be a maximum induced matching in G_0. Then M_0 is an induced matching in G. Moreover, if there exists a maximum induced matching M in G which does not saturate any vertex in $N_G[x_0]$ then $|M| = |M_0|$. In other words, M_0 is a maximum induced matching in G.*

Proof. The proof is omitted due to space constraint.   □

Case 2: x_0 is saturated by M, that is $x_0 y \in M$ for some $y \in N_G(x_0)$.
We construct a convex bipartite graph G_0^y in the following way:

Construction 3: $G_0^y = G[V \setminus (N_G(x_0) \cup N_G(y))]$.

Lemma 7. *G_0^y is a convex bipartite graph.*

Proof. The proof is easy, and hence is omitted due to space constraint.   □

Lemma 8. *Let M_0 be a maximum induced matching in G_0^y. Then $M_0 \cup \{x_0 y\}$ is an induced matching in G. Moreover, if there exists a maximum induced matching M in G containing the edge $x_0 y$ then $|M| = |M_0| + 1$.*

Proof. The proof is omitted due to space constraint.   □

Case 3: M does not saturate x_0 but M saturates at most 3 neighbors of x_0.
Again the following three cases arise:

Subcase 3.1: M saturates exactly one neighbor y_i of x_0, that is, $x y_i \in M$ for some $x \in N_G(y_i) \setminus \{x_0\}$.
We construct a convex bipartite graph $G_x^{y_i}$ in the following way:

Construction 4: First remove all the neighbors of x_0 other than y_i from G. Let us call the resultant graph G'. Now define $G_x^{y_i} = G'[V(G') \setminus (N_{G'}(x) \cup N_{G'}(y_i))]$.

Lemma 9. *$G_x^{y_i}$ is a convex bipartite graph.*

Proof. The proof is easy, and hence is omitted due to space constraint.   □

Lemma 10. *Let M_i be a maximum induced matching in $G_x^{y_i}$. Then $M_i \cup \{x y_i\}$ is an induced matching in G. Moreover, if there exists a maximum induced matching M in G such that M does not saturate x_0, and M saturates exactly one neighbor y_i of x_0, and $x y_i \in M$, then $|M| = |M_i| + 1$. In other words, $M_i \cup \{x y_i\}$ is a maximum induced matching in G.*

Proof. The proof is omitted due to space constraint.   □

Subcase 3.2: M saturates exactly two neighbors y_i, y_j of x_0, that is, $x_r y_i, x_s y_j \in M$ where $x_r \in N_G(y_i) \setminus N_G(y_j)$, and $x_s \in N_G(y_j) \setminus N_G(y_i)$.
We construct a convex bipartite graph $G_{x_r x_s}^{y_i y_j}$ in the following way:

Algorithm 2: INDUCED-M-TRIAD-CONVEX()

Input: A triad-convex bipartite graph $G = (X, Y, E)$ with a triad $T = (X, E_X)$, where $X = \{x_0\} \cup X_1 \cup X_2 \cup X_3$, and $|X_1| = n_1$, $|X_2| = n_2$, $|X_3| = n_3$.

Output: A maximum induced matching M^* in G.

$M = \emptyset$, $M^* = \emptyset$;

Construct G_0 using Construction 2;

Find a maximum induced matching M in G_0;

$M^* = M$;

foreach $y \in N_G(x_0)$ **do**
  Construct G_0^y using Construction 3;
  Find a maximum induced matching M in G_0^y;
  Update $M = M \cup \{x_0 y\}$;
  if $|M^*| < |M|$ **then**
    $M^* = M$;

foreach $y_i \in N_G(x_0)$ **do**
  foreach $x \in N_G(y_i) \setminus \{x_0\}$ **do**
    Construct $G_x^{y_i}$ using Construction 4;
    Find a maximum induced matching M in $G_x^{y_i}$;
    Update $M = M \cup \{x y_i\}$;
    if $|M^*| < |M|$ **then**
      $M^* = M$;

foreach $y_i, y_j \in N_G(x_0)$ such that $N_G(y_i) \setminus N_G(y_j) \neq \emptyset$ and $N_G(y_j) \setminus N_G(y_i) \neq \emptyset$ **do**
  foreach $x_r \in N_G(y_i) \setminus N_G(y_j)$ **do**
    foreach $x_s \in N_G(y_j) \setminus N_G(y_i)$ **do**
      Construct $G_{x_r x_s}^{y_i y_j}$ using Construction 5;
      Find a maximum induced matching M in $G_{x_r x_s}^{y_i y_j}$;
      Update $M = M \cup \{x_r y_i, x_s y_j\}$;
      if $|M^*| < |M|$ **then**
        $M^* = M$;

foreach $y_i, y_j, y_k \in N_G(x_0)$ such that $N_G(y_i) \setminus (N_G(y_j) \cup N_G(y_k)) \neq \emptyset$ and $N_G(y_j) \setminus (N_G(y_i) \cup N_G(y_k)) \neq \emptyset$ and $N_G(y_k) \setminus (N_G(y_i) \cup N_G(y_j)) \neq \emptyset$ **do**
  foreach $x_r \in N_G(y_i) \setminus (N_G(y_j) \cup N_G(y_k))$ **do**
    foreach $x_s \in N_G(y_j) \setminus (N_G(y_i) \cup N_G(y_k))$ **do**
      foreach $x_t \in N_G(y_k) \setminus (N_G(y_i) \cup N_G(y_j))$ **do**
        Construct $G_{x_r x_s x_t}^{y_i y_j y_k}$ using Construction 6;
        Find a maximum induced matching M in $G_{x_r x_s x_t}^{y_i y_j y_k}$;
        Update $M = M \cup \{x_r y_i, x_s y_j, x_t y_k\}$;
        if $|M^*| < |M|$ **then**
          $M^* = M$;

return M^*.

Fig. 6. Algorithm to compute a maximum induced matching in a triad-convex bipartite graph.

Construction 5: First remove all the neighbors of x_0 other than y_i, y_j from G. Let us call the resultant graph G'. Now define $G_{x_r x_s}^{y_i y_j} = G'[V(G') \setminus (N_{G'}(x_r) \cup N_{G'}(x_s) \cup N_{G'}(y_i) \cup N_{G'}(y_j))]$.

Lemma 11. $G_{x_r x_s}^{y_i y_j}$ is a convex bipartite graph.

Proof. The proof is omitted due to space constraint. □

Lemma 12. *Let M_{ij} be a maximum induced matching in $G_{x_r x_s}^{y_i y_j}$. Then $M_{ij} \cup \{x_r y_i, x_s y_j\}$ is an induced matching in G. Moreover, if there exists a maximum induced matching M in G such that M does not saturate x_0, and M saturates exactly two neighbors y_i, y_j of x_0, and $x_r y_i, x_s y_j \in M$, then $|M| = |M_{ij}| + 2$. In other words, $M_{ij} \cup \{x_r y_i, x_s y_j\}$ is a maximum induced matching in G.*

Proof. The proof is omitted due to space constraint. □

Subcase 3.3: M saturates exactly three neighbors y_i, y_j, y_k of x_0, that is, $x_r y_i, x_s y_j, x_t y_k \in M$ where $x_r \in N_G(y_i) \setminus (N_G(y_j) \cup N_G(y_k))$, and $x_s \in N_G(y_j) \setminus (N_G(y_i) \cup N_G(y_k))$, and $x_t \in N_G(y_k) \setminus (N_G(y_i) \cup N_G(y_j))$. We construct a convex bipartite graph $G_{x_r x_s x_t}^{y_i y_j y_k}$ in the following way:

Construction 6: First remove all the neighbors of x_0 other than y_i, y_j, y_k from G. Let us call the resultant graph G'. Now define $G_{x_r x_s x_t}^{y_i y_j y_k} = G'[V(G') \setminus (N_{G'}(x_r) \cup N_{G'}(x_s) \cup N_{G'}(x_t) \cup N_{G'}(y_i) \cup N_{G'}(y_j) \cup N_{G'}(y_k))]$.

Lemma 13. $G_{x_r x_s x_t}^{y_i y_j y_k}$ *is a convex bipartite graph.*

Proof. The proof is easy, and hence is omitted due to space constraint. □

Lemma 14. *Let M_{ijk} be a maximum induced matching in $G_{x_r x_s x_t}^{y_i y_j y_k}$. Then $M_{ijk} \cup \{x_r y_i, x_s y_j, x_t y_k\}$ is an induced matching in G. Moreover, if there exists a maximum induced matching M in G such that M does not saturate x_0, and M saturates exactly three neighbors y_i, y_j, y_k of x_0, and $\{x_r y_i, x_s y_j, x_t y_k\} \subseteq M$, then $|M| = |M_{ijk}| + 3$. In other words, $M_{ijk} \cup \{x_r y_i, x_s y_j, x_t y_k\}$ is a maximum induced matching in G.*

Proof. The proof is omitted due to space constraint. □

Based on the above discussion, the detailed algorithm for finding a maximum induced matching in a triad-convex bipartite graph G is shown in Fig. 6.

The following theorem directly follows from Lemmas 6, 8, 10, 12, 14 and the algorithm INDUCED-M-TRIAD-CONVEX.

Theorem 5. *A maximum induced matching in a triad-convex bipartite graph can be computed in $O(n^8)$ time.*

6 Conclusion

In this paper, we showed that the MAXIMUM INDUCED MATCHING problem is polynomial time solvable for circular-convex bipartite graphs and triad-convex bipartite graphs. We also proved some NP-completeness results. It will be an interesting problem to propose algorithms with better time complexity for the MAXIMUM INDUCED MATCHING problem in triad-convex bipartite graphs.

References

1. Stockmeyer, L.J., Vazirani, V.V.: NP-completeness of some generalizations of the maximum matching problem. Inf. Process. Lett. **15**, 14–19 (1982)
2. Cameron, K.: Induced matchings. Discret. Appl. Math. **24**, 97–102 (1989)
3. Kobler, D., Rotics, U.: Finding maximum induced matchings in subclasses of claw-free and P_5-free graphs, and in graphs with matching and induced matching of equal maximum size. Algorithmica **37**, 327–346 (2003)
4. Zito, M.: Induced matchings in regular graphs and trees. In: Widmayer, P., Neyer, G., Eidenbenz, S. (eds.) WG 1999. LNCS, vol. 1665, pp. 89–101. Springer, Heidelberg (1999). doi:10.1007/3-540-46784-X_10
5. Duckworth, W., Manlove, D.F., Zito, M.: On the approximability of the maximum induced matching problem. J. Discret. Algorithms **3**, 79–91 (2005)
6. Lozin, V.V.: On maximum induced matchings in bipartite graphs. Inf. Process. Lett. **81**, 7–11 (2002)
7. Cameron, K., Sritharan, R., Tang, Y.: Finding a maximum induced matching in weakly chordal graphs. Discret. Math. **266**, 133–142 (2003)
8. Golumbic, M., Lewenstein, M.: New results on induced matchings. Discret. Appl. Math. **101**, 157–165 (2000)
9. Liang, Y.D., Blum, N.: Circular convex bipartite graphs: maximum matching and hamiltonial circuits. Inf. Process. Lett. **56**, 215–219 (1995)
10. Liu, T.: Restricted bipartite graphs: comparison and hardness results. In: Gu, Q., Hell, P., Yang, B. (eds.) AAIM 2014. LNCS, vol. 8546, pp. 241–252. Springer, Heidelberg (2014). doi:10.1007/978-3-319-07956-1_22
11. Liu, T., Lu, M., Lu, Z., Xu, K.: Circular convex bipartite graphs: feedback vertex sets. Theoret. Comput. Sci. (2014). doi:10.1016/j.tcs.2014.05.001
12. Lu, Z., Liu, T., Xu, K.: Tractable connected domination for restricted bipartite graphs (extended abstract). In: Du, D.-Z., Zhang, G. (eds.) COCOON 2013. LNCS, vol. 7936, pp. 721–728. Springer, Heidelberg (2013). doi:10.1007/978-3-642-38768-5_65
13. Pandey, A., Panda, B.S.: Domination in some subclasses of bipartite graphs. In: Ganguly, S., Krishnamurti, R. (eds.) CALDAM 2015. LNCS, vol. 8959, pp. 169–180. Springer, Heidelberg (2015). doi:10.1007/978-3-319-14974-5_17
14. Chen, H., Lei, Z., Liu, T., Tang, Z., Wang, C., Xu, K.: Complexity of domination, hamiltonicity and treewidth for tree convex bipartite graphs. J. Comb. Optim. **32**, 95–110 (2016)
15. Liu, W.J.T., Wang, C., Xu, K.: Feedback vertex sets on restricted bipartite graphs. Theoret. Comput. Sci. **507**, 41–51 (2013)
16. Song, Y., Liu, T., Xu, K.: Independent domination on tree convex bipartite graphs. In: Snoeyink, J., Lu, P., Su, K., Wang, L. (eds.) AAIM/FAW -2012. LNCS, vol. 7285, pp. 129–138. Springer, Heidelberg (2012). doi:10.1007/978-3-642-29700-7_12
17. Brandstädt, A., Eschen, E.M., Sritharan, R.: The induced matching and chain subgraph cover problems for convex bipartite graphs. Inf. Process. Lett. **381**, 260–265 (2007)
18. Zhang, Y., Bao, F.: A review of tree convex sets test. Comput. Intell. **28**, 358–372 (2012)
19. Golumbic, M.C., Goss, C.F.: Perfect elimination and chordal bipartite graphs. J. Graph Theory **2**, 155–163 (1978)

Hamiltonicity in Split Graphs - A Dichotomy

P. Renjith[(⊠)] and N. Sadagopan

Indian Institute of Information Technology Design and Manufacturing
Kancheepuram, Chennai, India
{coe14d002,sadagopan}@iiitdm.ac.in

Abstract. In this paper, we investigate the well-studied Hamiltonian cycle problem, and present an interesting dichotomy result on split graphs. T. Akiyama, T. Nishizeki, and N. Saito [23] have shown that the Hamiltonian cycle problem is NP-complete in planar bipartite graph with maximum degree 3. Using this reduction, we show that the Hamiltonian cycle problem is NP-complete in split graphs. In particular, we show that the problem is NP-complete in $K_{1,5}$-free split graphs. Further, we present polynomial-time algorithms for Hamiltonian cycle in $K_{1,3}$-free and $K_{1,4}$-free split graphs. We believe that the structural results presented in this paper can be used to show similar dichotomy result for Hamiltonian path and other variants of Hamiltonian cycle problem.

1 Introduction

The Hamiltonian cycle (path) problem is a well-known decision problem which asks for the presence of a spanning cycle (path) in a graph. Hamiltonian problems play a significant role in various research areas such as operational research, physics and genetic studies [3,7,9]. This well-known problem has been studied extensively in the literature, and is NP-complete in general graphs. Various sufficient conditions for the existence of Hamiltonian cycle were introduced by G.A. Dirac, O. Ore, J.A. Bondy, and A. Ainouche, which were further generalized by A. Kemnitz [2]. A general study on the sufficient conditions were produced by H.J. Broersma and R.J. Gould [10,19,20]. Hamiltonicity has been looked at with respect to various structural parameters, the popular one is toughness. A relation between graph toughness, introduced by V. Chvatal [24] and Hamiltonicity has been well studied. A detailed survey on Hamiltonicity and toughness is presented in [5,19]. In split graphs, it is proved by D. Kratsch, J. Lehel, and H. Muller [8] that $\frac{3}{2}$ tough split graphs are Hamiltonian and due to Chvatal's result [24], Hamiltonian graphs are 1-tough. Therefore, the split graphs are Hamiltonian if and only if the toughness is in the range $1..\frac{3}{2}$.

On algorithmic front, the Hamiltonian cycle problem is NP-complete in chordal [1], chordal bipartite [11], planar [14], and bipartite [23] graphs. Further, T. Akiyama [23] have shown that the problem is NP-complete in bipartite graph with maximum degree 3. There is a simple reduction for the Hamiltonian cycle problem in bipartite graphs of maximum degree 3 to the Hamiltonian cycle problem in split graphs which we show as part of our dichotomy. Inspite of the

© Springer International Publishing AG 2017
D. Gaur and N.S. Narayanaswamy (Eds.): CALDAM 2017, LNCS 10156, pp. 320–331, 2017.
DOI: 10.1007/978-3-319-53007-9_28

hardness of the Hamiltonian problem in various graph classes, nice polynomial-time algorithms have been obtained in interval, circular arc, 2-trees, and distance hereditary graphs [12,21,22,25].

Split graphs are a popular subclass of chordal graphs and on which R.E. Burkard and P.L. Hammer presented a necessary condition [18]. Subsequently, N.D. Tan and L.X. Hung [15] have shown that the necessary condition of [18] is sufficient for some special split graphs. In this paper, we shall revisit Hamiltonicity restricted to split graphs and present a dichotomy result. We show that Hamiltonian cycle is NP-complete in $K_{1,5}$-free split graphs and polynomial-time solvable in $K_{1,4}$-free split graphs. It is important to note that a very few NP-complete problems have dichotomy results in the literature [4,13,16].

We use standard basic graph-theoretic notations. Further, we follow [6]. All the graphs we mention are simple, and unweighted. Graph G has vertex set $V(G)$ and edge set $E(G)$ which we denote using V, E, respectively, once the context is unambiguous. For *minimal vertex separator*, *maximal clique*, and *maximum clique* we use the standard definitions. A graph G is 2-connected if every minimal vertex separators are of size at least two. Split graphs are $C_4, C_5, 2K_2$-free graphs and the vertex set of a split graph can be partitioned into a clique K and an independent set I. For a split graph with vertex partition K and I, we assume K to be a maximum clique. For every $v \in K$ we define $N^I(v) = N(v) \cap I$, where $N(v)$ denotes the neighborhood of vertex v. $d^I(v) = |N^I(v)|$ and $\Delta^I = \max\{d^I(v) : v \in K\}$. We define an n-book H as follows; $V(H) = \{u, v, v_1, \ldots, v_{n-2}\}$ and $E(H) = \{uv\} \cup \bigcup_{1 \le i \le n-2} \{uv_i, vv_i\}$. For a cycle or a path C, we use $\overrightarrow{C}$ to represent an ordering of the vertices of C in one direction (forward direction) and $\overleftarrow{C}$ to represent the ordering in the other direction. $u\overrightarrow{C}u$ represents the ordered vertices from u to v in C. For two paths P and Q, $P \cap Q$ denotes $V(P) \cap V(Q)$. For simplicity, we use P to denote the underlying set $V(P)$.

2 Hamiltonian Problem in Split Graphs - Polynomial Results

In this section we shall present structural results on some special split graphs using which we can find Hamiltonian cycle in such graphs. In particular we explore the structure of $K_{1,3}$-free split graphs and $K_{1,4}$-free split graphs.

Lemma 1. [13] *For a claw-free split graph G, if $\Delta^I = 2$, then $|I| \le 3$.*

Observation 1. *2-connected split graph G with $\Delta^I = 1$ has a Hamiltonian cycle.*

Lemma 2. *Let G be a $K_{1,3}$-free split graph. G contains a Hamiltonian cycle if and only if G is 2-connected.*

Proof. Necessity is trivial. For the sufficiency, we consider the following cases.
Case 1: $|I| \geq 4$. As per Lemma 1, $\Delta^I = 1$ and by Observation 1, G has a Hamiltonian cycle.
Case 2: $|I| \leq 3$. If $\Delta^I = 1$, then by Observation 1, G has a Hamiltonian cycle. When $\Delta^I > 1$, note that either $|I| = 2$ or $|I| = 3$.
(a) $|I| = 2$, i.e., $I = \{s, t\}$. Since $\Delta^I = 2$, there exists a vertex $v \in K$ such that $d^I(v) = 2$ and $N^I(v) = \{s, t\}$. Let $S = N(s)$ and $T = N(t)$. Clearly, $K = S \cup T$. Suppose the set $S \setminus T$ is empty, then the vertices $T \cup \{t\}$ induces a clique, larger in size than K, contradicting the maximality of K. Therefore, the set $S \setminus T$ is non-empty. Similarly, $T \setminus S \neq \emptyset$. It follows that, $|K| \geq 3$; let $x, v, w \in K$ such that $x \in T \setminus S$ and $w \in S \setminus T$. Further, $(w, s, v, t, x, v_1, v_2, \ldots, v_k, w)$ is a Hamiltonian cycle in G where $\{v_1, v_2, \ldots, v_k\} = K \setminus \{v, w, x\}$.
(b) If $|I| = 3$ and let $I = \{s, t, u\}$. Since G is a 2-connected claw free graph, clearly $|K| \geq 3$. Since $\Delta^I = 2$, there exists $v \in K$ such that $N^I(v) = \{s, t\}$. Further, since G is claw free, for every vertex $w \in K$, $N^I(w) \cap \{s, t\} \neq \emptyset$. Let $S = N(s)$ and $T = N(t)$. Suppose the set $S \setminus T$ is empty, then the vertices $T \cup \{t\}$ induces a clique, larger in size than K, contradicting the maximality of K. Therefore, the set $S \setminus T \neq \emptyset$. Similarly, $T \setminus S \neq \emptyset$. Note that $|N(u)| \geq 2$ as G is 2-connected. If u is adjacent to a vertex $v \in S \cap T$, then $\{v\} \cup N^I(v)$ induces a claw. Therefore, $N(u) \cap (S \cap T) = \emptyset$. It follows that u is adjacent to some vertices in $S \setminus T$ or $T \setminus S$. Suppose $N(u) \cap (S \setminus T) = \emptyset$ and $N(u) \cap (T \setminus S) \neq \emptyset$, then there exists a vertex $z \in T \setminus S$ such that $N^I(z) = \{t, u\}$. Since, $S \setminus T \neq \emptyset$, there exists $z' \in S \setminus T$ and $\{z, u, t, z'\}$ induces a claw, a contradiction. Therefore, $N(u) \cap (S - T) \neq \emptyset$ and similarly, $N(u) \cap (T - S) \neq \emptyset$. It follows that there exists a vertex $x \in (S - T)$, $y \in (T - S)$ such that $x, y \in N(u)$. If $|K| = 3$, then (x, s, v, t, y, u, x) is a Hamiltonian cycle in G. Further, if $|K| \geq 3$, then $(x, s, v, t, y, u, x, w_1, \ldots, w_l)$ is the desired Hamiltonian cycle where $w_1, \ldots, w_l \in K$. This completes the case analysis, and the proof of Lemma 2. □

$K_{1,4}$-free Split Graphs

Now we shall present structural observations in $K_{1,4}$-free split graphs. The structural results in turn gives a polynomial-time algorithm for the Hamiltonian cycle problem, which is one part of our dichotomy. From Observation 1, it follows that a split graph G with $\Delta^I = 1$ has a Hamiltonian cycle if and only if G is 2-connected. It is easy to see that for $K_{1,4}$-free split graphs, $\Delta^I \leq 3$, and thus we analyze such a graph G in two variants, $\Delta_G^I = 2$, and $\Delta_G^I = 3$.
Claim A: For a split graph G with $\Delta^I = 3$, let $v \in K, d^I(v) = 3$, and $U = N^I(v)$. If G is $K_{1,4}$-free, then $N(U) = K$.

Proof. If not, let $w \in K$ such that $w \notin N(U)$. Clearly, $N^I(v) \cup \{w, v\}$ induces a $K_{1,4}$, a contradiction. □

Claim B: For a $K_{1,4}$-free split graph G with $\Delta^I = 3$, let $v \in K$ such that $d^I(v) = 3$, and the split graph $H = G - N^I(v)$. Then, $\Delta_H^I \leq 2$.

Proof. Otherwise, if there exists $x \in K, d_H^I(x) = 3$, then $N^I(v) \cup \{x, v\}$ induces a $K_{1,4}$ in G, a contradiction. □

The next theorem shows a necessary and sufficient condition for the existence of Hamiltonian cycle in split graphs with $\Delta^I = 2$. We define the notion short cycle in a $K_{1,4}$-free split graph G. Consider the subgraph H of G where $V_a = \{u \in I : d(u) = 2\}$, $V_b = N(V_a)$, $V(H) = V_a \cup V_b$ and $E(H) = \{uv : u \in V_a, v \in V_b\}$. Clearly, H is a bipartite subgraph of G. Let C be an induced cycle in H such that $V(K) \setminus V(C) \neq \emptyset$. We refer to C in H as a *short cycle* in G.

Theorem 1. *Let G be a $K_{1,4}$-free split graph with $\Delta^I = 2$. G has a Hamiltonian cycle if and only if there are no short cycles in G.*

Proof. Necessity is trivial as, if there exists a short cycle D, then $c(G - S) > |S|$ where $S = V(D) \cap K$. For sufficiency: if $|I| = |K|$ and G has no short cycles, then clearly, H is a spanning cycle of G. Further, if G has no short cycles with $|K| > |I|$, then note that H is a collection of paths $P_1, \ldots, P_i$, all of them are having end vertices in K. Let $V' = I \setminus V_a$. If $V' = \emptyset$, then it is easy to join the paths using clique edges to get a Hamiltonian cycle of G. Otherwise we partition the vertices in V' into three sets V_2, V_1, V_0 where $V_2 = \{u \in V' : N(u) \cap P_j \neq \emptyset$ and $N(u) \cap P_k \neq \emptyset, 1 \leq j \neq k \leq i\}$, $V_1 = \{u \in V' : N(u) \cap P_j \neq \emptyset, 1 \leq j \leq i$ and $u \notin V_2\}$, $V_0 = \{u \in V' : N(u) \cap P_j = \emptyset, 1 \leq j \leq i\}$. From the definitions, vertices in V_2 are adjacent to the end vertices of at least two paths, vertices in V_1 are adjacent to the end vertices of exactly one path and that of V_0 are not adjacent to the end vertices of any paths. Now we obtain two graphs H_1, and H_2 from H and finally, we see that H_2 is a collection of paths containing all the vertices of I, all those paths having end vertices in K. We iteratively add the vertices in V_2 and V_1 into H to obtain H_1, based on certain preferences till $V_2 = V_1 = \emptyset$. We pick a vertex (from V_2 if $V_2 \neq \emptyset$, otherwise from V_1) and add it to H. If we add a vertex u from V_2, then we join two arbitrary paths each has its end vertex adjacent to u. Therefore, the addition of a vertex from V_2 reduces the number of paths in H by one. If $u \in V_1$, then u is added to H in such a way that one of its end vertex is an end vertex of a path and the other is not an end vertex of any paths. Clearly, such two vertices are possible due to the fact that $d(u) \geq 3$. Note that in this case, one of the paths in H gets its size increased, still in both cases all the paths have their end vertices in K. We add the vertices in such a way that the vertices in V_2 gets preference over that of V_1. Also, after every addition of a vertex, we add the new vertex to V' and re-compute the partitions V_2, V_1 and V_0. Observe that once the set V_2 is empty, and a vertex from V_1 is added, the re-computation may result in a case where $V_2 \neq \emptyset$. Once $V_2 = V_1 = \emptyset$, the graph H_1 obtained is a collection of one or more paths having end vertices in K.

Now we continue the addition by including the vertices of V_0. A vertex $u \in V_0$ is added in such a way that it forms a P_3 with two of its arbitrary neighbors in K. Evidently, the addition of vertices from V_0 increases the number of paths by one. Note that the addition of vertices from V_0 may result in $V_2 \neq \emptyset$ or $V_1 \neq \emptyset$. Therefore, when we iteratively add the vertices to H_1, we give first preference to the vertices in V_2, then to the vertices in V_1 (if $V_2 = \emptyset$) and finally to that in V_0 (if $V_2 = V_1 = \emptyset$). Here also, after each addition of a vertex u, we add u to V' and re-compute the partitions. H_2 is the graph obtained by iteratively adding all the vertices left in I. It is interesting to see that the number of paths in H_2 is at least the number of paths in H_1. Finally, in H_2 we have a collection of one or more paths with end vertices in K. It is easy to see that the paths could be joined using clique edges to get a Hamiltonian cycle in G. This completes the proof.   □

Having presented a characterization for the Hamiltonian cycle problem in split graphs with $\Delta^I \leq 2$, we shall now present our main result, which is a necessary and sufficient condition for the existence of Hamiltonian cycle in $K_{1,4}$-free split graphs. Note that for a $K_{1,4}$-free split graph G, $\Delta^I \leq 3$ and thus the left over case to analyze is when $\Delta^I = 3$. When $\Delta^I = 3$, there exists a vertex $v \in C$ with $d^I(v) = 3$. We obtain $H = G - N^I(v)$, and from Claim B, $\Delta^I_H \leq 2$. Now we shall observe some characteristics of H.

Let G be a 2-connected $K_{1,4}$-free split graph with $\Delta^I = 3$, $|K| \geq |I| \geq 8$ and $H = G - N^I(v)$ where $v \in K, N^I(v) = \{v_1, v_2, v_3\}$. If there are no induced short cycles in G, then by the constructive proof of Theorem 1, in H there exists a collection of vertex disjoint paths $\mathbb{C}$. Note that each path in $\mathbb{C}$ alternates between an element in K and an element in I, and all the paths are having the end vertices in K. Therefore the paths are having odd number of vertices. Thus, $\mathbb{C} = \mathbb{P}_1 \cup \mathbb{P}_3, \ldots, \mathbb{P}_{2i+1}$, where $\mathbb{P}_j$ is the set of maximal paths of size j where for every $P \in \mathbb{P}_j$, there does not exists $P' \in \mathbb{C}$ such that $E(P) \subset E(P')$. A path $P_a \in \mathbb{C}$ is defined on the vertex set $V(P_a) = \{w_1, \ldots, w_j, x_1, \ldots, x_{j-1}\}$, $E(P_a) = \{w_i x_i : 1 \leq i \leq j-1\} \cup \{w_k x_{k-1} : 2 \leq k \leq j\}$ such that $\{w_1, \ldots, w_j\} \subseteq K$, $\{x_1, \ldots, x_{j-1}\} \subseteq I$. We denote such a path as $P_a = P(w_1, \ldots, w_j; x_1, \ldots, x_{j-1})$ We shall now present our structural observations on paths in $\mathbb{C}$.

Claim 1. *If there exists a path $P_a \in \mathbb{P}_i, i \geq 11$ such that $P_a = P(w_1, \ldots, w_j; x_1, \ldots, x_{j-1}), j \geq 6$, then there exists $v_1 \in N^I(v)$ such that $v_1 w_i \in E(G), 2 \leq i \leq j-1$.*

Proof. First we show that for every non-consecutive $2 \leq i, k \leq j-1$, for the pair of vertices w_i, w_k, $v_1 w_i \in E(G)$ and $v_1 w_k \in E(G)$. Suppose, if exactly one of w_i, w_k is adjacent to v_1, say $w_i v_1 \in E(G)$, then $N^I(w_i) \cup \{w_i, w_k\}$ has an induced $K_{1,4}$. If $v_1 w_i \notin E(G)$ and $v_1 w_k \notin E(G)$, then by Claim A, there exists an adjacency for w_i, w_k in v_2, v_3. Further, if either v_2 or v_3 is adjacent to both w_i, w_k, then $v_1 = v_2$, or $v_1 = v_3$, and the claim is true. Therefore, we shall assume without loss of generality, $v_2 w_i \in E(G)$ and $v_2 w_k \notin E(G)$. This implies that $N^I(w_i) \cup \{w_i, w_k\}$ has an induced $K_{1,4}$, a contradiction. Since the above observation is true for all such pair of vertices in $W = \{w_i\}$, $2 \leq i \leq j-1$ and $|W| \geq 4$, it follows that there exists $v_1 \in N^I(v)$ such that $v_1 w_i \in E(G), 2 \leq i \leq j-1$.   □

Claim 2. $\mathbb{P}_i = \emptyset, i \geq 13$.

Proof. Assume for a contradiction that there exists $P_a \in \mathbb{P}_i, i \geq 13$. Let $P_a = P(w_1, \ldots, w_j; x_1, \ldots, x_{j-1}), j \geq 7$. From Claim 1 there exists $v_1 \in N^I_G(v)$ such that $v_1 w_k \in E(G), 2 \leq k \leq j-1$. Since the clique is maximum in G, there exists

$s \in K$ such that $v_1 s \notin E(G)$. Further, there exists at least three vertices in $x_1, \ldots, x_{j-1}$ adjacent to s, otherwise, for some $2 \leq r \leq j - 1$, $N_G^I(w_r) \cup \{w_r, s\}$ induces a $K_{1,4}$. Finally, from Claim A, either $v_2 s \in E(G)$ or $v_3 s \in E(G)$. It follows that $\{s\} \cup N_G^I(s)$ has an induced $K_{1,4}$, a contradiction. $\qquad \square$

Claim 3. *Let* $P_a = P(w_1, \ldots, w_i; x_1, \ldots, x_{i-1}), i \geq 3$, *and* $P_b = P(s_1, \ldots, s_j; t_1, \ldots, t_{j-1}), j \geq 3$ *be arbitrary paths in* $\mathbb{C}$. *There exists* $v_1 \in N^I(v)$ *such that* $\forall\, 2 \leq l \leq i - 1, v_1 w_l \in E(G)$, *and* $\forall\, 2 \leq m \leq j - 1, v_1 s_m \in E(G)$.

Proof. We shall consider every pair of vertices w_l, s_m and show that $v_1 w_l, v_1 s_m \in E(G)$. Suppose, if exactly one of w_l, s_m is adjacent to v_1, say $w_l v_1 \in E(G)$, then $N^I(w_l) \cup \{w_l, s_m\}$ has an induced $K_{1,4}$. If $v_1 w_l \notin E(G)$ and $v_1 s_m \notin E(G)$, then by Claim A, there exists an adjacency for w_l, s_m in v_2, v_3. Further, if either v_2 or v_3 is adjacent to both w_l, s_m, then $v_1 = v_2$, or $v_1 = v_3$, and the claim is true. Therefore, we shall assume without loss of generality, $v_2 w_l \in E(G)$ and $v_2 s_m \notin E(G)$. This implies that $N^I(w_l) \cup \{w_l, s_m\}$ has an induced $K_{1,4}$, a contradiction. $\qquad \square$

Corollary 1. *Let* $P_a = P(w_1, \ldots, w_i; x_1, \ldots, x_{i-1}), i \geq 3$, $P_b = P(s_1, \ldots, s_j; t_1, \ldots, t_{j-1}), j \geq 3$ *and* $P_c = P(y_1, \ldots, y_k; z_1, \ldots, z_{k-1}), k \geq 3$ *be arbitrary paths in* $\mathbb{C}$. *There exists* $v_1 \in N^I(v)$ *such that* $\forall\, 2 \leq l \leq i - 1, v_1 w_l \in E(G)$, $\forall\, 2 \leq m \leq j - 1, v_1 s_m \in E(G)$, *and* $\forall\, 2 \leq n \leq k - 1, v_1 y_n \in E(G)$.

Proof. From Claim 3, $\forall\, 2 \leq l \leq i - 1, v_1 w_l \in E(G)$, and $\forall\, 2 \leq m \leq j - 1, v_1 s_m \in E(G)$. Similarly, $\forall\, 2 \leq l \leq i - 1, v_1 w_l \in E(G)$, and $\forall\, 2 \leq n \leq k - 1, v_1 y_n \in E(G)$. Thus the corollary follows from Claim 3. $\qquad \square$

Claim 4. *If there exists* $P_a \in \mathbb{P}_{11}$, *then* $\mathbb{P}_j = \emptyset, j \neq 11, j \geq 5$.

Proof. Assume for a contradiction that there exists such a path $P_b \in \mathbb{P}_j, j \geq 5$. Let $P_a = (w_1, \ldots, w_6; x_1, \ldots, x_5)$ and $P_b = (s_1, \ldots, s_r; t_1, \ldots, t_{r-1}), r \geq 3$. From Claim 1, there exists a vertex $v_1 \in N^I(v)$, such that $v_1 w_i \in E(G), 2 \leq i \leq 5$ and from Claim 3, $v_1 s_j \in E(G), 2 \leq j \leq r - 1$. Now we claim $v_1 w_1 \in E(G)$. Otherwise, by Claim A, $v_2 w_1$ or $v_3 w_1$ is in $E(G)$. Observe that either $w_1 x_2 \in E(G)$ or $w_1 x_3 \in E(G)$, otherwise $N^I(w_3) \cup \{w_3, w_1\}$ induces a $K_{1,4}$. Similarly, either $w_1 x_4 \in E(G)$ or $w_1 x_5 \in E(G)$. Now $\{w_1\} \cup N^I(w_1)$ induces a $K_{1,4}$. Using similar argument, we establish $v_1 w_6 \in E(G)$. Since the clique is maximal, there exists a vertex $w' \in K$ such that $v_1 w' \notin E(G)$. We see the following cases.
Case 1: $w' = s_1$. By Claim A, $s_1 v_2 \in E(G)$ or $s_1 v_3 \in E(G)$. Further, $s_1 x_2 \in E(G)$, otherwise $N^I(w_2) \cup \{w_2, s_1\}$ induces a $K_{1,4}$ or $N^I(w_3) \cup \{w_3, s_1\}$ induces a $K_{1,4}$. Similarly, $s_1 x_4 \in E(G)$. Now $\{s_1\} \cup N^I(s_1)$ induces a $K_{1,4}$. Similarly, we could establish a contradiction if $w' = s_r$. *Case 2:* $w' \notin P_b$. By Claim A, $w' v_2 \in E(G)$ or $w' v_3 \in E(G)$. Also due to the similar reasoning for s_1, $w' x_2, w' x_4 \in E(G)$. Now, either $t_1 w' \in E(G)$ or $t_2 w' \in E(G)$, otherwise $N^I(s_2) \cup \{s_2, w'\}$ induces a $K_{1,4}$. Finally, $\{w'\} \cup N^I(w')$ induces a $K_{1,4}$, a contradiction. Therefore, P_b does not exist. This completes the case analysis and the proof. $\qquad \square$

Claim 5. *If there exists $P_a \in \mathbb{P}_{11}$, then G has a Hamiltonian cycle.*

Proof. Let $P_a = (w_1, \ldots, w_6; x_1, \ldots, x_5)$. From Claim 1, there exists a vertex say $v_1 \in N_G^I(v)$, such that $v_1 w_i \in E(G)$, $2 \leq i \leq 5$. From the proof of the previous claim, $v_1 w_1, v_1 w_6 \in E(G)$. Since the clique is maximal, there exists $w' \in K$, such that $w' v_1 \notin E(G)$. By Claim A, $w' v_2 \in E(G)$ or $w' v_3 \in E(G)$. Without loss of generality, let $w' v_2 \in E(G)$. We claim $w' x_2 \in E(G)$ and $w' x_4 \in E(G)$, otherwise for some $2 \leq i \leq 5$, $N^I(w_i) \cup \{w_i, w'\}$ induces a $K_{1,4}$. One among v_2, x_2, x_4 is adjacent to w_1, otherwise $N^I(w') \cup \{w_1, w'\}$ induces a $K_{1,4}$. Similar argument holds good with respect to the vertex w_6. Note that for every $t \in \{v, w', w_1, \ldots, w_6\}$, $d^I(t) = 3$ and there exists a vertex $w'' \in K$, $w'' \neq t$, where $w'' v_3 \in E(G)$. Now we claim $w'' v_1 \in E(G)$. If not, for some $1 \leq j \leq 6$, $N^I(w_j) \cup \{w_j, w''\}$ induces a $K_{1,4}$. Finally $(w_1 \overrightarrow{P_a} w_6, v_1, w'', v_3, v, v_2, w')$ is a (w_1, w') path containing all the vertices of $P_a \cup \{v, w', w''\} \cup N^I(v)$, which could be easily extended to a Hamiltonian cycle in G using clique edges to join other vertex disjoint paths. $\square$

Claim 6. *If there exists $P_a \in \mathbb{P}_9$, then $\mathbb{P}_j = \emptyset, j \neq 9, j \geq 5$.*

Due to page constraints, the detailed proof is included in [17]. In the following claims to show the existence of Hamiltonian cycle, we shall do a constructive approach in which we produces a (u, v)-path where $u, v \in K$. The path is obtained by joining some paths in $\mathbb{C}$ using the vertices in $N^I(v)$. Therefore such a *desired path* is sufficient to show that G has a Hamiltonian cycle, which is in turn obtained by joining all such vertex disjoint paths using clique edges.

Claim 7. *If there exists $P_a \in \mathbb{P}_9$ and G has no short cycles, then G has a Hamiltonian cycle.*

Due to page constraints, the detailed proof is included in [17].

Claim 8. $|\mathbb{P}_7| \leq 2$. *Further, if $|\mathbb{P}_7| = 2$, then $\mathbb{P}_5 = \emptyset$.*

Due to page constraints, the detailed proof is included in [17].

Claim 9. *If $|\mathbb{P}_7| = 1$, then $|\mathbb{P}_5| \leq 1$.*

Due to page constraints, the detailed proof is included in [17].

Claim 10. *If there exists $P_a, P_b \in \mathbb{P}_7$, then G has a Hamiltonian cycle.*

Proof. Let $P_a, P_b \in \mathbb{P}_7$ such that $P_a = (w_1, \ldots, w_4; x_1, \ldots, x_3)$, $P_b = (s_1, \ldots, s_4; t_1, \ldots, t_3)$. Similar to the arguments in the proof of Claim 8, there exists $v_1 \in N^I(v)$ such that $v_1 w_i, v_1 s_i \in E(G), 1 \leq i \leq 4$. Since K is a maximal clique, there exists $w' \in K$ such that $w' v_1 \notin E(G)$. From Claim A, either $w' v_2 \in E(G)$ or $w' v_3 \in E(G)$. Without loss of generality, let $w' v_3 \in E(G)$. Note that $w' x_2 \in E(G)$, otherwise, either $N^I(w_2) \cup \{w_2, w'\}$ induces a $K_{1,4}$ or $N^I(w_3) \cup \{w_3, w'\}$ induces a $K_{1,4}$. Similarly, $w' t_2 \in E(G)$. Note that the vertices $w_i, s_i, i \in \{1, 4\}$ is adjacent to one of the vertices in $\{v_3, t_2, x_2\}$, if not, say $w_1 v_3, w_1 t_2, w_1 x_2 \notin E(G)$, then $N^I(w') \cup \{w', w_1\}$ induces a $K_{1,4}$. Similar arguments hold for w_2, s_1, s_2. It follows that for every $s' \in S = \{w_1 \ldots, w_4, s_1, \ldots, s_4\}$, $d^I(s') = 3$. Since G is two connected, there exists $w'' \in K \setminus S$ such that $w''v_2 \in E(G)$. Observe that $(w'', v_2, v, v_3, w', w_1 \overrightarrow{P_a} w_4, v_1, s_1 \overrightarrow{P_b} s_4)$ is a path containing $N^I(v)$ which could be easily extended to a Hamiltonian cycle in G.   □

Claim 11. *If there exists $P_a \in \mathbb{P}_7, P_b \in \mathbb{P}_5$ and G has no short cycle, then G has a Hamiltonian cycle.*

Proof. Let $P_a \in \mathbb{P}_7$, $P_b \in \mathbb{P}_5$, such that $P_a = (w_1, \ldots, w_4; x_1, \ldots, x_3)$, $P_b = (s_1, \ldots, s_3; t_1, \ldots, t_2)$. From Claim 3, there exists $v_1 \in N^I(v)$ such that $v_1 w_i, v_1 s_2 \in E(G), i \in \{2, 3\}$. Now we claim that $v_1 w_1 \in E(G)$. If not, observe that either $v_2 w_1 \in E(G)$ or $v_3 w_1 \in E(G)$. Also note that $w_1 x_2 \in E(G)$ or $w_1 x_3 \in E(G)$, otherwise $N^I(w_3) \cup \{w_3, w_1\}$ induces a $K_{1,4}$. Further, $w_1 t_1 \in E(G)$ or $w_1 t_2 \in E(G)$, otherwise $N^I(s_2) \cup \{s_2, w_1\}$ induces a $K_{1,4}$. Clearly, $\{w_1\} \cup N^I(w_1)$ induces a $K_{1,4}$, contradicting $v_1 w_1 \notin E(G)$. Similarly, $v_1 w_4 \in E(G)$. Since the clique is maximal, there exists $w' \in K$ such that $v_1 w' \notin E(G)$. We see the following cases.

Case 1: $w' \notin \{s_1, s_3\}$, thus $v_1 s_1, v_1 s_3 \in E(G)$. From the previous claims, it is easy to see that w' is an end vertex of a path P_c in $\mathbb{P}_3 \cup \mathbb{P}_1$. From Claim A, there exists $v_3 \in N^I(v)$ such that $v_3 w' \in E(G)$. Now we claim $w' x_2 \in E(G)$, otherwise, either $N^I(w_2) \cup \{w_2, w'\}$ induces a $K_{1,4}$ or $N^I(w_3) \cup \{w_3, w'\}$ induces a $K_{1,4}$. Also observe that either $w' t_1$ or $w' t_2$ is in $E(G)$, otherwise $N^I(s_2) \cup \{s_2, w'\}$ induces a $K_{1,4}$. Without loss of generality, let $w' t_2 \in E(G)$. Now note that all the vertices in $\{w_1, w_4, s_1\}$ has an adjacency in $\{t_2, x_2, v_3\}$. Clearly, $d^I(w_j) = d^I(s_1) = d^I(s_2) = 3$, $1 \le j \le 4$ and since the graph is 2-connected, $v_2 w'' \in E(G)$ where w'' is the end vertex of a path P_d in $\mathbb{P}_3 \cup \mathbb{P}_1$. Here we obtain $(\overrightarrow{P_d} w'', v_2, v, v_3, w' \overrightarrow{P_c}, w_1 \overrightarrow{P_a} w_4, v_1, s_1 \overrightarrow{P_b} s_3)$ as a desired path.

Case 2: $w' \in \{s_1, s_3\}$. Without loss of generality, let $w' = s_1$, i.e., $v_1 s_1 \notin E(G)$. From Claim A, there exists $v_3 \in N^I(v)$ such that $v_3 s_1 \in E(G)$. Also note that $s_1 x_2 \in E(G)$, otherwise either $N^I(w_2) \cup \{w_2, s_1\}$ induces a $K_{1,4}$ or $N^I(w_3) \cup \{w_3, s_1\}$ induces a $K_{1,4}$. Now we claim that w_1 and w_4 are adjacent to one of the vertices in $S = \{v_3, t_1, x_2\}$. Suppose $w_1 v_3, w_1 t_1, w_1 x_2 \notin E(G)$, then $N^I(s_1) \cup \{s_1, w_1\}$ induces a $K_{1,4}$. Similar arguments hold for w_4. It follows that $d^I(w_j) = d^I(s_k) = 3$, $1 \le j \le 4$, $k \in \{1, 2\}$. Since G is 2-connected, there exists $w^* \in K$ such that $w^* v_2 \in E(G)$. We see the following sub cases based on the possibility of w^*.

Case 2.1: $w^* = s_3$. i.e., $v_2 s_3 \in E(G)$. In this sub case we claim that there exists a vertex $w'' \ne v \in K$ such that $w'' \notin P_a \cup P_b$ and $w'' x' \in E(G)$ where $x' \in \{v_2, v_3, t_1, t_2\}$. Suppose such a w'' does not exist, then observe that, in the set $S = \{t_1, t_2, y, z\}$, $d(t_1) = d(t_2) = d(y) = d(z) = 2$, and $S \cup N(S)$ has a short cycle, a contradiction. Note that, w'' is an end vertex of a path P_d in $\mathbb{P}_3 \cup \mathbb{P}_1$. Now depending on the adjacency of w'', we obtain the following paths.

If $w'' v_2 \in E(G)$, then we obtain $(\overrightarrow{P_d} w'', v_2, s_3 \overleftarrow{P_b} s_1, v_3, v, v_1, w_1 \overrightarrow{P_a} w_4)$ as a desired path.

If $w''v_3 \in E(G)$, then we obtain $(\overrightarrow{P_d}w'', v_3, s_1\overrightarrow{P_b}s_3, v_2, v, v_1, w_1\overrightarrow{P_a}w_4)$ as a desired path.

If $w''t_1 \in E(G)$, then we obtain $(\overrightarrow{P_d}w'', t_1, s_1, v_3, v, v_2, s_3\overleftarrow{P_b}s_2, v_1, w_1\overrightarrow{P_a}w_4)$ as a desired path.

If $w''t_2 \in E(G)$, then we obtain $(\overrightarrow{P_d}w'', t_2, s_3, v_2, v, v_3, s_1\overrightarrow{P_b}s_2, v_1, w_1\overrightarrow{P_a}w_4)$ as a desired path.

Case 2.2: $w^* \neq s_3$. Note that w^* is an end vertex of a path P_e in $\mathbb{P}_3 \cup \mathbb{P}_1$. We see the following sub cases to complete our argument.

Case 2.2.1: $v_1s_3 \in E(G)$. Here we obtain $(\overrightarrow{P_e}w^*, v_2, v, v_3, s_1\overrightarrow{P_b}s_3, v_1, w_1\overrightarrow{P_a}w_4)$ as a desired path.

Case 2.2.2: $v_1s_3 \notin E(G)$. Clearly, from Claim A either $s_3v_2 \in E(G)$ or $s_3v_3 \in E(G)$. We now claim that $s_3x_2 \in E(G)$. Otherwise either $N^I(w_2) \cup \{w_2, s_3\}$ induces a $K_{1,4}$ or $N^I(w_3) \cup \{w_3, s_3\}$ induces a $K_{1,4}$. Here we obtain $(\overrightarrow{P_e}w^*, v_2, v, v_3, s_1\overrightarrow{P_b}s_3, x_2\overrightarrow{P_a}w_4, v_1, w_2, x_1, w_1)$ as a desired path.

This completes the case analysis and the proof. □

Claim 12. *If there exists $P_a \in \mathbb{P}_7, P_b, P_c \in \mathbb{P}_3$ and $\mathbb{P}_5 = \emptyset$ and G has no short cycles, then G has a Hamiltonian cycle.*

Due to page constraints, the detailed proof is included in [17].

Claim 13. *If $\mathbb{P}_j = \emptyset, j \geq 7$, and $\mathbb{P}_5 \neq \emptyset$, then $|\mathbb{P}_5| \leq 2$.*

Proof. For a contradiction assume that there exists paths $P_a, P_b, P_c \in \mathbb{P}_5$ such that $P_a = (w_1, w_2, w_3; x_1, x_2)$, $P_b = (s_1, s_2, s_3; t_1, t_2)$, and $P_c = (q_1, q_2, q_3; r_1, r_2)$. From Claim 3, there exists $v_1 \in N^I(v)$ such that $v_1w_2, v_1s_2, v_1q_2 \in E(G)$. Since the clique K is maximal, there exists $w' \in K$ such that $w'v_1 \notin E(G)$. Therefore, w' is an end vertex of some path in $\mathbb{P}_i, i \in \{1, 3, 5\}$. We see the following cases.

Case 1: w' is an end vertex of a P_5.

Without loss of generality assume $w' = w_1$, i.e., $w_1v_1 \notin E(G)$. Note that $w_1v_2 \in E(G)$ or $w_1v_3 \in E(G)$. Observe that either $w_1t_1 \in E(G)$ or $w_1t_2 \in E(G)$, otherwise, $N^I(s_2) \cup \{s_2, w_1\}$ induces a $K_{1,4}$. Further, either $w_1r_1 \in E(G)$ or $w_1r_2 \in E(G)$, otherwise, $N^I(q_2) \cup \{q_2, w_1\}$ induces a $K_{1,4}$. Clearly, $\{w_1\} \cup N^I(w_1)$ induces a $K_{1,4}$, and thus $w' \notin \mathbb{P}_5$.

Case 2: $w' \in \mathbb{P}_3 \cup \mathbb{P}_1$.

Note that $w'v_2 \in E(G)$ or $w'v_3 \in E(G)$. Observe that either $w'x_1 \in E(G)$ or $w'x_2 \in E(G)$, otherwise, $N^I(w_2) \cup \{w_2, w'\}$ induces a $K_{1,4}$. Similarly, either $w't_1 \in E(G)$ or $w't_2 \in E(G)$, otherwise, $N^I(s_2) \cup \{s_2, w'\}$ induces a $K_{1,4}$. Further, either $w'r_1 \in E(G)$ or $w'r_2 \in E(G)$, otherwise, $N^I(q_2) \cup \{q_2, w'\}$ induces a $K_{1,4}$. It follows that, $\{w'\} \cup N^I(w')$ has an induced $K_{1,4}$. This contradicts the assumption that there exists three such paths P_a, P_b, P_c. This completes the cases analysis and a proof. □

Claim 14. *If there exists $P_a, P_b \in \mathbb{P}_5, P_c \in \mathbb{P}_3$ and G has no short cycles, then G has a Hamiltonian cycle.*

Due to page constraints, the detailed proof is included in [17].

Observation: In the following claims, we shall produce a cycle C containing all the vertices of $N^I(v)$. Consider a path $P_m \in \mathbb{C}$ which is not a subpath of C. It is easy to observe that P_m is adjacent to at least one vertex in $\{v_1, v_2, v_3\}$, say v_1. Further, $(\overrightarrow{P_m}, v_1 \overrightarrow{C} v_1^-)$ is a desired path in G, where the vertices v_1^-, v_1 occur consecutively in $\overrightarrow{C}$.

Claim 15. *Let $P_a, P_b, P_c \in \mathbb{P}_3$, and $v_1 \in N^I(v)$ such that v_1 is adjacent to end vertex of at least two paths in P_a, P_b, P_c, $|\mathbb{P}_5| \leq 1$, and $\mathbb{P}_j = \emptyset$, $j \geq 7$. If G has no short cycles, then G has a Hamiltonian cycle.*

Due to page constraints, the detailed proof is included in [17].

Claim 16. *If there exists $P_a \in \mathbb{P}_5, P_b, P_c, P_d \in \mathbb{P}_3$ and G has no short cycles, then G has a Hamiltonian cycle.*

Proof. Note that there are four paths and thus there exists a vertex in $N^I(v)$, adjacent to at least two different paths. Without loss of generality, let us assume that the end vertices of two different paths are adjacent to $v_1 \in N^I(v)$. If two paths, say $P_b, P_c \in \mathbb{P}_3$ are adjacent to v_1, then by Claim 15, G has a Hamiltonian cycle. On the other hand, we shall assume that no such two paths exists in $\mathbb{P}_3$. Therefore, we could assume that the end vertices of P_b, P_c, P_d are adjacent to v_1, v_2, v_3, respectively. Now end vertices of P_a is adjacent to a vertex in $N^I(v)$, say v_1. Observe that $P = (\overrightarrow{P_a}, v_1, \overrightarrow{P_b}, \overrightarrow{P_c}, v_2, v, v_3, \overrightarrow{P_d})$ is a desired path in G. This completes the proof.   □

Claim 17. *If $\mathbb{P}_j = \emptyset, j \geq 5$ and there exists $P_a, P_b, P_c, P_d, P_e \in \mathbb{P}_3$ and G has no short cycles, then G has a Hamiltonian cycle.*

Proof. Note that there exists five paths and at least two those paths are adjacent to one of v_1, v_2, v_3. Observe that the premise of Claim 15 is satisfied, and therefore, G has a Hamiltonian cycle.   □

Theorem 2. *Let G be a 2-connected, $K_{1,4}$-free split graph with $|K| \geq |I| \geq 8$. G has a Hamiltonian cycle if and only if there are no induced short cycles in G. Further, finding such a cycle is polynomial-time solvable.*

Proof. Necessity is trivial. Sufficiency follows from the previous claims.

3   Hardness Result

T. Akiyama, T. Nishizeki, and N. Saito [23] proved the NP-completeness of Hamiltonian cycle in planar bipartite graphs with maximum degree 3. Here we give a reduction from Hamiltonian cycle problem in planar bipartite graphs with maximum degree 3 to Hamiltonian cycle problem in $K_{1,5}$-free split graph.

Theorem 3. *Hamiltonian cycle problem in $K_{1,5}$-free split graph is NP-complete.*

Proof. For NP-hardness result, we present a deterministic polynomial-time reduction that reduces an instance of planar bipartite graph with maximum degree 3 to the corresponding instance in split graphs Consider a planar bipartite graph G with maximum degree 3, and let A, B be the partitions of $V(G)$. We construct two graphs H_1, H_2 from G as follows.

$$V(H_1) = V(H_2) = V(G), E(H_i) = E(G) \cup E_i, i \in \{1,2\}$$
$$\text{where } E_1 = \{uv \ : \ u, v \in A\}, E_2 = \{uv \ : \ u, v \in B\}$$

Clearly, the reduction is a polynomial-time reduction and H_1, H_2 are split graphs with maximal cliques A, B, respectively. We now show that there exists a Hamiltonian cycle in G if and only if there exists a Hamiltonian cycle in H_1 and there exists a Hamiltonian cycle in H_2. Note that, if G has a Hamiltonian cycle, then $|A| = |B|$.

Necessity: If there exists a Hamiltonian cycle C in G, then C is a Hamiltonian cycle in H_1 and H_2, since G is a strict subgraph of H_1 and similarly, that of H_2.

Sufficiency: Since there exists a Hamiltonian cycle in H_1, $|A| \geq |B|$ and since H_2 has a Hamiltonian cycle, $|A| \leq |B|$. It follows that $|A| = |B|$. Now we claim that any Hamiltonian cycle C in H_1 is also a Hamiltonian cycle in G. If not, there exists at least one edge $uv \in E(C)$ where $u, v \in A$. It follows that at least one vertex in B is not in C, which contradicts the Hamiltonicity of H_1. Hence the sufficiency follows. We now show that the constructed graphs H_1 and H_2 are $K_{1,5}$-free. Suppose there exists a $K_{1,5}$ in H_1 or H_2 induced on vertices $\{u, v, w, x, y, z\}$, centered at v. At most two vertices (say u, v) of $K_{1,5}$ belongs to the clique K. Therefore, $w, x, y, z \in I$ and this implies $d^I_{H_i}(v) \geq 4, i = 1, 2$. It follows that $d_G(v) = 4$, which is a contradiction to the maximum degree of the bipartite graph G. Since a given instance of Hamiltonian problem in $K_{1,5}$-free split graphs can be verified in deterministic polynomial time, the problem is in class NP. It follows that the Hamiltonian cycle problem in $K_{1,5}$-free split graphs is NP-complete and the theorem follows.   □

References

1. Bertossi, A.A., Bonuccelli, M.A.: Hamiltonian circuits in interval graph generalizations. Inf. Process. Lett. **23**, 195–200 (1986)
2. Kemnitz, A., Schiermeyer, I.: Improved degree conditions for Hamiltonian properties. Discret. Math. **312**(14), 2140–2145 (2012)
3. Malakis, A.: Hamiltonian walks and polymer configurations. Stat. Mech. Appl. Phys. (A) **84**, 256–284 (1976)
4. de Figueiredo, C.M.H.: The P versus NP-complete dichotomy of some challenging problems in graph theory. Discret. Appl. Math. **160**(18), 2681–2693 (2012)
5. Bauer, D., Broersma, H.J., Heuvel, J., Veldman, H.J.: Long cycles in graphs with prescribed toughness and minimum degree. Discret. Math. **141**(1), 1–10 (1995)
6. West, D.B.: Introduction to Graph Theory, 2nd edn. (2003)
7. Dorninger, D.: Hamiltonian circuits determining the order of chromosomes. Discret. Appl. Math. **50**, 159–168 (1994)

8. Kratsch, D., Lehel, J., Muller, H.: Toughness, Hamiltonicity and split graphs. Discret. Math. **150**(1), 231–245 (1996)
9. Irina, G., Halskau, O., Laporte, G., Vlcek, M.: General solutions to the single vehicle routing problem with pickups and deliveries. Euro. J. Oper. Res. **180**, 568–584 (2007)
10. Broersma, H.J.: On some intriguing problems in Hamiltonian graph theory - a survey. Discret. Math. **251**, 47–69 (2002)
11. Muller, H.: Hamiltonian circuits in chordal bipartite graphs. Discret. Math. **156**, 291–298 (1996)
12. Keil, J.M.: Finding Hamiltonian circuits in interval graphs. Inf. Process. Lett. **20**, 201–206 (1985)
13. Illuri, M., Renjith, P., Sadagopan, N.: Complexity of steiner tree in split graphs - dichotomy results. In: Govindarajan, S., Maheshwari, A. (eds.) CALDAM 2016. LNCS, vol. 9602, pp. 308–325. Springer, Heidelberg (2016). doi:10.1007/978-3-319-29221-2_27
14. Garey, M.R., Johnson, D.S., Tarjan, R.E.: Planar Hamiltonian circuit problem is NP-complete. SIAM J. Comput. **5**, 704–714 (1976)
15. Tan, N.D., Hung, L.X.: On the Burkard-Hammer condition for Hamiltonian split graphs. Discret. Math. **296**(1), 59–72 (2005)
16. Narayanaswamy, N.S., Sadagopan, N.: Connected (s, t)-vertex separator parameterized by chordality. J. Graph Algorithms Appl. **19**(1), 549–565 (2015)
17. Renjith, P., Sadagopan, N.: Hamiltonicity in split graphs - a dichotomy. https://arxiv.org/abs/1610.00855
18. Burkard, R.E., Hammer, P.L.: A note on Hamiltonian split graphs. J. Comb. Theory Ser. B **28**(2), 245–248 (1980)
19. Gould, R.J.: Updating the Hamiltonian problem - a survey. J. Graph Theory **15**, 121–157 (1991)
20. Gould, R.J.: Advances on the Hamiltonian problem - a survey. Graphs Comb. **19**, 7–52 (2003)
21. Hung, R.W., Chang, M.S.: Linear-time algorithms for the Hamiltonian problems on distance-hereditary graphs. Theoret. Comput. Sci. **341**, 411–440 (2005)
22. Hung, R.W., Chang, M.S., Laio, C.H.: The Hamiltonian cycle problem on circular-arc graphs. In: Proceedings of the International Conference of Engineers and Computer Scientists (IMECS, Hong Kong), pp. 18–20 (2009)
23. Akiyama, T., Nishizeki, T., Saito, N.: NP-completeness of the Hamiltonian cycle problem for bipartite graphs. J. Inf. Process. **3**(2), 73–76 (1980)
24. Chvátal, V.: Tough graphs and Hamiltonian circuits. Discret. Math. **5**, 215–228 (1973)
25. Shih, W.K., Chern, T.C., Hsu, W.L.: An O(n^2log n) algorithm for the Hamiltonian cycle problem on circular-arc graphs. SIAM J. Comput. **21**, 1026–1046 (1992)

Finding Large Independent Sets in Line of Sight Networks

Pavan Sangha and Michele Zito[✉]

University of Liverpool, Liverpool, UK
{P.Sangha,M.Zito}@liverpool.ac.uk

Abstract. Line of Sight (LoS) networks provide a model of wireless communication which incorporates visibility constraints. Vertices of such networks can be embedded in finite d-dimensional grids of size n, and two vertices are adjacent if they share a line of sight and are at distance less than ω. In this paper we study large independent sets in LoS networks. We prove that the computational problem of finding a largest independent set can be solved optimally in polynomial time for one dimensional LoS networks. However, for $d \geq 2$, the (decision version of) the problem becomes NP-hard for any fixed $\omega \geq 3$ and even if ω is chosen to be a function of n that is $O(n^{1-\epsilon})$ for any fixed $\epsilon > 0$. In addition we show that the problem is also NP-hard when $\omega = n$ for $d \geq 3$. This result extends earlier work which showed that the problem is solvable in polynomial time for gridline graphs when $d = 2$. Finally we describe simple algorithms that achieve constant factor approximations and present a polynomial time approximation scheme for the case where ω is constant.

1 Introduction

Geometric graphs have become a popular tool for reasoning about wireless networks. Typically wireless devices positioned in some physical space can be represented by a collection of vertices. A graph can then be constructed by representing communication between pairs of vertices by edges.

The disk intersection model is a commonly used model for representing wireless sensor networks [16]. Sensors are modelled as vertices in some topological setting and their communication ranges are represented by circles having some prescribed radius. Overlapping circles then represent communication between pairs of vertices and makes it possible to construct a graph. Unfortunately in many real world applications of wireless networks, the environments often come with a large number of obstacles which impose line of sight constrictions on vertices. These obstacles are often difficult to incorporate in the geometric models described above.

Frieze et al. [7] developed the notion of a (random) line of sight network to provide a model of wireless networks which can incorporate line of sight constraints. For positive integers d and n, let $\mathbb{Z}_n^d = \{(x_1, x_2, \ldots, x_d) : x_i \in \{0, 1, \ldots, n-1\}, 1 \leq i \leq d\}$ and $\mathbb{Z}_+^d = \cup_{n=1}^{\infty} \mathbb{Z}_n^d$. In the rest of the paper the distance between points $\boldsymbol{x} = (x_1, x_2, \ldots, x_d)$ and $\boldsymbol{x}' = (x_1', x_2', \ldots, x_d')$ in $\mathbb{Z}_n^d$ is the quantity $\sum_{i=1}^d |x_i - x_i'|$.

© Springer International Publishing AG 2017
D. Gaur and N.S. Narayanaswamy (Eds.): CALDAM 2017, LNCS 10156, pp. 332–343, 2017.
DOI: 10.1007/978-3-319-53007-9_29

We say that two distinct points $\boldsymbol{x} = (x_1, x_2, \ldots, x_d), \boldsymbol{x}' = (x_1', x_2', \ldots, x_d')$ in $\mathbb{Z}_n^d$ share a line of sight if there exists a $j \in \{1, \ldots, d\}$ such that $x_i = x_i'$ for all $i \in \{1, \ldots, d\} \setminus \{j\}$, moreover in this case we say that $\boldsymbol{x}$ and $\boldsymbol{x}'$ share a j-line.

Definition 1. *A graph $G = (V, E)$ is said to be a* Line of Sight (LoS) *network with parameters d, n and ω, if there exists an embedding $f_G : V \to \mathbb{Z}_n^d$, such that $\{u, v\} \in E$ if and only if $f_G(u)$ and $f_G(v)$ share a line of sight and the distance between them is strictly less than ω.*

We denote the set of embedded vertices of $G = (V, E)$ by $f_G(V) \in \mathbb{Z}_n^d$. In what follows, because all graphs will be embedded we will abuse notation so that u will denote both a vertex in V and the corresponding embedded vertex $f_G(u) \in f_G(V)$. The parameter ω, called the network *range parameter*, can be used to model the usual proximity constraint in a wireless network. The line of sight visibility constraint is the mechanism that allows the modelling of obstacles.

Definition 2. *We say that a LoS network G with parameters d, n and ω is d-dimensionally spanning if for each $j \in \{1, 2, \ldots, d\}$ there exists an edge $\{u, v\} \in E$ such that $f_G(u)$ and $f_G(v)$ share a j-line.*

Let $L_{n,\omega}^d$ denote the set of all graphs G which are d-dimensionally spanning LoS networks with parameters n, d and ω. Let $L_n^d = \cup_{n \in \mathbb{N}} L_{n,\omega}^d$. Note that ω might be as large as n. To study the properties of LoS networks with large range parameter sometimes we are interested in the properties of $L_{n,g}^d$ (or L_g^d) the set of LoS networks with range parameter $\omega = g(n)$, if $g : \mathbb{N} \to \mathbb{N}$ is a monotone increasing function.

LoS networks generalise other well known geometric graph models. For example, a LoS network with parameter $\omega = 2$ is known as a grid graph [3], where each vertex can only share an edge with the $2d$ other vertices at distance one in $\mathbb{Z}_n^d$. On the other hand the elements of L_n^d are known as gridline graphs [15]. So far LoS networks have been studied with respect to their typical connectivity properties [4,7].

In this paper we investigate the well known maximum independent set problem (MIS) for both 2-dimensional and higher dimensional LoS networks as the range parameter ω varies. Large independent sets in graphs have been the subject of significant study in various branches of Mathematics as they provide a measure of network dispersion and have a strong connection with other important graph measures such as vertex covers, cliques and colourings [6]. It is well known that finding a largest independent set in a graph is an NP-hard problem [8], and even good approximate solutions are hard to find [9]. We show that the problem can be solved optimally in polynomial time for $d = 1$, for any ω, using a straightforward greedy strategy. For $\omega = 2$ and any d the problem can be solved in polynomial time because the LoS network is a bipartite graph. For $d = 2$ the problem can be solved in polynomial time for the case $\omega = n$. In higher dimensions ($d \geq 3$) the problem becomes more difficult and the overall picture is less clear cut. We prove the following (here IS is the decision version of MIS)

Theorem 1. IS($L_{n,\omega}^d$) *is NP-hard, for each fixed* $d \geq 2$ *and fixed integer* $\omega \geq 3$. *Additionally, for any given* $\epsilon > 0$, IS(L_g^d) *is NP-hard for any choice of* g *such that* $g(n) = O(n^{1-\epsilon})$.

The proof of Theorem 1 cannot be extended to cover the case of very large range parameters. However a different reduction allows us to prove the following:

Theorem 2. IS(L_n^d) *is NP-hard for each fixed* $d \geq 3$.

Note the statement of Theorem 2 is a generalizion to higher dimensions of a result by Peterson on 2-dimensional gridline graphs [15]. Finally, we complement these negative results by describing two heuristics that achieve constant factor approximations and an efficient polynomial time approximation scheme (EPTAS), as defined in [2], for the case when ω is a fixed constant, for any d.

The layout of this paper is as follows. We start our investigation in Sect. 2 by studying a natural greedy heuristic for the MIS problem in LoS networks. The algorithm is optimal for $d = 1$ and represents a base-line benchmark for approximation strategies in higher dimensions. In Sects. 3 and 4 we study the problem complexity for $d \geq 2$. The final part of the paper is devoted to the remaining algorithmic results mentioned above.

In what follows if Π is a computational problem and $\mathcal{I}$ is a particular set of instances for it, then $\Pi(\mathcal{I})$ will denote the restriction of Π to instances belonging to $\mathcal{I}$. If Π_1 and Π_2 are decision problems, then $\Pi_1 \leq_p \Pi_2$ will denote the fact that Π_1 is polynomial-time reducible to Π_2. Unless otherwise stated we follow [6] for all our graph-theoretic notations.

2   The Corner Greedy Algorithm

We start off our investigation of the MIS problem for LoS networks by describing and analysing a very simple, natural heuristic for tackling this problem. The strategy is an adaptation of an algorithm described in [12] in the context of unit disk graphs. Given a graph $G = (V, E)$ and $u \in V$ we denote by $N(u)$ the set of neighbours of u in V. For a d-dimensional LoS network $G = (V, E)$ and a vertex $u = (u_1, u_2 \ldots, u_d) \in V$ we denote the set of vertices $v = (v_1, v_2, \ldots, v_d)$, such that u and v share a j-line and $v \in N(u)$ as $N_j(u)$ (note that if $v \in N_j(u)$ then $u \in N_j(v)$) and we say that $u <_j v$ if $u_j < v_j$. Corners in LoS networks are vertices that are extremes w.r.t. the $<_j$ relationship for all j. More precisely we give the following definition:

Definition 3. *Given a d-dimensional LoS network* $G = (V, E)$, *we say that* $u \in V$ *is a* corner *if for each* $j \in \{1, 2, \ldots, d\}$, $u <_j v$ *for all* $v \in N_j(u)$ *or* $v <_j u$ *for all* $v \in N_j(u)$.

If we lexicographically order all vertices in a LoS network w.r.t their d-tuples the first vertex in such an ordering is a corner. The corner greedy Algorithm can be described as follows:

Corner Greedy Algorithm: Initially $S = \emptyset$

1. Add a corner u to S and remove the closed neighbourhood of $N(u)$ from G.
2. Repeat 1 until the graph G is empty.
3. Return S.

For $d = 1$ the process is clearly optimal and in general we can prove (here $\alpha(G)$ is the *independence number* of G, the size of the largest independent sets in G):

Theorem 3. *For any d-dimensional LoS network G the size of the set of independent vertices returned by the corner greedy algorithm is at least $\frac{\alpha(G)}{d}$.*

Proof. In each iteration the algorithm picks a corner vertex, adds it to S and then removes the vertex and all its neighbours from the graph. Let S^* be an independent set of maximum cardinality in G. Each vertex $v \in S^*$ belongs to exactly one closed neighbourhood identified in step 2 of the Corner Greedy algorithm. Furthermore since each corner vertex has neighbours in at most d different line of sights, d different elements of S^* may belong to a particular closed neighbourhood. The result follows.   □

Thus, in general, the corner greedy algorithm described in this section represents a d-approximation algorithm for the MIS problem in d-dimensional LoS networks. It is then natural to ask whether this can be improved. Perhaps LoS networks are sufficiently close to grid graphs and the MIS problem can indeed be solved in polynomial time. Unfortunately in the sections that follow we will give a negative answer to such conjecture.

3 Hardness for Small Ranges

In this section we prove Theorem 1. For $d = 2$ (resp. $d \geq 3$) we describe explicit embeddings in $\mathbb{Z}_n^d$ of graphs which are subdivisions of planar graphs with maximum degree four (resp. subdivisions of bounded degree graphs). In both instances we start by embedding the given graph $G = (V, E)$ orthogonally (see Sect. 3.1). We then add further vertices to obtain (the embedding of) a d-dimensionally spanning LoS network. Once this is done to prove NP-hardness we show the existence of a linear relationship between the size of an independent set in G and the size of an independent set in the resulting LoS network, and since the IS problem is NP-hard for both planar graphs of maximum degree four and bounded degree graphs the result follows.

3.1 Embeddings

Graph embedding has been an active research area for quite some time (the interested reader is referred to reviews like [5], or the more recent one [13], for additional bibliographic details). Here in particular we will be interested in so called *orthogonal* embeddings of bounded degree graphs $G = (V, E)$ in $\mathbb{Z}_n^d$.

Define a path in $\mathbb{Z}_n^d$ to be any sequence of distinct points in $\mathbb{Z}_n^d$ such that any two consecutive points in the sequence have distance one. An orthogonal embedding of a graph $G = (V, E)$ in $\mathbb{Z}_n^d$ denoted by $\Gamma(G)$ is an embedding where the vertices $v \in V$ are mapped to points in $\mathbb{Z}_n^d$ and the edges $\{u, v\} \in E$ to paths with end points $\Gamma(u), \Gamma(v)$. Paths representing distinct edges can only intersect at their end-points. It is well known that an orthogonal embedding of G in $\mathbb{Z}_n^d$ requires $d \geq \lceil \Delta(G)/2 \rceil$ and, for general graphs, the bound is tight except for $\Delta(G) \leq 4$. In this work we will use the following two results, dealing with the case $d = 2$ and $d \geq 3$ respectively. Moreover the embeddings described in the following two theorems can be constructed in polynomial time.

Theorem 4. [17] *Any planar graph* $G = (V, E)$ *with* $\Delta(G) \leq 4$ *and* $|E| = m$, *admits an orthogonal embedding* $\Gamma(G)$ *in* $\mathbb{Z}_{3m}^2$.

Theorem 5. [18] *Any simple graph* $G = (V, E)$ *with* $\Delta(G) \geq 5$ *admits an orthogonal embedding* $\Gamma(G)$ *in* $\mathbb{Z}_{k|V|}^d$ *where* $d = \lceil \Delta(G)/2 \rceil$ *and* k *is a positive integer constant.*

3.2   Padding

Given a bounded degree graph $G = (V, E)$ and an arbitrary function r mapping the edges of G to positive integers, let $F(G, r)$ be a new graph obtained by replacing each edge $e \in E$ by a path containing a $2r(e)$ additional vertices between it's end points. The graph $F(G, r)$ is also known as a *subdivision* of G (see [6, Chap. 1]). Our reduction then uses the following well-known result.

Lemma 1. *Let* $G = (V, E)$ *be a graph, and* r *an arbitrary function mapping the edges of* G *to positive integers. Then* $\alpha(G) \geq k$ *if and only if* $\alpha(F(G, r)) \geq k + \sum_{e \in E} r(e)$.

3.3   Reduction

We now sketch the main construction in the proof of Theorem 1. Given a bounded degree graph $G = (V, E)$ we use Theorem 4 (for $d = 2$) or Theorem 5 (for $d > 2$) to define the appropriate orthogonal embedding $\Gamma(G)$ in the appropriate d-dimensional finite grid. We then *stretch* the embedding $\Gamma(G)$ to obtain a new orthogonal embedding $\Gamma'(G)$ by mapping each embedded vertex $\Gamma(u) = (u_1, \ldots, u_d) \in \Gamma(G)$ to the embedded vertex $\Gamma'(u) = (4(\omega - 1)u_1, \ldots, 4(\omega - 1)u_d) \in \Gamma'(G)$. The paths P_{uv} in $\Gamma(G)$ with end points $\Gamma(u), \Gamma(v)$ get mapped to the corresponding paths $P'_{uv} \in \Gamma'(G)$ with end points $\Gamma'(u), \Gamma'(v)$. Finally we pad the orthogonal embedding $\Gamma'(G)$ with additional *sensor* vertices to obtain a LoS network embedding of G which is also a subdivision $F(G, r)$, for a suitable choice of r. It is easy to check from the orthogonal embedding of G that the LoS network G' is d-dimensionally spanning. The additional sensor vertices are placed along the paths in $\Gamma'(G)$ corresponding to the edges of G in a way that satisfies the following constraints:

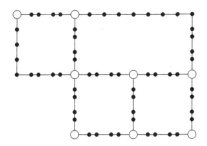

Fig. 1. A 2-dimensionally spanning LoS network $G' = F(G,r)$ with sensor vertices in black, constructed from an orthogonal embedding of the 4-planar graph G with 8 white vertices.

1. the total number of padding vertices placed on each path P'_{uv} is even;
2. a padding vertex must be placed on every corner (bend in the path);
3. for any set of three consecutive sensor vertices $\{v_s, v'_s, v''_s\}$ placed on a path P'_{uv}, v''_s and v_s are at a distance at least ω, but v'_s is at a distance at most $\omega - 1$ from v_s and v''_s, maintaining a connected line of sight path structure.

Figure 1 sketches an example of the construction for $d = 2$. The correctness of the reduction follows from Lemma 1, where for each $\{u, v\} \in E$, $4 \times |\Gamma(P_{uv})|$ additional sensor vertices are added to the path P'_{uv}.

The proof of the second part of the statement of Theorem 1 is obtained by noticing that the construction above works if ω is bounded above by a function $g(n) = O(n^{1-\epsilon})$ for any fixed $\epsilon > 0$.

4 Hardness for (Very) Long Ranges

The outcome of Sect. 3 is a proof that for any $d \geq 2$, $\mathrm{IS}(L^d_\omega)$ is NP-hard for small values of ω. The analysis of the problem's complexity for $\omega = n$ while different to the sublinear case is quite interesting in its own right. In this section we prove Theorem 2. Let $G = (V, E) \in L^2_n$ then for each vertex $u \in V$ in a LoS embedding of G, $v \in N(u)$ if and only if u and v share a line of sight since ω is as large as possible. A simple reduction shows that $\mathrm{MIS}(L^2_n)$ can be solved via a single bipartite matching computation [11,15].

When $d \geq 3$ things become more complicated. Each element of L^d_n can still be mapped to a d-partite graph. However the independent sets of the LoS network correspond to structures that are computationally less tractable than matchings. In what follows MAX 2-SAT(3) is the problem of finding a truth assignment to the variables of a 2-CNF boolean propositional formula that maximizes the number of satisfied clauses, restricted to instances in which each variable occurs in at most three clauses [1]. The proof of Theorem 2 is completed in the case $d \geq 3$ using the following result:

Theorem 6. MAX 2-SAT(3) $\leq_p \mathrm{IS}(L^d_n)$, for every $d \geq 3$.

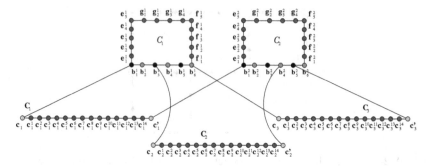

Fig. 2. The construction of the graph $\mathcal{E}_{2,3}^5(F)$ corresponding to the formula $F = (X_1 \vee X_2) \wedge (X_1 \vee \overline{X}_2) \wedge (\overline{X}_1 \vee X_2)$ and the assignment $X_1 = \text{TRUE}$, $X_2 = \text{TRUE}$ satisfying all three clauses.

The proof of this result consists of two steps. We first describe how to translate a given 2-CNF formula into a graph and then we embed such a graph into a d-dimensionally spanning LoS network where ω is maximised.

4.1   From Formulae to Graphs

Consider an instance of MAX 2-SAT(3) consisting of a formula $F = C_1 \vee C_2 \vee \ldots \vee C_m$ on n variables and m clauses, each formed by two literals. We construct a graph denoted $\mathcal{E}_{n,m}^d(F)$ as follows. For each variable X_j we construct a variable gadget $\mathcal{X}_j$ which is a cycle of even length $2m + 3d - (d \bmod 2)$ in $\mathcal{E}_{n,m}^d(F)$. The vertices in C_j are split into four sets: a set of $2m$ *base vertices*, $\{\mathbf{b}_1^j, \mathbf{b}_2^j, \ldots, \mathbf{b}_{2m}^j\}$ and $3d - (d \bmod 2)$ additional vertices which are further split into three sets. There are vertices $\{\mathbf{e}_1^j, \mathbf{e}_2^j, \ldots, \mathbf{e}_{d+1-d \bmod 2}^j\}$, $\{\mathbf{f}_1^j, \mathbf{f}_2^j, \ldots, \mathbf{f}_d^j\}$ and $\{\mathbf{g}_1^j, \mathbf{g}_2^j, \ldots, \mathbf{g}_{d-1}^j\}$. Cycle $\mathcal{X}_j$ is then given by

$$\mathbf{b}_1^j \ldots \mathbf{b}_{2m}^j \mathbf{f}_1^j \ldots \mathbf{f}_d^j \mathbf{g}_{d-1}^j \ldots \mathbf{g}_1^j \mathbf{e}_{d+1-d \bmod 2}^j \cdots \mathbf{e}_1^j \mathbf{b}_1^j.$$

For each clause C_i in F we construct a clause gadget $\mathbf{C}_i$ containing two *clause vertices* $\mathbf{c}_i$ and $\mathbf{c}_i'$ which are connected by a path of $3d - (d \bmod 2)$ *dummy vertices*. Thus the clause gadget consists of the path $\mathbf{c}_i \mathbf{c}_i^1 \mathbf{c}_i^2 \ldots \mathbf{c}_i^{3d-(d \bmod 2)} \mathbf{c}_i'$. If C_i in F contains the variable X_j (resp. $\overline{X}_j$) we select odd (resp. even) parity vertex $\mathbf{b}_{2i-1}^j$ (resp. $\mathbf{b}_{2i}^j$) on the cycle $\mathcal{X}_j$ and we connect it to one of the clause vertices in $\mathbf{C}_i$. Note that because of the variable occurrence constraint, in each variable gadget at most three base vertices are selected. Figure 2 shows an example of this construction.

Claim. For any instance of MAX 2-SAT(3) with m clauses on n variables, there is an assignment that satisfies r clauses if and only if the graph $\mathcal{E}_{n,m}^d(F)$ has an independent set of size

$$r + \lfloor 3d/2 \rfloor \times m + n \times (m + (3d - (d \bmod 2)))/2.$$

Each variable gadget $\mathcal{X}_j$ has an independent set of size $m+(3d-(d \bmod 2))/2$ and in fact there's always exactly two sets of this size, one using all odd indexed $\mathbf{b}_h^j$, the other using all the even indexed ones. The former of these corresponds to setting X_j to TRUE, the other one to FALSE. Also, each clause gadget $\mathbf{C}_i$ is a path of even length, the largest independent sets of these paths are of size $\lfloor 3d/2 \rfloor + 1$ and include one of $\mathbf{c}_i$ or $\mathbf{c}_i'$. If clause C_i is satisfied at least one of its literals is set to TRUE and that implies that an independent set can be picked in the corresponding variable gadget that leaves at least one of $\mathbf{c}_i$ or $\mathbf{c}_i'$ free to be added to an independent set. Conversely if C_i is not satisfied, then neither $\mathbf{c}_i$ nor $\mathbf{c}_i'$ are free and the corresponding clause gadgets will only contribute $\lfloor 3d/2 \rfloor$ vertices to the independent set. The first part of the claim follows.

For the opposite direction let $\mathcal{I}$ be an independent set in $\mathcal{E}_{n,m}^d(F)$. Two cases arise. If all the variable gadgets contain half of their vertices in $\mathcal{I}$ then the 2^n possible settings correspond to the possible ways to assign a truth value to the variables of F. In each of these cases a clause gadget will contain $\lfloor 3d/2 \rfloor + 1$ elements of $\mathcal{I}$ if and only if at least one of the clause vertices is one of them. The number r of clause gadgets containing $\lfloor 3d/2 \rfloor + 1$ elements of $\mathcal{I}$ satisfies

$$r = |\mathcal{I}| - \lfloor 3d/2 \rfloor \times m - n \times (m + (3d - (d \bmod 2))/2).$$

We call an independent set of this type *full*. If $\mathcal{I}$ is not full then wlog let $\mathcal{X}_k$ be a variable gadget such that less than half the vertices of $\mathcal{X}_k$ belong to $\mathcal{X}_k \cap \mathcal{I}$. Call $\mathbf{b}^k(x)$, $\mathbf{b}^k(y)$, $\mathbf{b}^k(z)$ the three base variables from $\mathcal{X}_k$ that are adjacent to clause gadgets $\mathbf{C}_x$, $\mathbf{C}_y$ and $\mathbf{C}_z$. We replace the set $\mathcal{X}_k \cap \mathcal{I}$ with a maximum independent set of $\mathcal{X}_k$ containing half it's vertices according to the parity of $\mathbf{b}^k(x)$, $\mathbf{b}^k(y)$, and $\mathbf{b}^k(z)$: if $\mathbf{b}^k(x)$, $\mathbf{b}^k(y)$, and $\mathbf{b}^k(z)$ all have the same, say, odd parity, we replace $\mathcal{X}_k \cap \mathcal{I}$ with the set of all even vertices in $\mathcal{X}_k$. If two of them are of the same parity we pick the opposite to the majority. During this process we may remove at most one clause vertex from $\mathcal{I}$. For instance in the case where a base vertex not in the majority and not in the independent set $\mathcal{X}_k \cap \mathcal{I}$ is added to the independent set during the process. In this case it's adjacent clause vertex must be removed once it is added. However in replacing $\mathcal{X}_k \cap \mathcal{I}$ with an independent set in $\mathcal{X}_k$ consisting of half of the vertices we add at least one extra vertex from $\mathcal{X}_k$ to our independent set, thus negating the potential loss of a clause vertex. The process terminates in at most n steps with an independent set $|\mathcal{I}'| \geq |\mathcal{I}|$.

4.2   Embedding

To complete the reduction we need to show that there exists an integer N, polynomial in n and m for fixed d, such that $\mathcal{E}_{n,m}^d(F)$ has a d-dimensionally spanning LoS network embedding in $\mathbb{Z}_N^d$ satisfying $\omega = N$, thus showing $\mathcal{E}_{n,m}^d(F) \in L_{N,N}^d$. In what follows we use $N = 7(m+d)n$.

In order to explain the embedding of $\mathcal{E}_{n,m}^d(F)$, we first discuss how to embed paths and cycles in $\mathbb{Z}^{+d}$ satisfying LoS network constraints with the range parameter maximized. We use methods called *path rotation* and *path connection* to embed, respectively, long and short paths. Given a path $P = v_1, \ldots, v_k$ the process starts by assigning (arbitrarily or according to some rule) a d-tuple $(x_1, x_2, \ldots, x_d)$ to v_1. After that, the path rotation method assigns d-tuples iteratively so that the d-tuple assigned to v_i differs from the d-tuple assigned to v_{i-1} in co-ordinate position $(i-1)$ mod d. Furthermore the integer value assigned to the d-tuple v_i in the co-ordinate position $(i-1)$ mod d is one more than the integer value assigned to the d-tuple v_{i-1} in co-ordinate position $(i-1)$ mod d. Note for two adjacent vertices v_{i-1}, v_i on the path, there remains an edge between their assigned d-tuples as these differ in exactly one co-ordinate position (i.e. share a line of sight). The path connection method is used to specifically embed a path $P' = x, c_1, c_2, \ldots c_{d-1}, y$ in $\mathbb{Z}^{+d}$ of length d where the d-tuples $x = (x_1, x_2, \ldots, x_d)$ and $y = (y_1, y_2, \ldots, y_d)$ are pre-assigned and distinct ($x_i \neq y_i$ for $i \in \{1, 2, \ldots, d\}$). We can denote $x = c_0$ and $y = c_d$. We then embed c_i for $i \in \{1, \ldots, d-1\}$ as follows. Let S_d denote the set of all permutations of the set $\{1, \ldots, d\}$ and pick $\sigma \in S_d$. Then the pair (c_i, c_{i-1}) differ in co-ordinate position $\sigma(i)$ and the value assigned to c_i in this position is $y_{\sigma(i)}$.

We can use a combination of the path rotation and path connection methods to embed a cycle $v_1, \ldots, v_k, v_1$ where $k \geq 3d$ in $\mathbb{Z}^{+d}$. We embed the path $v_1, v_2, \ldots, v_{k-(d-1)}$ using the path rotation method, we then embed $v_{k-(d-1)+1}, \ldots, v_k$ using the path connection method. Note that the resulting embedding is d-dimensionally spanning. We embed each variable gadget $\mathcal{C}_i$ in $\mathcal{E}^d_{n,m}(F)$ according to the method described above, embedding the vertices $\mathbf{e}^j_{d+1-d \bmod 2}, \ldots, \mathbf{e}^j_1$, $\mathbf{b}^j_1, \ldots, \mathbf{b}^j_{2m}, \mathbf{f}^j_1, \ldots, \mathbf{f}^j_d$ using the path rotation method initialising $\mathbf{e}^j_{d+1-d \bmod 2} = ((j-1)(B_{m,d}+1), (j-1)(B_{m,d}+1), \ldots, (j-1)(B_{m,d}+1))$ where $B_{m,d} = \lceil \frac{2m+2d-1}{d} \rceil$. The vertices $\mathbf{g}^j_1, \ldots, \mathbf{g}^j_{d-1}$ are then embedded using the path connection method. We next embed the clause gadgets $\mathbf{C}_j$ for $j = 1, \ldots, m$. For each j we first embed the clause vertices $\mathbf{c}_j$ and $\mathbf{c}'_j$. For each clause vertex we choose the co-ordinate position that it will differ in from it's adjacent base vertex carefully. Without loss of generality assume that clause vertex $\mathbf{c}_j$ is adjacent to some base vertex denoted $\mathbf{b}^i$ (we do not distinguish between $\mathbf{b}^i = \mathbf{b}^i_{2j-1}$ and $\mathbf{b}^i = \mathbf{b}^i_{2j}$). Then we ensure that for each base vertex $\mathbf{b}^i_* \in N(\mathbf{b})$, the co-ordinate positions that the pairs of d-tuples $(\mathbf{b}^i, \mathbf{c}_j)$ and $(\mathbf{b}^i, \mathbf{b}^i_*)$ differ in are not the same. The value assigned in the co-ordinate position in which $(\mathbf{b}^i, \mathbf{c}_j)$ differ is then unique for each $\mathbf{c}_j$ ensuring we do not add unwanted edges. Once $\mathbf{c}_j$ and $\mathbf{c}'_j$ have been assigned their d-tuples the remaining dummy vertices $\mathbf{c}^1_j, \mathbf{c}^2_j, \ldots, \mathbf{c}^{3d-d \bmod 2}_j$ are embedded using a combination of path rotation and path connection methods fully embedding $\mathbf{C}_j$.

5   Approximation Algorithms

In this section we further extend our understanding of the computational properties of the MIS problem for d-dimensional LoS networks with constant range parameter $\omega \in \mathbb{N}$. The hardness proofs of Sects. 3 and 4 will be complemented by two additional algorithmic results. First, we define a polynomial approximation scheme that works for fixed ω. Then we describe a local improvement strategy that beats the approximation guarantee of the corner greedy heuristic described in Sect. 2 for the extreme case $\omega = n$.

5.1 An Efficient Polynomial Time Approximation Scheme

In this section we describe an algorithm that accepts as input any d-dimensional LoS network $G = (V, E)$ with constant range parameter $\omega \in \mathbb{N}$ and returns a $(1 + \epsilon)$-approximate independent set in the given graph. The process mimics an approximation scheme proposed by Nieberg *et al.* [14] for the MIS problem in *unit disk graphs*. The algorithm in this section however has a running time that, for fixed ω, is linear in the number of vertices of the input graph (although the constant hidden in the big-Oh notation depends exponentially on ϵ^{-1}). Algorithms of this type have been referred to as *Efficient* polynomial time approximation schemes [2].

Let $\epsilon > 0$ and let $\rho = 1 + \epsilon$ denote the desired approximation guarantee. Given a LoS network $G = (V, E)$ we seek to construct an independent set of size at least $\alpha(G)/\rho$. Let r be a non-negative integer and for $u \in V$ define the *bounding box* $B^r(u)$ *centered at* $u = (u_1, u_2, \ldots, u_d)$ as the region in $\mathbb{Z}_n^d$ containing the set of points

$$\{y = (y_1, y_2, \ldots, y_d) \in \mathbb{Z}_n^d \mid \max_{i=1,\ldots,d} |y_i - u_i| < r \cdot \omega\}.$$

Let $G[B^r(u)]$ denote the induced subgraph of G in the region $B^r(u)$ in the embedding of G. The proposed algorithm is an iterative process that removes vertices from G. At each iteration it picks a corner vertex $u \in G$ among the surviving vertices, builds $G[B^0(u)], G[B^1(u)], \ldots, G[B^{\hat{r}}(u)]$, computing a maximum independent set I_r of $G[B^r(u)]$ and removes $G[B^{\hat{r}+1}(u)]$ from G (note $G[B^0(u)]$ consists of just the single vertex u). The value $\hat{r}$ is the smallest positive r for which

$$|I_{r+1}| < \rho \cdot |I_r|. \tag{1}$$

The independent set returned by the algorithm is the union of the sets $I_{\hat{r}}$ produced by each iteration of the algorithm. Note that a maximum cardinality independent set can be found in $B^r(u)$ in time $O((r \cdot \omega)^d \, r^d \omega^{d-1})$. It follows from (1) and by the definition of $\hat{r}$ in addition to the fact that $|I_0| = 1$ that for each $r \leq \hat{r}$,

$$|I_r| \geq (1 + \epsilon)^r. \tag{2}$$

In addition, using the pigeon hole principle it can be shown that

$$|I_r| \leq r \cdot (r \cdot \omega)^{d-1}. \tag{3}$$

Hence, for each $r \leq \hat{r}$,

$$(1 + \epsilon)^r \leq |I_r| \leq r^d \cdot \omega^{d-1}.$$

The value $\hat{r}$ is therefore upper bound by the smallest positive integer r for which $r^d \omega^{d-1} < (1+\epsilon)^r$ and the process is indeed an EPTAS for the MIS problem for LoS networks in the case where $\omega \in \mathbb{N}$. The correctness of the algorithm is entailed by the following result, which mirrors a similar statement in [14]:

Theorem 7. *Suppose inductively that we can compute a ρ-approximate independent set $I' \subset V \setminus N^{\hat{r}+1}(v)$ for G'. Then $I \equiv I_{\hat{r}} \cup I'$ is a ρ-approximate independent set for G.*

5.2   An Improved Approximation Algorithm

In this section we show that for the case $\omega \geq n$ it is possible to improve the guarantee of the greedy heuristic provided in Sect. 2. We provide a local improvement strategy that approximately doubles the guarantee from $\frac{1}{d}$ to $\frac{2}{d} - \epsilon$ where $\epsilon > 0$ is a fixed constant. The algorithm uses the following well known result of Hurkens and Schrijver on set systems [10].

Theorem 8. *[10] Let $E_1, \ldots, E_m$ be subsets of a set T of n elements. Suppose that:*

1. *Each element of T is contained in at most $k \geq 3$ of the $E_1, \ldots, E_m$ sets;*
2. *For any $p \leq t$, any p sets among $E_1, \ldots, E_m$ cover at least p elements of T.*

Then:

$$\frac{m}{n} \leq \frac{k(k-1)^r - k}{2(k-1)^r - k} \text{ if } t = 2r - 1, \qquad \frac{m}{n} \leq \frac{k(k-1)^r - 2}{2(k-1)^r - 2} \text{ if } t = 2r.$$

Consider a d-dimensional LoS network $G = (V, E)$ with $\omega \geq n$, the algorithm works as follows. Start with any collection of independent vertices $S = \{s_1, s_2, \ldots, s_r\}$ and fix a $t \geq 1$ in G. We now perform the following iterative local search algorithm H_t which takes any set of $p \leq t$ vertices from S and replaces them with a collection of $p + 1$ vertices from $V \setminus S$ so that the new collection remains pairwise independent. By repeating this algorithm it will terminate with a set of disjoint embedded vertices $S' = \{s'_1, s'_2, \ldots, s'_n\}$ such that for each set of $p + 1 \leq t + 1$ pairwise independent vertices amoung $V \setminus S'$ they intersect at least $p + 1$ vertices amoung $s'_1 \ldots, s'_n$, otherwise we could run the algorithm for another step. Let $U = \{u_1, \ldots, u_{\alpha(G)}\}$ be an independent set in G of maximum size. Then we claim that $\frac{\alpha(G)}{|S'|}$ satisfies Theorem 9 conditions.

Theorem 9. *Let t be a positive integer, $G = (V, E)$ a LoS network with $\omega \geq n$ and S' the independent set returned by the algorithm H_t. Then*

$$\frac{\alpha(G)}{|S'|} \leq \frac{d - c/(d-1)^{\lfloor t/2 \rfloor}}{2 - c/(d-1)^{\lfloor t/2 \rfloor}}$$

where $c = d - (d-2)(1 - (t \bmod 2))$.

In our application $T = S'$. Also, if $U = \{u_1, \ldots, u_{\alpha(G)}\}$ is a maximum independent set in G, we may define $E_i = \{v \in S' : \{v, u_i\} \in E(G)\}$ for each $i \in \{1, \ldots, \alpha(G)\}$. Finally since U is an independent set it follows that any vertex $s'_j \in S'$ can only belong to at most d sets E_i because s'_j has exactly d different line of sights in an embedding of G. Thus $k = d$ in our application, note that condition 2 in Theorem 8 is satisfied by construction when the algorithm terminates.

References

1. Ausiello, G., Crescenzi, P., Gambosi, G., Kann, V., Marchetti-Spaccamela, A., Protasi, M.: Complexity and Approximation. Combinatorial Optimization Problems and their Approximability Properties. Springer, Berlin (1999)
2. Cesati, M., Trevisan, L.: On the efficiency of polynomial time approximation schemes. Inf. Process. Lett. **64**, 165–171 (1997)
3. Clark, B.N., Colbourn, C.J., Johnson, D.S.: Unit disk graphs. Discret. Math. **86**, 165–177 (1990)
4. Devroye, L., Farczadi, L.: Connectivity for line-of-sight-networks in higher dimensions. Discret. Math. Theor. Comput. Sci. **15**, 71–86 (2013)
5. Di Battista, G., Eades, P., Tamassia, R., Tollis, I.: Algorithms for drawing graphs. An annotated bibliography. Comput. Geom.: Theory Appl. **4**, 235–282 (1994)
6. Diestel, R.: Graph Theory. Springer, New York (1999)
7. Frieze, A., Kleinberg, J., Ravi, R., Debani, W.: Line of sight networks. Comb. Probab. Comput. **18**, 142–163 (2009)
8. Garey, M.R., Johnson, D.S.: Computer and Intractability: A Guide to the Theory of NP-Completeness. Freeman and Company, New York (1979)
9. Håstad, J.: Clique is hard to approximate within $n^{1-\varepsilon}$. Acta Math. **182**, 105–142 (1999)
10. Hurkens, C.A., Schrijver, A.: On the size of sets every t of which have an SDR, with an application to the worst case ratio of heuristics for packing problems. SIAM J. Discret. Math. **2**, 68–72 (1989)
11. Lovasz, L.: Matching Theory. North Holland, Amsterdam (1986)
12. Marathe, M.V., Breu, H., Hunt, H.B., Ravi, S.S., Rosenkrantz, D.J.: Simple heuristics for unit disk. Graphs Netw. **25**, 59–68 (1995)
13. Mohar, B., Thomassen, C.: Graphs on Surfaces. John Hopkins University Press, Baltimore (2001)
14. Nieberg, T., Hurink, J., Kern, W.: A robust PTAS for maximum weight independent sets in unit disk graphs. In: Hromkovič, J., Nagl, M., Westfechtel, B. (eds.) WG 2004. LNCS, vol. 3353, pp. 214–221. Springer, Heidelberg (2004). doi:10.1007/978-3-540-30559-0_18
15. Peterson, D.: Gridline graphs: a review in two dimensions and an extension to higher dimensions. Discret. Appl. Math. **126**, 223–239 (2003)
16. Stoyan, D., Kendall, W.S., Mecke, J., Ruschendorf, L.: Stochastic Geometry and Its Applications, vol. II. Wiley, Chichester (1995)
17. Valiant, L.: Universality considerations in VLSI circuits. IEEE Trans. Comput. **C–30**, 135–140 (1981)
18. Wood, D.R.: On higher-dimensional orthogonal graph drawing. In: Australian Computer Science Communications - CATS 1997 Proceedings of the Computing: The Australasian Theory Symposium, pp. 3–8. Australian Computer Science Association, Melbourne (1997)

A Lower Bound of the cd-Chromatic Number and Its Complexity

M.A. Shalu, S. Vijayakumar, and T.P. Sandhya[✉]

Indian Institute of Information Technology, Design & Manufacturing (IIITD&M),
Kancheepuram, Chennai 600127, India
{shalu,vijay,mat11d001}@iiitdm.ac.in

Abstract. The cd-coloring is motivated by the super-peer architecture in peer-to-peer resource sharing technology. A vertex set partition of a graph G into k independent sets $V_1, V_2, \ldots, V_k$ is called a k-color domination partition (k-cd-coloring) of G if there exists a vertex $u_i \in V(G)$ such that u_i dominates V_i in G for $1 \leq i \leq k$ and the smallest integer k for which G admits a k-cd-coloring is called the cd-chromatic number of G, denoted by $\chi_{cd}(G)$. A *subclique* is a set S of vertices of a graph G such that for any $x, y \in S$, $d(x,y) \neq 2$ in G and the cardinality of a maximum subclique in G is denoted by $\omega_s(G)$. Clearly, $\omega_s(G) \leq \chi_{cd}(G)$ for a graph G.

In this paper, we explore the complexity status of SUBCLIQUE: for a given graph G and a positive integer k, SUBCLIQUE is to decide whether G has a subclique of size at least k. We prove that SUBCLIQUE is NP-complete for (i) bipartite graphs, (ii) chordal graphs, and (iii) the class of H-free graphs when H is a fixed graph on 5-vertices. In addition, we prove that SUBCLIQUE for the class of H-free graphs is polynomial time solvable only if H is an induced subgraph of P_4; otherwise the problem is NP-complete. Moreover, SUBCLIQUE is polynomial time solvable for trees, split graphs, and co-bipartite graphs.

Keywords: Clique · Subclique · cd-coloring

1 Introduction

A vertex coloring is a partition of the vertex set of a graph into independent sets and many practical problems are modeled using coloring. For example, coloring is used in problems such as scheduling, register allocation, and mobile radio frequency assignment [5]. Another well studied parameter in graph theory is domination which asks for a subset S of the vertex set of the graph such that every vertex outside of S has a neighbour in S. It is employed in applications such as facility location problems and monitoring communication or electrical networks [10]. Combining coloring and domination, two variations of coloring/domination have been studied; (i) d-coloring [8] and (ii) cd-coloring [20]. A k-vertex coloring $V_1, V_2, \ldots, V_k$ of a graph G is called a k-d-coloring if for every vertex $v \in V(G)$, there exists an i, $1 \leq i \leq k$ such that $V_i \subseteq N[v]$. It is known that the d-coloring

© Springer International Publishing AG 2017
D. Gaur and N.S. Narayanaswamy (Eds.): CALDAM 2017, LNCS 10156, pp. 344–355, 2017.
DOI: 10.1007/978-3-319-53007-9_30

of bipartite, planar, and split graphs are NP-complete [3] while it is polynomial time solvable for trees [15]. On the other hand, a k-vertex coloring $V_1, V_2, \ldots, V_k$ of a graph G is called a k-cd-coloring if for every i, $1 \leq i \leq k$, there exists a vertex x_i such that $V_i \subseteq N[x_i]$. The minimum number of colors needed to cd-color a graph is called its cd-chromatic number. It is proved that the cd-colorability of split graphs and P_4-free graphs are polynomial time solvable [19]. In [13], the authors described an $O(2^n n^4 log\ n)$-time algorithm for finding the cd-chromatic number of a graph and proved that the problem is FPT when parameterized by the number of colors q on graphs of girth at least 5 and on chordal graphs. We also note that d-coloring of split graphs is NP-complete [3] while the cd-coloring of split graphs is in P [19].

Engineering applications such as peer-to-peer resource sharing technologies [4] and distributed control system in a smart grid [1,17] demand (i) sharing (coloring) and (ii) monitoring usage (domination) of its limited resources. A peer-to-peer software (like Skype, Freenet, and BitTorrent) helps computers to act both as clients and servers, and bandwidth utilization is a serious concern which can be managed by the super-peer architecture [2,14] which assigns a set of trusted computers (administrators) as monitors/super-peers. The cd-coloring captures two essential features (i) bandwidth utilization (independent sets/coloring) and (ii) monitoring by super-peers (u_i/domination) of super-peer architecture.

In this paper we study a lower bound of the cd-chromatic number. For this, we note that two vertices x, y of a graph G are in the same cd-color class of a cd-coloring of G only if $d(x, y) = 2$ in G. So if we consider a subset S of $V(G)$ whose elements are mutually not at a distance two, then any two vertices in S belong to different color classes in any cd-coloring of G and the cardinality of such a set is a lower bound of $\chi_{cd}(G)$. The set S is called a subclique and next, we analyze the algorithmic complexity of subclique problem.

SUBCLIQUE

Instance: A graph G and a positive integer k

Question: Does G contain a subclique of size k?

The main results of this paper are

- SUBCLIQUE is NP-complete for (i) bipartite graphs, (ii) chordal graphs, and (iii) the class of H-free graphs when H is a graph on 5-vertices,
- SUBCLIQUE for the class of H-free graphs is polynomial time solvable only if H is an induced subgraph of P_4; otherwise the problem is NP-complete, and
- SUBCLIQUE is polynomial time solvable for trees, split graphs, and co-bipartite graphs.

2   Preliminaries

We consider finite, undirected, and unweighted simple graphs only. For graph terminologies, we refer [21]. The *components* of a graph G are its maximal connected subgraphs. A *clique (An independent set)* is a subset of vertices which are pairwise adjacent (respectively, non-adjacent) in a graph G. The size of a maximum clique (independent set) in G is denoted by $\omega(G)$ (respectively, $\alpha(G)$). A k-vertex coloring of a graph G is a partition $V_1, V_2, \ldots, V_k$ of $V(G)$ such that V_i is an independent set in G for $1 \leq i \leq k$. The chromatic number of G is defined as $\chi(G) = \min\{k: G \text{ admits a } k\text{-vertex coloring}\}$. In a graph G, a subset $S \subseteq V(G)$ is a dominating set if every vertex not in S has a neighbour in S. For a pair of vertices x, y of a graph G, $d(x, y)$ denotes the length of a shortest x-y path in G. A graph G is a *split graph* if there exists a vertex partition $V(G) = V_1 \cup V_2$ where V_1 is a clique and V_2 is an independent set in G. For a graph H, G is said to be H-*free* if no induced subgraph of G is isomorphic to H. A graph is a *chordal graph* if it is C_n-free for every $n \geq 4$. For a set $X \subseteq V(G)$, $G[X]$ denotes the graph induced by X in G. For a set $A \subseteq V(G)$, we denote $G[V(G) \setminus A]$ as $G \setminus A$. Define $[A, B] = \{\{a, b\} : a \in A, b \in B\}$ where A and B are two non-empty disjoint sets. The *join* $G_1 + G_2$ of two vertex disjoint graphs G_1 and G_2 is a graph with vertex set $V(G_1) \cup V(G_2)$ and edge set $E(G_1) \cup E(G_2) \cup [V(G_1), V(G_2)]$. The *union* $G_1 \cup G_2 \cup \ldots \cup G_k$ of pairwise vertex disjoint graphs $G_1, G_2, \ldots, G_k$ is a graph with vertex set $V(G_1) \cup V(G_2) \cup \ldots \cup V(G_k)$ and edge set $E(G_1) \cup E(G_2) \cup \ldots \cup E(G_k)$. For a vertex v of a graph G, $N(v) = \{u \in V(G) : uv \in E(G)\}$, $N[v] = \{v\} \cup N(v)$, and $deg(v) = |N(v)|$. Often we denote an edge $\{a, b\}$ in a graph as ab or ba. Given a graph G and a positive integer k, CLIQUE is to decide whether G has a clique of size at least k.

Let G be a graph and $V_1, V_2, \ldots, V_k$ be a partition of $V(G)$ into k independent sets. We say $V_1, V_2, \ldots, V_k$ is a k-*color domination partition* (k-cd-coloring) of G if there exists a vertex $u_i \in V(G)$ such that $V_i \subseteq N[u_i]$ for $1 \leq i \leq k$. We define the cd-chromatic number of a graph G, denoted by $\chi_{cd}(G)$, as

$$\chi_{cd}(G) = \min\{k: G \text{ admits a } k\text{-cd-coloring}\}.$$

We note that [20] $\chi_{cd}(K_{1,n}) = 2$ and

$$\chi_{cd}(P_n) = \chi_{cd}(C_n) = \begin{cases} \lceil \frac{n}{2} \rceil & \text{if } n \equiv 0, 1, 3 \pmod 4 \\ \frac{n}{2} + 1 & \text{if } n \equiv 2 \pmod 4 \end{cases}$$

where $K_{1,n}$ is the star graph on $n+1$ vertices and C_n and P_n respectively denote the cycle and the path on n vertices. So $\chi_{cd} - \chi$ is very large even for paths and cycles.

A *subclique* is a set S of vertices of a graph G such that for any $x, y \in S$, $d(x, y) \neq 2$ in G. The cardinality of a maximum subclique in G is denoted by $\omega_s(G)$ and $\omega_s(G) \leq \chi_{cd}(G)$. For a graph G with components $G_1, G_2, \ldots, G_k$,

$$\omega_s(G) = \sum_{i=1}^{k} \omega_s(G_i).$$

We note that the difference between χ_{cd} and ω_s can be arbitrarily large in general. For example, if $G_k = \underbrace{C_5 + C_5 + \ldots + C_5}_{k \text{ copies}}$, then $\chi_{cd}(G_k) = 3k$ and $\omega_s(G_k) = 2k$.

Next, we note that a clique is also a subclique and thus $\omega(G) \leq \omega_s(G)$. Hence for a graph G, $\omega(G) \leq \omega_s(G) \leq \chi_{cd}(G)$. For a connected split graph G, $\chi_{cd}(G) = \omega(G)$ [19] and thus $\omega_s(G) = \omega(G)$. A similar result holds for a connected P_4-free graph G since $\chi_{cd}(G) = \omega(G)$ [19]. So we have the following proposition.

Proposition 1. (i) If G is a connected split graph, then an optimal subclique of G can be found in $O(n^2)$ time where n denotes the number of vertices of G and $\omega_s(G) = \omega(G)$.

(ii) If G is a connected P_4-free graph, then an optimal subclique of G can be found in $O(n+m)$ time where n and m denote the numbers of vertices and edges of G respectively and $\omega_s(G) = \omega(G)$.   □

The paper is organized as follows. In Sects. 3 and 4, we analyze the time complexity of SUBCLIQUE for bipartite graphs and chordal graphs respectively. In Sect. 5, we do an exhaustive study of the complexity of SUBCLIQUE for H-free graphs where H is a graph on at most 5 vertices. In Sect. 6, we explore the time complexity of SUBCLIQUE for some subclasses of graphs such as trees and co-bipartite graphs.

Next we consider SUBCLIQUE for chordal graphs and bipartite graphs. All our reductions are from the NP-complete set packing problem [12], denoted SETPACKING. Let $X = \{x_1, x_2, \ldots, x_n\}$ be any finite set and let $\mathcal{S}$ be a collection of subsets of X. Given $(X, \mathcal{S})$ and a positive integer k, SETPACKING asks whether there are k disjoint subsets in the collection $\mathcal{S}$.

3 Bipartite Graphs

Theorem 1. SUBCLIQUE for bipartite graphs is NP-complete.

Proof. Given a subset S of the vertex set of a bipartite graph, it can be verified in polynomial time that the distance between any pair of vertices in S is not two. So, SUBCLIQUE for bipartite graphs is in NP.

We now provide a reduction from SETPACKING to SUBCLIQUE for bipartite graphs that is polynomial time computable. Let $((X, \{S_1, S_2, \ldots, S_m\}), k)$ be an instance of SETPACKING where $X = \{x_1, x_2, \ldots, x_n\}$. We transform it to an instance of SUBCLIQUE for bipartite graphs $((A, B, E), k')$ as follows. The bipartite graph (A, B, E) we construct has vertex sets $A = \{s_1, s_2, \ldots, s_m\} \cup \{s_0\}$, $B = X$, and edge set $E = \{\{s_i, x\} \mid x \in X, 1 \le i \le m, \text{and } x \in S_i\} \cup \{\{s_0, x_i\} \mid i = 1, 2, \ldots, n\}$ (Fig. 1). The parameter k' is set to $k + 1$. We will argue that $((X, \{S_1, S_2, \ldots, S_m\}), k)$ is a YES instance of SETPACKING if and only if $((A, B, E), k + 1)$ is a YES instance of SUBCLIQUE for bipartite graphs.

Claim 1. $(X, \{S_1, S_2, \ldots, S_m\})$ has a set packing of size k if and only if the bipartite graph $G = (A, B, E)$ has a subclique of size $k + 1$.

Suppose $(X, \{S_1, S_2, \ldots, S_m\})$ has a set packing of size k say $S_1, S_2, \ldots, S_k$. Since $S_i \cap S_j = \emptyset$, $d(s_i, s_j) = 4$ for $1 \le i < j \le k$ in G (because of the path $s_i x_i s_0 x_j s_j$ where $x_i \in S_i$ and $x_j \in S_j$). Let $x \in S_1$. Note that $d(x, s_i) = 3$ for $2 \le i \le k$ (because of the path $x s_0 x_i s_i$ where $x_i \in S_i$ and the fact that $S_1 \cap S_i = \emptyset$). So $\{x, s_1, s_2, \ldots, s_k\}$ is a subclique of size $k + 1$ in G.

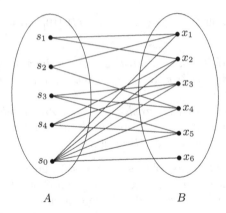

Fig. 1. The bipartite graph $G = (A, B, E)$ constructed from the set packing instance $((X, \{S_1, S_2, S_3, S_4\}), k)$, where $X = \{x_1, x_2, x_3, x_4, x_5, x_6\}$, $S_1 = \{x_1, x_2\}$, $S_2 = \{x_1, x_4\}$, $S_3 = \{x_3, x_4, x_5\}$, and $S_4 = \{x_2, x_3, x_5\}$.

Conversely, suppose that $G = (A, B, E)$ has a subclique S of size $k + 1$. Note that $|S \cap B| \leq 1$ since $d(x_i, x_j) = 2$ for $1 \leq i < j \leq n$ (because of the path $x_i s_0 x_j$). Hence $|S \cap A| \geq k$. Now there are two cases. Case (1) $s_0 \notin S$. So $S \cap A \subseteq \{s_1, s_2, \ldots, s_m\}$. W.l.o.g. let $\{s_1, s_2, \ldots, s_k\} \subseteq S \cap A$. Since S is a subclique, $d(s_i, s_j) \neq 2$ and thus $S_i \cap S_j = \emptyset$ for $1 \leq i < j \leq k$. Hence $S_1, S_2, \ldots, S_k$ is a set packing of size k. Case (2) $s_0 \in S$. Since $d(s_0, s_i) = 2$, $s_i \notin S$ for all i, $1 \leq i \leq m$. Also $|S \cap B| \leq 1$ and thus $k + 1 = |S| \leq 2$. So S_1 is a set packing of size $k = 1$. From Claim 1, it can be deduced that $((X, \{S_1, S_2, \ldots, S_m\}), k)$ is a YES instance of SETPACKING if and only if $((A, B, E), k + 1)$ is a YES instance of SUBCLIQUE for bipartite graphs. $\qquad \square$

In the previous theorem, we provided a polynomial time reduction from SETPACKING to SUBCLIQUE for bipartite graphs. This reduction is optimum preserving in that the optimum increases by exactly one. This implies that if the optimization version of SETPACKING has no $\alpha(n)$-approximation algorithm unless P $\neq$ NP, then the optimization version of SUBCLIQUE for bipartite graphs also does not have an $\alpha(n)$-approximation algorithm unless P $\neq$ NP. It is known that the set packing problem does not have a constant factor approximation algorithm unless P $\neq$ NP [9,11]. So, we have that SUBCLIQUE for bipartite graphs also does not have a constant factor approximation algorithm unless P $\neq$ NP.

4   Chordal Graphs

SUBCLIQUE for split graphs is polynomial time solvable (Proposition 1). Next, we consider chordal graphs, a super class of split graphs. In this section, we show that SUBCLIQUE for chordal graphs is NP-complete by reducing SETPACKING to SUBCLIQUE for chordal graphs. Every chordal graph G constructed by our reduction is essentially a split graph and a complete graph, with one common vertex.

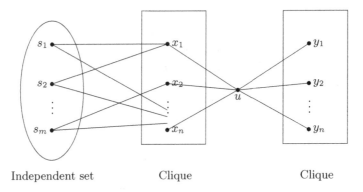

Fig. 2. The chordal graph G constructed from a set packing instance $((X, \{S_1, S_2, \ldots, S_m\}), k)$

Theorem 2. SUBCLIQUE *for chordal graphs is NP-complete.*

Proof. It is easy to prove that SUBCLIQUE for chordal graphs is in NP.

We now provide a reduction from SETPACKING to SUBCLIQUE for chordal graphs that is polynomial time computable. Let $((X, \{S_1, S_2, \ldots, S_m\}), k)$ be an instance of SET PACKING where $X = \{x_1, x_2, \ldots, x_n\}$. It is transformed into an instance of SUBCLIQUE for chordal graphs (G, k') as follows. The graph $G = (V, E)$ we construct has $V = X \cup \{s_1, s_2, \ldots, s_m\} \cup \{y_1, y_2, \ldots, y_n\} \cup \{u\}$ and $E = \{\{x_i, x_j\} \mid i, j = 1, 2, \ldots, m \text{ and } i \neq j\} \cup \{\{y_i, y_j\} \mid i, j = 1, 2, \ldots, m \text{ and } i \neq j\} \cup \{\{s_i, x\} \mid x \in X, 1 \leq i \leq m, \text{ and } x \in S_i\} \cup \{\{u, x_i\} \mid i = 1, 2, \ldots, m\} \cup \{\{u, y_i\} \mid i = 1, 2, \ldots, m\}$. That is, G is obtained from a split graph and a complete graph by identifying (merging) a vertex of the split graph and a vertex of the complete graph; therefore it is chordal (Fig. 2). The parameter k' is set to $n + k$. We now argue that $((X, \{S_1, S_2, \ldots, S_m\}), k)$ is a YES instance of SETPACKING if and only if $(G, n + k)$ is a YES instance of SUBCLIQUE for chordal graphs.

Claim 2. $(X, \{S_1, S_2, \ldots, S_m\})$ *has a set packing of size k if and only if G has a subclique of size $n + k$.*

Suppose $(X, \{S_1, S_2, \ldots, S_m\})$ has a set packing of size k say $S_1, S_2, \ldots, S_k$. Since $S_i \cap S_j = \emptyset$, $d(s_i, s_j) = 3$ in G for $1 \leq i < j \leq k$ (because of the path $s_i x_i x_j s_j$ where $x_i \in S_i$ and $x_j \in S_j$). In addition, $d(s_i, y_j) = 3$ in G for $1 \leq i \leq k$ and $1 \leq j \leq n$ (because of the path $s_i x_i u y_j$ where $x_i \in S_i$). Also $d(y_i, y_j) = 1$ in G for $1 \leq i < j \leq n$. So $\{s_1, s_2, \ldots, s_k, y_1, y_2, \ldots, y_n\}$ is a subclique of size $n + k$ in G.

Conversely, suppose that G has a subclique S of size $n + k$. Note that $|S \cap (\{x_1, x_2, \ldots, x_n\} \cup \{y_1, y_2, \ldots, y_n\})| \leq n$ since $d(x_i, y_j) = 2$ in G for $1 \leq i, j \leq n$. Hence $|S \cap \{s_1, s_2, \ldots, s_m, u\}| \geq k$. Now there are two cases. Case (1) $u \notin S$. Since $|S \cap \{s_1, s_2, \ldots, s_m\}| \geq k$, w.l.o.g. let $\{s_1, s_2, \ldots, s_k\} \subseteq S$. Since S is a subclique in G, $d(s_i, s_j) \neq 2$ in G and thus $S_i \cap S_j = \emptyset$ for $1 \leq i < j \leq k$. Hence $S_1, S_2, \ldots, S_k$ is a set packing of size k. Case (2) $u \in S$. Since $d(u, s_i) = 2$

in G for $1 \leq i \leq m$, $s_i \notin S$ and thus $S \subseteq \{x_1, x_2 \ldots, x_n, y_1, y_2, \ldots, y_n, u\}$ and $n + k = |S| \leq n + 1$. So S_1 is a set packing of size $k = 1$.

From Claim 2, it can be deduced that $((X, \{S_1, S_2, \ldots, S_m\}), k)$ is a YES instance of SETPACKING if and only if $(G, n+k)$ is a YES instance of SUBCLIQUE for chordal graphs.   □

Next, we consider SUBCLIQUE for the class of H-free graphs and prove some important dichotomy results.

5   Dichotomy Results on H-free Graphs

Proposition 2. *For a fixed graph H, let the family*
$\mathcal{G} = \{G + K_1 : G \text{ is a } H\text{-free graph}\}$. *If $\mathcal{G}$ is H-free (that is, every member of $\mathcal{G}$ is H-free) and* CLIQUE *for the class of H-free graphs is NP-complete, then* SUBCLIQUE *for the class of H-free graphs is NP-complete.*

Proof. Given a subset S of the vertex set of a H-free graph, it can be verified in polynomial time that the distance between any pair of vertices in S is not two. So, SUBCLIQUE for the class of H-free graphs is in NP.

We now provide a reduction from CLIQUE of H-free graphs to SUBCLIQUE for graphs in $\mathcal{G}$ that is polynomial time computable. For a H-free graph G, let $G' \cong K_1 + G$, where $V(G') = \{x\} \cup V(G)$ and $E(G') = E(G) \cup \{xu : u \in V(G)\}$. Then $G' \in \mathcal{G}$. Note that, G and G' are H-free. Next, we prove that G contains a clique of size k if and only if G' contains a subclique of size $k + 1$. Suppose that $\{v_1, v_2, \ldots, v_k\}$ is a clique of G. Then $\{v_1, v_2, \ldots, v_k, x\}$ is a subclique of size $k + 1$ in G'. Next, suppose that G' contains a subclique of size $k + 1$, say S'. Let $S = S' \setminus \{x\}$. Then $|S| \geq k$. We claim that S is a clique in G. If not, there exist $u, v \in S$ such that $uv \notin E(G)$ and hence $d(u, v) = 2$ in G', a contradiction since S' is a subclique in G'. Hence S is a clique in G. Since CLIQUE for H-free graphs is NP-complete, SUBCLIQUE for $\mathcal{G}$ is NP-complete and thus SUBCLIQUE for H-free graphs is NP-complete since $\mathcal{G}$ is a subclass of the class of all H-free graphs.   □

Note that if $H \cong 3K_1$, then the family $\mathcal{G} = \{G + K_1 : G \text{ is a } H\text{-free graph}\}$ is a subclass of $3K_1$-free graphs. A similar result holds when $H \cong 2K_2$. Since CLIQUE is NP-complete for $3K_1$-free graphs [18], and $2K_2$-free graphs [18], by Proposition 2, we have the following corollary.

Corollary 1. SUBCLIQUE *is NP-complete for the class of (i) $3K_1$-free graphs, and (ii) $2K_2$-free graphs.*   □

Next, we prove two important dichotomy results (Propositions 3 and 4).

Proposition 3. *For a graph H on 3 vertices,* SUBCLIQUE *is NP-complete for the class of H-free graph if $H \ncong P_3$ or $K_2 \cup K_1$ and the problem is polynomial time solvable for the class of (i) P_3-free graphs and (ii) $(K_2 \cup K_1)$-free graphs.*

Proof. Let H be a graph on 3 vertices. Then there are four cases. Case (1) $H \cong 3K_1$. By Corollary 1 (i), SUBCLIQUE for H-free graphs is NP-complete. Case (2) $H \cong K_3$. By Theorem 1, SUBCLIQUE for bipartite graphs is NP-complete and hence the problem is NP-complete for K_3-free graphs. Case (3) $H \cong P_3$. Note that if a graph G is P_3-free, then it is a union of cliques and hence $\omega_s(G) = |V(G)|$. Case (4) $H \cong K_2 \cup K_1$. Then the class of H-free graphs is a subclass of P_4-free graphs and hence by Proposition 1, $\omega_s(G) = \sum_{i=1}^{k} \omega_s(G_i) = \sum_{i=1}^{k} \omega(G_i)$ where $G_1, G_2, \ldots, G_k$ are components of G. Clearly, in Cases (3) and (4), SUBCLIQUE is polynomial time solvable. $\square$

Proposition 4. *For a graph H on 4 vertices, SUBCLIQUE is NP-complete for the class of H-free graphs if $H \not\cong P_4$ and the problem is polynomial time solvable for the class of P_4-free graphs.*

Proof. For a graph H on 4 vertices, there are three cases. Case (1) H is $\{3K_1, K_3\}$-free. Then it can be verified that $H \cong 2K_2$ or P_4 or C_4. If $H \cong 2K_2$, then by Corollary 1 (ii), SUBCLIQUE is NP-complete. If $H \cong C_4$, then by Theorem 2, SUBCLIQUE for chordal graphs is NP-complete and hence the problem is NP-complete for C_4-free graphs. If $H \cong P_4$, then by Proposition 1, SUBCLIQUE is polynomial time solvable. Case (2) H contains a $3K_1$ as an induced subgraph. Then the class of H-free graphs contains the class of all $3K_1$-free graphs and by Proposition 3, SUBCLIQUE for the class of H-free graphs is NP-complete. Case (3) H contains a K_3 as an induced subgraph. Then the class of H-free graphs contains the class of all K_3-free graphs and by Proposition 3, SUBCLIQUE for the class of H-free graphs is NP-complete. $\square$

Proposition 5. *For a fixed graph H on five vertices, SUBCLIQUE is NP-complete for the class of H-free graphs.*

Proof. Let H be a graph on 5 vertices. By Propositions 3 and 4, it is clear that if H has an induced subgraph isomorphic to $3K_1$, K_3, or $2K_2$, then SUBCLIQUE is NP-complete. So we consider the remaining case; that is H is $\{3K_1, 2K_2, K_3\}$-free. Then it can be verified that $H \cong C_5$. Since SUBCLIQUE is NP-complete for bipartite graphs by Theorem 1, the problem is NP-complete for C_5-free graphs. $\square$

By summarizing Propositions 3, 4, and 5 we have the following theorem.

Theorem 3. *For a fixed graph H, SUBCLIQUE for the class of H-free graphs is polynomial time solvable only if H is an induced subgraph of P_4; otherwise the problem is NP-complete.* $\square$

6 Cycles, Trees, and Co-bipartite Graphs

For a given G, construct a graph G^* where $V(G^*) = V(G)$ and $xy \in E(G^*)$ if and only if $d(x, y) = 2$ in G. In other words, end vertices of every induced P_3 in

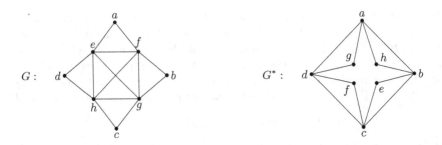

Fig. 3. An example of a graph G and its corresponding graph G^*

G is made adjacent in G^*. An example for this construction is given in Fig. 3. We note that, for a 3-tuple of vertices in a graph G, we need to check whether it induces a P_3 in G and it takes $O(1)$ time. So to construct G^* from G, we need $O(n^3)$ time where $n = |V(G)|$. Throughout this section, G^* denotes the graph constructed from a given graph G using the above construction.

Proposition 6. *A set S is a subclique in G if and only if S is an independent set in G^*.*

Proof. Assume that S is a subclique in G. We claim that S is an independent set in G^*. If not, there exist $x, y \in S$ such that $xy \in E(G^*)$. Then $d(x, y) = 2$ in G, a contradiction since S is a subclique in G. Conversely, suppose that S is an independent set in G^*. We claim that S is a subclique in G, else there exist $x, y \in S$ such that $d(x, y) = 2$ in G and $xy \in E(G^*)$, a contradiction since S is an independent set in G^*. □

From Proposition 6, we can deduce the following observations.

Observation 1. *For a graph G, $\omega_s(G) = \alpha(G^*)$.* □

Observation 2. *(i) $C^*_{2n+1} = C_{2n+1}$ for $n \geq 2$,*
(ii) $C^*_{2n} = C_n \cup C_n$ for $n \geq 3$, and
(iii) $P^*_n = P_{\lfloor \frac{n}{2} \rfloor} \cup P_{\lceil \frac{n}{2} \rceil}$ for $n \geq 3$. □

Corollary 2. *(i) $\omega_s(C_{2n+1}) = \alpha(C^*_{2n+1}) = \alpha(C_{2n+1}) = n$,*
(ii) $\omega_s(C_{2n}) = \alpha(C^*_{2n}) = \alpha(C_n \cup C_n) = 2\alpha(C_n) = 2\lfloor \frac{n}{2} \rfloor$, and
(iii) $\omega_s(P_n) = \alpha(P_{\lfloor \frac{n}{2} \rfloor}) + \alpha(P_{\lceil \frac{n}{2} \rceil})$
$$= \begin{cases} 2k, & \text{if } n=4k \\ 2k+1, & \text{if } n=4k+1 \\ 2k+2, & \text{if } n=4k+2 \text{ or } 4k+3 \end{cases}$$
□

The above corollary proves that the difference between ω and ω_s can be arbitrarily large since $\omega(P_n) = \omega(C_n) = 2$.

Proposition 7. *If G is a tree, then G^* is a chordal graph.*

Proof. If not, let C be an induced cycle in G^* with vertex set $\{v_1, v_2, \ldots, v_k\}$ and edges $v_1v_2, v_2v_3, \ldots, v_{k-1}v_k, v_kv_1$, of length $k \geq 4$. Note that, for each $v_iv_{i+1} \in E(G^*)$, there exists a vertex $u_i \in V(G)$ such that $v_iu_i, u_iv_{i+1} \in E(G)$ and $v_iv_{i+1} \notin E(G)$ for $1 \leq i \leq k$, $(i \bmod k)$. Let $X = V(C) \cup \bigcup_{i=1}^{k} \{u_i\}$. Note that (a) $v_i \neq u_i$, else $v_iv_{i+1} \in E(G)$, a contradiction and (b) $v_{i+1} \neq u_i$, else $v_iv_{i+1} \in E(G)$, a contradiction. So $deg(u_i) \geq 2$ in $G[X]$ since $v_iu_i, u_iv_{i+1} \in E(G)$ and $v_i, v_{i+1} \in X$. Next, we claim that $u_{i-1} \neq u_i$. If not, $v_{i-1}u_i(= u_{i-1})v_{i+1}$ is the unique path from v_{i-1} to v_{i+1} in G (since G is a tree) and thus $d(v_{i-1}, v_{i+1}) = 2$ in G. Hence $v_{i-1}v_{i+1} \in E(G^*)$, a contradiction. Hence $deg(v_i) \geq 2$ in $G[X]$ since $u_{i-1}v_i, u_iv_i \in E(G)$. So $G[X]$ contains a cycle as a subgraph since the degree of every vertex in $G[X]$ is at least two, a contradiction to the assumption that G is a tree. Hence G^* is a chordal graph.   □

Theorem 4. SUBCLIQUE *Problem for trees can be solved in $O(n^3)$ time where n denotes the number of vertices of a given tree.*

Proof. From a given tree G, G^* can be constructed in time $O(n^3)$ where n denotes the number of vertices of G. By Proposition 7, G^* is a chordal graph and by Observation 1, $\omega_s(G) = \alpha(G^*)$. Since the maximum independent set problem is solvable in linear time for a chordal graph [7], SUBCLIQUE for a tree can be solved in $O(n^3)$ time.   □

We note that if G is a split graph, G^* may not be a chordal graph. For example, consider G, a split graph, in Fig. 3. But G^* is not a chordal graph. It is an interesting problem to characterize the class of graphs G for which G^* is a chordal graph.

Theorem 5. SUBCLIQUE *for a co-bipartite graph can be solved in $O(n^3)$ time where n denotes the number of vertices of the given co-bipartite graph.*

Proof. Let G be a co-bipartite graph with $V(G) = V_1 \cup V_2$ where V_1 and V_2 are cliques in G. Since V_1 and V_2 are subcliques in G, they are independent sets in G^* by Proposition 6 and hence G^* is a bipartite graph. In addition, by Observation 1, $\omega_s(G) = \alpha(G^*)$ and the maximum independent set problem of a bipartite graph can be solved in $O(\sqrt{n}m)$ time [16]. So SUBCLIQUE for a co-bipartite graph can be solved in $O(n^3)$ time, since the construction of G^* takes $O(n^3)$ time.   □

7   Conclusion

We proved that SUBCLIQUE is polynomial time solvable for trees, split graphs, and co-bipartite graphs, and the problem is NP-complete for bipartite graphs and chordal graphs. In addition, we explored the complexity of SUBCLIQUE for the class of H-free graphs, for a fixed graph H and proved that SUBCLIQUE

Table 1. Comparison between CLIQUE and SUBCLIQUE

Graph class	CLIQUE	SUBCLIQUE
Split	P	P
Chordal	P	NP-c
$3K_1$-free	NP-c	NP-c

is polynomial time solvable only if H is an induced subgraph of P_4; otherwise the problem is NP-complete. Table 1 describes the comparison between the time complexity of CLIQUE and SUBCLIQUE for some subclasses of graphs and they are 'different' in time complexity perspective.

Open problems: To find the time complexity of SUBCLIQUE of (i) interval graphs, (ii) planar graphs, and (iii) graphs with bounded degree.

Acknowledgment. The authors wish to thank the anonymous referees whose suggestions improved the presentation of this paper.

References

1. Amin, S.M., Wollenberg, B.F.: Toward a smart grid. IEEE Power Energy Mag. **3**, 34–41 (2005)
2. Androutsellis-Theotokis, S., Spinellis, D.: A survey of peer-to-peer content distribution technologies. ACM Comput. Surv. **36**, 335–371 (2004)
3. Arumugam, S., Chandrasekar, K.R., Misra, N., Philip, G., Saurabh, S.: Algorithmic aspects of dominator colorings in graphs. In: Combinatorial Algorithms, pp. 19–30 (2011)
4. Canali, C., Renda, M.E., Santi, P., Burresi, S.: Enabling efficient peer-to-peer resource sharing in wireless mesh networks. IEEE Trans. Mob. Comput. **9**, 333–347 (2010)
5. Formanowicz, P., Tanaś, K.: A survey of graph coloring - its types, methods and applications. Found. Comput. Decis. Sci. **37**, 223–238 (2012)
6. Garey, M.R., Johnson, D.S.: Computers and Intractability; A Guide to the Theory of NP-Completeness. W. H. Freeman and Co., New York (1990)
7. Gavril, F.: Algorithms for minimum coloring, maximum clique, minimum covering by cliques, and maximum independent set of a chordal graph. SIAM J. Comput. **1**, 180–187 (1972)
8. Gera, R., Horton, S., Rasmussen, C.: Dominator colorings and safe clique partitions. Congr. Numer. **181**, 19–32 (2006)
9. Håstad, J.: Clique is hard to approximate within $n^{1-\epsilon}$. Acta Math. **182**, 105–142 (1999)
10. Haynes, T.W., Hedetniemi, S.T., Slater, P.J.: Fundamentals of Domination in Graphs. Marcel Dekker, New York (1998)
11. Hazan, E., Safra, S., Schwartz, O.: On the complexity of approximating k-set packing. Comput. Complex. **15**, 20–39 (2006)
12. Karp, R.M.: Reducibility among combinatorial problems. In: Miller, R.E., Thatcher, J.W. (eds.) Complexity of Computer Computations. Plenum Press, New York (1972)

13. Krithika, R., Rai, A., Saurabh, S., Tale, P.: Parameterized and Exact Algorithms for Class Domination Coloring. http://www.imsc.res.in/~ashutosh/papers/cd_coloring.pdf (manuscript)
14. Löser, A., Naumann, F., Siberski, W., Nejdl, W., Thaden, U.: Semantic overlay clusters within super-peer networks. In: Aberer, K., Koubarakis, M., Kalogeraki, V. (eds.) DBISP2P 2003. LNCS, vol. 2944, pp. 33–47. Springer, Heidelberg (2004). doi:10.1007/978-3-540-24629-9_4
15. Merouane, H.B., Chellali, M.: On the dominator colorings in trees. Discuss. Math. Gr. Theory **32**(4), 677–683 (2012)
16. Micali, S., Vazirani, V.V.: An $O(\sqrt{|V|}\,|E|)$ algorithm for finding maximum matchings in general graphs. In: Proceedings of the 21st IEEE Symposium on Foundations of Computer Science, pp. 17–27 (1980)
17. Monti, A., Ponci, F., Benigni, A., Liu, J.: Distributed intelligence for smart grid control. In: International School on Nonsinusoidal Currents and Compensation, 15–18 June 2010, Łagów, Poland (2010)
18. Poljak, S.: A note on the stable sets and coloring of graphs. Commun. Math. Univ. Carolin. **15**, 307–309 (1974)
19. Shalu, M.A., Sandhya, T.P.: The cd-coloring of graphs. In: Govindarajan, S., Maheshwari, A. (eds.) CALDAM 2016. LNCS, vol. 9602, pp. 337–348. Springer, Heidelberg (2016). doi:10.1007/978-3-319-29221-2_29
20. Venkatakrishnan, Y.B., Swaminathan, V.: Color class domination numbers of some classes of graphs. Algebra Discret. Math. **18**, 301–305 (2014)
21. West, D.B.: Introduction to Graph Theory, 2nd edn. Prentice-Hall, USA (2000)

Stability Number and k-Hamiltonian $[a, b]$-factors

Sizhong Zhou[1]($\boxtimes$), Yang Xu[2], and Lan Xu[3]

[1] School of Mathematics and Physics, Jiangsu University of Science and Technology,
Mengxi Road 2, Zhenjiang 212003, Jiangsu, People's Republic of China
zsz_cumt@163.com
[2] Department of Mathematics, Qingdao Agricultural University,
Qingdao 266109, Shandong, People's Republic of China
[3] Department of Mathematics, Changji University,
Changji 831100, Xinjiang, People's Republic of China

Abstract. Let a, b and k be three integers with $b > a \geq 2$ and $k \geq 0$, and let G be a graph. If $G - U$ contains a Hamiltonian cycle for any $U \subseteq V(G)$ with $|U| = k$, then G is called a k-Hamiltonian graph. An $[a, b]$-factor F of a graph G is Hamiltonian if F admits a Hamiltonian cycle. If $G - U$ includes a Hamiltonian $[a, b]$-factor for every subset $U \subseteq V(G)$ with $|U| = k$, then we say that G has a k-Hamiltonian $[a, b]$-factor. In this paper, we prove that if G is a k-Hamiltonian graph with

$$1 \leq \alpha(G) \leq \frac{4(b-2)(\delta(G) - a - k + 1)}{(a + k + 1)^2},$$

then G admits a k-Hamiltonian $[a, b]$-factor. Furthermore, it is shown that this result is sharp.

Keywords: Graph · Stability number · Minimum degree · k-Hamiltonian graph · k-Hamiltonian $[a, b]$-factor

1 Introduction

The graphs considered in this work will be finite and undirected graphs which have neither loops nor multiple edges. Let $G = (V(G), E(G))$ be a graph, where $V(G)$ and $E(G)$ denote its vertex set and edge set, respectively. For each $x \in V(G)$, the neighborhood $N_G(x)$ of x is the set of vertices of G adjacent to x, and the degree $d_G(x)$ of x is $|N_G(x)|$. For $X \subseteq V(G)$, we write $N_G(X) = \bigcup_{x \in X} N_G(x)$ and $N_G[X] = N_G(X) \cup X$. We use $G[X]$ to denote the subgraph of G induced by X, and $G - X = G[V(G) \setminus X]$. If X and Y are two disjoint subsets of $V(G)$,

Supported by the National Natural Science Foundation of China (Grant Nos. 11371009, 11501256, 61503160) and the National Social Science Foundation of China (Grant No. 14AGL001) and the Natural Science Foundation of Xinjiang Province of China (Grant No. 2015211A003), and sponsored by 333 Project of Jiangsu Province.

D. Gaur and N.S. Narayanaswamy (Eds.): CALDAM 2017, LNCS 10156, pp. 356–361, 2017.
DOI: 10.1007/978-3-319-53007-9_31

then $e_G(X, Y)$ denotes the number of edges that join a vertex in X and a vertex in Y. A vertex set $X \subseteq V(G)$ is called independent if $G[X]$ has no edges. The minimum degree and the stability number of G are denoted by $\delta(G)$ and $\alpha(G)$, respectively. Given a real number r, recall that $\lfloor r \rfloor$ is the greatest integer such that $\lfloor r \rfloor \leq r$.

Let a and b be two integers with $0 \leq a \leq b$. Then a spanning subgraph F of G is called an $[a, b]$-factor if $a \leq d_F(x) \leq b$ for all $x \in V(G)$. If $a = b = r$, then an $[a, b]$-factor is an r-factor, which is a regular factor. If $G - U$ includes a Hamiltonian cycle for any $U \subseteq V(G)$ with $|U| = k$, then G is called a k-Hamiltonian graph. An $[a, b]$-factor F is Hamiltonian if F admits a Hamiltonian cycle. If $G - U$ has a Hamiltonian $[a, b]$-factor for any subset $U \subseteq V(G)$ with $|U| = k$, then we say that G admits a k-Hamiltonian $[a, b]$-factor. In particular, a 0-Hamiltonian graph is said to be a Hamiltonian graph; a 0-Hamiltonian $[a, b]$-factor is called a Hamiltonian $[a, b]$-factor.

Many authors have investigated graph factors [2–7, 11–13, 15]. The following results on Hamiltonian k-factors, Hamiltonian $[k, k+1]$-factors and Hamiltonian $[a, b]$-factors are known.

Theorem 1 (Wei and Zhu [10]). Let $k \geq 2$ be an integer and G be a graph of order n with $n \geq 8k - 4$, kn being even and $\delta(G) \geq \frac{n}{2}$. Then G has a Hamiltonian k-factor.

Theorem 2 (Cai et al. [1]). Let $k \geq 2$ be an integer and let G be a graph of order $n \geq 3$ with minimum degree at least k, $n \geq 8k - 16$ for even n and $n \geq 6k - 13$ for odd n. If for each pair of nonadjacent vertices x and y of G,

$$d_G(x) + d_G(y) \geq n,$$

then G has a Hamiltonian $[k, k+1]$-factor.

Theorem 3 (Matsuda [9]). Let $k \geq 2$ be an integer. Suppose that G is a 2-connected graph of order $n \geq 3$ with $n \geq 8k - 16$ for even n and $n \geq 6k - 13$ for odd n. If $\delta(G) \geq k$ and

$$\max\{d_G(x), d_G(y)\} \geq \frac{n}{2}$$

for each pair of nonadjacent vertices x and y of $V(G)$, then G has a Hamiltonian $[k, k+1]$-factor.

Theorem 4 (Zhou [14]). Let a and b be two integers with $2 \leq a < b - 1$, and let G be a Hamiltonian graph of order n with $n \geq a + 2$. If $t(G) \geq a - 1 + \frac{a-1}{b-2}$, then G has a Hamiltonian $[a, b]$-factor.

In this paper, we study k-Hamiltonian $[a, b]$-factors in graphs and obtain a sufficient condition for the existence of k-Hamiltonian $[a, b]$-factors in graphs. Our main result is the following theorem.

Theorem 5. Let a, b and k be three integers with $b > a \geq 2$ and $k \geq 0$, and let G be a k-Hamiltonian graph. If G satisfies

$$1 \leq \alpha(G) \leq \frac{4(b-2)(\delta(G) - a - k + 1)}{(a + k + 1)^2},$$

then G admits a k-Hamiltonian $[a, b]$-factor.

Note that a 0-Hamiltonian graph is said to be a Hamiltonian graph; a 0-Hamiltonian $[a, b]$-factor is called a Hamiltonian $[a, b]$-factor. Set $k = 0$ in Theorem 5. Then we obtain the following corollary.

Corollary 1. Let a and b be two integers with $b > a \geq 2$, and let G be a Hamiltonian graph. If G satisfies

$$1 \leq \alpha(G) \leq \frac{4(b-2)(\delta(G) - a + 1)}{(a + 1)^2},$$

then G admits a Hamiltonian $[a, b]$-factor.

2   The Proof of Theorem 5

The proof of Theorem 5 depends on the following lemma, which is a special case of Lovász's (g, f)-factor theorem [8].

Lemma 1 (Lovász [8]). Let G be a graph and let a and b be integers such that $1 \leq a < b$. Then G has an $[a, b]$-factor if and only if

$$\delta_G(S, T) = b|S| + d_{G-S}(T) - a|T| \geq 0$$

for all disjoint subsets S and T of $V(G)$, where $d_{G-S}(T) = \sum_{x \in T} d_{G-S}(x)$.

Proof of Theroem 5. We write $G' = G - U$, where U is an arbitrary subset of $V(G)$ with $|U| = k$. According to the condition of Theorem 5 and the definition of k-Hamiltonian graph, G' contains a Hamiltonian cycle C. For $a = 2$, Theorem 5 holds since C itself is a k-Hamiltonian $[2, b]$-factor of G. Hence we may assume that $a \geq 3$. Set $H = G' - E(C)$. Thus, we have $V(H) = V(G') = V(G) \setminus U$ and $\delta(H) = \delta(G') - 2 \geq \delta(G) - k - 2$. Obviously, G has a k-Hamiltonian $[a, b]$-factor if and only if H includes an $[a - 2, b - 2]$-factor. In order to prove Theorem 5 by contradiction, we assume that H has no $[a-2, b-2]$-factor. Then, from Lemma 1, there exist two disjoint subsets S and T of $V(H) = V(G') = V(G) \setminus U$ such that

$$\delta_H(S, T) = (b - 2)|S| + d_{H-S}(T) - (a - 2)|T| \leq -1. \tag{1}$$

We choose such subsets S and T so that $|T|$ is minimum. Clearly, $T \neq \emptyset$ by (1).

Then, we claim that $d_{H-S}(x) \leq a - 3$ for any $x \in T$. This follows since, if $d_{H-S}(x) \geq a - 2$ for some $x \in T$, then the subsets S and $T \setminus \{x\}$ satisfy (1), as $\delta_H(S,T \setminus \{x\}) \le \delta_H(S,T)$, contradicting the choice of S and T. Now define $h = \min\{d_{H-S}(x) : x \in T\}$. Note that we trivially have:

$$0 \le h \le a - 3. \tag{2}$$

Now we consider the subgraph $G[T]$ of G induced by T. Set $T_1 = G[T]$. Let x_1 be a vertex with minimum degree in T_1 and $N_1 = N_{T_1}[x_1]$. Moreover, for $i \ge 2$, let x_i be a vertex with minimum degree in $T_i = G[T] - \bigcup_{1 \le j < i} N_j$ and $N_i = N_{T_i}[x_i]$. We continue these procedures until we reach the situation in which $T_i = \emptyset$ for some i, say for $i = r + 1$. Then in view of the above definition we know that $\{x_1, x_2, \cdots, x_r\}$ is an independent set of G (and of $G[T]$). Clearly, $r \ge 1$ since $T \ne \emptyset$.

We denote the order of N_i by n_i. The following properties are easily verified.

$$\alpha(G[T]) \ge r \tag{3}$$

and

$$|T| = \sum_{1 \le i \le r} n_i. \tag{4}$$

Since our choice of x_i implies that all vertices in N_i have degree at least $n_i - 1$ in T_i, we have

$$\sum_{1 \le i \le r} \left(\sum_{x \in N_i} d_{T_i}(x) \right) \ge \sum_{1 \le i \le r} n_i(n_i - 1). \tag{5}$$

It follows from $G' = G - U$, $H = G' - E(C)$, (4) and (5) that

$$
\begin{aligned}
d_{H-S}(T) = d_{G'-E(C)-S}(T) &\ge d_{G'-S}(T) - 2|T| \\
&= d_{G-U-S}(T) - 2|T| \ge d_{G-S}(T) - k|T| - 2|T| \\
&= d_{G-S}(T) - (k+2)|T| \\
&\ge \sum_{1 \le i \le r} n_i(n_i - 1) + \sum_{1 \le i < j \le r} e_G(N_i, N_j) - (k+2) \sum_{1 \le i \le r} n_i \\
&\ge \sum_{1 \le i \le r} n_i(n_i - 1) - (k+2) \sum_{1 \le i \le r} n_i \\
&= \sum_{1 \le i \le r} n_i(n_i - k - 3),
\end{aligned}
$$

that is,

$$d_{H-S}(T) \ge \sum_{1 \le i \le r} n_i(n_i - k - 3). \tag{6}$$

Using (3), $\alpha(G) \le \frac{4(b-2)(\delta(G)-a-k+1)}{(a+k+1)^2}$ and the obvious inequality $\alpha(G) \ge \alpha(G[T])$, we have

$$r \le \alpha(G[T]) \le \alpha(G) \le \frac{4(b-2)(\delta(G) - a - k + 1)}{(a + k + 1)^2}. \tag{7}$$

It is easy to see that $\min\{d_H(x) : x \in T\}$ is both an upper bound on $\delta(H)$ and a lower bound on $|S|+h$, and so $\delta(H) \leq |S|+h$. Note that $\delta(H) \geq \delta(G) - k - 2$. According to (1), (2), (4), (6), (7) and the obvious inequality $n_i(n_i - a - k - 1) \geq -\frac{(a+k+1)^2}{4}$, we have

$$
\begin{aligned}
-1 \geq \delta_H(S,T) &= (b-2)|S| + d_{H-S}(T) - (a-2)|T| \\
&\geq (b-2)|S| + \sum_{1 \leq i \leq r} n_i(n_i - k - 3) - (a-2)\sum_{1 \leq i \leq r} n_i \\
&= (b-2)|S| + \sum_{1 \leq i \leq r} n_i(n_i - a - k - 1) \\
&\geq (b-2)|S| - \sum_{1 \leq i \leq r} \frac{(a+k+1)^2}{4} \\
&= (b-2)|S| - \frac{(a+k+1)^2}{4}r \\
&\geq (b-2)(\delta(H) - h) - \frac{(a+k+1)^2}{4} \cdot \frac{4(b-2)(\delta(G) - a - k + 1)}{(a+k+1)^2} \\
&= (b-2)(\delta(H) - h) - (b-2)(\delta(G) - a - k + 1) \\
&\geq (b-2)(\delta(G) - k - 2 - h) - (b-2)(\delta(G) - a - k + 1) \\
&= (b-2)(a - 3 - h) \geq 0,
\end{aligned}
$$

which is a contradiction. This completes the proof of Theorem 5.   □

3   Remark

If $\alpha(G) = 1$ in Corollary 1, then $\delta(G) - a + 1 \geq 1$, and hence G is a complete graph with $\delta(G) \geq a$. In this case, G contains a clique of order $a + 1$, which is a Hamiltonian a-factor of G. In the following, we assume that $\alpha(G) \geq 2$ in Corollary 1. Now we show that the condition

$$
\alpha(G) \leq \frac{4(b-2)(\delta(G) - a + 1)}{(a+1)^2}
$$

in Corollary 1 cannot be replaced by

$$
\alpha(G) \leq \frac{4(b-2)(\delta(G) - a + 1)}{(a+1)^2} + 1.
$$

Put $m = \lfloor \frac{4(b-2)(\delta(G)-a+1)}{(a+1)^2} \rfloor$. From $\alpha(G) \geq 2$, we have $m \geq 2$. Set $t = \lfloor \frac{am}{b} \rfloor$. We construct a graph $G = K_t \vee (m+1)K_a$, where $\vee$ means *join*. Obviously, $\frac{4(b-2)(\delta(G)-a+1)}{(a+1)^2} < \alpha(G) = m+1 = \lfloor \frac{4(b-2)(\delta(G)-a+1)}{(a+1)^2} \rfloor + 1 \leq \frac{4(b-2)(\delta(G)-a+1)}{(a+1)^2} + 1$.

Thus, if we take $S = V(K_t)$ and $T = V((m + 1)K_a)$, then

$$
\begin{aligned}
\delta_G(S, T) &= b|S| + d_{G-S}(T) - a|T| \\
&= bt + (m + 1)a(a - 1) - (m + 1)a^2 \\
&\leq b \cdot \frac{am}{b} + (m + 1)a(a - 1) - (m + 1)a^2 \\
&= -a < 0.
\end{aligned}
$$

According to Lemma 1, G has no $[a, b]$-factor, and so G has no Hamiltonian $[a, b]$-factor. In the above sense, the result in Corollary 1 is best possible. And so the result in Theorem 5 is also best possible.

Acknowledgments. The authors are grateful to the anonymous referees for their very helpful and detailed comments in improving this paper.

References

1. Cai, M., Li, Y., Kano, M.: A $[k, k + 1]$-factor containing given Hamiltonian cycle. Sci. China Ser. A **41**, 933–938 (1998)
2. Cai, J., Liu, G., Hou, J.: The stability number and connected $[k, k + 1]$-factor in graphs. Appl. Math. Lett. **22**, 927–931 (2009)
3. Ekstein, J., Holub, P., Kaiser, T., Xiong, L., Zhang, S.: Star subdivisions and connected even factors in the square of a graph. Discret. Math. **312**, 2574–2578 (2012)
4. Fourtounelli, O., Katerinis, P.: The existence of k-factors in squares of graphs. Discret. Math. **310**, 3351–3358 (2010)
5. Liu, H., Liu, G.: A neighborhood condition for graphs to have (g, f)-factors. Ars Comb. **93**, 257–264 (2009)
6. Liu, H., Liu, G.: Neighbor set for the existence of (g, f, n)-critical graphs. Bull. Malays. Math. Sci. Soc. **34**, 39–49 (2011)
7. Liu, G., Yu, Q., Zhang, L.: Maximum fractional factors in graphs. Appl. Math. Lett. **20**, 1237–1243 (2007)
8. Lovász, L.: Subgraphs with prescribed valencies. J. Comb. Theory **8**, 391–416 (1970)
9. Matsuda, H.: Degree conditions for the existence of $[k, k + 1]$-factors containing a given Hamiltonian cycle. Australas. J. Comb. **26**, 273–281 (2002)
10. Wei, B., Zhu, Y.: Hamiltonian k-factors in graphs. J. Graph Theory **25**, 217–227 (1997)
11. Zhou, S.: Binding numbers for fractional ID-k-factor-critical graphs. Acta Math. Sin. Engl. Ser. **30**, 181–186 (2014)
12. Zhou, S.: A new neighborhood condition for graphs to be fractional (k, m)-deleted graphs. Appl. Math. Lett. **25**, 509–513 (2012)
13. Zhou, S.: Independence number, connectivity and (a, b, k)-critical graphs. Discret. Math. **309**, 4144–4148 (2009)
14. Zhou, S.: Toughness and the existence of Hamiltonian $[a, b]$-factors of graphs. Util. Math. **90**, 187–197 (2013)
15. Zhou, S., Bian, Q.: Subdigraphs with orthogonal factorizations of digraphs (II). Eur. J. Comb. **36**, 198–205 (2014)

Subgraphs with Orthogonal $[0, k_i]_1^n$-Factorizations in Graphs

Sizhong Zhou[1($\boxtimes$)], Tao Zhang[2], and Zurun Xu[1]

[1] School of Mathematics and Physics, Jiangsu University of Science and Technology, Mengxi Road 2, Zhenjiang 212003, Jiangsu, People's Republic of China
zsz_cumt@163.com
[2] School of Economic and Management,
Jiangsu University of Science and Technology, Mengxi Road 2,
Zhenjiang 212003, Jiangsu, People's Republic of China

Abstract. Let m, n, r and k_i ($1 \leq i \leq m$) be positive integers with $n \leq m$ and $k_1 \geq k_2 \geq \cdots \geq k_m \geq 2r - 1$. Let G be a graph, and let $H_1, H_2, \cdots, H_r$ be vertex-disjoint n-subgraphs of G. It is verified in this article that every $[0, k_1 + k_2 + \cdots + k_m - n + 1]$-graph G includes a subgraph R such that R has a $[0, k_i]_1^n$-factorization orthogonal to every H_i, $1 \leq i \leq r$.

Keywords: Graph · Subgraph · Factor · $[0, k_i]_1^m$-Factorization · Orthogonal

1 Introduction

Many physical structures can conveniently be modelled by networks. Examples include a communication network with nodes and links modelling cities and communication channels, respectively, or a railroad network with nodes and links representing railroad stations and railways between two stations, respectively. The file transfer problem can be modelled as $(0, f)$-factorizations in graphs [12]. The designs of Latin squares and Room squares, combinatorial design, network design, circuit layout and so on are related to orthogonal factorizations of graphs [1], and so the research on orthogonal factorizations in graphs attracted a great deal of attention in recent years. It is well known that a network can be represented by a graph. Vertices and edges of the graph correspond to nodes and links between the nodes, respectively. Henceforth we use the term *graph* instead of *network*.

Supported by the National Natural Science Foundation of China (Grant No. 11371009), the National Social Science Foundation of China (Grant No. 14AGL001) and the Natural Science Foundation of the Higher Education Institutions of Jiangsu Province (Grant No. 14KJD110002), and sponsored by 333 Project of Jiangsu Province.

D. Gaur and N.S. Narayanaswamy (Eds.): CALDAM 2017, LNCS 10156, pp. 362–370, 2017.
DOI: 10.1007/978-3-319-53007-9_32

We only consider finite graphs which have neither loops nor multiple edges. Let G be a graph. We denote its vertex set and edge set by $V(G)$ and $E(G)$, respectively. For $x \in V(G)$, the degree of x in G is denoted by $d_G(x)$. Let $g, f : V(G) \rightarrow \mathbb{Z}$ be two functions satisfying $0 \leq g(x) \leq f(x)$ for any $x \in V(G)$. A spanning subgraph F of G with $g(x) \leq d_F(x) \leq f(x)$ for each $x \in V(G)$ is a (g, f)-factor. Specially, G is said to be a (g, f)-graph if G itself is a (g, f)-factor. Let a and b be two nonnegative integers with $a \leq b$. If $g(x) \equiv a$ and $f(x) \equiv b$ for all $x \in V(G)$, then a (g, f)-factor is an $[a, b]$-factor and a (g, f)-graph is an $[a, b]$-graph. A subgraph H of G is an m-subgraph if $|E(H)| = m$. A (g, f)-factorization of G is a decomposition of the edge set of G into edge-disjoint (g, f)-factors $F_1, F_2, \cdots, F_m$. Let H be an mr-subgraph. A (g, f)-factorization $F = \{F_1, F_2, \cdots, F_m\}$ of G is r-orthogonal to H if $|E(H) \cap E(F_i)| = r$ for $1 \leq i \leq m$. Let k_i be a positive integer, $1 \leq i \leq m$. A $[0, k_i]_1^m$-factorization $F = \{F_1, F_2, \cdots, F_m\}$ of G is a decomposition of G into edge-disjoint factors $F_1, F_2, \cdots, F_m$, where F_i is a $[0, k_i]$-factor, $1 \leq i \leq m$. A $[0, k_i]_1^m$-factorization $F = \{F_1, F_2, \cdots, F_m\}$ of G is r-orthogonal to H if $|E(H) \cap E(F_i)| = r$ for $1 \leq i \leq m$. Specially, 1-orthogonal is simply called orthogonal.

Many authors investigated orthogonal (g, f)-factorizations in $(mg + m - 1, mf - m + 1)$-graphs [4–7]. Wang [9] verified that there exists a subgraph R in an $(mg + k, mf - k)$-graph such that R admits a (g, f)-factorization orthogonal to n vertex-disjoint k-subgraphs. Feng showed that every $(0, mf - m + 1)$-graph has a $(0, f)$-factorization orthogonal to any given m-subgraph [2] and every $[0, k_1 + k_2 + \cdots + k_m - m + 1]$-graph has a $[0, k_i]_1^m$-factorization orthogonal to any given m-subgraph [3]. Wang [8] investigated subgraphs with orthogonal $[0, k_i]_1^n$-factorizations in $[0, k_1 + k_2 + \cdots + k_m - n + 1]$-graphs. Orthogonal (g, f)-factorizations in digraphs were obtained in [10,11]. The following results on orthogonal factorizations in graphs are known.

Theorem 1 [3]. Let G be a $[0, k_1 + k_2 + \cdots + k_m - m + 1]$-graph, where $m \geq 1$ is an integer and $k_1, k_2, \cdots, k_m$ are positive integers. Let H be an arbitrary m-subgraph of G. Then G has a $[0, k_i]_1^m$-factorization orthogonal to H.

Theorem 2 [8]. Let G be a $[0, k_1 + k_2 + \cdots + k_m - n + 1]$-graph, where m, n and $k_1, k_2, \cdots, k_m$ are positive integers with $n \leq m$ and $k_1 \geq k_2 \geq \cdots \geq k_m$. Let H be an arbitrary n-subgraph of G. Then there exists a subgraph R of G such that R admits a $[0, k_i]_1^n$-factorization orthogonal to H.

In this article, we proceed to investigate the problem on orthogonal factorizations in graphs and obtain a new result on orthogonal factorizations in $[0, k_1 + k_2 + \cdots + k_m - n + 1]$-graphs, which is shown in Sect. 3.

2 Preliminary Lemmas

Let G be a graph. For any $S \subseteq V(G)$, we denote by $G - S$ the subgraph of G by deleting the vertices in S together with the edges incident to vertices in S. Let S and T be two disjoint vertex subsets of G. We use $E_G(S, T)$ to denote the set of edges in G joining S and T, and write $e_G(S,T) = |E_G(S,T)|$. For any function φ defined on $V(G)$, we write $\varphi(S) = \sum_{x \in S} \varphi(x)$ and $\varphi(\emptyset) = 0$. Especially, $d_{G-S}(T) = \sum_{x \in T} d_{G-S}(x)$.

Let S and T be two disjoint vertex subsets of G, and E_1 and E_2 be two disjoint edge subsets of G. Set

$$U = V(G) - (S \cup T), \quad E(S) = \{xy \in E(G) : x, y \in S\},$$

$$E(T) = \{xy \in E(G) : x, y \in T\}.$$

We write

$$E_1' = E_1 \cap E(S), \quad E_1'' = E_1 \cap E_G(S, U),$$

$$E_2' = E_2 \cap E(T), \quad E_2'' = E_2 \cap E_G(T, U),$$

$$\alpha_G(S, T; E_1, E_2) = 2|E_1'| + |E_1''|,$$

$$\beta_G(S, T; E_1, E_2) = 2|E_2'| + |E_2''|.$$

If there is no ambiguity, we denote $\alpha_G(S, T; E_1, E_2)$ and $\beta_G(S, T; E_1, E_2)$ by α and β, respectively. It is easy to see that $\alpha \leq d_{G-T}(S)$ and $\beta \leq d_{G-S}(T)$.

Lam et al. [4] proved the following result, which is very useful in the proof of our main result.

Lemma 1 [4]. Let G be a graph, and let g, f be two integer-valued functions defined on $V(G)$ satisfying $0 \leq g(x) < f(x) \leq d_G(x)$ for each $x \in V(G)$, and E_1 and E_2 be two disjoint subsets of $E(G)$. Then G admits a (g, f)-factor F with $E_1 \subseteq E(F)$ and $E_2 \cap E(F) = \emptyset$ if and only if

$$\delta_G(S, T) = f(S) + d_{G-S}(T) - g(T) \geq \alpha_G(S, T; E_1, E_2) + \beta_G(S, T; E_1, E_2) \quad (1)$$

for any two disjoint subsets S and T of $V(G)$.

In the following, we always assume that G is a $[0, k_1 + k_2 + \cdots + k_m - n + 1]$-graph, where m, n and k_i $(1 \leq i \leq m)$ are positive integers with $1 \leq n \leq m$ and $k_1 \geq k_2 \geq \cdots \geq k_m$. For each $[0, k_i]$-factor F_i and each isolated vertex x of G, we have $d_{F_i}(x) = 0$. We write I for the set of all isolated vertices of G. If $G - I$ has a $[0, k_i]$-factor, then G has also a $[0, k_i]$-factor. Therefore, we may assume that G has no isolated vertices. For any $x \in V(G)$, we define

$$p(x) = \max\{0, d_G(x) - (k_1 + k_2 + \cdots + k_{m-1} - n + 2)\},$$

$$q(x) = \min\{k_m, d_G(x)\}.$$

In terms of the definitions of $p(x)$ and $q(x)$, we have $0 \leq p(x) < q(x) \leq k_m$ for each $x \in V(G)$.

Lemma 2. Let G be a $[0, k_1 + k_2 + \cdots + k_m]$-graph, and let $H_1, H_2, \cdots, H_r$ be independent edges of G, where m, r and k_i $(1 \leq i \leq m)$ are positive integers. Then G admits a $[0, k_1]$-factor including H_i $(1 \leq i \leq r)$.

Proof. Set $E_1 = \{H_1, H_2, \cdots, H_r\}$ and $E_2 = \emptyset$. We define α and β as before for two disjoint subsets S and T of $V(G)$. According to the definitions of α and β, we obtain

$$\alpha \leq \min\{2r, |S|\} \quad and \quad \beta = 0.$$

Thus, we have by $k_1 \geq 1$

$$\delta_G(S, T) = k_1|S| + d_{G-S}(T) - 0 \cdot |T| \geq |S| \geq \alpha = \alpha + \beta,$$

where $\delta_G(S, T)$ is defined by Equation (1) by replacing g and f by 0 and k_1. Then by using Lemma 1, G admits a $[0, k_1]$-factor including H_i $(1 \leq i \leq r)$. This completes the proof of Lemma 2.   □

3   Main Result and Its Proof

In the following, we show our main result in this paper.

Theorem 3. Let G be a $[0, k_1 + k_2 + \cdots + k_m - n + 1]$-graph, and let $H_1, H_2, \cdots, H_r$ be vertex-disjoint n-subgraphs of G, where k_i $(1 \leq i \leq m)$, m, n and r are positive integers with $1 \leq n \leq m$ and $k_1 \geq k_2 \geq \cdots \geq k_m \geq 2r - 1$. Then there exists a subgraph R of G such that R has a $[0, k_i]_1^n$-factorization orthogonal to every H_i, $1 \leq i \leq r$.

If $r = 1$ and $m = n$ in Theorem 3, then we obtain Theorem 1. Hence, Theorem 1 is a special case of Theorem 3. Furthermore, we partially solve the following question posed by Alspach et al. [1]: Given a subgraph H of a graph G, does there exist a factorization F of G with some fixed type orthogonal to H?

Proof of Theorem 3. According to Theorem 2, Theorem 3 holds for $r = 1$. Hence, we assume that $r \geq 2$.

We apply induction on m and n. In terms of Lemma 2, Theorem 3 holds for $n = 1$. Therefore, we assume that $n \geq 2$. For the inductive step, suppose that Theorem 3 holds for any $[0, k_1 + k_2 + \cdots + k_{m'} - n' + 1]$-graph G' with $m' < m$, $n' < n$ and $1 \leq n' \leq m'$, and any vertex-disjoint n'-subgraphs $H_1', H_2', \cdots, H_r'$ of G'. We now consider a $[0, k_1 + k_2 + \cdots + k_m - n + 1]$-graph G and any vertex-disjoint n-subgraphs $H_1, H_2, \cdots, H_r$ of G.

Define $p(x)$ and $q(x)$ the same as before. For $1 \leq i \leq r$, we put

$$A_{i1} = \{xy : xy \in E(H_i), p(x) \geq 1, p(y) \geq 1\};$$

$$A_{i2} = \{xy : xy \in E(H_i), p(x) \geq 1 \text{ or } p(y) \geq 1\};$$

$$A_i = \begin{cases} A_{i1}, & if \ A_{i1} \neq \emptyset, \\ A_{i2}, & if \ A_{i1} = \emptyset \text{ and } A_{i2} \neq \emptyset, \\ E(H_i), & otherwise. \end{cases}$$

Choose $u_i v_i \in A_i$ for $1 \leq i \leq r$. Set $E_1 = \{u_i v_i : 1 \leq i \leq r\}$, $E_2 = \bigcup_{i=1}^{r} E(H_i) \setminus E_1$. Thus, we have $|E_1| = r$ and $|E_2| = (n-1)r$. We define E_1', E_1'',

E_2', E_2'', α and β as before for two disjoint subsets S and T of $V(G)$. In terms of the definitions of α and β, we obtain

$$\alpha \leq \min\{2r, |S|\} \quad and \quad \beta \leq \min\{2(n-1)r, (n-1)|T|\}$$

Now we define $\delta_G(S,T)$ in Equation (1) by replacing g and f by p and q, and choose disjoint subsets S and T of $V(G)$ satisfying

(1) $\delta_G(S,T) - \alpha_G(S,T;E_1,E_2) - \beta_G(S,T;E_1,E_2)$ is minimum.
(2) $|S|$ is minimum subject to (1).

We now verify the following claim.

Claim 1. If $S \neq \emptyset$, then $q(x) \leq d_G(x) - 1$ for each $x \in S$, and so $q(x) = k_m$ for each $x \in S$.

Proof. We write $S_1 = \{x \in S : q(x) \geq d_G(x)\}$. In the following, we prove $S_1 = \emptyset$.
  Suppose that $S_1 \neq \emptyset$. Set $S_0 = S \setminus S_1$. Thus, we have

$$
\begin{aligned}
\delta_G(S,T) &= q(S) + d_{G-S}(T) - p(T) \\
&= q(S_0) + q(S_1) + d_G(T) - e_G(S_0,T) - e_G(S_1,T) - p(T) \\
&= q(S_0) + d_{G-S_0}(T) - p(T) + q(S_1) - e_G(S_1,T) \\
&\geq \delta_G(S_0,T) + d_G(S_1) - e_G(S_1,T) \\
&= \delta_G(S_0,T) + d_{G-T}(S_1).
\end{aligned}
$$

Note that

$$
\alpha_G(S,T;E_1,E_2) + \beta_G(S,T;E_1,E_2) \leq \alpha_G(S_0,T;E_1,E_2) + \beta_G(S_0,T;E_1,E_2) \\
+ \alpha_G(S_1,T;E_1,E_2)
$$

and

$$d_{G-T}(S_1) \geq \alpha_G(S_1,T;E_1,E_2).$$

Hence, we obtain

$$
\delta_G(S,T) - \alpha_G(S,T;E_1,E_2) - \beta_G(S,T;E_1,E_2) \geq \delta_G(S_0,T) - \alpha_G(S_0,T;E_1,E_2) \\
- \beta_G(S_0,T;E_1,E_2),
$$

which contradicts the choice of S. Therefore, $S_1 = \emptyset$. And so, if $S \neq \emptyset$, then $q(x) \leq d_G(x) - 1$ for each $x \in S$. Furthermore, we obtain $q(x) = k_m$ for each $x \in S$. This completes the proof of Claim 1. □

The remaining of the proof is dedicated to verifying that G admits a (p,q)-factor F_n satisfying $E_1 \subseteq E(F_n)$ and $E_2 \cap E(F_n) = \emptyset$. In terms of Lemma 1 and the choice of S and T, it is sufficient to prove that $\delta_G(S,T) \geq \alpha + \beta$.
  In the following, we write $\lambda = k_1 + k_2 + \cdots + k_{m-1} - n + 2$, $T_1 = \{x : d_G(x) - \lambda > 0, x \in T\}$ and $T_0 = T \setminus T_1$. Obviously, $p(x) = 0$ for any $x \in T_0$ and

$p(x) = d_G(x) - \lambda$ for any $x \in T_1$. According to the definition of $\beta_G(S, T; E_1, E_2)$, we have

$$\beta_G(S, T_0; E_1, E_2) + \beta_G(S, T_1; E_1, E_2) = \beta_G(S, T; E_1, E_2). \qquad (2)$$

According to the definitions of α and β, we derive $\alpha \leq \min\{2r, |S|\} \leq |S|$ and $\beta \leq d_{G-S}(T)$. If $T_1 = \emptyset$, then we obtain

$$\begin{aligned} \delta_G(S, T) &= q(S) + d_{G-S}(T) - p(T) \\ &= q(S) + d_{G-S}(T) - p(T_0) - p(T_1) \\ &\geq |S| + d_{G-S}(T) \geq \alpha + \beta. \end{aligned}$$

If $S = \emptyset$, then $\alpha = 0$. In view of Equation (2), $r \geq 2$, $2 \leq n \leq m$ and $k_i \geq 2r - 1$, $1 \leq i \leq m - 1$, we have

$$\begin{aligned} \delta_G(S, T) &= q(S) + d_{G-S}(T) - p(T) = d_G(T) - p(T_1) \\ &= d_G(T_0) + d_G(T_1) - (d_G(T_1) - \lambda|T_1|) = d_G(T_0) + \lambda|T_1| \\ &\geq d_G(T_0) + ((m-1)(2r-1) - n + 2)|T_1| \geq d_G(T_0) + (n-1)|T_1| \\ &\geq \beta_G(\emptyset, T_0; E_1, E_2) + \beta_G(\emptyset, T_1; E_1, E_2) \\ &= \beta_G(\emptyset, T; E_1, E_2) = \beta = \alpha + \beta. \end{aligned}$$

In the following, we always assume that $S \neq \emptyset$ and $T_1 \neq \emptyset$. In order to verify Theorem 3, we discuss two cases.

Case 1. $|S| \geq |T_1|$.
Using Claim 1 and the definition of T_1, we have

$$\begin{aligned} \delta_G(S, T) &= q(S) + d_{G-S}(T) - p(T) \\ &= q(S) + d_{G-S}(T) - p(T_1) \\ &= k_m|S| + d_{G-S}(T) - d_G(T_1) + \lambda|T_1| \\ &= k_m(|S| - |T_1|) + (\lambda + k_m)|T_1| + d_{G-S}(T) - d_G(T_1) \\ &\geq k_m(|S| - |T_1|) + d_G(T_1) + |T_1| + d_{G-S}(T) - d_G(T_1) \\ &= (k_m - 1)(|S| - |T_1|) + |S| + d_{G-S}(T) \\ &\geq \alpha + \beta. \end{aligned}$$

Case 2. $|S| \leq |T_1| - 1$.
In terms of Claim 1 and the definitions of T_0 and T_1, we obtain

$$\begin{aligned} \delta_G(S, T) &= q(S) + d_{G-S}(T) - p(T) \\ &= q(S) + d_{G-S}(T_0) + d_{G-S}(T_1) - p(T_1) \\ &= k_m|S| + d_{G-S}(T_0) + d_G(T_1) - p(T_1) - e_G(S, T_1) \\ &= k_m|S| + d_{G-S}(T_0) + \lambda|T_1| - e_G(S, T_1), \end{aligned}$$

that is,

$$\delta_G(S, T) = k_m|S| + d_{G-S}(T_0) + \lambda|T_1| - e_G(S, T_1). \qquad (3)$$

Subcase 2.1. $|T_1| \leq k_m - 1$.
Using Equations (2) and (3), $r \geq 2$, $2 \leq n \leq m$ and $k_i \geq 2r - 1$, $1 \leq i \leq m - 1$, we have

$$
\begin{aligned}
\delta_G(S, T) &= k_m|S| + d_{G-S}(T_0) + \lambda|T_1| - e_G(S, T_1) \\
&\geq k_m|S| + d_{G-S}(T_0) + \lambda|T_1| - |S||T_1| \\
&\geq k_m|S| + d_{G-S}(T_0) + ((m-1)(2r-1) - n + 2)|T_1| - (k_m - 1)|S| \\
&\geq |S| + d_{G-S}(T_0) + (n-1)|T_1| \\
&\geq \alpha + \beta_G(S, T_0; E_1, E_2) + \beta_G(S, T_1; E_1, E_2) \\
&= \alpha + \beta_G(S, T; E_1, E_2) = \alpha + \beta.
\end{aligned}
$$

Subcase 2.2. $|T_1| \geq k_m$.
Subcase 2.2.1. $|S| \leq 2n - 4$.
According to $r \geq 2$, $2 \leq n \leq m$ and $k_i \geq 2r - 1$, $1 \leq i \leq m - 1$, we have

$$
\lambda \geq (m-1)(2r-1) - n + 2 \geq 3(n-1) - n + 2 = 2n - 1,
$$

and so,

$$
\lambda - |S| \geq (2n - 1) - (2n - 4) = 3 > 0.
$$

Combining this with Equations (3), $|T_1| \geq k_m$, $r \geq 2$, $2 \leq n \leq m$ and $k_i \geq 2r - 1$, $1 \leq i \leq m$, we obtain

$$
\begin{aligned}
\delta_G(S, T) &= k_m|S| + d_{G-S}(T_0) + \lambda|T_1| - e_G(S, T_1) \\
&\geq k_m|S| + \lambda|T_1| - |S||T_1| = k_m|S| + (\lambda - |S|)|T_1| \\
&\geq k_m|S| + (\lambda - |S|)k_m = \lambda k_m \\
&\geq ((m-1)(2r-1) - n + 2)(2r-1) \\
&\geq (3(n-1) - n + 2)(2r-1) \\
&= 2nr + (2n-2)r - 2n + 1 \\
&\geq 2nr + 2(2n-2) - 2n + 1 = 2nr + 2n - 3 \\
&> 2nr = 2r + 2(n-1)r \geq \alpha + \beta.
\end{aligned}
$$

Subcase 2.2.2. $|S| \geq 2n - 3$.
Note that $k_1 \geq k_2 \geq \cdots \geq k_m \geq 2r - 1$, $\lambda = k_1 + k_2 + \cdots + k_{m-1} - n + 2$ and G is a $[0, k_1 + k_2 + \cdots + k_m - n + 1]$-graph. Hence, we derive $\lambda \geq (m-1)(2r-1) - n + 2$ and $d_G(S) \leq (k_1 + k_2 + \cdots + k_m - n + 1)|S| = (\lambda + k_m - 1)|S|$. In terms of Claim 1, the definition of T_1, $|S| \leq |T_1| - 1$, $r \geq 2$ and $2 \leq n \leq m$, we have

$$
\begin{aligned}
\delta_G(S, T) &= q(S) + d_{G-S}(T) - p(T) \\
&= q(S) + d_G(T) - e_G(S, T) - p(T_1) \\
&= k_m|S| + d_G(T) - e_G(S, T) - d_G(T_1) + \lambda|T_1| \\
&\geq k_m|S| - e_G(S, T) + \lambda|T_1| \\
&= \lambda(|T_1| - |S|) + (\lambda + k_m)|S| - e_G(S, T) \\
&\geq \lambda + |S| + d_G(S) - e_G(S, T)
\end{aligned}
$$

$$\geq (m-1)(2r-1) - n + 2 + 2n - 3 + d_{G-T}(S)$$
$$\geq (n-1)(2r-1) + n - 1 + \alpha = \alpha + 2(n-1)r$$
$$\geq \alpha + \beta.$$

In conclusion, $\delta_G(S,T) \geq \alpha_G(S,T; E_1, E_2) + \beta_G(S,T; E_1, E_2)$. In terms of the choice of S and T, we obtain $\delta_G(S', T') \geq \alpha_G(S', T'; E_1, E_2) + \beta_G(S', T'; E_1, E_2)$ for any disjoint subsets S' and T' of $V(G)$. According to Lemma 1, G admits a (p,q)-factor F_n satisfying $E_1 \subseteq E(F_n)$ and $E_2 \cap E(F_n) = \emptyset$, and F_n is also a $[0, k_n]$-factor of G. According to the definitions of $p(x)$ and $q(x)$, we obtain

$$d_{G-F_n}(x) = d_G(x) - d_{F_n}(x) \geq d_G(x) - q(x) \geq 0$$

and

$$d_{G-F_n}(x) = d_G(x) - d_{F_n}(x) \leq d_G(x) - p(x)$$
$$\leq d_G(x) - (d_G(x) - (k_1 + k_2 + \cdots + k_{m-1} - n + 2))$$
$$= k_1 + k_2 + \cdots + k_{m-1} - (n-1) + 1.$$

Hence, $G - F_n$ is a $[0, k_1 + k_2 + \cdots + k_{m-1} - (n-1) + 1]$-graph. Set $H_i' = H_i - u_i v_i$ for $1 \leq i \leq r$. Clearly, $H_1', H_2', \cdots, H_r'$ are vertex-disjoint $(n-1)$-subgraphs of $G - F_n$. In terms of the induction hypothesis, there exists a subgraph R' of $G - F_n$ such that R' has a $[0, k_i]_{i=1}^{n-1}$-factorization orthogonal to every H_i', $1 \leq i \leq r$. Write R for the subgraph of G induced by $E(R') \cup E(F_n)$. Therefore, R is a subgraph of G such that R admits a $[0, k_i]_{i=1}^{n}$-factorization orthogonal to every H_i, $1 \leq i \leq r$. This completes the proof of Theorem 3.   □

Acknowledgments. The authors are grateful to the anonymous referees for their very helpful and detailed comments in improving this paper.

References

1. Alspach, B., Heinrich, K., Liu, G.: Contemporary Design Theory: A Collection of Surveys. Wiley, New York (1992)
2. Feng, H.: On orthogonal $(0, f)$-factorizations. Acta Math. Sci., Englis Ser. **19**(3), 332–336 (1999)
3. Feng, H., Liu, G.: Orthogonal factorizations of graphs. J. Graph Theory **40**, 267–276 (2002)
4. Lam, P.C.B., Liu, G., Li, G., Shiu, W.: Orthogonal (g, f)-factorizations in networks. Networks **35**(4), 274–278 (2000)
5. Liu, G.: Orthogonal (g, f)-factorizations in graphs. Discret. Math. **143**, 153–158 (1995)
6. Liu, G., Long, H.: Randomly orthogonal (g, f)-factorizations in graphs. Acta Appl. Math. Sin. **18**(3), 489–494 (2002). English Ser.
7. Liu, G., Zhu, B.: Some problems on factorizations with constrants in bipartite graphs. Discret. Appl. Math. **128**, 421–434 (2003)

8. Wang, C.: Orthogonal factorizations in networks. Int. J. Comput. Math. **88**(3), 476–483 (2011)
9. Wang, C.: Subgraphs with orthogonal factorizations and algorithms. Eur. J. Combin. **31**, 1706–1713 (2010)
10. Wang, C.: Subdigraphs with orthogonal factorizations of digraphs. Eur. J. Combin. **33**, 1015–1021 (2012)
11. Zhou, S., Bian, Q.: Subdigraphs with orthogonal factorizations of digraphs (II). Eur. J. Combin. **36**, 198–205 (2014)
12. Zhou, X., Nishizeki, T.: Edge-coloring and f-coloring for various classes of graphs. Lect. Notes Comput. Sci. **834**, 199–207 (1994)

Author Index

Printed in the United States
By Bookmasters